Methods of
Experimental Physics

VOLUME 19

ULTRASONICS

METHODS OF EXPERIMENTAL PHYSICS:

L. Marton and C. Marton, *Editors-in-Chief*

Volume 19

Ultrasonics

Edited by

PETER D. EDMONDS

Bioengineering Research Center
SRI International
Menlo Park, California

1981

ACADEMIC PRESS

A Subsidiary of Harcourt Brace Jovanovich, Publishers

New York   London   Toronto   Sydney   San Francisco

ACADEMIC PRESS, INC.
111 Fifth Avenue, New York, New York 10003

United Kingdom Edition published by
ACADEMIC PRESS, INC. (LONDON) LTD.
24/28 Oval Road, London NW1   7DX

Library of Congress Cataloging in Publication Data
Main entry under title:

Ultrasonics.

   (Methods of experimental physics; v. 19)
   Includes bibliographical references and index.
   1. Ultrasonics. I. Edmonds, Peter D. II. Series.
QC244.U43   534.5'5   80-28270
ISBN 0-12-475961-0

PRINTED IN THE UNITED STATES OF AMERICA

81 82 83 84   9 8 7 6 5 4 3 2 1

CONTENTS

2. Ultrasonic Wave Velocity and Attenuation Measurements
by M. A. BREAZEALE, JOHN H. CANTRELL, JR., AND JOSEPH S. HEYMAN

3. Dynamic Viscosity Measurement
by GILROY HARRISON AND A. JOHN BARLOW

CONTRIBUTORS

Numbers in parentheses indicate the pages on which the authors' contributions begin.

ROBERT E. APFEL, *Department of Engineering and Applied Science, Yale University, New Haven, Connecticut 06520* (355)

A. JOHN BARLOW, *Department of Electronics and Electrical Engineering, The University, Glasgow, G12 8QQ, Scotland* (137)

M. A. BREAZEALE, *Department of Physics, The University of Tennessee, Knoxville, Tennessee 37916* (67)

L. J. BUSSE,* *Department of Physics, Washington University, St. Louis, Missouri 63130* (29)

JOHN H. CANTRELL, JR., *National Aeronautics and Space Administration, Langley Research Center, Hampton, Virginia 23665* (67)

F. DUNN, *Bioacoustics Research Laboratory, Department of Electrical Engineering, University of Illinois, Urbana, Illinois 61801* (1)

PETER D. EDMONDS, *Bioengineering Research Center, SRI International, Menlo Park, California 94025* (1)

JAMES F. GREENLEAF, *Department of Physiology and Biophysics, Biodynamics Research Unit, Mayo Clinic/Mayo Foundation, Rochester, Minnesota 55901* (563)

GILROY HARRISON, *Department of Electronics and Electrical Engineering, The University, Glasgow, G12 8QQ, Scotland* (137)

JOSEPH HEISERMAN, *Edward L. Ginzton Laboratory, Stanford University, Stanford, California 94305* (413)

JOSEPH S. HEYMAN, *National Aeronautics and Space Administration, Langley Research Center, Hampton, Virginia 23665* (67)

B. P. HILDEBRAND,† *Battelle Memorial Institute, Pacific Northwest Laboratories, Richland, Washington 99352* (533)

---

* Present address: Battelle Northwest, Richland, Washington 99352.

† Present address: Spectron Development Laboratories, Inc., Costa Mesa, California 92626.

J. G. MILLER, *Department of Physics, Washington University, St. Louis, Missouri 63130* (29)

MATTHEW O'DONNELL,* *Department of Physics, Washington University, St. Louis, Missouri 63130* (29)

EMMANUEL P. PAPADAKIS, *Ford Motor Company, Manufacturing Processes Laboratory, Detroit, Michigan 48239* (237)

JAMES A. ROONEY, *Department of Physics and Astronomy, University of Maine, Orono, Maine 04473* (299)

LEON J. SLUTSKY, *Department of Chemistry, University of Washington, Seattle, Washington 98195* (179)

G. I. A. STEGEMAN,† *Department of Physics, University of Toronto, Toronto, Ontario M5S 1A7, Canada* (455)

RICHARD M. WHITE, *Department of Electrical Engineering and Computer Sciences, University of California, Berkeley, California 94720* (495)

* Present address: Signal Electronics Laboratory, Corporate Research and Development, General Electric Company, Schenectady, New York 12345.

† Present address: Optical Sciences Center, University of Arizona, Tucson, Arizona 85721.

FOREWORD

This volume, edited by Dr. Peter Edmonds, is the first of the Methods to be devoted to acoustics. Future volumes will deal with the more classical aspects of acoustics, a field that has been adopted by our colleagues in engineering as well.

Ultrasonics plays many roles, ranging from physical and bioengineering applications to the study of the fundamental properties of materials. Dr. Edmonds and his contributors cover these areas in a manner that should make this volume a definitive reference work on the subject. We expect that researchers in a given specialization will find much useful information and have their imaginations stimulated by going through the book as a whole.

L. MARTON
C. MARTON

PREFACE

This volume offers detailed and comprehensive treatments of a number of important topics in the broad field of ultrasonics. It is intended to serve the needs of graduate students and also of specialists in other fields who may desire an assessment of the capabilities of ultrasonics as a technique with the potential for solving specific problems.

Ultrasonics interfaces with many fields, including optics, low temperature and solid state physics, chemical kinetics, cavitation, viscoelasticity, lubrication, nondestructive evaluation, medical diagnostic imaging, signal processing, and materials processing. The authors of one or more of the following parts discuss these fields. However, other important topics have been omitted, e.g., ultrasonics in gaseous media, plasma- and magneto-acoustics, and phonon phenomena in general. Ultrasonic scattering in noncrystalline media proved to be insufficiently developed for treatment in this treatise . (Seekers of information on these topics should consult the excellent treatise "Physical Acoustics," edited by W. P. Mason and R. N. Thurston, published by Academic Press.)

I wish to thank all authors for their cooperation and hard work in writing and many supplementary tasks. The essential contributions made by the secretarial assistants to the authors and by their institutions are also acknowledged.

I am grateful to several anonymous reviewers of parts of this volume, whose excellent advice has been freely given and usually heeded. Valuable support provided by the management and staff of SRI International is acknowledged with thanks.

All who have contributed to this volume profoundly regret that one of its editors-in-chief, Dr. Ladislaus Marton, did not live to see its publication. In his absence, the functions of editor-in-chief have been admirably fulfilled by Mrs. Claire Marton.

PETER D. EDMONDS

METHODS OF EXPERIMENTAL PHYSICS

Editors-in-Chief
L. Marton   C. Marton

Volume 1. Classical Methods
Edited by Immanuel Estermann

Volume 2. Electronic Methods. Second Edition (in two parts)
Edited by E. Bleuler and R. O. Haxby

Volume 3. Molecular Physics, Second Edition (in two parts)
Edited by Dudley Williams

Volume 4. Atomic and Electron Physics—Part A: Atomic Sources
and Detectors, Part B: Free Atoms
Edited by Vernon W. Hughes and Howard L. Schultz

Volume 5. Nuclear Physics (in two parts)
Edited by Luke C. L. Yuan and Chien-Shiung Wu

Volume 6. Solid State Physics (in two parts)
Edited by K. Lark-Horovitz and Vivian A. Johnson

Volume 7. Atomic and Electron Physics—Atomic Interactions (in
two parts)
Edited by Benjamin Bederson and Wade L. Fite

Volume 8. Problems and Solutions for Students
Edited by L. Marton and W. F. Hornyak

Volume 9. Plasma Physics (in two parts)
Edited by Hans R. Griem and Ralph H. Lovberg

Volume 10. Physical Principles of Far-Infrared Radiation
By L. C. Robinson

Volume 11. Solid State Physics
Edited by R. V. Coleman

Volume 12. Astrophysics—Part A: Optical and Infrared
Edited by N. Carleton
Part B: Radio Telescopes, Part C: Radio Observations
Edited by M. L. Meeks

Volume 13. Spectroscopy (in two parts)
Edited by Dudley Williams

Volume 14. Vacuum Physics and Technology
Edited by G. L. Weissler and R. W. Carlson

Volume 15. Quantum Electronics (in two parts)
Edited by C. L. Tang

Volume 16. Polymers (in three parts)
Edited by R. A. Fava

Volume 17. Accelerators in Atomic Physics
Edited by P. Richard

Volume 18. Fluid Dynamics (in two parts)
Edited by R. J. Emrich

Volume 19. Ultrasonics
Edited by Peter D. Edmonds

0. INTRODUCTION: PHYSICAL DESCRIPTION OF ULTRASONIC FIELDS†

By Peter D. Edmonds and F. Dunn

List of Symbols

a	subscript denoting amplitude
A	magnitude of α/f^2 contributed by a relaxation process (except viscosity) at $f \ll f_r$
B	magnitude of α/f^2 contributed by the viscosity relaxation process at $f \ll f_v$
C_p	specific heat of medium at constant pressure
C_p'	specific heat per unit mass of medium at constant pressure
C_v	specific heat of medium at constant volume
D	logarithmic decrement
E_0	acoustic energy density = energy stored per unit volume
ΔE_0	energy loss per cycle due to absorption
f	frequency
f_v	relaxation frequency for viscosity
$g_1(\tau)$	distribution function for relaxation times
G	shear modulus of elasticity
$G^*(j\omega)$	complex shear modulus of elasticity
$G'(\omega)$, $G''(\omega)$	real and imaginary parts of G^*
G^∞	infinite frequency asymptote of G'
I	intensity
I_{1+}, I_{1-}, I_{2+}	intensities of incident, reflected, and transmitted waves
j	$\sqrt{-1}$
k	wave vector
$\mathscr{K}$	magnitude of **k**
K	bulk modulus of elasticity
n	integer
p	acoustic pressure
P	amplitude of p
P_0	ambient pressure

---

† Portions of this introduction have been adapted with permission from W. J. Fry and F. Dunn, Ultrasound: Analysis and experimental methods in biological research, *in* ''Physical Techniques in Biological Research,'' (W. L. Nastuk, ed.), Vol. 4, pp. 265–275. Academic Press, New York, 1962; and from F. Dunn, P. D. Edmonds, and W. J. Fry, Absorption and dispersion of ultrasound in biological media, *in* ''Biological Engineering'' (H. P. Schwan, ed.), pp. 207–233. McGraw-Hill, New York, 1969.

METHODS OF EXPERIMENTAL PHYSICS, VOL. 19

P_+, P_- P_{1+}, P_{1-}, P_{2+}, P_{2-} P_{max}, P_{min} P_{1i}, P_{1r}, P'_{1i}, P'_{1r} P_{2t}, P'_{2t}	see Table IV
q	general field variable
Q	amplitude of q
Q_m	quality factor $= \pi/D$
$\mathbf{e}$	general spatial coordinate
$r_{2/1}$, $r_{3/1}$, $r_{2/3}$ $r'_{1/1}$, $r'_{2/1}$	see Table IV
$\mathscr{R}_a$	amplitude reflection coefficient
$\mathscr{R}_I$	intensity reflection coefficient
s	condensation $= (\rho - \rho_0)/\rho_0$
S	amplitude of condensation; entropy (as subscript)
SWR	standing wave ratio
t	time
t_D	decay time of field amplitude parameter in an absorbing medium
T	absolute temperature
$\mathbf{T}$	shear stress
$\mathbf{T}_i$	initial value of shear stress
$\mathbf{T}^0$	asymptotic final value of shear stress
$\hat{\mathbf{T}}$	amplitude of sinusoidal shear stress
$\hat{\mathbf{T}}^0$	amplitude of asymptotic final value of shear stress
$\mathscr{T}_a$	amplitude transmission coefficient
$\mathscr{T}_I$	intensity transmission coefficient
v	wave propagation velocity; sound speed
v^0	limiting sound speed at zero frequency (compressional wave)
v_l	compressional wave speed
v_s^0	limiting shear wave speed at zero frequency
v_1, v_2, v_3 v_{l1}, v_{l2}, v_{s1}, v_{s2}	see Table IV
Δv	velocity dispersion
x	spatial coordinate
x_3	see Table IV
Z_0, Z_1, Z_2	characteristic acoustic impedance
α	amplitude absorption coefficient
α_r	generalized relaxational contribution to α
α_v	contribution to α from viscosity
β_S	adiabatic compressibility
β_T	isothermal compressibility
γ	ratio of specific heats $= C_p/C_v$
δ	phase lag between acoustic pressure and particle velocity in an absorbing medium
$\eta(\omega)$	shear viscosity coefficient
η^0	limiting shear viscosity as frequency tends to zero
$\eta^*(j\omega)$	complex shear viscosity
$\eta'(\omega)$	real part of η^*
θ	isobaric thermal expansion coefficient
θ_1, θ_2, θ'_1, θ'_2	see Table IV

Θ	amplitude of temperature peturbation
λ	wavelength
λ_3	see Table IV
ξ	"particle" (elemental volume) displacement
ξ_x	particle displacement in x direction
$\dot{\xi}$	particle velocity
$\dot{\xi}_x$	particle velocity in x direction
$\ddot{\xi}_x$	particle acceleration in x direction
$\Xi_{\mp}$	amplitudes of particle displacement for waves in the positive and negative directions
$\dot{\Xi}$	amplitude of particle velocity
$\ddot{\Xi}$	amplitude of particle acceleration
ρ	density
ρ_0	mean density
ρ_1, ρ_2, ρ_3	see Table IV
τ	relaxation time
τ_a, τ_b	limits of relaxation time distribution
τ_v	relaxation time for viscosity
Y	instantaneous temperature increment in medium
ϕ	scalar displacement potential (irrotational)
Φ	vector displacement potential (rotational)
ψ	scalar velocity potential
ω	angular frequency
(dot over symbol)	differentiation with respect to time

0.1. Development of Propagation Relations

The propagation of an acoustic disturbance or the presence of an acoustic field in an elastic medium is characterized by changes in a number of the physical variables that describe the state of the system or medium. Examples of these variables are pressure, temperature, and density.

For a traveling, sinusoidal, plane wave propagating in the positive direction of the x axis (when no attenuation of the waves occurs because we assume absorption of energy by the medium is absent), the changes in the physical variables can each be expressed in the form of Eq. (0.1.1), provided that the medium responds linearly to the stresses imposed upon it.

$$q = Q \cos \omega(t - x/v) \quad \text{or} \quad q = \text{Re}\{Q \exp[j\omega(t - x/v)]\}. \quad (0.1.1)$$

In this equation q designates any one of the variables that undergoes sinusoidal change owing to the presence of the disturbance in the medium and Q designates the amplitude of the cyclic change in that variable; t and x are the time and space coordinates, respectively, ω is the angular frequency ($\omega = 2\pi f$), f the frequency, and v the free-field sound speed, i.e., the propagation speed of a plane wave traveling through a liquid medium of infinite extent. Equation (0.1.1) is one solution, namely, that representing a wave traveling in the positive x direction, of the one-dimensional elastic wave equation as it applies to an ideal, linear, homogenous, perfectly elastic (dissipationless), fluid medium

$$\partial^2 q/\partial t^2 = (1/v^2)\, \partial^2 q/\partial x^2. \tag{0.1.2}$$

In this equation q could represent the instantaneous displacement ξ of an element of volume of the medium. This approximation to the more general hydrodynamical equation is valid under conditions that permit linearization, that is, when the velocity amplitude $\ddot{\Xi} = (\partial \xi/\partial t)_{max}$ of the elementary volume is small in comparison with the speed of sound v and when the adiabatic compressibility β_S, which is the reciprocal of the adiabatic elastic bulk modulus K, is not significantly dependent on pressure over the range of pressure variations present in the acoustic field.

Since sound propagation is very close to an adiabatic process at most frequencies of interest, the adiabatic compressibility is a significant parameter in the description of sound propagation. It is related to the free-field sound speed for compressional waves as follows:

$$v^2 = v_l^2 = \frac{1}{\rho_0 \beta_S} = \frac{\gamma}{\rho_0 \beta_T} = \frac{C_p/C_v}{\rho_0 \beta_T}, \tag{0.1.3}$$

where β_S is the adiabatic compressibility of the medium and ρ_0 the mean density of the medium. The sound speed can be expressed, as indicated in Eq. (0.1.3), in terms of the isothermal compressibility β_T by introducing the ratio of specific heats $\gamma = C_p/C_v$, where C_p and C_v are the specific heats of the medium at constant pressure and constant volume, respectively. Clearly, a measurement of the speed of a plane compressional wave can be interpreted immediately to yield the adiabatic compressibility of the medium if the density is known; and if the value of γ is also known, the isothermal compressibility can be determined.

Equation (0.1.2) is a special case of the more general wave equation that is applicable to three-dimensional propagation:

$$\frac{\partial^2 \xi}{\partial t^2} = \frac{1}{\rho_0 \beta_S} \nabla^2 \xi. \tag{0.1.4}$$

Solutions of Eq. (0.1.4) include not only waves propagating in the positive r direction away from the origin but also those propagating in the negative r direction toward the origin. All are represented when the $\pm$ sign is placed in the exponent for one-dimensional propagation, e.g.,

$$\xi = \Xi_{\mp}(r) \exp[j(\omega t \pm \mathbf{k} \cdot \mathbf{r})]. \tag{0.1.5}$$

The wave vector $\mathbf{k}$ that appears in the solution is related to the angular frequency and the sound speed as

$$\mathbf{k} = \mathcal{K}\mathbf{n}; \qquad \mathcal{K} = -\omega/v = 2\pi/\lambda; \qquad v = f\lambda. \qquad (0.1.6)$$

Equation (0.1.4) is itself a specialization, applicable to fluids of the type indicated, of the following wave equation describing propagation of disturbances in a dissipationless, isotropic, elastic solid:

$$\frac{\partial^2 \boldsymbol{\xi}}{\partial t^2} = \frac{K + 4G/3}{\rho_0} \nabla\nabla \cdot \boldsymbol{\xi} - \frac{G}{\rho_0} \nabla \times \nabla \times \boldsymbol{\xi}, \qquad (0.1.7)$$

where K and G are, respectively, the bulk and shear moduli of elasticity of the medium.

It is possible to express the displacement vector as the sum of terms involving a scalar potential ϕ and a vector potential Φ as

$$\boldsymbol{\xi} = \nabla\phi + \nabla \times \boldsymbol{\Phi} \qquad (0.1.8)$$

For irrotational motion, such as in a spherical wave, the vector potential $\Phi = 0$ and only the scalar displacement potential ϕ remains; that is,

$$\boldsymbol{\xi} = \nabla\phi. \qquad (0.1.9a)$$

The time derivative of the displacement potential is the velocity potential ψ, i.e.,

$$\partial\phi/\partial t = \psi; \qquad \dot{\boldsymbol{\xi}} = \nabla\psi. \qquad (0.1.9b)$$

These potentials are fundamental functions (analogous to electric field potentials) in terms of which acoustic field parameters may be expressed. The specialization of Eq. (0.1.7) for fluids is obtained when the modulus of shear rigidity G is set equal to zero, which is true for lossless fluids, since the latter are characterized by an inability to support an elastic shear strain, and $\boldsymbol{\Phi} = 0$.

Returning to a consideration of the simple plane wave propagating in an ideal isotropic elastic medium in the positive x direction, we can express the sinusoidally varying acoustic parameters in terms of the displacement potential or velocity potential and in terms of one another.

$$p = -\rho_0 \, \partial\psi/\partial t, \qquad \dot{\xi}_x = (\nabla\psi)_x, \qquad (0.1.10)$$

$$s = (\rho - \rho_0)/\rho_0 = \beta_S p, \qquad (0.1.11)$$

$$Y = (T\theta/\rho_0 C_p')p = (\gamma - 1)(\beta_S/\theta)p, \qquad (0.1.12)$$

where s is the condensation or the fractional change in density, ρ the instantaneous density, Y the instantaneous temperature increment resulting from adiabatic compression of the medium, T the absolute temper-

TABLE I.

Parameter	Parameter symbol q	Amplitude symbol Q	P	S
Pressure	p	P	—	$\mp\rho_0 v^2$
Condensation	s	S	$\pm\dfrac{1}{\rho_0 v^2}$	—
Particle displacement	ξ	Ξ	$\pm\dfrac{1}{j\omega\rho_0 v}$	$\pm\dfrac{v}{j\omega}$
Particle velocity	$\dot{\xi}$	$\dot{\Xi}$	$\pm\dfrac{1}{\rho_0 v}$	$\pm v$
Particle acceleration	$\ddot{\xi}$	$\ddot{\Xi}$	$\pm\dfrac{j\omega}{\rho_0 v}$	$\pm j\omega v$
Temperature	Y	Θ	$\pm\dfrac{1}{\theta}\left(\beta_T - \dfrac{1}{\rho_0 v^2}\right)$	$\pm\dfrac{\rho_0 v^2}{\theta}\left(\beta_T - \dfrac{1}{\rho_0 v^2}\right)$

[a] Multiply expression in the table by the column heading to obtain the relations equal to the amplitude quantities tabulated in the amplitude symbol column. Note that $j = \sqrt{-1}$. The relations apply to plane waves traveling in either direction. The upper sign applies to waves traveling in the positive direction and the lower sign to the negative direction [see Eq. (0.1.5)]. The amplitude of a change in any one physical parameter is equal to the amplitude of the change in any other physical parameter multiplied by the absolute value of the appropriate quantity in the table. A self-consistent set of units is used throughout the table (e.g., mks or cgs).

ature of the medium, θ the isobaric thermal expansion coefficient, and C_p' the heat capacity at constant pressure per unit mass. The interrelation of the acoustic field parameters is shown in Table I.

The method of detection and description of the field, in any specific case, may depend on the measurement of one or several of these parameters. The quantity $\rho_0 v$, the product of density and sound speed, which appears in many relations in the table, is known as the characteristic acoustic impedance of the medium Z_0; that is,

$$Z_0 = \rho_0 v. \qquad (0.1.13)$$

For plane traveling waves, Z_0 is numerically equal to the specific acoustic impedance, which is defined as the ratio of the pressure p to the particle velocity $\dot{\xi}$ at any point in the field. For other field configurations, including plane standing waves, the specific acoustic impedance differs numerically from $\rho_0 v$ and is, in general, a function of position. It should also be noted that the characteristic acoustic impedance is dependent on

Relations between Amplitudes of the Various Physical Parameters[a]

Ξ	$\dot\Xi$	$\ddot\Xi$	θ
$\pm j\omega\rho_0 v$	$\pm\rho_0 v$	$\pm\dfrac{\rho_0 v}{j\omega}$	$\pm\dfrac{\theta}{(\beta_T - (1/\rho_0 v^2))}$
$\pm\dfrac{j\omega}{v}$	$\pm\dfrac{1}{v}$	$\pm\dfrac{1}{j\omega v}$	$\pm\dfrac{\theta}{\rho_0 v^2[\beta_T - (1/\rho_0 v^2)]}$
—	$\pm\dfrac{1}{j\omega}$	$\mp\dfrac{1}{\omega^2}$	$\pm\dfrac{\theta}{j\omega\rho_0 v[\beta_T - (1/\rho_0 v^2)]}$
$\pm j\omega$	—	$\pm\dfrac{1}{j\omega}$	$\pm\dfrac{\theta}{\rho_0 v[\beta_T - (1/\rho_0 v^2)]}$
$\mp\omega^2$	$\pm j\omega$	—	$\pm\dfrac{j\omega\theta}{\rho_0 v[\beta_T - (1/\rho_0 v^2)]}$
$\pm\dfrac{j\omega\rho_0 v}{\theta}\left(\beta_T - \dfrac{1}{\rho_0 v^2}\right)$	$\pm\dfrac{\rho_0 v}{\theta}\left(\beta_T - \dfrac{1}{\rho_0 v^2}\right)$	$\pm\dfrac{\rho_0 v}{j\omega\theta}\left(\beta_T - \dfrac{1}{\rho_0 v^2}\right)$	—

the type of wave that is propagating, since the speed of shear waves is different from that of compressional waves.

The intensity I of the sound wave is defined as the time average of the rate of propagation of energy through unit area normal to the direction of propagation; for plane traveling waves, I is related to field-parameter amplitudes by

$$I = P^2/2Z_0 = P\ddot\Xi/2 = Z_0\ddot\Xi^2/2. \qquad (0.1.14)$$

The energy density E_0 of the wave motion at a specific position in the field is the sum of the kinetic energy per unit volume of the moving volume element and the potential energy per unit volume of compression (or expansion) of the element. For plane traveling waves, it is equal to the ratio of the intensity to the sound speed, i.e.,

$$E_0 = \rho_0\ddot\Xi^2/2 = I/v. \qquad (0.1.15)$$

Root mean square (rms) quantities are not employed in the majority of publications in acoustics, and consequently the symbols in Eqs. (0.1.14) and (0.1.15) are the amplitudes of the acoustic field parameters. If rms values had been used, the factors 2 would have been eliminated from the equations.

As stated previously, linearizing of the hydrodynamical equations depends on two assumptions which can now be expressed symbolically as

$$\ddot\Xi/v \ll 1; \qquad [(\beta_S)_{P_0+P} - (\beta_S)_{P_0-P}]/(\beta_S)_{P_0} \ll 1, \qquad (0.1.16)$$

TABLE II. Numerical Example of Physical Parameters for Water

Material	f	T	P_0	I	P	S	Ξ	$\dot{\Xi}$	$\ddot{\Xi}$	Θ	$\ddot{\Xi}/v$
To obtain results in:	MHz	°C	atm	W/cm²	atm		cm	cm/sec	cm/sec²	°C	
Multiply figures in table by:	1	1	1	1	1	10^{-5}	10^{-6}	1	10^{6}	10^{-4}	10^{-5}
To obtain results in:			N/m²^a	W/m²	N/m²^a		m	m/s	m/s²		
Multiply figures in table by:			1.013×10^5	10^4	1.013×10^5		10^{-8}	10^{-2}	10^4		
Water											
degassed and distilled	1	30	1	0.01	0.171	0.762	0.183	1.15	7.22	3.82	0.762
				1	1.71	7.62	1.83	11.5	72.2	38.2	7.62
				100	17.1	76.2	18.3	115	722	382	76.2

^{a} N/m² ≡ Pascal (Pa).

where P_0 represents the ambient pressure in the absence of a sound wave. Nonlinear or second-order effects still may be of importance for values of $\ddot{\Xi}/v$ smaller, for example, than 0.01, but the linearized equations constitute a good first approximation for calculating values of the physical parameters when this numerical limit is placed on the interpretation of the symbol $\ll 1$.

Table II shows values of the numerical magnitudes of the acoustic field parameters for a plane traveling wave, when the propagation medium is water, for representative intensity values of the wave spanning four orders of magnitude. It may be noted in particular that the temperature excursion in water is small and that this parameter is entirely unrelated to the monotonic rise in temperature of the specimen that occurs when energy is absorbed by the specimen. However, even for low-amplitude ultrasonic waves, which may be used as a probe to measure the response of a system to an extremely small perturbation, the pressure amplitude may be comparable to one atmosphere, and the amplitude of the particle acceleration can be exceedingly high and give rise to significant local stresses.

Table III lists values for the various characteristic constants of a number of materials of general utility. These data may be used in connection with the relations appearing in Table I to obtain numerical values of field parameters such as those listed in Table II. It is usually convenient to express the intensity in watts per square centimeter and the acoustic pressure amplitude in atmospheres. However, for calculations using the expressions of Table I, the intensity should be expressed in ergs per square centimeter per second and the pressure amplitude in dynes per square centimeter if the other parameters are expressed in the indicated units. Equivalent mks units may also be used.

0.2. Reflection and Refraction

Reflection and refraction of acoustic waves occur in a manner analogous to that for electromagnetic waves, and many of the concepts that arise in the theory of transmission lines are applicable in "one-dimensional" situations. The formulas listed in Table IV are for media within which no acoustic absorption occurs and for which the normals to the planar wave fronts and the normals to the interfaces lie in the same plane.

Case 1. Reflection and transmission occur at a single interface between two media. The reflection coefficient $\mathcal{R}_a$, the transmission coefficient $\mathcal{T}_a$, and the standing wave ratio (SWR) for waves incident on the interface

TABLE IIIA. Physical Constants of Various Materials

Material	T	P_0	ρ_0	v	$\rho_0 v$	C_p/C_v	β_T	θ	α
To obtain results in:	°C	atm	g/cm³	cm/sec	g/(cm² sec)		cm²/d	(°C)⁻¹	Np/cm
Multiply figures in table by:	1	1	1	10⁵	10⁵	1	10⁻¹²	10⁻⁵	1
To obtain results in:		N/m²ᵃ	kg/m³	m/s	kg/(m² s)		m²/Nᵃ		Np/m
Multiply figures in table by:		1.013 × 10⁵	10³	10³	10⁶		10⁻¹¹		10²
Water									
Degassed, distilled	0	1	0.999841	1.4027	1.4025	1.000583	50.86	−5.89	
α proportional to f^{2b}	10	1	0.999701	1.4476	1.4472	1.001085	47.79	+9.45	
	20	1	0.998207	1.4827	1.4800	1.00656	45.86	21.19	25 × 10⁻⁵
	30	1	0.995651	1.5094	1.5028	1.01526	44.76	30.75	
	40	1	0.992220	1.5292	1.5173	1.02575	44.20	38.93	
	0	136	0.9941	1.4245	1.4161	1.00012	49.58	2.01	
	10	136	0.9946	1.4700	1.4621	1.00356	46.69	15.09	
	20	136	0.9961	1.5057	1.4998	1.01041	44.74	25.10	
	30	136	0:9986	1.5329	1.5308	1.01827	43.40	34.05	
	40	136	1.0019	1.5531	1.5560	1.02672	42.48	40.92	
Water Solutions									
0.9% normal salineᶜ	0	1	1.00668	1.4134	1.4228			1.98	
α proportional to f^{2b}	10	1	1.00631	1.4582	1.4674			8.46	
	20	1	1.00460	1.4932	1.5001			23.89	25 × 10⁻⁵
	30	1	1.00189	1.5198	1.5268			29.94	
	40	1	0.99837	1.5394	1.5369			40.07	

Oils					
Castor, at 30°C					
α proportional to $f^{5/3\,b,d}$					
0	1	0.972	1.580	1.536	0.26
10	1	0.960	1.536	1.474	0.16
20	1	0.952	1.494	1.422	0.096
30	1	0.946	1.452	1.374	0.057
40	1	0.941	1.411	1.328	0.037
Phenylated silicone					
Dow-Corning No. 710					
α proportional to f^2 to ~20 MHz^{b,e}					
0	1	1.124	1.446	1.625	0.135
10	1	1.112	1.409	1.567	0.070
20	1	1.102	1.378	1.518	0.040
30	1	1.095	1.349	1.477	0.024
40	1	1.089	1.321	1.438	
Aluminum (rolled)		2.70	6.42	17.3	
Ceramics (approximate range)		2.5–3.4	4.6–6.8	12–18	
Glasses					
Borate crown (light)		2.24	5.10	11.4	
Pyrex (702)		2.32	5.64	13.1	
Silicate flint (heavy)		3.88	3.98	15.4	
Silica (fused)		2.2	5.97	13.1	
Stainless steel (347)		7.91	5.79	45.8	

TABLE IIIB. Physical Constants of Biological Media[f]

Material	T	P_0	ρ_0	v	$\rho_0 v$	α
To obtain results in:	°C	atm	g/cm³	cm/sec	g/(cm² sec)	Np/cm
Multiply figures in table by:	1	1	1	10⁵	10⁵	1
To obtain results in:		N/m²[a]	kg/m³	m/sec	kg/(m² sec)	Np/m
Multiply figures in table by:		1.013×10^5	10³	10³	10⁶	10²
Central nervous system[g-i] Brain (average)	37	1	1.03	1.51	1.56	See Table IIIC
Soft parenchymal tissues, e.g., liver, kidney (average)	37	1	1.05	1.56	1.64	
Muscle (skeletal)[b,g,h]	37	1	1.07	1.57	1.68	0.13
Fat[b,h]	37	1	0.97	1.44	1.40	0.05
Bone						
Skull (human)[j]	37	1	1.7	3.36	6.0	
Frequency (MHz) 0.6		1				0.4
0.8		1				0.9
1.2		1				1.7
1.6		1				3.2
1.8		1				4.2
2.25		1				5.3
3.5		1				7.8

TABLE IIIC. Ultrasonic Absorption in Biological Tissues[j]

Tissue	Frequency f (MHz)						Regression analysis fit	
	0.5	0.7	1	3	4	7	α	R
Brain	—	0.014 ± 0.003	0.029 ± 0.004	—	—	0.23 ± 0.09	$0.024f^{1.18}$	0.993
Heart	—	0.018 ± 0.009	0.033 ± 0.006	—	—	0.21 ± 0.03	$0.028f^{1.04}$	0.995
Kidney	—	0.017 ± 0.007	0.033 ± 0.004	—	—	0.20 ± 0.002	$0.028f^{1.02}$	0.994
Liver	0.010 ± 0.006	0.020 ± 0.003	0.023 ± 0.004	—	0.14 ± 0.03	0.24 ± 0.02	$0.026f^{1.17}$	0.995
Tendon	0.050 ± 0.03	0.16 ± 0.1	0.11 ± 0.04	0.53 ± 0.2	0.75 ± 0.4	1.4 ± 0.5	$0.14f^{1.17}$	0.973
Testis	0.0078 ± 0.002	0.0085 ± 0.001	0.015 ± 0.003	—	0.079 ± 0.02	0.12 ± 0.02	$0.015f^{1.11}$	0.995

[a] $N/m^2 \equiv$ Pascal (Pa); $(m^2/N) \equiv (Pa^{-1})$.
[b] Values of α for 1 MHz.
[c] Measurements of W.D. Wilson, U.S. Naval Ordnance Laboratory.
[d] Indicated power dependence holds over entire range of measurements from 400 kHz to 500 MHz at 30°C.
[e] Measurements at 26°C over the frequency range 1 to 2000 MHz indicate (assuming negligible velocity dispersion) the presence of a single relaxation process centered at 40 MHz.
[f] Extensive tabular and graphical data are given by Goss et al.,[1] Goss et al.,[2] Chivers and Parry,[3] and Bamber and Hill.[4]
[g] α varies with direction of sound propagation relative to fiber orientation.
[h] α proportional to frequency.
[i] Absorption coefficient listed for bone includes effects of reflections at interfaces within the bone structure. More extensive data are given by Fry and Barger.[5]
[j] Goss et al.[2] The absorption coefficient $\alpha \pm$ standard deviation is given in nepers per centimeter at 37°C.

TABLE IIID.   Ultrasonic Attenuation in Biological Tissues[a]

Tissue	Frequency f (MHz)						Regression analysis fit	
	0.5	0.7	1	3	4	7	A	R
Brain	0.032	0.047	0.07	0.24	0.34	0.64	$0.07f^{1.14}$	0.822
Heart	0.060	0.086	0.13	0.41	0.56	1.0	$0.13f^{1.07}$	0.98
Kidney	0.049	0.070	0.10	0.34	0.47	0.87	$0.10f^{1.09}$	0.973
Liver	0.038	0.055	0.08	0.29	0.40	0.75	$0.08f^{1.13}$	0.934
Tendon	0.33	0.42	0.56	1.3	1.6	2.5	$0.56f^{0.763}$	0.998

[a] Goss et al.[2] The attenuation coefficient A is given in nepers per centimeter at 37°C.

from medium 1 are functions only of the ratio of the characteristic acoustic impedances of the two media, $r_{2/1} = \rho_2 v_2 / \rho_1 v_1$.

For the partial reflection at normal incidence, the complete expression for the pressure variation for a sinusoidal disturbance of infinite extent can be represented by the summation of two waves, one traveling in the positive and the second in the negative x direction:

$$p = P_+ \exp[j\omega(t - x/v)] + P_- \exp[j\omega(t + x/v)], \qquad (0.2.1)$$

where P_+ is the amplitude of the pressure wave traveling in the positive direction and P_- the amplitude of a similar wave traveling in the negative direction.   The standing wave ratio in either medium may be defined as

$$\text{SWR} = \frac{|P_{max}|}{|P_{min}|} = \frac{1 + |P_-/P_+|}{1 - |P_-/P_+|}, \qquad (0.2.2)$$

where P_{max} is the maximum value of the pressure amplitude in the field of interference of the incident and reflected waves and P_{min} the minimum value of the pressure amplitude.

A distinction is required between reflection and transmission coefficients referring to the amplitude of the disturbance and those referring to the power carried by the acoustic waves.   The coefficients are defined, respectively, as the ratios of the amplitudes or the intensities of the re-

[1] S. A. Goss, R. L. Johnston, and F. Dunn, J. Acoust. Soc. Am. 64, 423–467 (1978).

[2] S. A. Goss, L. A. Frizzell, and F. Dunn, Ultrasound Med. Biol. 5, 181–186 (1979).

[3] R. C. Chivers and R. J. Parry, J. Acoust. Soc. Am. 63, 940–953 (1978).

[4] J. C. Bamber and C. R. Hill, Ultrasound Med. Biol. 5, 149–157 (1979).

[5] F. J. Fry and J. E. Barger, J. Acoust. Soc. Am. 63, 1576–1590 (1978).

TABLE IV. Pressure Amplitude of Reflected and Transmitted Waves for Various Combinations of Media[a]

Configuration	Definition	Formula

Case 1. Wave in medium 1 at normal incidence on boundary between medium 1 and medium 2. No energy returned to interface from medium 2. No absorption in media.

$\rho_1, v_1 \quad \rho_2, v_2$

$\xrightarrow{\quad} P_{1+}$

$\xrightarrow{\quad} P_{2+}$

$\xleftarrow{\quad} P_{1-}$

$P_{2-} = 0$

$\xrightarrow{\quad} +$

$$r_{2/1} = \frac{\rho_2 v_2}{\rho_1 v_1}$$

$$\mathscr{R}_a = \left| \frac{P_{1-}}{P_{1+}} \right| = \left| \frac{1 - r_{2/1}}{1 + r_{2/1}} \right|$$

$$\mathscr{T}_a = \left| \frac{P_{2+}}{P_{1+}} \right| = \left| \frac{2r_{2/1}}{1 + r_{2/1}} \right|$$

$$(SWR)_1 = \begin{cases} r_{2/1} & \text{when } r_{2/1} > 1 \\ 1/r_{2/1} & \text{when } r_{2/1} < 1 \end{cases}$$

Case 2. Wave in medium 1 at normal incidence. Slab of medium 3 interposed between media 1 and 2. No energy returned to interface from medium 2. No absorption in media.

$\rho_1, v_1 \quad | \quad \rho_3, v_3 \quad | \quad \rho_2, v_2$

$\xrightarrow{\quad} P_{1+}$

$\xleftarrow{\ x_3\ \rightarrow}$

$\xrightarrow{\quad} P_{2+}$

$\xleftarrow{\quad} P_{1-}$

$\xleftarrow{\quad} P_{2-} = 0$

$\xrightarrow{\quad} +$

$$r_{2/1} = \frac{\rho_2 v_2}{\rho_1 v_1}$$

$$r_{3/1} = \frac{\rho_3 v_3}{\rho_1 v_1}$$

$$r_{2/3} = \frac{\rho_2 v_2}{\rho_3 v_3}$$

$$\mathscr{R}_a = \left| \frac{P_{1-}}{P_{1+}} \right| = \left[\frac{4r_{2/1}}{(r_{2/1} + 1)^2 \cos^2(\omega x_3 / v_3) + (r_{3/1} + r_{2/3})^2 \sin^2(\omega x_3 / v_3)} \right]^{1/2}$$

$$\mathscr{T}_a = \left| \frac{P_{2+}}{P_{1+}} \right| = \left[\frac{(4r_{2/1})^2}{(r_{2/1} + 1)^2 \cos^2(\omega x_3 / v_3) + (r_{3/1} + r_{2/3})^2 \sin^2(\omega x_3 / v_3)} \right]^{1/2}$$

[a] Only the ratios of the magnitudes of the pressure amplitudes are shown in the table, i.e., phase factors are not shown.

TABLE IV (Continued)

Configuration	Definition	Formulas
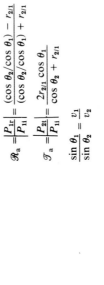	$$r_{2/1} = \frac{\rho_2 v_2}{\rho_1 v_1}$$	$$\mathscr{R}_a = \left|\frac{P_{1r}}{P_{1i}}\right| = \left|\frac{(\cos\theta_2/\cos\theta_1) - r_{2/1}}{(\cos\theta_2/\cos\theta_1) + r_{2/1}}\right|$$ $$\mathscr{T}_a = \left|\frac{P_{2t}}{P_{1i}}\right| = \left|\frac{2r_{2/1}\cos\theta_1}{\cos\theta_2 + r_{2/1}}\right|$$ $$\frac{\sin\theta_1}{\sin\theta_2} = \frac{v_1}{v_2}$$

Case 3. Wave in medium 1 incident at angle θ_1 to normal to boundary between media 1 and 2. Fluid media. No energy returned to interface from medium 2. No absorption in media.

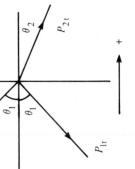

Case 4. Longitudinal or shear wave in medium 1 incident at angle θ_1 to normal to boundary. Solid or viscoelastic media. Mode conversion with generation of shear and longitudinal waves in media 1 and 2. No energy returned to interface from medium 2. Absorption in media sufficiently small for impedances of media to be approximated by their real parts.[b]

$$r_{2/1} = \frac{\rho_2 v_{l2}}{\rho_1 v_{l1}}$$

$$r'_{2/1} = \frac{\rho_2 v_{s2}}{\rho_1 v_{l1}}$$

$$r'_{1/1} = \frac{\rho_1 v_{s1}}{\rho_1 v_{l1}}$$

$$\frac{\sin \theta_1}{\sin \theta_2} = \frac{v_{l1}}{v_{l2}}$$

$$\frac{\sin \theta_1}{\sin \theta'_2} = \frac{v_{l1}}{v_{s2}}$$

$$\frac{\sin \theta'_1}{\sin \theta'_1} = \frac{v_{l1}}{v_{s1}}$$

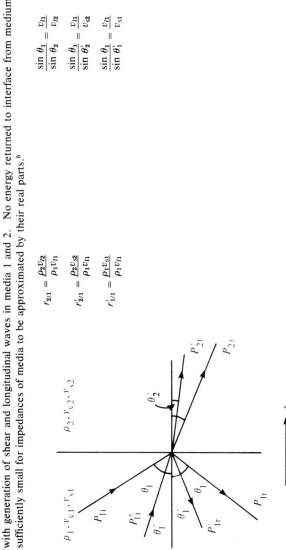

[b] See Muskat and Meres[6] for expressions for ratios of normal components of wave amplitudes.

[6] M. Muskat and M. W. Meres, *Geophysics* **5**, 115 (1940).

flected and transmitted waves to the amplitude or intensity of the incident wave, i.e.,

$$\mathcal{R}_a = \frac{P_{1-}}{P_{1+}} = \frac{Z_2 - Z_1}{Z_2 + Z_1}, \qquad \mathcal{T}_a = \frac{P_{2+}}{P_{1+}} = \frac{2Z_2}{Z_2 + Z_1}, \qquad (0.2.3)$$

$$\mathcal{R}_I = \frac{I_{1-}}{I_{1+}} = \left(\frac{P_{1-}}{P_{1+}}\right)^2 = \mathcal{R}_a{}^2, \qquad \mathcal{T}_I = \frac{I_{2+}}{I_{1+}} = \frac{Z_1}{Z_2} \mathcal{T}_a{}^2. \qquad (0.2.4)$$

In these expressions the subscript a designates amplitude coefficients and the subscript I power or intensity coefficients. Conservation of energy requires that the sum of the power reflection and transmission coefficients should always equal unity whereas the sum of the amplitude coefficients is not in general equal to unity.

$$\mathcal{R}_I + \mathcal{T}_I = 1. \qquad (0.2.5)$$

The coefficients described here as transmission coefficients are frequently described as absorption coefficients in experiments involving irradiation of a specimen. This difference in viewpoint arises from the fact that in irradiation experiments interest is confined to measuring the standing wave ratio in the medium situated in front of the specimen. Any energy transmitted into the specimen is therefore effectively lost or appears to be absorbed. From the point of view of the properties of the interface between two media, this "lost" energy is merely transmitted into the second medium. The term "absorption coefficient" will be reserved for later use when a study is made of the attenuation of the transmitted wave amplitude mechanisms operative within the second medium.

Case 2. A slab of a third medium of thickness x_3 is interposed between two media. The reflection coefficient $\mathcal{R}_a$ and the transmission coefficient $\mathcal{T}_a$ are functions of the ratios of the characteristic impedances and of the quantity $\omega x_3/v_3$ (equal to $2\pi x_3/\lambda_3$), which is determined by the ratio of the thickness of medium 3 and the wavelength in it. If the characteristic impedance of medium 3 is intermediate between those of media 1 and 2, then the transmission coefficient can be maximized by choosing thickness x_3 to satisfy the relation

$$x_3/\lambda_3 = (2n - 1)/4, \qquad n = 1, 2, 3, \ldots . \qquad (0.2.6)$$

The transmission coefficient then becomes

$$\mathcal{T}_a = |2r_{2/1}/(r_{3/1} + r_{2/3})|; \qquad (0.2.7)$$

that is, for maximum transmission, the best choice of thickness that can be made for any interposed material (if its characteristic acoustic impedance has any value between those of the other media) is one-quarter wavelength or odd multiples thereof. In addition, if one is free to choose the interposed material so that its characteristic acoustic impedance is optimum for transmitting the acoustic energy, then the reflected wave in medium 1 can be eliminated by choosing the intermediate material so that

$$(\rho_3 v_3)^2 = (\rho_1 v_1)(\rho_2 v_2). \tag{0.2.8}$$

If the characteristic acoustic impedance of medium 3 does not have a value between those of the other two media, then the optimum choice of thickness for the slab to obtain the maximum value to the transmission coefficient is an integral multiple of a half-wavelength, i.e.,

$$x_3/\lambda_3 = n(\tfrac{1}{2}), \qquad n = 1, 2, 3, \ldots . \tag{0.2.9}$$

The transmission coefficient then becomes identical with Eq. (0.2.3). If media 1 and 2 have nearly equal characteristic acoustic impedances that are less than that of medium 3 and if the thickness of the interposed slab satisfies the relation $r_{3/1}(\omega x_3/v_3) \leq \tfrac{1}{10}$, then the transmission coefficient does not differ from that of case 1 by more than 1%. If the characteristic acoustic impedance of medium 3 is less than that of media 1 and 2, then $r_{2/3}$ should be used in place of $r_{3/1}$ in the foregoing inequality.

Case 3. A plane wave is incident at any angle θ_1 on the plane interface between two fluid media. The angle of refraction θ_2 is a function of the angle of incidence and the ratio of the velocities of sound in the two media. The pressure transmission and reflection coefficients also involve the ratio of characteristic impedances. If $\sin \theta_1 > v_1/v_2$, then the incident wave is totally reflected, and there is no propagation of a refracted wave in medium 2. It should also be observed from the form of the reflection coefficient that there is no reflected wave if the ratio of velocities satisfies either of the relations $\rho_2/\rho_1 > v_1/v_2 > 1$ or $\rho_2/\rho_1 < v_1/v_2 < 1$ and if the angle of incidence satisfies the relation

$$\sin \theta_1 = \frac{r_{2/1}^2 - 1}{r_{2/1}^2 - (v_2/v_1)^2}. \tag{0.2.10}$$

Case 4. When the waves are incident obliquely and both media are solid or viscoelastic, the effects of shear rigidity are exhibited. The boundary conditions to be satisfied are: continuity of pressure and continuity of the normal and parallel components of particle displacement. The condition on the parallel component is satisfied by the occurrence of shear waves in one or both media. In the configuration of Table IV, the direction of polarization of the shear waves is in the plane of the diagram. An obliquely incident longitudinal wave generates reflected and refracted longitudinal waves and, in addition, reflected and refracted shear waves. An incident shear wave polarized in the plane of the diagram generates a similar set of four waves. A shear wave obliquely incident as indicated and polarized perpendicularly to the diagram, that is, parallel to the interface, will generate only refracted and reflected shear waves of the same polarization since there is no component of motion perpendicular to the interface.[7] If either medium behaves as an ideal fluid, then it does not support shear wave propagation.[8] Muskat and Meres[6] derived expressions for the ratios of the components of displacements perpendicular to an interface between two perfectly elastic solids; their results may be used not only for perfectly elastic solids but also, with caution, for those viscoelastic solids exhibiting small absorption of energy. Absorption can be regarded as the result of independent processes influencing the wave amplitudes during propagation toward and away from the interface.

The formulas given in this section (Cases 1–4) are important, for example, in calculating, at least approximately, sound speed values from standing wave data, the magnitude of the effect of the reflected acoustic energy on driving transducers, the amplitude of the waves reflected at tissue interfaces, the accuracy of geometric placement or localization of a beam focus deep in tissue, etc. The formulas are also useful in the design of ultrasonic instruments where considerations of energy transfer from the transducer to the material of interest arise. More complicated configurations of materials and interfaces may arise in practice. The effects on the field of absorption within a medium will be considered in the following sections dealing with the physical mechanisms of absorption.

0.3. Absorption

When an ultrasonic wave propagates through any real medium, energy is absorbed from the wave and converted into heat. The rate of heat production in a selected volume of a medium in which such a field exists is determined by the amplitude, frequency, and spatial distribution of the field parameters. A variety of different mechanisms may play a role in the conversion of sonic energy into heat.

The occurrence of absorption modifies the phenomenological description of lossless plane wave propagation by the introduction of an absorption coefficient into Eq. (0.1.1), i.e.,

$$q = Q \exp(-\alpha x) \, \text{Re}\{\exp[j\omega(t - x/v)]\}, \qquad (0.3.1)$$

where α is the amplitude absorption coefficient per unit distance. The in-

[7] M. R. Redwood, "Mechanical Waveguides." Pergamon, Oxford, 1960.
[8] W. G. Mayer, *IEEE Trans. Sonics Ultrason.* **SU-11,** 1 (1964).

tensity absorption coefficient per unit distance is equal to 2α. The fractional energy *loss* per unit volume per cycle is

$$\frac{\Delta E_0}{E_0} = \frac{1}{E_0} \int_0^{1/f} (P_0 + p)\, dV = 2\alpha\lambda, \tag{0.3.2}$$

where E_0 is the energy stored per unit volume. This quantity may also be expressed in terms of a quality factor Q_m or the logarithmic decrement D of a field parameter per cycle, defined by Eq. (0.3.3), both of which are commonly used to describe the behavior of acoustic or electrical resonators:

$$\frac{\Delta E_0}{\pi E_0} = \frac{1}{Q_m} = \frac{D'}{\pi} = \tan \delta = \frac{1}{\pi f t_D} = \frac{\alpha\lambda}{\pi} \qquad \text{for} \quad \alpha \ll \frac{\omega}{v}, \tag{0.3.3}$$

where δ is the angle of lag between a perturbation applied to the medium and an appropriate response parameter and t_D is the decay constant of a field-amplitude parameter. Absorption occurs in a homogeneous medium when the changes in density are not in time phase with the changes in pressure, i.e., when the time at which the maximum pressure occurs differs from the time at which the maximum density occurs. This type of behavior is produced by a variety of mechanisms classified under two general categories: relaxation and hysteresis for homogeneous media; relative motion and bubble mechanisms, for inhomogeneous media.

0.3.1. Relaxation Processes

It will be convenient to discuss relaxation phenomena first in terms of a specific example. The relaxation mechanism that is related to the shear viscosity of the medium is chosen for this purpose. If viscosity is the only mechanism responsible for absorption of a traveling, plane, compressional wave, then the absorption coefficient is given by

$$\alpha_v = \frac{2\pi^2 f^2}{\rho_0 v_l^3} \cdot \frac{4}{3} \eta \equiv B f^2 \qquad \text{for} \quad \frac{\alpha_v \lambda}{2\pi} \ll 1, \tag{0.3.4}$$

where η is the shear viscosity coefficient of the medium and $B = 8\pi^2\eta/3\rho_0 v_l^3$ is the classical absorption parameter related to viscosity. In many nonmetallic liquids, particularly those which are associated and thus exhibit appreciable viscosity, it is found that the measured absorption coefficient is approximately described by the classical absorption expression (0.3.4) within a factor of about 3. In other cases the classical and measured absorption coefficient values differ by orders of magnitude. Consider first a hypothetical liquid for which Eq. (0.3.4) accurately describes the measured absorption coefficient at lower frequencies. At higher frequencies it appears that the absorption coefficient should increase in proportion to the square of the frequency, assuming that the viscosity remains constant while the frequency is allowed to increase. This prediction is approximately true over only a limited frequency range for which the effective value of the viscosity coefficient is the same as the value at low frequencies, that is, under "static" conditions. As the frequency increases, the effective viscosity decreases monotonically toward zero, owing to the finite time required for the transfer of momentum between adjacent regions of the medium.

Under nonequilibrium conditions the instantaneous shear stress $\mathbf{T}$ across a planar element at any position in a medium is not equal to the "static" value given by the product of the "static" shear viscosity coefficient and the space gradient of the particle velocity, but this product constitutes an asymptotic value $\mathbf{T}^0$ toward which $\mathbf{T}$ tends as time increases. The simplest assumption regarding the approach to the "static" value is that the rate is proportional to the difference between the instantaneous value and the "static" value; i.e.,

$$\frac{\partial \mathbf{T}}{\partial t} = \frac{1}{\tau_v} \left(\eta^0 \frac{\partial \dot{\boldsymbol{\xi}}}{\partial x} - \mathbf{T} \right), \qquad (0.3.5)$$

where τ_v is the proportionality constant which is in the nature of a time constant, the relaxation time, and η^0 is the low-frequency viscosity coefficient.

Consider a step function change of velocity to be imposed on the x boundary of the system. Since transfer of momentum in the x direction is necessary to change the internal stress conditions, the shear stress $\mathbf{T}(t)$ will not rise to $\mathbf{T}^0$ immediately but will tend asymptotically to this limit. An approximate solution to Eq. (0.3.5) is obtained by equating the instantaneous viscous and inertial forces and by regarding $\eta^0 \, \partial \dot{\boldsymbol{\xi}} / \partial x$ as the time-independent stress $\mathbf{T}^0$; Eq. (0.3.5) becomes

$$\frac{\partial \mathbf{T}}{\partial t} = \frac{1}{\tau_v} (\mathbf{T}^0 - \mathbf{T}), \qquad (0.3.6)$$

yielding the solution

$$\mathbf{T} = \mathbf{T}^0 - (\mathbf{T}^0 - \mathbf{T}_i) \exp(-t/\tau_v),$$

where $\mathbf{T}_i$ was the initial value of $\mathbf{T}$. Consequently the shear stress in this hypothetical experiment increases approximately exponentially toward $\mathbf{T}^0$ with time constant τ_v. A time delay is exhibited in the response of the liquid, where "response" refers to the changes in the time-dependent stress $\mathbf{T}(t)$, after imposition of the step function.

The concept of $\mathbf{T}^0$ as a stress toward which the instantaneous stress in the liquid tends, even though it may never reach it, can be helpful in visualizing the response of a viscous liquid to a *sinusoidal* change in the imposed strain rate. In this case it is evident that a time delay in the response of the liquid will result in a phase delay between stress and imposed strain rate. Such a phase delay is characteristic of relaxing systems subjected to sinusoidal perturbation.

When the ultrasonic perturbation is a sinusoidal change in the rate of strain, the following substitutions are made in Eqs. (0.3.5) and (0.3.6):

$$\mathbf{T} = \mathbf{T}(x)e^{j(\omega t + \delta)}, \qquad \dot{\xi} = \dot{\Xi}(x)e^{j\omega t},$$
$$\mathbf{T}^0 = \hat{\mathbf{T}}^0(x)e^{j\omega t}, \tag{0.3.7}$$

where δ is the phase delay between stress and particle velocity. The substitution of these expressions leads to a description of the response of the system in terms of a frequency-dependent effective viscosity:

$$\frac{\mathbf{T}}{\partial \xi / \partial x} = \frac{\mathbf{T}_0 e^{j\delta}}{\partial \dot{\Xi} / \partial x} = \frac{\eta^0}{1 + j\omega\tau_v} \equiv \eta^*(j\omega). \tag{0.3.8}$$

The effective viscosity obtained here is a complex number and contains a contribution [the imaginary part of $\eta^*(j\omega)$] which implies that such a medium has the property of a dynamic shear modulus $G^*(j\omega) = G' + jG''$ with a nonzero real part. The real part of the complex viscosity coefficient at any frequency decreases uniformly from the low-frequency value η^0 to zero as the frequency increases.

$$\eta'(\omega) = \eta^0 / [1 + (\omega\tau_v)^2]. \tag{0.3.9}$$

It is appropriate to define the equivalent complex shear modulus $G^*(j\omega) \equiv G'(\omega) + jG''(\omega)$ by

$$G^*(j\omega) = \frac{\mathbf{T}_0 e^{j\delta}}{\partial \dot{\Xi} / \partial x} = \frac{j\omega\eta^0}{1 + j\omega\tau_v} = \frac{G^\infty(\omega\tau_v)^2}{1 + (\omega\tau_v)^2} + j\frac{G^\infty\omega\tau}{1 + (\omega\tau_v)^2}, \tag{0.3.10}$$

where $G^\infty \equiv \eta^0/\tau_v$. The real part of the complex shear modulus (shown in Fig. 2b of Part 3) varies from zero at zero frequency to an asymptotic value G^∞ at frequencies very much greater than $f_v = 1/2\pi\tau_v$. The frequency f_v is thus called the *relaxation frequency* for the viscous mechanism. The imaginary part of the complex shear modulus increases from zero at zero frequency to a maximum at the relaxation frequency and falls again to zero at indefinitely high frequencies. As the frequency increases, the liquid exhibits an effective nonzero real part of the shear modulus and this property allows the propagation of heavily damped shear waves to occur. The propagation velocity as well as the absorption coefficient will be strongly dependent on frequency:

$$v_s = v_s^0 \frac{[1 + (G''/G')^2]^{1/4}}{\cos(G''/2G')}, \qquad \alpha_s = \frac{2\pi f}{v_s^0} = \frac{\sin(G''/2G')}{[1 + (G''/G')^2]^{1/4}}, \quad (0.3.11)$$

where $v_s^0 = (G'/\rho_0)^{1/2}$, the limiting shear wave speed as the frequency approaches zero. We therefore observe a drastic change in the behavior of a liquid medium of moderate viscosity as the frequency of an applied acoustic perturbation is varied. At low frequencies it behaves like a viscous liquid that does not allow the propagation of shear waves. At frequencies considerably higher than the relaxation frequency, the equations predict that the absorption coefficient due to viscosity should approach a constant nonzero value and that the velocity of propagation of such waves should approach a constant value which is determined by G^∞. It is found in the case of some hydrocarbon oils[9] that G^∞ is of the order of magnitude 10^{10} dyn/cm^2, i.e., within two orders of magnitude of the values characteristic of metals. In other words, the liquid is behaving much like a glass. The first known example of materials exhibiting behavior that is determined predominately by a single viscous relaxation mechanism of the type just described was molten zinc chloride.[10] Many other liquids behave as if several such viscous relaxation processes, described by different relaxation times and magnitudes, were superposed, as described in Part 3.

As one might expect, the observed behavior of most materials, including those of biological interest, is by no means so simple that it can be described adequately by a single relaxation time. It is found that sonic parameter magnitudes generally vary less drastically with frequency than the predictions of a single relaxation process require. Such behavior can be encompassed within the theory of relaxation processes by supposing that a discrete number of such processes are operative at the same frequency (each process may be described by a different value of the relaxation time) or, alternatively, by supposing that a continuous distribution of relaxation times exists. Since the latter possibility is more general, attention here will be confined to it. For the viscosity relaxation mechanism, Eq. (0.3.10) would be replaced by

$$G^*(j\omega) = G^\infty \int_{\tau_a}^{\tau_b} g_1(\tau) \frac{(\omega\tau)^2 \, d\tau}{1 + (\omega\tau)^2} + jG^\infty \int_{\tau_a}^{\tau_b} g_1(\tau) \frac{\omega\tau \, d\tau}{1 + (\omega\tau)^2}, \quad (0.3.12)$$

where τ_a and τ_b are bounds of the distribution of relaxation times. The distribution function $g_1(\tau)$, bounded by the values zero and one, expresses the contribution to the complex shear modulus which is derived from pro-

[9] A. J. Barlow and J. Lamb, *Proc. R. Soc. London Ser. A* **253**, 52 (1959).
[10] G. Gruber and T. A. Litovitz, *J. Chem. Phys.* **40**, 13 (1964).

cesses having relaxation times between τ and $\tau + d\tau$. These concepts are considered in detail by Harrison and Barlow in Part 3.

Equations (0.3.4) and (0.3.9) may be combined and represented by

$$\alpha_v v/f = \alpha_v \lambda = 2(\alpha_v \lambda)_{max} \omega \tau_v /[1 + (\omega \tau_v)^2], \qquad (0.3.13)$$

where η' is identified with η and $(\alpha_v \lambda)_{max} = 4\pi^2 \eta^0 f_v/3\rho_0 v_l^2$ is the maximum value of the relaxational absorption in unit wavelength, attained at the relaxation frequency $f_v = 1/2\pi\tau_v$, i.e., when $f/f_v = \omega\tau = 1$ or $\log(\omega\tau) = 0$. An alternative formulation of Eq. (0.3.13) is

$$\alpha_v/f^2 = B/[1 + (f/f_v)^2], \qquad (0.3.14)$$

where $B = 8\pi^2\eta/3\rho_0 v_l^3 = 2(\alpha_v \lambda)_{max}/vf_v$.

The relaxational concept can now be extended to include processes that respond with a time delay to local changes of pressure or temperature. Examples of such processes are structural changes responding to pressure and redistribution of energy among vibrational degrees of freedom at the molecular level responding to temperature. Any process involving a change of molar volume or molar enthalpy will respond to the perturbations of local pressure [as given by Eq. (0.1.10)] or local temperature [as given by Eq. (0.1.12)]. If the reactants and products of the process are initially present in approximately equal mole fractions (i.e., if the process is approximately in equilibrium), then its contribution to the absorption of ultrasonic energy can be significant and expressions (0.3.13) and (0.3.14) apply with values of A, $(\alpha_r \lambda)_{max}$, and τ governed by the process

$$\alpha_r v/f = \alpha_r \lambda = 2(\alpha_r \lambda)_{max} \omega \tau /[1 + (\omega\tau)^2] \qquad (0.3.15)$$

$$\alpha_r/f^2 = A/[1 + (f/f_r)^2], \qquad (0.3.16)$$

where α_r, τ, and A are generalizations of α_v, τ_v, and B, respectively. Consequently, ultrasonic absorption measurements as a function of frequency offer a technique for exploring the kinetics of certain fast physicochemical reactions. In general, both volumetric and enthalpic changes accompany a reaction and complexities arise early in the interpretation of such measurements. The subject is discussed in detail by Slutsky in Part 4.

In liquids of low viscosity the relaxation times $\tau = 1/2\pi f_r$ for the volumetric and enthalpic processes are frequently several orders of magnitude greater than τ_v for the viscosity relaxation process. Therefore, the viscous process contributes the frequency-independent amount B to the quantity α/f^2 in the range of interest for studying many physicochemical reactions.

Three options for representing such typical relaxational behavior of

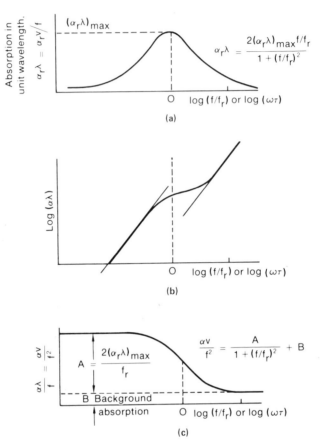

FIG. 1. (a) Relaxational contribution to the ultrasonic absorption per wavelength $\alpha_r\lambda$ for a single relaxation process. (b) Absorption in unit wavelength $\alpha_r\lambda$ for a single relaxation process added to absorption due to viscosity $\alpha_v = Bf^2$. (c) Absorption parameter α_r/f^2 for a single relaxation process and background absorption due to viscosity.

media are shown in Fig. 1. Figure 1a in analogy with Fig. 2b of Part 3 shows a peak in the absorption in unit wavelength for a general relaxation process other than viscosity (which would give rise to another peak at much higher frequencies). Figure 1b shows the same quantities with the ordinate plotted on a logarithmic scale. A relaxation process causes the curve to transfer from one line of slope 2 to another displaced to higher frequencies. Figure 1c shows the measured value of the absorption parameter α/f^2 where the viscosity contributes the background absorption B (which tends to zero only at much higher frequencies):

$$\left(\frac{\alpha}{f^2}\right)_{\text{meas}} = \frac{A}{1 + (f/f_r)^2} + B. \tag{0.3.17}$$

The occurrence of relaxation is shown by the sigmoid form of α/f^2 plotted against frequency. Two measurements of α/f^2 (for example, at the highest and lowest frequencies available) are sufficient to show whether at least one relaxation region lies between the frequencies employed. Methods of measuring α are discussed in detail by Breazeale, Cantrell, and Heyman in Part 2.

The occurrence of relaxation is not only associated with a peak in the relation between the absorption in unit wavelength and the frequency but it is also accompanied by dispersion of the sound speed in the same frequency range, i.e.,

$$v^2 = (v^0)^2 + 2v^0 \, \Delta v \, \frac{\omega\tau}{1 + (\omega\tau)^2}, \tag{0.3.18}$$

where v^0 is the limiting sound speed as frequency tends to zero. The amplitude of the absorption peak and the increment of sound dispersion Δv are related as

$$\Delta v/v \approx (\alpha_r\lambda)_{\text{max}}/\pi, \tag{0.3.19}$$

provided $(\alpha_r\lambda)_{\text{max}}/\pi \ll 1$, which is the case for *compressional* waves in most materials of interest.

Since the magnitude of the velocity dispersion is very small (usually less than 1%), very accurate measurement of velocity is required if quantitative deductions are to be made. Observation of velocity dispersion in connection with an absorption maximum provides valuable confirmation that relaxational behavior is being observed. If only part of the relaxational spectrum is accessible experimentally, the following equation provides a useful relationship between $\alpha_r\lambda$ and the slope of the velocity dispersion curve at the same frequency:

$$\frac{\alpha_r\lambda}{dv/df} = \frac{1}{2v\tau}(1 + \omega\tau). \tag{0.3.20}$$

0.4. Attenuation

Any process that removes energy from a traveling acoustic wave but does not dissipate that energy as heat contributes to the attenuation of the acoustic wave but not the absorption by the medium. The primary contribution to attenuation in addition to absorption is acoustic scattering by inhomogeneities in the medium.

Scattering of ultrasound is governed by the same principles as scattering of electromagnetic radiation, e.g., radar signals in the atmosphere or light in a turbid liquid. Three cases are distinguished, depending on the ratio of the wavelength of the radiation to the linear dimension of the inhomogeneities in "refractive index"; i.e., much less than unity, much greater than unity, or comparable in magnitude. When the ratio is much less than unity, behavior is approximately described by the theory for reflection and refraction at plane interfaces. When the ratio is much greater than unity (small scatterers), Rayleigh's theory is applicable. The greatest complexity arises when wavelength and linear dimension of the inhomogeneities are comparable in magnitude.

In all three cases, however, the acoustic problem is inherently more complicated than the electromagnetic problem, because elastic media support both bulk compressional and shear waves, in general, and surface waves can be generated at interfaces. Furthermore the acoustic "refractive index" will vary in response to variations of either density or elastic modulus or both. Further difficulties arise when the absorption by the medium is not negligible and when the locations and properties of the inhomogeneities are time dependent. It is therefore hardly surprising that the theory of scattering of ultrasound is in an early stage of development.

Scattering in polycrystalline media is discussed in detail by Papadakis in Part 5. Scattering by noncrystalline media, such as biological tissues, is a subject of ongoing research on which a consensus is not yet available. Readers seeking information on this subject may consult original articles by Sigelmann and Reid,[11] Shung *et al.*,[12] and Waag *et al.*[13] and a review by Chivers.[14] Texts by Morse and Ingard[15] and Chernow[16] provide theoretical bases for the two major aspects of the problem, i.e., scattering from discrete entities or in inhomogeneous continua. Twersky's[17] work provides a foundation for studies of multiple scattering.

[11] R. A. Sigelmann and J. M. Reid, *J. Acoust. Soc. Am.* **53**, 1351 (1973).

[12] K. K. Shung, R. A. Sigelmann, and J. M. Reid, *IEEE Trans. Biomed. Eng.* **BME-24**, 460 (1976).

[13] R. C. Waag, R. M. Lerner, and R. Gramiak, "Seminar on Tissue Characterization," NBS Special Publication 453, pp. 213–228, U.S. Government Printing Office, Washington, D.C., 1976.

[14] R. C. Chivers, *Ultrasd. Med. Biol.* **3**, 1 (1977).

[15] P. M. Morse and K. U. Ingard, "Theoretical Acoustics," McGraw-Hill, New York, 1968.

[16] L. A. Chernow, "Wave Propagation in a Random Medium," Dover, New York, 1960.

[17] V. Twersky, *J. Res. Nat. Bur. Stand.* **64D**, 715 (1960).

1. PIEZOELECTRIC TRANSDUCERS

By Matthew O'Donnell, L. J. Busse, and J. G. Miller

List of Symbols

A	cross-sectional area	t	time
c	elastic stiffness constant	$\mathbf{T}$	stress
c^D	elastic stiffness constant at constant electric displacement	$\mathbf{u}$	particle velocity
		v	speed of sound
c^E	elastic stiffness constant at constant electric field	v^D	speed of sound at constant electric displacement
C	equivalent electrical (mechanical) capacitance (compliance)	v^E	speed of sound at constant electric field
C_0	plate capacitance at constant strain (i.e., clamped)	V	voltage
		Z	equivalent electrical or mechanical impedance
d	piezoelectric strain constant	Z_0	characteristic impedance
$\mathbf{D}$	electric displacement	α	attenuation coefficient
e	piezoelectric stress constant	ϵ	dielectric permittivity
$\mathbf{E}$	electric field	ϵ_0	dielectric permittivity of free space
f	frequency	ϵ^S	dielectric permittivity at constant strain
$\mathbf{F}$	force	
g, h	piezoelectric constants	ϵ^T	dielectric permittivity at constant stress
i	$\sqrt{-1}$	
I	current	θ	propagation constant $[\alpha + i(\omega/v)]$
k	wave number	λ	wavelength
k_T	electromechanical coupling efficiency	ξ	particle displacement
k_T^2	electromechanical coupling constant	ρ	density
q	electric charge	σ	surface charge density
Q	mechanical or electrical quality factor	ϕ	ideal transformer turns ratio
R	equivalent electrical or mechanical resistance	ω	angular frequency $(2\pi f)$
$\mathbf{S}$	strain	

1.1. Introduction

Transducers are used as both transmitters and receivers of mechanical vibrations, converting electrical energy to acoustical energy and vice versa. For ultrasonic applications, transducers constructed of piezoelectric materials are used most commonly. Consequently, the goal of this part is to present a coherent discussion of piezoelectric transducers for use in experimental ultrasonics.

29

METHODS OF EXPERIMENTAL PHYSICS, VOL. 19

The design or selection of a particular transducer must be dictated, ultimately, by the specific application. In light of the vastly divergent applications of ultrasonic methods that will be described in subsequent
chapters of this book, a coherent treatment can be achieved only by iden-
tifying the underlying physical principles common to all transducer applications. Thus in the following sections the constitutive relations appropriate to piezoelectric materials, the physical basis for the generation and detection of ultrasonic signals in piezoelectric materials, and equivalent circuits describing the electrical and mechanical behavior of piezoelectric
plates will be described. Following an elementary discussion of these underlying physical principles, we shall summarize certain specific features
of commonly used piezoelectric materials and address some problems associated with the design and construction of ultrasonic transducers. The
part will conclude with a description of the acoustical and electrical properties of several transducers which serves to illustrate the use of the underlying physical principles with practical examples.

An overview of piezoelectric transducer design and construction
follows to provide a framework in the context of which subsequent specific discussions can be understood.

The term *piezoelectricity* describes the generation of an electrical polarization in a substance by the application of a mechanical stress and, conversely, a change in the shape of a substance when an electric field is applied.[1] An essential feature of piezoelectricity is the validity of a linear
relationship between applied electric field and mechanical stress or strain
that occurs only in materials exhibiting the absence of a center of symmetry. Although a large number of solids satisfy this criterion, practical
piezoelectric transducers are fabricated from a modest number of materials that exhibit a favorable combination of mechanical, electrical, and
piezoelectric properties. Among these are naturally occurring crystals,
such as quartz, and certain man-made ceramic materials, such as barium
titanates, lead zirconate–titanates, and lead metaniobates. Crystals such
as quartz are inherently piezoelectric, with properties determined by their
crystallographic features. In contrast, ferroelectric ceramics are initially
isotropic and are subsequently polarized above the Curie temperature by
the application of strong electric fields to induce the anisotropy responsible for their strong piezoelectric properties.

Although piezoelectricity is a bulk property, the conversion of electrical to mechanical energy occurs principally at the surfaces of piezoelectric devices. The physical basis for this is related to the driving term for
electrical to mechanical conversion, which involves a spatial gradient of

---

[1] H. Jaffe, Piezoelectricity, *in* Encyclopædia Brittanica, Chicago, Illinois, 1961.

certain electrical and piezoelectric parameters. Consequently, substantial contributions to this driving term occur only at the discontinuities in material properties represented by surfaces.

The active element in a typical ultrasonic transducer is a thin plate fabricated from a piezoelectric material. Such a plate functions as a resonator in a thickness expander mode for the generation and detection of longitudinal waves or in a thickness shear mode for the generation and detection of transverse waves. A plate in the form of a disk of diameter approximately 10 mm and thickness of an order of 1 mm is typical for operation in the low megahertz frequency range. Electrical contact to the piezoelectric element is usually accomplished by vacuum depositing or electroplating metal electrodes. The mechanical coupling to the piezoelectric plate and the electrical coupling to the appropriate electrical transmitting or receiving circuitry depends on the specific application. Many applications fall into either of two categories: narrowband or broadband.

In a number of narrowband applications, the transducer is used to measure the mechanical properties of a specimen that has been prepared in the form of an ultrasonic resonator. Transducers chosen for these applications are usually coupled directly to the specimen of interest and become part of a composite ultrasonic resonator consisting of specimen plus transducer. To determine the ultrasonic properties of the specimen from measurements carried out on the composite resonator, the ultrasonic losses in the transducer should be small compared to those in the specimen. Furthermore, the mechanical properties of the composite resonator must be isolated sufficiently from the electrical system so that only an insignificant amount of energy is dissipated in the external circuitry. Electrical isolation can be accomplished by using "weakly coupled" piezoelectric materials. Transducers made from quartz, for example, satisfy these requirements since quartz exhibits low ultrasonic attenuation (high mechanical Q) and low electromechanical coupling. Additional isolation can be obtained by using electrical matching schemes designed to mismatch purposely the electrical properties of the transducer and the electrical properties of the transmitting or receiving electrical circuitry.

In Fig. 1 we illustrate the pulse-echo and standing wave responses in the neighborhood of 90 MHz of a composite resonator consisting of a single crystal vanadium specimen and two piezoelectric plates. Determination of the ultrasonic attenuation and phase velocity can be made from either of these composite resonator responses. For example, the ultrasonic attenuation can be related to either the time constant of the exponential decay of the pulse-echo pattern or the frequency width of the individual mechanical resonances in the standing wave pattern. The ultra-

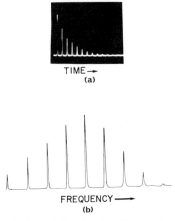

TIME ⟶

(a)

FREQUENCY ⟶

(b)

FIG. 1. Response patterns in the neighborhood of 90 MHz for a composite resonator consisting of a single crystal of vanadium with AT-cut quartz transducers bonded to opposite faces: (a) the pulse-echo pattern is the time-domain response; (b) the cw standing-wave pattern is the frequency-domain response. (Adapted from Miller.[2])

sonic phase velocity can be related to either the temporal spacing of the echoes in the pulse-echo pattern or the frequency spacing of the mechanical resonances in the standing wave pattern. Methods used to extract the ultrasonic attenuation and phase velocity from narrowband measurements such as those presented in Fig. 1 are described in Part 2.

Broadband transducers are used in applications ranging from imaging with short ultrasonic pulses to quantitative measurements of the phase velocity and attenuation over a continuous range of frequencies. Consequently, quite different criteria apply to the choice of piezoelectric elements for use in broadband as opposed to narrowband applications. To achieve acceptable signal-to-noise ratios over a broad bandwidth, high electromechanical conversion efficiencies are usually required. In addition, to achieve satisfactory time-domain resolution in imaging applications, short-duration pulses are usually desired. The use of "strongly coupled," relatively low mechanical Q, ceramic, piezoelectric elements is consistent with these needs.

For broadband applications, the transducer itself often takes the form of a composite resonator with a low mechanical Q. One method frequently used to achieve a low mechanical Q is the bonding of a well-matched material to the back of the piezoelectric plate. This backing material is fabricated to exhibit very high ultrasonic attenuation and thus to appear as an infinite transmission line representing a purely resistive

[2] J. G. Miller, J. Acoust. Soc. Am. **53**, 710 (1973).

mechanical load.   Such an approach leads to a relatively broadband frequency-domain response and a compact time-domain response. These characteristics are achieved at the expense of a substantial insertion loss since a large fraction of the ultrasonic energy is absorbed in the lossy backing layer.   Time-domain and frequency-domain response characteristics of a transducer of this design radiating through a quarter-wave matching layer into a water load are shown in Fig. 2.   Transducers of this sort can be excited with an electromagnetic impulse of several hundred volts amplitude and broadband ultrasonic measurements carried out using spectral analysis techniques.   An alternative approach, which achieves broadband frequency characteristics without the high insertion loss associated with the lossy backing layer, consists of the use of an air-backed piezoelectric element coupled to the load through several intermediate quarter-wave matching layers.   Improved insertion losses are achieved using this approach since energy which would have been dissipated in the lossy backing material is now reflected at the piezoelectric-to-air interface and directed toward the specimen.

The preceding two examples serve to indicate the range of considerations that enter into the design or selection of an appropriate piezoelec-

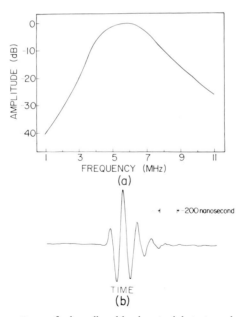

Fig. 2. Response patterns of a broadband lead metaniobate transducer: (a) the response as a function of frequency over the range 1–11 MHz; (b) the time-domain response to an "impulse" stimulus.

tric transducer. The next several sections will be devoted to those under-lying physical principles common to all applications of piezoelectric trans-ducers. The electrical and mechanical properties of piezoelectric mate-rials in common use will be discussed following the review of the general principles. This part will conclude with a description of a number of practical transducers for broadband or narrowband ultrasonic applica-tions.

1.2. Physical Principles of Piezoelectricity

1.2.1. Piezoelectric Constitutive Relations

The mechanical constitutive relations for a nonpiezoelectric, linear, elastic solid express the proportionality between stress and strain. This generalization of Hooke's law takes the form[3]

$$T_{ij} = c_{ijkl} S_{kl}, \qquad i, j, k, l = 1, 2, 3, \qquad (1.2.1)$$

where T_{ij} is the second-rank stress tensor, S_{kl} the second-rank strain tensor, and c_{ijkl} the fourth-rank tensor of elastic stiffness constants. The Einstein summation convention for repeated indices is implied. In an elastic solid both the stress and strain tensors are symmetric. Conse-quently, at most 36 of the 81 components of c_{ijkl} are independent. These 36 components can be reorganized into a 6×6 matrix, which is the basis of the condensed Voigt notation.[3,4] Since the symmetric stress and strain tensors have at most six independent elements each, the two-index Carte-sian notation is replaced by a single-index notation with range of 1–6. The relationship between the two classification schemes is illustrated by the following expressions for the strain:

$$S = \begin{bmatrix} S_{xx} & S_{xy} & S_{xz} \\ S_{xy} & S_{yy} & S_{yz} \\ S_{xz} & S_{yz} & S_{zz} \end{bmatrix} = \begin{bmatrix} S_1 & \frac{1}{2}S_6 & \frac{1}{2}S_5 \\ \frac{1}{2}S_6 & S_2 & \frac{1}{2}S_4 \\ \frac{1}{2}S_5 & \frac{1}{2}S_4 & S_3 \end{bmatrix}. \qquad (1.2.2)$$

The factor of $\frac{1}{2}$ is introduced in Eq. (1.2.2) to simplify the general expres-sion for Hooke's law in the Voigt classification scheme. Using this con-densed notation, the relationship between stress and strain takes the form

$$T_I = c_{IJ} S_J, \qquad I, J = 1, 2, \ldots, 6, \qquad (1.2.3)$$

[3] B. A. Auld, "Acoustic Fields and Waves in Solid," Vols. I and II. Wiley, New York, 1973.
[4] D. A. Berlincourt, D. R. Curran, and H. Jaffe, Piezoelectric and piezomagnetic mate-rials and their function in transducers, in "Physical Acoustics" (W. P. Mason, ed.), Vol. 1A, Chapter 3. Academic Press, New York, 1964.

or in matrix notation,

$$
\begin{bmatrix} T_1 \\ T_2 \\ T_3 \\ T_4 \\ T_5 \\ T_6 \end{bmatrix} = (c_{IJ}) \begin{bmatrix} S_1 \\ S_2 \\ S_3 \\ S_4 \\ S_5 \\ S_6 \end{bmatrix},
\tag{1.2.4}
$$

where c_{IJ} is the 6×6 matrix of elastic stiffness constants. In Voigt notation, uppercase letters are used as subscripts to denote summation over the six indices. If parameters described by Voigt notation and parameters described by Cartesian subscripts appear in the same equation, the Cartesian subscripts are written as lowercase letters and the Voigt subscripts as uppercase letters.

In piezoelectric materials the elastic and electrical properties are coupled, so a further generalization of both the mechanical and electrical constitutive relations is necessary.[3,4] To describe this coupling it is usually adequate to consider the relationship among the mechanical parameters stress T and strain S and the electrical parameters field E and displacement D. The piezoelectric constitutive relations involving these four parameters can take four representations depending on which set of two variables is chosen as independent. In Table I the commonly used symbol for the appropriate piezoelectric constant is listed for each choice of independent and dependent variables. As an illustration of the use of Table I, the piezoelectric constitutive relations for the case where the electric field and the mechanical strain are taken as the independent variables are

$$
T_J = c_{JI}^E S_I - e_{Jj} E_j,
\tag{1.2.5a}
$$

$$
D_i = \epsilon_{ij}^S E_j + e_{iI} S_I.
\tag{1.2.5b}
$$

In Eqs. (1.2.5), c_{JI}^E is the elastic stiffness tensor for constant applied electric field and ϵ_{ij}^S the permittivity matrix for constant strain. The piezo-

TABLE I. Piezoelectric Constants

Independent variable	Dependent variable	Piezoelectric coupling constant
E,S	D,T	e
E,T	D,S	d
D,S	E,T	h
D,T	E,S	g

electric coupling constants e_{Jj} and e_{iI} are third-rank tensors relating second-rank tensors T_J and S_I to first-rank quantities, i.e., vectors, E_j and D_i. In the absence of piezoelectricity (i.e., for $e_{Jj} = e_{iI} = 0$), Eqs. (1.2.5) reduce to the familiar constitutive relations for isolated mechanical and electrical systems.

Constitutive relations such as those of Eqs. (1.2.5) involve complicated combinations of tensor quantities. In practice, the complexity can be reduced substantially by an appropriate choice of plate orientation relative to either the crystalline lattice or poling axis so that a single piezoelectric coupling coefficient is dominant. Under these conditions, a one-dimensional representation of the constitutive relationships is frequently adequate. An example is the case of a piezoelectric ceramic with ultrasonic waves propagating either parallel or perpendicular to the axis of polarization. The piezoelectric coupling constants in the one-dimensional representation are scalar quantities describing the magnitude of the piezoelectric effect for a particular configuration. Numerical values of the piezoelectric constants in a number of materials are presented in Table V in Section 1.4.

1.2.2. Propagation of Ultrasound in Piezoelectric Materials

Equations governing the propagation of acoustic waves in a piezoelectric material are obtained by combining an equation of motion (Newton's second law) with the appropriate constitutive relations. We consider the case of longitudinal waves propagating in a piezoelectric material. Newton's second law takes the form

$$\rho \, \partial^2 \xi / \partial t^2 = \nabla \cdot \mathbf{T}, \qquad (1.2.6)$$

where ξ is the particle displacement. The strain $\mathbf{S} \equiv \nabla \xi$, stress $\mathbf{T}$, and electric field $\mathbf{E}$ are related by a constitutive equation [Eq. (1.2.5a)], which can be combined with Eq. (1.2.6) to yield[5,6]

$$\rho \, \partial^2 \xi / \partial t^2 = c^E \nabla^2 \xi - \nabla(eE). \qquad (1.2.7)$$

For one-dimensional propagation, Eq. (1.2.7) can be written to yield an inhomogeneous wave equation for the particle displacement

$$\frac{\partial^2 \xi}{\partial x^2} - \frac{1}{v^2} \frac{\partial^2 \xi}{\partial t^2} = \frac{1}{c^E} \frac{\partial}{\partial x}(eE), \qquad (1.2.8)$$

where $v = (c^E/\rho)^{1/2}$ is the velocity of propagation.

To specify the problem further, we assume that the piezoelectric mate-

[5] A. R. Hutson and D. L. White, *J. Appl. Phys.* **33**, 40 (1962).
[6] E. H. Jacobsen, *J. Acoust. Soc. Am.* **32**, 949 (1960).

rial is prepared in the form of a plate with metal electrodes on each face and we use Maxwell's equations in conjunction with Eq. (1.2.8). If the electrodes on opposite faces are short circuited, the electric field E is forced to be zero and Eq. (1.2.8) reduces to a homogeneous wave equation exhibiting an ultrasonic wave propagation velocity of $(c^E/\rho)^{1/2}$. In contrast, if the electrodes are open circuited, the electric field is nonzero and the value of E on the right-hand side of Eq. (1.2.8) must be obtained from the one-dimensional representation of the constitutive relation given in Eq. (1.2.5b).[4] Using this substitution, Eq. (1.2.8) can be written

$$\frac{\partial^2 \xi}{\partial x^2} - \frac{1}{v^2}\frac{\partial^2 \xi}{\partial t^2} = \frac{e}{\epsilon^S c^E}\frac{\partial D}{\partial x} - \frac{e^2}{\epsilon^S c^E}\frac{\partial^2 \xi}{\partial x^2}. \tag{1.2.9}$$

Since there are no free charges, Maxwell's equations require that $\nabla \cdot D = 0$, which eliminates the first term on the right side of Eq. (1.2.9). Thus for an open-circuited plate, Eq. (1.2.9) also reduces to a homogeneous wave equation

$$\frac{\partial^2 \xi}{\partial x^2} - \frac{1}{v^2(1 + e^2/\epsilon^S c^E)}\frac{\partial^2 \xi}{\partial t^2} = 0. \tag{1.2.10}$$

Ultrasonic waves propagate with a higher velocity $(c^D/\rho)^{1/2}$ and the material is described as being piezoelectrically stiffened. The elastic stiffness constant for the open-circuited plate c^D is related to that for the short-circuited plate c^E by

$$c^D = c^E(1 + e^2/\epsilon^S c^E). \tag{1.2.11}$$

The quantity $e^2/\epsilon^S c^E$ is the "electromechanical coupling constant" and is written k_T^2. The magnitude of the electromechanical coupling constant is a useful index of the strength of the piezoelectric effect in a particular material.

1.2.3. Piezoelectric Generation and Detection of Ultrasound

Ultrasonic waves are generated by the application of an external electric field to a piezoelectric material. The inhomogeneous wave equation represented by Eq. (1.2.8) indicates that a gradient in the product of the piezoelectric constant of the material and the electric field serves as the source term for the generation of mechanical disturbances. In practice, the surfaces of a piezoelectric material offer the sharpest discontinuity in both e and E, and thus represent the strongest source of sound.[6]

The phenomenon of surface generation may be illustrated with the following simplified picture. Figure 3 shows a parallel plate capacitor containing a piezoelectric material of dielectric constant ϵ. In Fig. 3b and c,

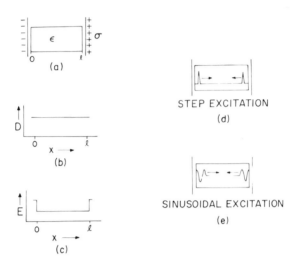

STEP EXCITATION

(d)

SINUSOIDAL EXCITATION

(e)

FIG. 3. Simplified illustration of surface generation in a piezoelectric plate: (a) parallel plate capacitor containing a piezoelectric material of dielectric constant ϵ; (b) and (c), respectively, the electrical displacement D and electric field E depicted at a particular time shortly after a surface charge density has been applied to the capacitor; (d) the spatial distribution of mechanical disturbance shortly after a voltage step is applied; (e) the response of the crystal to a harmonic voltage excitation. (Adapted in part from Redwood.[7])

respectively, the electrical displacement D and electric field E are depicted a particular time t shortly after a charge density σ has been applied to the capacitor. Since there are no free charges between the plates of the capacitor, $\nabla \cdot \mathbf{D} = 0$ and hence D is constant as illustrated in (b). Because $\mathbf{D} \cong \epsilon\mathbf{E}$ and D is constant, the magnitude of E is smaller inside the piezoelectric where the dielectric constant is larger, as illustrated in (c), from which it is clear that the only gradient in the electric fields occurs at the surfaces of the material. In addition, there is a large discontinuity in e at the surface of the crystal. Consequently, the surfaces of the piezoelectric material are the predominate sources for the generation of ultrasound. Figure 3d illustrates the spatial distribution of mechanical disturbance shortly after a voltage impulse is applied. Short acoustic transients are launched from both surfaces of the crystal.[7,8] The response of the crystal to a harmonic voltage excitation is illustrated in Fig. 3e. If a voltage at frequency ω is applied to the piezoelectric material, both surfaces of the material launch mechanical waves at frequency ω.

At microwave frequencies, high Q reentrant electromagnetic cavities

[7] M. Redwood, J. Acoust. Soc. Am. **33**, 527 (1961).
[8] E. F. Carome, P. E. Parks, and S. J. Mraz, J. Acoust. Soc. Am. **36**, 946 (1964).

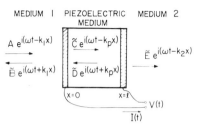

FIG. 4. The process of piezoelectric detection when an ultrasonic wave of amplitude A impinges upon a piezoelectric plate bounded on each side by nonpiezoelectric media.

are frequently employed to concentrate large electric fields near the surface of the piezoelectric material in order to launch ultrasonic waves.[6,9-12] At frequencies in the megahertz range, mechanically resonant structures are usually used to enhance the process of launching sound waves illustrated in Fig. 3. Particular mechanical modes of oscillation are selectively enhanced by bringing the surfaces of the piezoelectric material close together. The ultrasonic properties of the resulting thickness-mode resonant plates are best described in terms of acoustic transmission line theory using an approach discussed in the next section.

Piezoelectric detection of ultrasonic waves is reciprocal to the process of wave generation. That is, the conversion of mechanical energy into electrical energy is also a phenomenon dominated by the behavior at the surfaces of the piezoelectric material.[13] To illustrate this surface detection process, we consider a simple example of an ultrasonic wave of amplitude A impinging from medium 1 upon a piezoelectric plate that is bounded on each side by nonpiezoelectric media, as depicted in Fig. 4. A wave of (complex) amplitude $\tilde{B}$ is reflected from the piezoelectric material and a wave of amplitude $\tilde{E}$ is transmitted into medium 2. In addition, two waves of amplitude $\tilde{C}$ and $\tilde{D}$ propagate in opposite directions within the piezoelectric material. The voltage measured across the plate under these conditions is the integral of the electric field over the thickness of the crystal

$$V = \int_0^l E \, dx. \qquad (1.2.12)$$

[9] H. E. Bommel and K. Dransfeld, *Phys. Rev. Lett.* **1**, 234 (1958).

[10] K. N. Baranskii, *Dokl. Akad. Nauk SSSR* **114**, 517 (1957) [*English transl.: Sov. Phys. Dokl.* **24**, 237 (1958)].

[11] H. E. Bommel and K. Dransfeld, *Phys. Rev.* **119**, 1245 (1960).

[12] J. Lamb and H. Sequin, *J. Acoust. Soc. Am.* **39**, 752 (1966).

[13] R. T. Beyer and S. V. Letcher, "Physical Ultrasonics." Academic Press, New York, 1969.

Using the constitutive relation relating the electric field to the strain and the electric displacement, the voltage is given by

$$V \equiv \int_0^l - hS \, dx + \int_0^l \frac{D}{\epsilon^S} \, dx. \qquad (1.2.13)$$

From Gauss's law the electric displacement D is independent of x and is equal to the surface charge density σ, where $\sigma = q/A$ and q is the total charge on area A. Consequently, Eq. (1.2.13) reduces to[7]

$$V = - h \int_0^l \frac{\partial \xi}{\partial x} \, dx + \frac{ql}{\epsilon^S A} = - h[\xi(l) - \xi(0)] + \frac{q}{C_0}$$

$$= - h[\xi(l) - \xi(0)] + \frac{I}{i\omega C_0}, \qquad (1.2.14)$$

where C_0 is the clamped (i.e., constant strain) plate capacitance of the piezoelectric plate and I the current available to the external load. Equation (1.2.13) indicates that the open-circuit voltage ($I = 0$) developed across the piezoelectric plate is proportional to the relative displacements of the front and back surfaces of the material. The magnitude of the net displacement can be calculated from the boundary conditions placed on the five acoustic waves at the surfaces of the piezoelectric plate. Provided that medium 1 and medium 2 do not "clamp" the piezoelectric material, then the relative displacement of the front and back surfaces, $\xi(l) - \xi(0)$, is largest if the thickness of the plate corresponds to an odd integral number of half-wavelengths of the ultrasonic wave impinging upon the material, in which case the two surfaces oscillate 180° out of phase. If the thickness of the crystal is an even number of half-wavelengths, then the amplitude of oscillation of the two surfaces is in phase and, therefore, $\xi(l) - \xi(0) = 0$. In the next section we shall examine the properties of piezoelectric plate resonators using transmission line and lumped element equivalent circuits.

1.3. Distributed and Lumped Element Equivalent Circuits

The propagation of plane acoustic waves in a material can be described with the aid of transmission line theory. In this way, the time delays, reflection properties, and dispersion and attenuation effects that take place in materials can be modeled. The transmission line model of acoustic wave propagation has proved useful in exploring the mechanical properties of ultrasonic resonators and delay lines.[4,14-17] Furthermore, this model can be generalized to include the electrical as well as the mechanical properties of piezoelectric materials. In this section we shall use the transmission line model of acoustic wave propagation to explore equivalent circuits for a piezoelectric plate operating in the thickness expander or thickness shear mode.

1.3.1. Transmission Line Model

For one-dimensional propagation the equation of motion, Eq. (1.2.6), takes the form

$$\partial T/\partial x = \rho \partial^2 \xi/\partial t^2 = \rho \partial u/\partial t, \tag{1.3.1}$$

where T is the external stress and u the particle velocity. For disturbances that can be described in terms of a harmonic time dependence of the form $e^{i\omega t}$, Eq. (1.3.1) leads to the relationship

$$u = (i/\omega\rho) \, \partial T/\partial x. \tag{1.3.2}$$

It is sometimes convenient to appeal to an analogy between mechanical and electrical variables in terms of which Eq. (1.3.2) corresponds to Ohm's law, wherein "current" (particle velocity u) is proportional to the "voltage difference" (gradient of the stress $\partial T/\partial x$). We are primarily interested in the propagation of plane acoustic waves that can be described in terms of harmonic space and time dependences of the form $\exp[i(\omega t - kx)]$. The wave number $k \equiv 2\pi/\lambda$ is related to the frequency ω by an algebraic expression, usually referred to as a dispersion relationship. The dispersion relationship is obtained by combining an equation of motion [Eq. (1.3.1)] with an appropriate constitutive relationship such as $T = cS$ (i.e., Hooke's law), where c is the elastic stiffness constant. The dispersion relationship for a lossless medium is $\omega = vk$, where $v = (c/\rho)^{1/2}$ is the velocity of plane wave propagation. For a plane wave propagating in a lossless (nonattenuating) medium in which no reflections occur (i.e., an "infinite transmission line"), the ratio of stress T to particle velocity u defines an impedance per unit cross-sectional area A that is spe-

[14] W. P. Mason, "Electromechanical Transducers and Wave Filters," 2nd ed. Van Nostrand-Reinhold, Princeton, New Jersey, 1948.

[15] W. P. Mason, Use of piezoelectric crystals and mechanical resonators in filters and oscillators, in "Physical Acoustics" (W. P. Mason, ed.) Vol. 1A, Chapter 5. Academic Press, New York, 1964.

[16] J. E. May, Guided wave ultrasonic delay lines, in "Physical Acoustics" (W. P. Mason, ed.) Vol. 1A, Chapter 6. Academic Press, New York, 1964.

[17] E. K. Sittig, IEEE Trans. Sonics Ultrason. SU-14, 167 (1967).

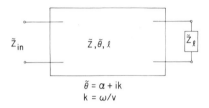

$$\tilde{\theta} = \alpha + ik$$
$$k = \omega/v$$

FIG. 5. General representation of a one-dimensional acoustic transmission line.

cific to the mechanical transmission line. From Eq. (1.3.2) and the dispersion relationship $\omega = vk$, the characteristic impedance Z_0 of a lossless line is

$$Z_0 = \rho vA. \qquad (1.3.3)$$

For a lossy (attenuating) transmission line, the dispersion relationship $\omega = v\tilde{k}$ can be satisfied if the wave vector is allowed to be complex ($\tilde{k}$). Therefore, defining ω/v as the real part of the wave vector and α as the imaginary part of the wave vector (i.e., the attenuation coefficient), the characteristic impedance of a lossy transmission line is

$$\tilde{Z}_0 = \rho vA \left[\frac{1}{1 + r^2} - \frac{ir}{1 + r^2} \right], \qquad (1.3.4)$$

where $r = \alpha v/\omega$. In the usual case, where the real part of the wave vector is much larger than the imaginary part (i.e., $r \ll 1$), Eq. (1.3.4) reduces approximately to Eq. (1.3.3).

For an acoustic transmission line of finite length l, the formalism must take into account reflection and transmission effects that depend on how the line is terminated. The (complex) specific acoustic impedance is a function of position along the transmission line and is defined as the ratio of stress to the particle velocity at that position. A well-known result from transmission line theory establishes that the (complex) input impedance $\tilde{Z}_{in}$ of an acoustical line with length l, real part of the wave number ω/v, imaginary part of the wave number α, and characteristic impedance $\tilde{Z}_0$, which is terminated at l with a complex impedance $\tilde{Z}_l$, is given by[18]

$$\tilde{Z}_{in} = \tilde{Z}_0 \left[\frac{\tilde{Z}_l + \tilde{Z}_0 \tanh(\tilde{\theta}l)}{\tilde{Z}_0 + \tilde{Z}_l \tanh(\tilde{\theta}l)} \right], \qquad (1.3.5)$$

where $\tilde{\theta} = \alpha + i\omega/v$. The two-port network corresponding to Eq. (1.3.5) is illustrated in Fig. 5. The result expressed in Eq. (1.3.5) can be used to describe the behavior of a composite mechanical system consisting of an arbitrary number of segments each of which exhibits arbitrary length,

[18] F. E. Terman, "Radio Engineers Handbook," Sect. 3. McGraw-Hill, New York, 1943.

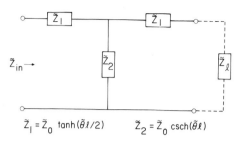

$$\tilde{Z}_1 = \tilde{Z}_0 \tanh(\tilde{\theta}\ell/2) \qquad \tilde{Z}_2 = \tilde{Z}_0 \operatorname{csch}(\tilde{\theta}\ell)$$

FIG. 6. Equivalent T network of a one-dimensional acoustic transmission line. (Adapted in part from Terman[18] and Mason.[14])

propagation parameters, and characteristic acoustic impedance. The input impedance for the composite structure is evaluated in a sequential fashion, beginning with the terminal segment whose input impedance is obtained directly from Eq. (1.3.5). The input impedance of the next-to-last segment is then evaluated using a form of Eq. (1.3.5) in which the terminating impedance is set equal to the input impedance previously calculated for the terminal segment. This process is continued until an expression for the input impedance of the entire composite structure is obtained.

Although the transmission line model presented in Fig. 5 represents a complete description of the acoustic properties of a sample of arbitrary length, it is often useful to replace this general transmission line with an equivalent T network when the length of the specimen becomes comparable to the acoustic wavelengths of interest.[14] The equivalent T network is presented in Fig. 6. The input impedance presented by the network shown in this figure is identical to that shown in Fig. 5 [i.e., $\tilde{Z}_{in}$ is given by Eq. (1.3.5)] provided that the T section is terminated in an impedance $\tilde{Z}_l$. A mechanical resonance of an acoustic transmission line occurs whenever the magnitude of the mechanical impedance reaches a minimum. The condition for mechanical resonance of a length of transmission line terminated in a low impedance (e.g., a free surface) is that the length of the transmission line be equal to an integral number of half-wavelengths of ultrasound, i.e., kl equals $n\pi$ for all integral values of n.

Starting with a transmission line model, a number of equivalent circuits have been derived for a piezoelectric plate operating in the thickness expander or thickness shear mode.[14,19-23] In the next section we shall

[19] R. Krimholtz, D. Leedom, and G. Matthaei, *Electron. Lett.* **6**, 398 (1970).

[20] D. Leedom, R. Krimholtz, and G. Matthaei, *IEEE Trans. Sonics Ultrason.* **SU-18**, 128 (1971).

[21] T. R. Meeker, *Ultrasonics* **10**, 26 (1972).

[22] E. K. Sittig, *IEEE Trans. Sonics Ultrason.* **SU-16**, 2 (1969).

[23] T. L. Rhyne, *IEEE Trans. Sonics Ultrason.* **SU-25**, 98 (1978).

discuss the Mason equivalent circuit,[14] a widely used model for characterizing the behavior of piezoelectric plates. Another model, derived by
Krimholtz *et al.*,[19,20] which retains the intrinsic transmission line character of the mechanical system but permits a lumped circuit analysis of the
electrical system, will be discussed in Section 1.3.3.

1.3.2. Mason Equivalent Circuit

In a piezoelectric material the mechanical and electrical properties are
coupled. Consequently, the equivalent circuit for a piezoelectric plate
might be expected to consist of elements exhibiting complicated mixtures
of electrical and mechanical parameters. However, Mason developed an
equivalent circuit in which mechanical and electrical properties are coupled only through an ideal transformer[14] with turns ratio $1:\phi$, where
$\phi = gC_0c^D$. The Mason equivalent circuit for a piezoelectric plate
operating in a thickness expander or thickness shear mode is presented in
Fig. 7. A comparison of Figs. 6 and 7 illustrates that the Mason equivalent circuit is a generalization of the equivalent T network describing an
arbitrary mechanical transmission line of length (i.e., thickness) l. In this
generalization the mechanical impedances are obtained by using open-
circuit electrical boundary conditions [i.e., $v = v^D = (c^D/\rho)^{1/2}$ and $\bar{\theta} =
\alpha + i\omega/v^D$]. The Mason equivalent circuit consists of two mechanical
ports, corresponding to the front and back surfaces of the plate, and one
electrical port.

Usually we are interested in coupling a piezoelectric element to other
materials and transmitting or receiving mechanical forces across the faces

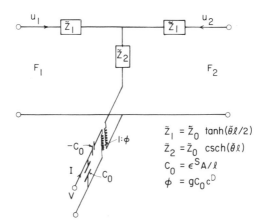

FIG. 7. Mason equivalent circuit of a piezoelectric plate operating in a thickness mode.
(Adapted in part from Mason.[14])

of the crystal. In general, both the front and back surfaces of the crystal are loaded by mechanical impedances. The circuit describing arbitrary loading on both faces of the crystal is illustrated in Fig. 8a. Here $\tilde{Z}_F$ is the load on the front face of the crystal, $\tilde{Z}_B$ the load on the back face of the crystal, F_F the mechanical force imparted to the front load, and F_B the force imparted to the back load.

Mason demonstrated[14] that the equivalent circuit presented in Fig. 8a

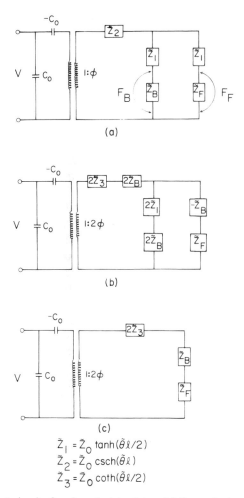

$$\tilde{Z}_1 = \tilde{Z}_0 \tanh(\tilde{\theta}\ell/2)$$
$$\tilde{Z}_2 = \tilde{Z}_0 \operatorname{csch}(\tilde{\theta}\ell)$$
$$\tilde{Z}_3 = \tilde{Z}_0 \coth(\tilde{\theta}\ell/2)$$

FIG. 8. Equivalent circuits for piezoelectric plates: (a) the equivalent circuit of a piezoelectric plate that is mechanically loaded on both the front and back faces; (b) the equivalent circuit of (a) redrawn using a circuit identity first developed by Norton; (c) an approximation to the circuit of (b). (Adapted in part from Mason.[14])

can be cast into a more useful form by the application of a circuit identity developed by Norton. Figure 8b shows the equivalent circuit for a piezoelectric plate loaded by complex impedances $\tilde{Z}_F$ and $\tilde{Z}_B$ after the application of the Norton identity. The circuit presented in Fig. 8b offers the advantage that the impedance of the branch containing the element $2\tilde{Z}_1$ is very large compared to that of the parallel branch for frequencies in the neighborhood of a mechanical resonance. Thus the branch containing $2\tilde{Z}_1$ can be neglected and the circuit can be simplified to yield Fig. 8c. Formally, Fig. 8c represents an approximate form of Figs. 8a and 8b; however, it is adequate for describing the operation of piezoelectric plates near resonance driving a wide variety of mechanical loads (e.g., solids, liquids, and gases). In the limit of an unloaded piezoelectric plate ($\tilde{Z}_B = \tilde{Z}_F = 0$) with no losses [$\tilde{\theta} = i(\omega/v^D)$], electrical resonances of the plate occur for $l \cong n(\lambda/2)$ for odd values of n. That is, electrical energy is optimally converted to mechanical energy only near odd mechanical resonances of the plate. This behavior was described physically in conjunction with the discussion of Eq. (1.2.14). At even mechanical resonances of the plate (i.e., for $l = n(\lambda/2)$ for even values of n), no "current" flows through the mechanical portion of the equivalent circuit and so the electrical impedance presented by the plate to external circuitry is simply that of the clamped plate capacitance C_0.

Near any odd mechanical resonance, the equivalent circuit of the plate can be simplified further. For example, near the fundamental plate resonance ($n = 1$, $f_0 = v^D/2l$), the impedance $2\tilde{Z}_3$ in Fig. 8c can be approximately represented by a series resistance–inductance–capacitance (RLC) circuit exhibiting a resonance at frequency f_0. This situation is illustrated in Fig. 9a. The values for the equivalent capacitance (compliance), inductance (mass), and resistance are

$$C_m = (\pi^2 f_0 Z_0)^{-1}, \qquad (1.3.6a)$$

$$L_m = Z_0/(4f_0), \qquad (1.3.6b)$$

$$R_m = \alpha l Z_0. \qquad (1.3.6c)$$

The equivalent circuit of Fig. 9a can be redrawn as illustrated in Fig. 9b, where the effects of the transformer have been included in the circuit elements. The resonant frequency f_0' for the circuit illustrated in Fig. 9b is

$$f_0' = f_0[1 - 4\phi^2(C_m/C_0)]^{1/2}, \qquad (1.3.7)$$

which is decreased from the nominal resonant frequency $f_0 = v^D/2l$ by an amount related to the strength of the piezoelectric coupling coefficient g, where $\phi = gC_0c^D$. We note, however, from Eq. (1.2.11), that the velocity of sound v^D in a piezoelectric plate increases with the strength of elec-

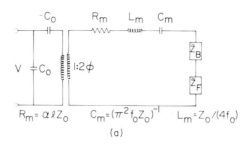

$$R_m = \alpha l Z_0 \qquad C_m = (\pi^2 f_0 Z_0)^{-1} \qquad L_m = Z_0/(4f_0)$$

(a)

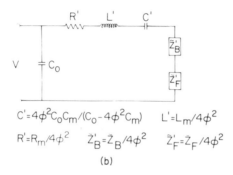

$$C' = 4\phi^2 C_0 C_m/(C_0 - 4\phi^2 C_m) \qquad L' = L_m/4\phi^2$$
$$R' = R_m/4\phi^2 \qquad \tilde{Z}'_B = \tilde{Z}_B/4\phi^2 \qquad \tilde{Z}'_F = \tilde{Z}_F/4\phi^2$$

(b)

FIG. 9. Equivalent circuits for piezoelectric plates: (a) the equivalent circuit of a piezo-electric plate operating near a mechanical resonance of the plate; (b) the effects of the transformer of (a) have been included in the lumped component equivalents. (Adapted in part from Mason.[14])

tromechanical coupling. The net effect of Eqs. (1.2.11) and (1.3.7) is to produce a resonant frequency of approximately $v^E/2l$.

1.3.3. KLM Equivalent Circuit

Although the Mason equivalent circuit has proved useful in many applications, an alternative equivalent circuit appears to be better suited for determining the optimum electrical and mechanical matching for broad-band operation. The Krimholtz et al.[19,20] (KLM) equivalent circuit, illustrated in Fig. 10, models the piezoelectric transducer as a lossless mechanical transmission line of length l coupled at the midpoint $l/2$ to a lumped element electrical network. The mechanical and electrical systems are coupled through an ideal transformer of turns ratio $1:\phi$, where ϕ is a function of frequency as well as other physical parameters:

$$\phi = [k_T(\pi/\omega_0 C_0 Z_0)^{1/2} \, \text{sinc}(\omega/2\omega_0)]^{-1}, \qquad (1.3.8)$$

and $\text{sinc}(x)$ equals $\sin(\pi x)/\pi x$. In Eq. (1.3.8), $\omega_0 \, (= 2\pi f_0)$ is the mechanical resonant frequency of the transmission line of length l and k_T^2 the

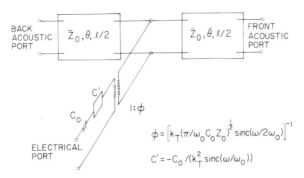

FIG. 10. KLM equivalent circuit of a piezoelectric plate operating in a thickness mode. (Adapted in part from Krimholtz *et al.*,[19] Leedom *et al.*,[20] and Auld.[3])

electromechanical coupling constant. The lumped element electrical network consists of two capacitors in series with the transformer. One capacitor represents the clamped capacitance of the transducer C_0 and the other capacitor is an explicit function of frequency and electromechanical coupling coefficients, written

$$C' = -C_0/[k_T^2 \ \text{sinc}(\omega/\omega_0)] \qquad (1.3.9)$$

The distinction between the ways in which mechanical and electrical components are treated in the KLM model is quite natural for describing the operation of broadband transducers designed to operate in the low-megahertz frequency range. Acoustical matching techniques over this frequency range require a transmission line approach whereas electrical matching can be accomplished with lumped components. DeSilets *et al.*[24] have used this approach in an attempt to design the optimal electrical and mechanical matching schemes for transferring mechanical energy from a high-impedance piezoelectric plate to a low-impedance mechanical load such as water. These authors assert that the condition for maximum bandwidth of a piezoelectric transducer is that the Q of the mechanical branch of the KLM circuit Q_m be matched to the Q of the electrical branch Q_e.

The Q of the mechanical branch of the KLM equivalent circuit can be estimated as follows. We consider the case of a lossless mechanical plate of high characteristic acoustic impedance $\bar{Z}_0$ with mechanical impedances $\bar{Z}_F$ and $\bar{Z}_B$. The transmission line theory developed earlier tells us that if $|\bar{Z}_F| \ll |\bar{Z}_0|$ and $|\bar{Z}_B| \ll |\bar{Z}_0|$, then the plate will resonate at frequencies

[24] C. S. DeSilets, J. D. Fraser, and G. S. Kino, *IEEE Trans. Son. Ultrason.* **SU-25**, 115 (1978).

f_n, where $f_n = nv^{D}/2l$ and n is an integer. If one further assumes that the mechanical loads can be represented as purely resistive loads (i.e., $\hat{Z}_F \rightarrow \hat{R}_F$ and $\hat{Z}_B \rightarrow \hat{R}_B$), then the Q of any particular mechanical resonance of the plate is determined by these mechanical loads. Thus Q_m, the mechanical Q of the fundamental plate resonance, is

$$Q_m \cong \frac{\pi}{2} \frac{Z_0}{R_F + R_B}. \tag{1.3.10}$$

The Q of the electrical branch of the circuit Q_E can be estimated as follows. Mechanical transmission line theory is used to represent the external mechanical loads R_F and R_B as a single equivalent resistor R, placed across the ideal transformer at the center of the mechanical transmission line. The value of this resistor is determined by using Eq. (1.3.5) to write expressions for R_F and R_B as seen at the center tap of the mechanical transmission line looking through the lossless transmission lines of length $l/2$. We note that at a plate resonance the segments of length $l/2$ constitute quarter-wavelength sections. The value of the equivalent mechanical resistance R is the parallel combination of these transformed resistances; at the resonant frequency it is given by

$$R = Z_0^2/(R_F + R_B). \tag{1.3.11}$$

In the electrical branch of the equivalent circuit, C_0 is in series with C'. Over the typical frequency range of interest in broadband applications, C' is very large (i.e., $|C'| \gg C_0$). Consequently, the effects of C' on the electrical branch are negligible. To maximize current flow near the resonant frequency, an external inductor L_0 can be inserted in series with the electrical branch of the transducer, with a value chosen to resonate with the plate capacitance at the center frequency (i.e., $L_0 = C_0/\omega_0^2$). The resulting expression for the Q of the electrical branch can be written as

$$Q_E = \frac{\phi^2}{\omega_0 C_0 R} = \frac{\pi}{4k_T^2} \left(\frac{R_F + R_B}{Z_0} \right), \tag{1.3.12}$$

where R/ϕ^2, the transformed value of the equivalent mechanical resistance R, represents the effects of the mechanical loads upon the electrical Q.

Imposing the condition that $Q_E = Q_m$ and making use of Eqs. (1.3.10) and (1.3.12), one obtains

$$R_F + R_B = \sqrt{2} Z_0 k_T. \tag{1.3.13}$$

Equation (1.3.13) specifies the optimum value of $R_F + R_B$ to achieve maximum bandwidth using a piezoelectric plate of characteristic impedance Z_0 and electromechanical coupling constant k_T^2. Using Eq. (1.3.11) and

TABLE II.[a]   Matching Formulas[b]

Number of matching sections	R	Z_{01}	Z_{02}	Z_{03}	$\cdots$	Z_{0n}	$\cdots$
1	Z_0^2/R_F	Z_0	—	—			
2	$Z_0^{4/3}/R_F^{1/3}$	Z_0	$Z_0^{1/3}R_F^{2/3}$	—			
3	$Z_0^{8/7}/R_F^{1/7}$	Z_0	$Z_0^{4/7}R_F^{3/7}$	$Z_0^{1/7}R_F^{6/7}$			
$\vdots$	$\vdots$	$\vdots$	$\vdots$	$\vdots$		$\vdots$	
N							

[a] Adapted in part from DeSilets et al.[24]
[b] For the nth matching layer in an N-layer system $Z_{0n} = Z_{0(n+1)}[Z_0/R_F]^x$, where $x = N!/[(2^N - 1)n!(N - n)!]$.

writing R_{opt} to specify this optimum value of $R_F + R_B$, one obtains[24]

$$R_{opt} = Z_0/\sqrt{2}\,k_T. \tag{1.3.14}$$

For the case in which a high characteristic acoustic impedance piezoelectric material is radiating into a low impedance load such as water, the value of R determined from Eq. (1.3.11) is usually much less than the optimum value R_{opt} obtained from Eq. (1.3.14). Consequently, to optimize the bandwidth, the equivalent mechanical resistance R is adjusted upwards by inserting quarter-wave matching layers between the transducer faces and the ultimate loads.

Quarter-wave matching schemes designed to yield maximally flat transmission properties have been formulated by Collins,[25] Riblet,[26] and Young[27] for microwave transmission lines and were adapted by DeSilets et al.[24] to the case of the acoustic transmission line. The goal is to insert the minimum number of quarter-wave matching layers necessary to transform the impedance of the terminal mechanical load up to a value approximately equal to the optimum value specified by Eq. (1.3.14). An inspection of the KLM model shown in Fig. 10 indicates that the front half of the transducer constitutes the first quarter-wave matching layer. In Table II we indicate the values of R for a small number N of quarter-wave matching layers with characteristic impedance Z_{0n}. By knowing the impedances of the transducer Z_0 and the ultimate load R_F, the value of R can be determined from Table II for as many as three quarter-wave matching layers. The appropriate number of matching layers is determined by comparing the value of R obtained from Table II for a fixed

[25] R. E. Collins, Proc. IRE 43, 179 (1955).
[26] H. Riblet, IRE Trans. MTT-5, 233 (1957).
[27] L. Young, IRE Trans. MTT-7, 223 (1959).

number of layers to R_{opt} calculated in Eq. (1.3.14). If R differs substantially from R_{opt}, then additional layers must be added. In Section 1.4.2 we shall show how this scheme is used to fabricate high-sensitivity broadband transducers.

1.4. Design Considerations for Practical Devices

1.4.1. Material Properties

Piezoelectric transducers are fabricated from a number of crystalline and ceramic solids that are usually prepared in the form of plate resonators. Although plate resonators exhibiting a variety of modes of vibration can be used, the most common are the thickness expander mode and thickness shear mode, which are illustrated in Fig. 11.[15,28] Transducers fabricated from piezoelectric plates exhibiting these modes of vibration are used to launch longitudinal and transverse ultrasonic waves, respectively.

In the case of crystalline piezoelectrics, the mode of vibration is determined by the orientation of the plate relative to the crystalline axes. The crystallographic axes of quartz are illustrated in Fig. 12a and frequently used plate orientations ("cuts") are shown in Figs. 12b and c.[29,30] Transducers for the generation of longitudinal waves are made from X-cut quartz. Transducers for the generation of transverse waves are made from Y-cut plates or rotated Y-cut plates such as the AT-cut. The desig-

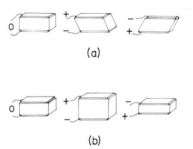

(a)

(b)

FIG. 11. Mechanical displacement for a piezoelectric quartz plate operating in the thickness shear mode (a) and the thickness expander mode (b). Part (a) shows a transverse AT-cut and part (b) a longitudinal X-cut. (Adapted in part from Posakony.[28])

[28] G. J. Posakony, *Ultrason. Symp. Proc.* IEEE Cat. #75 CHO 994-4SU (1975).

[29] "An Introduction to Piezoelectric Transducers," 6th Printing Valpey-Fisher Corp., Holliston, Massachusetts 1972.

[30] V. Bottoms, "The Theory and Design of Quartz Crystal Units." McMurray Press, Abilene, Texas, 1968.

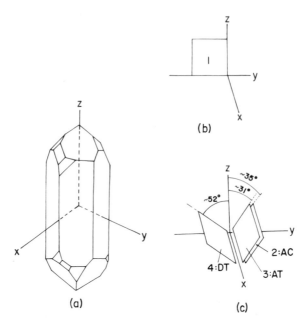

FIG. 12. Crystallographic axes (part a) and frequently used plate orientations. Part (b) shows an X-cut in quartz and part (c) a rotated Y-cut. (Adapted in part from Mason[15] and Bottoms.[30])

nation of a number of standard cuts of quartz along with the corresponding modes of vibration and useful frequency range are presented in Table III.

In the case of ceramic piezoelectrics, the mode of vibration is determined by the orientation of the plate relative to an axis that is imposed on the material by a process known as poling.[1,4,31,32] Ceramic piezoelectric

TABLE III.[a]   Quartz Plate Orientations

Plate number	Reference	Mode
1	X-cut	Thickness expander
2	AC-cut	Thickness shear
3	AT-cut	Thickness shear
4	DT-cut	Plate shear

[a] Adapted in part from Mason.[15]

[31] W. P. Mason, *Phys. Rev.* **74**, 1134 (1948).
[32] P. W. Forsbergh, Piezoelectricity, electrostriction and ferroelectricity, "Handbuch der Physik" (S. Flugge, ed.), Vol. 17, Sect. 2. Springer-Verlag, Berlin and New York, 1956.

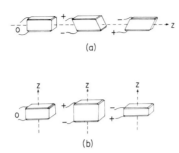

Fig. 13. Mechanical displacements relative to the poling axis z in ferroelectric ceramics for two modes: (a) the thickness shear mode with transverse orientation is poled perpendicular to the applied field; (b) the thickness expander mode with longitudinal orientation is poled parallel to the applied field. (Adapted in part from Posakony.[28])

elements are fabricated from polycrystalline ferroelectric materials. Initially, local regions ("domains") within the material exhibit electrical polarization, but the orientations of adjacent domains in a plate of such a material are random. Consequently, there is no net polarization and the plate is at most weakly piezoelectric. A preferred orientation for the domains is achieved by raising the temperature of the plate above the Curie point for the ceramic and imposing a large dc electric field ("poling"). The applied electric field causes a significant fraction of the domains to align parallel to the field, and the resulting net polarization is maintained if the ceramic is cooled before it is removed from the poling field. The direction of poling is conventionally designated as the z axis. Figure 13 illustrates that ceramic plates can be operated in thickness expander or thickness shear modes depending on the orientation of the plate axis relative to the axis of polarization.

The choice of a piezoelectric material is ultimately dictated by the specific application.[28,33] Certain distinct differences exist between crystalline piezoelectrics (most notably quartz) and ceramic piezoelectrics. In general, ceramics exhibit higher electromechanical coupling factors and lower mechanical Qs (i.e., higher mechanical losses) than crystalline piezoelectrics. Consequently, crystalline piezoelectrics are frequently used in narrowband applications whereas ceramics are commonly used in broadband applications. The principal parameters that are useful in choosing a particular material for a specific application are listed in Table IV. In Table V these properties are listed for a variety of crystalline and ceramic transducer materials operating in the thickness expander mode (longitudinal) or the thickness shear mode (transverse).

[33] J. Callerame, R. H. Tancrell, and D. T. Wilson, *Ultrason. Symp. Proc.* IEEE Cat. No. 78 CH 1344-ISU (1978).

TABLE IV. Acoustic and Piezoelectric Parameters

Symbol	Definition
d	Transmission constant − (strain out/field in)
g	Receiving constant − (field out/stress in)
ρ	Density
v^e	Ultrasonic velocity in a particular direction $[(c^E/\rho)^{1/2}]$
Z_0	Characteristic acoustic impedance (lossless approximation) ($= \rho v$)
ϵ^T	Free dielectric constant (unclamped)
k_T	Electromechanical coupling efficiency ($k_T{}^2 = e^2/\epsilon^S c^E$)
Q_m	Mechanical quality factor

TABLE V.[a] Material Properties

	Longitudinal					
	Quartz (0° X-cut)	PZT-4[b]	PZT-5[b]	PZT-5H[b]	PbNb$_2$O$_6$[b]	BaTiO$_3$[b]
d (10^{-12} m/V)	2	289	374	593	75	149
g (10^{-3} Vm/N)	50	26	25	20	35	14
ρ (kg/m^3)	2650	7600	7500	7500	5900	5700
v^E (m/sec)	5650	3950	3870	4000	2700	4390
Z_0 (10^6 kg/m^{2}sec)	15	30	29	30	16	25
ϵ^T/ϵ_0	4.5	1300	1700	3400	240	1700
k_T (%)	11	70	70	75	40	48
Q_m	>25000	<500	<75	<65	<5	<400

	Transverse					
	Quartz (0° Y-cut)	Quartz (AT-cut)	PZT-4[b]	PZT-5[b]	PZT-5H[b]	BaTiO$_3$[b]
d (10^{-12} m/V)	4.4	3.4	496	584	741	260
g (10^{-3} Vm/N)	110	80	38	38	27	20
ρ (kg/m^3)	2650	2650	7600	7500	7500	5700
v^E (m/sec)	3850	3320	1850	1680	1770	2725
Z_0 (10^6 kg/m^{2}sec)	10.2	8.8	14.0	12.6	13.3	15.5
ϵ^T/ϵ_0	4.5	4.6	1475	1730	3130	1450
k_T (%)	14	9	71	68	65	50
Q_m	>25000	>25000	<500	<75	<75	<300

[a] Adapted in part from Auld,[3] Posakony,[28] Mason,[34] and Berlincourt.[35]
[b] Values quoted are typical; some variations can occur.

[34] W. P. Mason," American Institute of Physics Handbook" (D. E. Gray, ed.), Chapter 3, pp. 118–129. McGraw-Hill, New York, 1972.
[35] D. Berlincourt, Piezoelectric crystals and ceramics, in "Ultrasonic Transducer Materials" (O. E. Mattiat, ed.), Chapter 2. Plenum Press, New York, 1971.

A large d constant indicates a material that would be an efficient transmitter of ultrasonic energy and a large g constant marks a material that would be a sensitive receiver of ultrasonic energy.[28] For applications where a single transducer is to be used as both transmitter and receiver, a material with a large dg product should be considered. The efficiency and sensitivity factors, however, must be weighed against other factors (such as mechanical Q and characteristic impedance) when practical devices are being designed. Design considerations will be discussed in Section 1.4.2.

1.4.2. Practical Transducers

Using the equivalent circuits of Chapter 1.3 with the electrical and mechanical properties of particular piezoelectric materials presented in Section 1.4.1, practical transducers can be designed for either narrowband (typically a few percent bandwidth) or broadband (typically 30–70% bandwidth) applications. Narrowband transducers are frequently used as components of a composite ultrasonic resonator to make precise measurements of phase velocity and attenuation near a specific frequency.[36] Broadband transducers are used in applications ranging from imaging,[37–39] using short ultrasonic pulses, to quantitative measurements of phase velocity[40,41] and attenuation[42–45] over a wide range of frequencies. In the next two sections we investigate some of the electrical and mechanical characteristics of practical transducers used for narrowband and broadband applications.

    1.4.2.1. Narrowband Transducers. In narrowband applications piezoelectric transducers usually form part of a composite resonator consisting of transducer plus specimen. To determine the ultrasonic properties of the specimen from measurements carried out on the composite resonator, the ultrasonic attenuation of the transducer must be small compared

[36] R. Truell, C. Elbaum, and B. B. Chick, "Ultrasonic Methods in Solid State Physics." Academic Press, New York, 1969.

[37] F. L. Thurstone, *IEEE Trans. Sonics Ultrason.* **SU-17**, 154 (1970).

[38] "Acoustical Holography," Vols. 1–7. Plenum Press, New York, 1967–1976.

[39] G. Kossoff, D. E. Robinson, and W. J. Garret, *IEEE Trans. Sonics Ultrason.* **12**, 31 (1965).

[40] E. H. Young, Jr., *IRE Trans. Ultrason. Eng.* **UE-9**, 13–21 (1962).

[41] W. Sachse and Y. H. Pao, *J. Appl. Phys.* **49**, 4320 (1978).

[42] E. P. Papadakis and K. A. Fowler, *J. Acoust. Soc. Am.* **50**, 729 (1971).

[43] J. G. Miller *et al., Ultrason. Symp. Proc.* IEEE Cat. No. 76 CH1120-5SU, 33 (1976).

[44] F. Lizzi, L. Katz, L. St. Louis, and P. S. Coleman, *Ultrasonics* **13**, 77–80 (1976).

[45] P. D. Lele, A. B. Mansfield, A. I. Murphy, J. Namery, and N. Senapati, *Proc. Seminar Ultrason. Tissue Characterization* Natl. Bur. Std. Publ. #453, pp. 167–196 (1976).

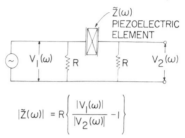

$$|\tilde{Z}(\omega)| = R\left\{\frac{|V_1(\omega)|}{|V_2(\omega)|} - 1\right\}$$

FIG. 14. A simple circuit for determining the magnitude of the electrical impedance of a piezoelectric plate. (Adapted in part from Bottoms.[30])

to that of the specimen.[42,46] Furthermore, the mechanical properties of the composite resonator must be sufficiently isolated from the electrical system that only an insignificant amount of energy is dissipated in the external circuitry.[46,47] Transducers made from crystalline materials such as quartz satisfy these requirements, exhibiting both low ultrasonic attenuation (high mechanical Q) and small electromechanical conversion efficiency.

For the design of narrowband transducers, the simple lumped equivalent circuit of Fig. 9 is usually adequate. The electrical properties can be characterized by measuring the frequency dependence of the impedance over a narrow range of frequencies in the neighborhood of a mechanical resonance of the transducer. One method for measuring the magnitude of the impedance is illustrated in Fig. 14.[30] Here the voltages $V_1(\omega)$ and $V_2(\omega)$ are measured over a range of frequencies near the mechanical resonance of the transducer and the impedance $|\tilde{Z}(\omega)|$ is calculated as shown.

An illustration of the response characteristics of a narrowband transducer is presented in Fig. 15. In Fig. 15a the magnitude of the electrical impedance is presented as a function of frequency in the neighborhood of the fundamental mechanical resonance of an unloaded, 0.2-cm-thick, X-cut quartz crystal operating in the thickness expander mode. The electrical impedance of the quartz plate exhibits a minimum at a frequency corresponding to the mechanical resonance frequency of the series mechanical RLC equivalent circuit of Fig. 9. At the resonant frequency the impedance measured consists of the parallel combination of R' and C_0. Using the data of Fig. 15a, R' is computed to be approximately 350 Ω for this quartz plate. The impedance of the quartz plate exhibits a maximum at a frequency slightly above the mechanical resonance frequency. This impedance maximum (sometimes described as an

[46] N. F. Foster and A. H. Meitzler, *J. Appl. Phys.* **39**, 4460 (1968).
[47] J. G. Miller and D. I. Bolef, *J. Appl. Phys.* **41**, 2282 (1970).

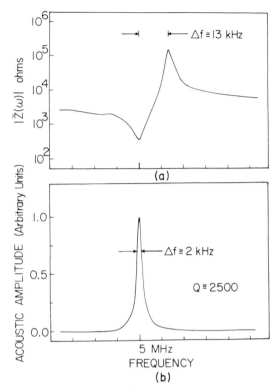

FIG. 15. Response characteristics of an isolated, 5-MHz, X-cut quartz plate operating near the fundamental frequency: (a) the frequency dependence of the magnitude of the electrical impedance; (b) the frequency-domain acoustic response with $Q \cong 2500$.

"antiresonance") corresponds to the parallel resonance of the plate capacitance C_0 with the net inductive reactance presented by the mechanical equivalent RLC network above the mechanical resonance frequency.

The mechanical response characteristics of a narrowband transducer can be measured using a number of time-domain or frequency-domain techniques discussed in Part 2. The sampled cw technique[2,48,49] described in Part 2 can be used to measure the frequency dependence of a signal related to the total mechanical energy stored in the resonator. With the sampled cw technique, a continuously running oscillator is gated on for a time sufficiently long that steady-state conditions are achieved in

[48] J. G. Miller and D. I. Bolef, *Rev. Sci. Instrum.* **40**, 915 (1969).
[49] D. I. Bolef and J. G. Miller, High frequency continuous wave ultrasonics, *in* "Physical Acoustics" (W. P. Mason, ed.), Vol. **8**, Chapter 3. Academic Press, New York, 1971.

the ultrasonic resonator. After steady-state conditions are achieved, the oscillator is gated off and the electrical response of the transducer is gated into a detector. Consequently, the transducer "samples" the steady-state ultrasonic signal that was established in the resonator during the oscillator-on interval. The square of the magnitude of the detected signal is proportional to the total mechanical energy stored in the resonator. In Fig. 15b we present the sampled cw frequency-domain response (the square of which is proportional to the acoustic energy stored in the un-loaded quartz resonator) as a function of frequency in the neighborhood of the fundamental mechanical resonance of the plate. The Q for this iso-lated resonator is estimated to be approximately 2500.

The effects of coupling a narrowband transducer to a specimen that approximates an infinite acoustical transmission line are illustrated in Fig. 16. The sampled cw frequency-domain response in the neighborhood of 5 MHz of the isolated quartz resonator is presented again in Fig. 16a. The response pattern resulting when the quartz resonator radiates into a

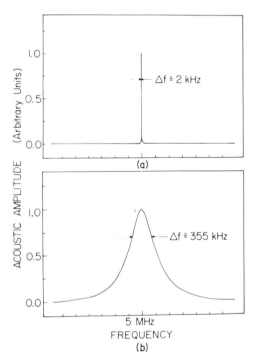

FIG. 16. Frequency-domain acoustic responses of the same quartz plate used in Fig. 15. This figure illustrates the alteration in the frequency domain acoustic response of the iso-lated quartz plate [part (a) with $Q \cong 2500$] when it is loaded on one side [part (b) with $Q \approx 14$] by a "purely resistive mechanical load" in the form of a long column of castor oil.

long column of castor oil is presented in Fig. 16b. The responses presented have been normalized to unit amplitude to facilitate a comparison of the widths of the resonance lines. The magnitude of the response of the castor oil loaded resonator is substantially lower than that of the unloaded resonator. Because castor oil is a highly attenuating substance, no ultrasonic energy transmitted into the castor oil is returned to the quartz plate. Thus the castor oil column represents an infinite transmission line. The equivalent circuit illustrated in Fig. 9 can be used to predict the extent to which the castor oil should lower the Q of the quartz resonator. The measured Q is reduced from the value of approximately 2500 shown in Fig. 16a to the value of approximately 14 shown in Fig. 16b. On the basis of the equivalent circuit of Fig. 9 and using handbook values for the material properties of X-cut quartz[3] and castor oil,[50] the Q is predicted to fall to 16, which is in reasonably good agreement with the experimental value of 14.

Transducers are frequently coupled to another mechanical resonator rather than to an infinite transmission line. The transducer–specimen combination then represents a composite ultrasonic resonator. The response characteristics of a composite resonator consisting of an AT-cut quartz transducer bonded to a 1.2-cm-thick sample of vanadium were presented in Fig. 1.[2] The sharp multiple resonances correspond approximately to half-wavelength mechanical resonances in the vanadium specimen. The spacing between the individual maxima is related to the phase velocity in the vanadium as discussed in Part 2. In addition, because the total acoustic loss in the sample is much larger than that in the transducer, the Q of the individual resonance lines is directly related to the attenuation in the sample. The envelope of the resonance pattern illustrated in Fig. 2 corresponds approximately to the response that would be exhibited by the quartz transducer if it were loaded by an infinite length of vanadium.

The use of a piezoelectric transducer to estimate the mechanical properties of a specimen from measurements carried out on a composite resonator is valid only in the limit where the mechanical properties of the composite resonator are approximately isolated from the external electrical system. If the electrical impedance of a narrowband transducer is well matched to the impedance of the electrical system, then the conversion of acoustical to electrical energy may result in nonnegligible dissipation of energy in the electrical circuit and a corresponding broadening of the

[50] F. Dunn, P. D. Edmonds, and W. J. Fry, Absorption and dispersion of ultrasound in biological media, "Biological Engineering" (H. Schwan, ed.), Chapter 3, p. 214. McGraw-Hill, New York, 1969.

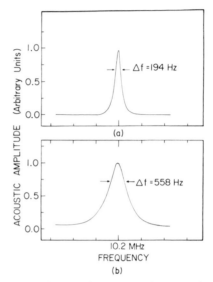

FIG. 17. The frequency-domain acoustic response of a composite resonator consisting of a 10-MHz, fundamental frequency, AT-cut quartz transducer bonded to a single crystal of InSb of thickness 1.2 cm is illustrated near a mechanical resonance for two conditions of electrical impedance matching. The response presented in part (a) was obtained with a large electrical mismatch, ($Q \cong 5300$), whereas the response presented in part (b) was obtained with reasonably good electrical match ($Q \cong 1800$). (Adapted in part from Miller and Bolef.[47])

frequency-domain response of the mechanical resonator.[47] As a consequence, the ultrasonic attenuation of the specimen might be overestimated. Effects arising from loading due to the electrical system can be minimized by deliberate electrical mismatch. In Fig. 17 the sampled cw frequency-domain response of a composite resonator is illustrated for two different states of impedance matching.[47] The composite resonator consists of a 10-MHz, fundamental frequency, AT-cut quartz plate bonded to a 1.2-cm-thick single crystal of InSb. The responses illustrated in Fig. 17 are centered at 10.2 MHz. The response presented in Fig. 17a was obtained with a large electrical mismatch whereas the response presented in Fig. 17b was obtained with a reasonably good electrical match. The mechanical resonance of the composite resonator has been artificially broadened from 194 Hz, corresponding to the true attenuation of the InSb specimen, to 558 Hz. As illustrated by these results, care must be taken to ensure that the mechanical properties of the composite resonator are adequately isolated from the external electrical system in order to determine correctly the attenuation of a specimen.

1.4.2.2. Broadband and Pulse-Echo Transducers. Broadband transducers are necessary in many applications including imaging[37-39] using short ultrasonic pulses and for rapid measurement of the phase velocity[40,41] and attenuation[42-45] over a continuous range of frequencies. Piezoelectric materials exhibiting large electromechanical coupling coefficients are necessary to achieve adequate sensitivity over a broad frequency range. For imaging applications, a short impulse response is desired whereas for attenuation and velocity measurements, it may be desirable to sacrifice a short impulse response for additional sensitivity over the same passband. The following three examples of broadband transducers are meant to illustrate the tradeoffs that must be made in the design of a practical ultrasonic system.

Transducers constructed for optimal impulse response usually consist of piezoelectric ceramics that are mechanically backed by high-loss materials[51] exhibiting mechanical impedances approximately equal to that of the piezoelectric element. The backing has the effect of "spoiling the Q" of the mechanical resonance of the piezoelectric plate and thus increasing the bandwidth. In Fig. 18 the electromechanical efficiency and impulse response of a commercially available lead metaniobate disk transducer backed by lossy tungsten-loaded epoxy are illustrated.[24] The disk transducer is 12.7 mm in diameter and has a resonance frequency of approximately 4.5 MHz. In Fig. 18a the two-way insertion loss of this transducer is illustrated as a function of frequency. The two-way insertion loss is a measure of the electromechanical efficiency of the transducer. It is defined as the ratio of the available electrical power generated by the device as a receiver to the electrical power dissipated in the device as a transmitter under conditions in which the acoustic wave produced is reflected from a perfectly reflecting interface and received by the same transducer.[52-57] The insertion loss plot indicates that this transducer is not highly efficient (35-dB insertion loss at the center of the passband) but exhibits a significant bandwidth and a smooth bandshape. The transducer also exhibits a very short impulse response, as is evident from Fig. 18b. In general, the spoiled Q resonator approach yields broad bandwidth and a compact impulse response, but at the expense of sensitivity.

[51] G. Kossoff, *IEEE Trans. Sonics Ultrason.* **SU-13**, 20 (1966).
[52] L. L. Foldy and H. Primakoff, *J. Acoust. Soc. Am.* **17**, 109 (1945).
[53] H. Primakoff and L. L. Foldy, *J. Acoust. Soc. Am.* **19**, 50 (1947).
[54] W. R. MacLean, *J. Acoust. Soc. Am.* **12**, 140 (1940).
[55] E. L. Carstensen, *J. Acoust. Soc. Am.* **19**, 961 (1947).
[56] G. A. Sabin, *J. Acoust. Soc. Am.* **36**, 168 (1964).
[57] J. M. Reid, *J. Acoust. Soc. Am.* **55**, 862 (1974).

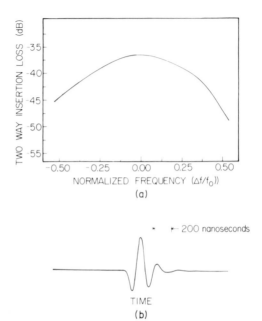

FIG. 18. Response of a broadband composite transducer consisting of a lead metaniobate disk backed by a lossy tungsten-loaded epoxy: (a) the two-way insertion loss as a function of frequency with $f_0 = 4.5$ MHz; (b) the time-domain response to an electrical impulse.

An approach for constructing broadband piezoelectric transducers with improved sensitivity combines the use of electrical tuning with mechanical backing layers bonded to the piezoelectric element. In Fig. 19, the two-way insertion loss and impulse response of this type of broadband transducer are illustrated. An inductor placed in parallel with the piezoelectric element is used as a tuning element for this system. The value of the inductor is chosen to resonate with the static capacitance C_0 of the piezoelectric plate at the operating frequency of the transducer. The insertion loss, illustrated in Fig. 19a, is reduced using this technique (15-dB insertion loss at band center). However, the improvement in sensitivity is gained at the expense of a compact time-domain response (Fig. 19b).

Broadband transducers possessing substantially lower insertion losses with bandwidths comparable to those of spoiled Q resonators can be constructed,[58] again, however, at the expense of a compact impulse response.[24] As discussed in Section 1.3.3, De Silets, Fraser, and Kino have used the KLM model (see Fig. 10) to design low-insertion-loss, broadband transducers using one or more quarter-wave matching schemes on

[58] J. H. Goll and B. A. Auld, *IEEE Trans. Sonics Ultrason.* **SU-22**, 52 (1975).

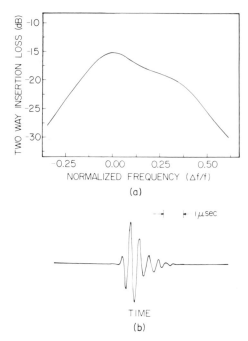

FIG. 19. Response of a composite transducer electrically tuned by the use of an inductor placed in parallel with the piezoelectric element: (a) the two-way insertion loss as a function of frequency with $f_0 \cong 2.4$ MHz; (b) the time-domain response to an electrical impulse.

the front (i.e., radiating) surface of an air-backed piezoelectric plate. In Fig. 20a the real and imaginary parts of the impedance measured by De Silets, Fraser, and Kino are illustrated for a 2-MHz, center frequency, lead metaniobate transducer. This transducer is air backed and has a single quarter-wave matching layer consisting of a thin layer of Dow Epoxy Resin 332, which was used to obtain the optimal acoustic match between lead metaniobate and water. For minimum insertion loss, the electrical reactance must be tuned out over the passband. Since the reactance of this transducer is negative, it can be tuned out approximately by an external series inductor. In Fig. 20b the two-way insertion loss of this transducer (including a 13.9-μH series inductor) is illustrated. The 3-dB bandwidth of this transducer is 40%, the bandshape is flat over the passband, and the two-way insertion loss is only 6.5 dB at the center of the passband. Figure 20c illustrates the impulse response of this transducer. The impulse response is notably less compact than that of the spoiled Q resonator response. The impulse response of an air-backed, multiple quarter-wave matching layer device can be anticipated on the basis of its

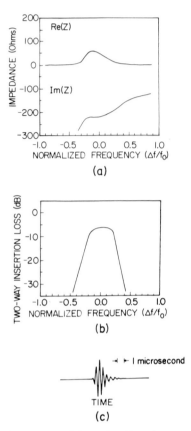

FIG. 20. The frequency dependence of the real and imaginary parts of the impedance [part (a) with $f_0 = 2.06$ MHz], the frequency dependence of the two-way insertion loss [part (b)], and the time-domain impulse response [part (c)] of a transducer consisting of an air-backed lead metaniobate disk bonded to a single quarter-wave matching layer. (Adapted from De Silets et al.[24])

frequency-domain response. Although the steady-state response of the air-backed, quarter-wave matching layer transducer exhibits a relatively broad bandwidth, the magnitude of the response exhibits relatively steep skirts which are accompanied by relatively rapid variations in phase. These frequency-domain characteristics mitigate against a compact response in the time domain. In contrast, the spoiled Q resonator design exhibits moderate roll off in the skirts, leading to a compact time-domain response.

All of the transducers considered must be appropriately matched to the electrical system for maximum bandwidth and minimum insertion loss.

To ensure minimum losses, the negative (i.e., capacitivelike) reactance of the transducer must be tuned out over the entire band. More sophisticated tuning networks than those mentioned above might be designed to compensate the specific frequency dependence of the negative reactance of a particular transducer. Once the reactance has been tuned out, the real part of the impedance at the center frequency can be matched to the electrical source impedance by using standard electrical impedance matching techniques.

1.5 Concluding Remarks

In this part, which dealt with piezoelectric transducers, we have surveyed many of the underlying physical principles of piezoelectricity as well as some practical details of transducer design and construction. By necessity, no one subject has been treated in the detail necessary for a thorough understanding. Consequently, we suggest the following general references as a starting point for further reading. The physics of piezoelectricity and its application to ultrasonic transducers is discussed in detail by Berlincourt et al.[4] A general account of wave propagation in elastic and piezoelectric solids is given by Auld.[3] For discussions of a number of practical aspects of ultrasonic transduction, the reader is referred to Hueter and Bolt,[59] Truell et al.,[36] and Sachse and Hsu.[60]

Acknowledgment

Pranoat Suntharothok-Priesmeyer was responsible for production of the text and illustrations.

[59] T. F. Hueter and R. H. Bolt, "Sonics: Techniques for the Use of Sound and Ultrasound in Engineering and Science." Wiley, New York, 1955.

[60] W. Sachse and N. N. Hsu, Characterization of ultrasonic transducers used in materials testing, in "Physical Acoustics," (W. P. Mason and R. N. Thurston, eds.) Vol. 14. Academic Press, New York, 1979.

2. ULTRASONIC WAVE VELOCITY AND ATTENUATION MEASUREMENTS

By M. A. Breazeale, John H. Cantrell, Jr., and Joseph S. Heyman

2.1. Introduction

During the past four decades the measurement of the velocity and attenuation of ultrasonic waves has been the basis of evaluation of a wide variety of physical properties of gases, liquids, and solids. In some cases the measurements have been made with great precision and accuracy. In other cases the ready availability of the result was more important than extreme precision. In developing the discussion that follows we have considered not only the precision achieved but also, wherever possible, the inherent accuracy of the system.

In writing this part we often encountered the fact that an author measured both velocity and attenuation and reported them in the same publication. Although this is appropriate in reporting scientific results, it nevertheless has created problems in our description of measurement of these two distinct phenomena. We found it possible to separate attenuation measurements from velocity measurements in the section on pulse techniques, but were not able to do this as completely in the section on continuous wave methods, because of the way the data are analyzed and reported. With optical techniques this is not a problem because, with certain exceptions, optical techniques are used primarily for velocity measurements only.

We recognize that other authors have addressed the task of writing a review of velocity and attenuation measurements. Notable among them are McSkimin,[1] who gave a thorough treatment of the fundamentals and of pulse techniques, and Papadakis, who described ultrasonic attenuation caused by scattering in polycrystalline media,[2] considered the effect of

---

[1] H. J. McSkimin, *in* "Physical Acoustics" (W. P. Mason, ed.), Vol. IA, pp. 271–417. Academic Press, New York, 1964.

[2] E. P. Papadakis, *in* "Physical Acoustics" (W. P. Mason, ed.), Vol IVB, pp. 269–328. Academic Press, New York, 1968.

diffraction,[3] and described pulse techniques with scientific and industrial applications.[4]

Our treatment of the subject of velocity and attenuation measurements is intended to cover a wider range of topics than has been attempted previously. The analysis covers optical techniques, pulse techniques, and continuous wave techniques in sufficient depth that the strong points, as well as some of the limitations, of each technique can be appreciated. The breadth of the discussion is intended to aid in the comparison of techniques at the time one is deciding which technique would be best to make a specific measurement or series of measurements.

2.1.1. Sources of Error

A number of sources of error need to be considered if one is measuring velocity and attenuation. In the first place, error is associated with the measurement of such quantities as time and distance. The effect of these errors can be evaluated directly. Thus, they need no further discussion here. Among the less-easily evaluated errors are those arising from velocity dispersion, from the transducer bond, from phase cancellation (resulting from lack of parallelism of the transducers), from material inhomogeneity, and from diffraction.

2.1.1.1. Dispersion. In general, measurement of velocity involves measurement either of the time required for the ultrasonic wave to travel a known distance or of the wavelength and frequency. In the first case, one is measuring what corresponds to group velocity; in the second, one is measuring the phase velocity. The difference between group velocity and phase velocity of ultrasonic waves in infinitely extended media most often is negligible, for, in general, the dispersion is quite small. On the other hand, in situations in which the sample is not effectively infinite in extent, the difference can be appreciable. In this case a correction of the data for velocity dispersion is essential.

2.1.1.2. Transducer Bond. An insidious source of error in velocity measurements in solids is the effect of the transducer producing the ultrasonic waves and of the bond coupling the transducer to the sample. With the optical technique one is measuring in the sample volume and, therefore, can neglect completely any complications caused by the transducers. With electronic pulse techniques and continuous wave techniques, however, effect of bonding the transducer to solid samples often is of the

---

[3] E. P. Papadakis, in "Physical Acoustics" (W. P. Mason and R. N. Thurston, eds.), Vol. XI, pp. 151–211. Academic Press, New York, 1975.

[4] E. P. Papadakis, in "Physical Acoustics" (W. P. Mason and R. N. Thurston, eds.), Vol. XII, pp. 277–374. Academic Press, New York, 1976.

order of 1% in the measured velocity. In this situation, one can correct the data, as described by McSkimin[5] and Papadakis.[6] Such corrections have been made in most of the measurements in solids we report in this study.

An alternative, which has appeal from the fundamental point of view, is the use of noncontact transducers. With this type of transducer the ends of the sample satisfy the free–free boundary conditions exactly. Data can be used without correction for phase shifts at the ends of the sample because the transducers do not load the ends of the sample appreciably. In a recent review article, Thompson[7] enumerates three general types of noncontact transducers: electromagnetic acoustic transducers (EMAT), capacitive transducers, and optical techniques.

With the EMAT a current is induced in a conducting sample by the Lorentz forces resulting from interaction with a static magnetic field.[8,9] A coil near the end of the sample couples the acoustical system to the electronic system. An EMAT can generate both longitudinal and transverse ultrasonic waves.[10] Also, an EMAT can be used to study surface waves,[11,12] angled bulk waves,[13] and other configurations.[14]

A capacitive transducer uses an electric field rather than a magnetic field. An electrode is spaced a distance of approximately 5 μm from the optically flat solid sample and is biased with a dc voltage. Vibration of the end of the sample induces an ac signal whose amplitude is proportional to the amplitude of the ultrasonic wave. Since the proportionality constant is known, the amplitude measurements are absolute. Gauster and Breazeale[15] describe a capacitive transducer capable of measurement of longitudinal wave amplitudes of the order of 10^{-10} cm at 30 MHz. Later, Peters and Breazeale[16] were able to measure amplitudes as low as 10^{-12} cm = 10^{-4} Å with a similar system. For measurements at low temperatures, differential expansion causes problems. To make measure-

[5] H. J. McSkimin, *J. Acoust. Soc. Am.* **33**, 12–16 (1961).

[6] E. P. Papadakis, *J. Acoust. Soc. Am.* **42**, 1045–1051 (1967).

[7] R. B. Thompson, *Proc. IEEE Ultrason. Symp., Phoenix,* p. 74 (1977).

[8] M. R. Gaerttner, W. D. Wallace, and B. W. Maxfield, *Phys. Rev.* **184**, 702 (1969).

[9] D. J. Meredith, R. J. Watts-Tobin, and E. R. Dobbs, *J. Acoust. Soc. Am.* **45**, 1393–1401 (1969).

[10] A. G. Betjemann, H. V. Bohm, D. J. Meredith, and E. R. Dobbs, *Phys. Lett.* **25A**, 753 (1967).

[11] R. B. Thompson, *IEEE Trans. Sonics Ultrason.* **SU-20**, 340 (1973).

[12] T. L. Szabo and A. M. Frost, *IEEE Trans. Sonics Ultrason.* **SU-23**, 323 (1976).

[13] T. J. Moran and R. M. Panos, *J. Appl. Phys.* **47**, 2225 (1976).

[14] C. M. Fortunko and R. B. Thompson, *Proc. IEEE Ultrason. Symp., Annapolis,* p. 12 (1976).

[15] W. B. Gauster and M. A. Breazeale, *Rev. Sci. Instrum.* **37**, 1544–1548 (1966).

[16] R. D. Peters and M. A. Breazeale, *Appl. Phys. Lett.* **12**, 106–108 (1968).

ments down to liquid helium temperatures, Peters *et al.*[17] developed a capacitive transducer in which the spacing could be adjusted between 3 and 10 μm by application of a pressure differential. Cantrell and Breazeale[18] developed a capacitive driver for use in ultrasonic wave velocity measurements, and later[19] they used two capacitive transducers for velocity measurements in which bond corrections were no longer necessary. More recently, a capacitive transducer for use in liquids has been constructed.[20]

As has been mentioned, bond corrections need not be made with optical techniques. Optical techniques traditionally have been used with compressional waves and, hence, traditionally have been limited to transparent media. However, in recent years there has been considerable emphasis on study of surface acoustic waves. In this case, even opaque materials can be studied by optical techniques since the light is reflected from the surface. Discussions of acousto-optical techniques and of surface acoustic waves are given in Parts 9 and 10.

2.1.1.3. **Phase Cancellation.** Fuller *et al.*[21] and Miller *et al.*[22] have pointed out that even with initially plane wave fronts phase cancellation of different portions of the wave fronts can occur and that this phase cancellation sets a limit on the measurement accuracy, especially at high frequencies. This is the reason for the use of optical tolerances in many of the measurements of solid samples made with quartz or other phase-sensitive transducers. The effect of lack of parallelism was carefully investigated by Truell and Oats.[23]

An interesting alternative from the fundamental point of view would be a phase-insensitive transducer. Southgate[24] reported a new type of transducer that is sensitive to acoustic flux rather than pressure. The device is based on phonon-charge carrier coupling in piezoelectric semiconductors.[25] A momentum transfer accompanies the coupling mechanism

[17] R. D. Peters, M. A. Breazeale, and V. K. Paré, *Rev. Sci. Instrum.* **39,** 1505–1506 (1968).

[18] J. H. Cantrell, Jr. and M. A. Breazeale, *Proc. IEEE Ultrason. Symp., Milwaukee,* p. 537 (1974).

[19] J. H. Cantrell, Jr. and M. A. Breazeale, *J. Acoust. Soc. Am.* **61,** 403–406 (1977).

[20] J. H. Cantrell, Jr., J. S. Heyman, W. T. Yost, M. A. Torbett, and M. A. Breazeale, *Rev. Sci. Instrum.* **50,** 31–33 (1979).

[21] E. R. Fuller, Jr., A. V. Granato, J. Holder, and E. R. Naimon, *in* "Methods of Experimental Physics" (R. V. Coleman, ed.), Vol. 11, pp. 371–441. Academic Press, New York, 1974.

[22] J. G. Miller, J. S. Heyman, D. E. Yuhas, and A. N. Weiss, *in* "Ultrasound in Medicine" (D. White, ed.), Vol. 1, pp. 447–453. Plenum, New York, 1975.

[23] R. Truell and W. Oates, *J. Acoust. Soc. Am.* **35,** 1382–1386 (1963).

[24] P. D. Southgate, *J. Acoust. Soc. Am.* **39,** 480 (1966).

[25] A. R. Hutson and D. L. White, *J. Appl. Phys.* **33,** 40 (1962).

and causes a net transport of charge carriers in the direction of wave propagation.[26] The physical process is called the "acoustoelectric effect," and the current resulting from it is proportional to the incident acoustic flux rather than the pressure, as is the case with other transducers. This means that the device is insensitive to phase in the acoustic wave. A practical device called an "acoustoelectric transducer" (AET) has been applied in medical measurements[27] and in material measurements.[28] Its advantages are demonstrated by Heyman.[29] With further testing, it appears likely that this transducer will make a fundamental improvement in precision velocity and attenuation measurements in which phase cancellation is present with conventional transducers.

2.1.1.4. Diffraction Corrections. The acoustic waves emitted by a transducer into a sample are not confined to a region defined by the area of the transducer and the normal to its emitting surface as often is assumed. Because of the finite size of the transducer, the acoustic beam spreads out into a diffraction field, a phenomenon that can introduce errors in both attenuation and velocity measurements. The diffraction error is related to the ratio of source dimension to acoustic wavelength and thus is especially large for low frequencies and small transducers. For attenuation measurements, it can be of the same order of magnitude as the measured value.

Investigations of the effect of diffraction on velocity and attenuation measurements have been made by a number of authors for the case of circular, axially concentric transmitting and receiving transducers of the same radius. The transmitting tranducer is treated as a finite piston source in an infinite rigid baffle radiating into a semi-infinite medium. The acoustic field is found at each point in the propagation medium and an integration is performed over the area in the field presented by the receiving transducer. The results show that the diffraction error involves two interference effects: one is intensity variation in the diffraction pattern and the other is the fact that the wave front is not strictly plane. Phase variations in the surface plane of the receiving transducer give rise to interference effects when integrated over a phase-sensitive transducer surface.

Huntington *et al.*[30] calculated diffraction corrections by numerical inte-

[26] G. Weinreich, *Phys. Rev.* **107,** 317 (1957).

[27] J. R. Klepper, G. H. Brandenburger, L. J. Busse, and J. G. Miller, *Proc. IEEE Ultrason. Symp., Phoenix,* p. 182 (1977).

[28] J. S. Heyman and J. H. Cantrell, Jr., *Proc. IEEE Ultrason. Symp., Phoenix,* p. 124 (1977).

[29] J. S. Heyman, *J. Acoust. Soc. Am.* **64,** 243–249 (1978).

[30] H. B. Huntington, A. G. Emslie, and V. W. Hughes, *J. Franklin Inst.* **245,** 1 (1948).

gration of an approximate expression for the piston field attributable to Lommel,[31] who had obtained the expressions, valid for large ka, by applying the Kirchhoff approximation to Fresnel diffraction of light by a circular hole in an opaque screen. Noting the similarity between Lommel's light diffraction problem and the acoustic diffraction problem for a piston source, Huntington *et al.* used Lommel's expression to approximate the piston field and obtained the so-called Lommel diffraction correction integral. The integral depends on a single-component variable $s = 2\pi z/ka^2$, where k is the wave number, a the transducer radius, and z the axial distance from the piston source. Numerically integrating the tabulated data of Lommel, Huntington *et al.* found the Lommel diffraction correction integral to be a monotonically decreasing function of s.

In a different approach to the problem, Williams[32] used an expression by King[33] to obtain an exact integral expression for the acoustic diffraction correction, which is a function of both s and ka. He also obtained an approximate expression, valid for large ka and z/a, which he numerically evaluated and found, as Huntington *et al.* did, to be a monotonically decreasing function of s. A better approximation to Williams's exact expression, valid for $z/a \geqslant 1$, was found by Bass[34] (with an error corrected by Williams[35]).

Seki *et al.*[36] noted that the monotonic decrease of the diffraction correction with increasing s, predicted by Huntington *et al.*[30] and Williams,[32] was inconsistent with experiment, particularly near $s = 1$, where they observed a maximum independent of ka. They surmised that the previous calculations used s intervals that were too large. Therefore they recalculated the diffraction correction using smaller s intervals and performed the integration graphically to find the experimentally observed peak at $s = 1.05$.

Khimunin[37] published extensive tables of the magnitude of the diffraction correction as a function of both ka and s, obtained by numerical integration of Williams's exact expression. Later Benson and Kiyohara[38] used a high-speed digital computer to evaluate Lommel's integral and published tables of both the magnitude and phase of the diffraction cor-

[31] E. Lommel, *Abh. Bayer. Akad. Wiss. Math. Naturwiss. Kl.* **15**, 233 (1886).
[32] A. O. Williams, *J. Acoust. Soc. Am.* **23**, 1–6 (1951).
[33] L. V. King, *Can. J. Res.* **11**, 135 (1934).
[34] R. Bass, *J. Acoust. Soc. Am.* **30**, 602–605 (1958).
[35] A. O. Williams, Jr., *J. Acoust. Soc. Am.* **48**, 285–289 (1970).
[36] H. Seki, A. Granato, and R. Truell, *J. Acoust. Soc. Am.* **28**, 230 (1956).
[37] A. S. Khimunin, *Acustica* **27**, 173 (1972).
[38] G. C. Benson and O. Kiyohara, *J. Acoust. Soc. Am.* **55**, 184 (1974).

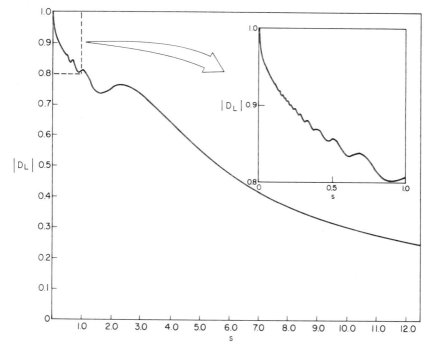

FIG. 1. Magnitude of the Lommel diffraction correction integral $|D_L|$ plotted as a function of $s = 2\pi z/ka^2$.[39]

rection to great precision. Comparison of the two calculations shows good agreement particularly in the range of large s and ka.

Rogers and Van Buren[39] showed that the required integral of Lommel's expression can be evaluated analytically to obtain a simple closed-form expression for the diffraction correction that is valid for all values of z/a, provided $(ka)^{1/2} \gg 1$. Their results are shown in Fig. 1.

Papadakis[40] derived an expression for the diffraction correction to pulse-echo attenuation measurements that takes into consideration which echoes are used in the measurement. He also studied the effect of anisotropy of the propagation medium on the diffraction.[41-43] Figure 2 shows curves of relative loss for various degrees of anisotropy represented by the parameter b. (The value $b = 0$ corresponds to the purely isotropic case.)

[39] P. H. Rogers and A. L. Van Buren, *J. Acoust. Soc. Am.* **55**, 724 (1974).
[40] E. P. Papadakis, *J. Acoust. Soc. Am.* **31**, 150–152 (1959).
[41] E. P. Papadakis, *J. Acoust. Soc. Am.* **35**, 490–494 (1963).
[42] E. P. Papadakis, *J. Acoust. Soc. Am.* **36**, 414–422 (1964).
[43] E. P. Papadakis, *J. Acoust. Soc. Am.* **40**, 863 (1966).

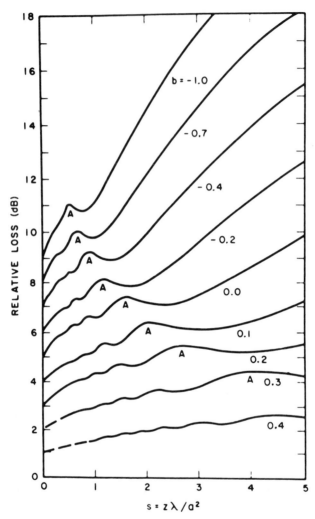

FIG. 2. Diffraction loss for a circular piston radiating longitudinal waves into an anisotropic medium along a direction of three-, four-, or fivefold symmetry.[43] Values of the anisotropy parameter b from -1 to $+0.4$ are shown. The upper limit of b is 0.5. The position of peak A is given by $s_A = 0.8/(0.5 - b)$.

The effect of diffraction on velocity measurements also was studied.[43] Pertinent to the diffraction corrections for velocity is the phase advance of the wave as a function of s. Figure 3 shows curves of relative phase for various degrees of anisotropy. Papadakis's results show that the effect of diffraction is to increase the measured velocity of propagation over the

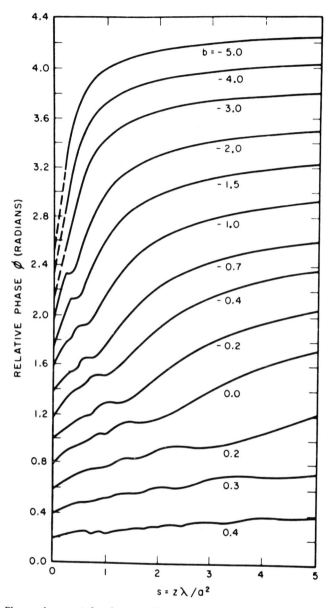

FIG. 3. Phase advance ϕ for the wave from a circular piston source.[43] This advance occurs as the secondary lobes leave the region of the main beam. The total advance from $s = 0$ to infinity is $\pi/2$ rad. The phase advance can be used in correcting both phase velocity and group velocity measurements. For group velocity it would be advantageous to place the echoes where $\partial\phi/\partial s = 0$.

plane wave value "only slightly," in agreement with the experimental work of McSkimin[44] and of Barshauskas *et al.*[45]   A correction of 0.005% is quoted by Cantrell and Breazeale[19] for velocity measurements in fused silica at 30 MHz. The effective diffraction loss and phase change in the field of broadband ultrasonic pulses were calculated[46] in terms of the *s* parameter at the center frequency of the pulse.

2.2. Systems for Making Measurements

2.2.1. Optical Systems

2.2.1.1. Light Diffraction (Fraunhofer Zone).
The interaction of light and ultrasonic waves has been investigated since the fundamental experiments of Debye and Sears[47] and Lucas and Biquard.[48] A theory describing the diffraction of light by ultrasonic waves was developed in a series of papers by Raman and Nath.[49-53] In the study of diffraction of light by ultrasonic waves, one distinguishes two regimes. Raman–Nath diffraction, named in honor of the two scientists who developed the fundamental theory, is analogous to the diffraction of light by a ruled grating. A theoretical treatment of Raman–Nath diffraction can be made by considering the effect of variations in refractive index caused by the ultrasonic wave. This produces a phase modulation of the light wave front and, hence, the diffraction pattern.

The second regime, Bragg diffraction, is more analogous to the diffraction of x rays by a crystalline lattice. A general treatment of both types of diffraction has been made by Bhatia and Noble.[54] Careful experimental

[44] H. J. McSkimin, *J. Acoust. Soc. Am.* **32**, 1401–1404 (1960).

[45] K. Barshauskas, V. Ilgunas, and O. Kubilynnene, *Sov. Phys. Acoust.* **10**, 21 (1964).

[46] E. P. Papadakis, *J. Acoust. Soc. Am.* **52**, 843–846, 847–849 (1972).

[47] P. Debye and F. W. Sears, *Proc. Natl. Acad. Sci. U.S.* **18**, 409 (1932).

[48] R. Lucas and P. Biquard, *J. Phys. Radium* **3**, 464 (1932).

[49] C. V. Raman and N. S. N. Nath, *Proc. Indian Acad. Sci. Sect. A Part I* **2**, 406–412 (1935).

[50] C. V. Raman and N. S. N. Nath, *Proc. Indian Acad. Sci. Sect. A Part II* **2**, 413–420 (1935).

[51] C. V. Raman and N. S. N. Nath, *Proc. Indian Acad. Sci. Sect. A Part III* **3**, 75–84 (1936).

[52] C. V. Raman and N. S. N. Nath, *Proc. Indian Acad. Sci. Sect. A Part IV* **3**, 119–125 (1936).

[53] C. V. Raman and N. S. N. Nath, *Proc. Indian Acad. Sci. Sect. A Part V* **3**, 459–465 (1936).

[54] A. B. Bhatia and W. J. Noble, *Proc. R. Soc. London Ser. A* **220**, 356–368, 369–385 (1953).

investigation of the overlap region of Raman–Nath diffraction and Bragg diffraction has been made by Nomoto.[55] Theoretical analysis of this region of overlap has been made by Klein et al.,[56] who define a dimensionless parameter

$$Q = K^{*2}L/\mu_0 K, \qquad (2.2.1)$$

where K^* is the ultrasonic propagation constant $2\pi/\lambda^*$, L the width of the ultrasonic beam, μ_0 the refractive index of the medium in which the ultrasonic wave propagates, and K the light propagation constant $2\pi/\lambda$. The magnitude of Q allows one to determine which type of diffraction is dominant. This parameter was experimentally investigated by Klein et al.[57] and by Martin et al.[58] For ultrasonic waves in water, $Q > 9$ indicates that the diffraction is totally Bragg diffraction. For $9 > Q > 1$, the diffraction is mixed. For $Q < 1$, the diffraction is Raman–Nath diffraction. Effectively, this means that one most readily uses Raman–Nath diffraction for velocity measurements at low frequencies and Bragg diffraction at high frequencies. Transition from one type of diffraction to the other occurs between 10 and 20 MHz in usual laboratory situations.

2.2.1.1.1. RAMAN–NATH DIFFRACTION. Figure 4 allows one to distin-

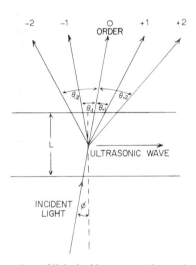

FIG. 4. Diffraction orders of light incident on an ultrasonic wave in a liquid.[58]

[55] O. Nomoto, Proc. Phys. Math. Soc. Jpn. 24, 380–400, 613–639 (1942).
[56] W. R. Klein, B. D. Cook, and W. G. Mayer, Acustica 15, 67–74 (1965).
[57] W. R. Klein, C. B. Tipnis, and E. A. Hiedemann, J. Acoust. Soc. Am. 38, 229–233 (1965).
[58] F. D. Martin, L. Adler, and M. A. Breazeale, J. Appl. Phys. 43, 1480–1487 (1972).

guish the optical arrangements. For nonnormal incidence, the diffraction orders are located at angles θ_n that satisfy

$$\sin(\theta_n + \phi) - \sin \phi = n\lambda/\lambda^*. \tag{2.2.2}$$

For Raman–Nath diffraction, $Q < 1$ implies that the ultrasonic frequency is typically less than 10 MHz. Good separation of the diffraction orders puts a lower limit on the frequency in the neighborhood of 1 MHz. In the frequency range, then, of 1–10 MHz, one can use normal incidence ($\phi = 0$) and reduce Eq. (2.2.2) to

$$\sin \theta_n = n\lambda/\lambda^*. \tag{2.2.3}$$

This equation is valid for both progressive and standing ultrasonic waves.

A very simple and direct measurement of λ^* and, hence, the ultrasonic wave velocity is suggested by Eq. (2.2.3). A monochromatic light source, an optical bench, a transducer, and a source of rf voltage constitute the necessary equipment. Figure 5 shows two optical systems ca-

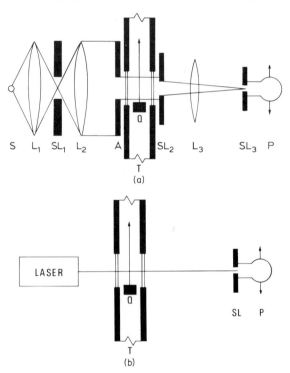

FIG. 5. Optical arrangements for studying the diffraction of light by ultrasonic waves: (a) with mercury vapor light source S; (b) with laser. Lenses are labeled L_i, slits SL_i, aperture A, quartz transducer Q, tank T, and photomultiplier P.

pable of measurement to accuracies of the order of 1%. Figure 5a is the optical system used with a mercury vapor or similar light source; Fig. 5b is the system used with a laser. A ground glass screen can be used in place of the photomultiplier P if one is interested in velocity measurement only.

Figure 6a is a diffraction pattern observed with a mercury vapor lamp and a source slit. The propagating medium is water. The number of diffraction orders increases with an increase of ultrasonic wave intensity, and the light intensity in the diffraction orders goes through maxima and minima as shown. For truly progressive ultrasonic waves, the minima are actually zeros of intensity. If the ultrasonic wave undergoes waveform distortion because of nonlinear effects, this diffraction pattern becomes

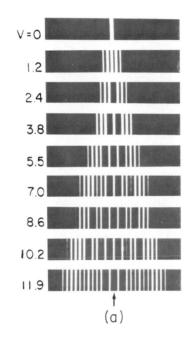

(a)

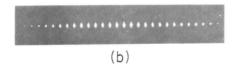

(b)

FIG. 6. Diffraction patterns produced: (a) by progressive ultrasonic waves in water[59]; (b) by standing waves.

asymmetrical.[59]   On the other hand, if the ultrasonic wave is a standing wave, the zeros of intensity in the lower diffraction orders disappear, as shown in Fig. 6b (a laser light source was used).   The diffraction of light by ultrasonic waves of various standing wave ratios was considered by Cook and Hiedemann,[60] who showed how the light intensity in the diffraction orders changes as the ultrasonic standing wave ratio changes from SWR = ∞ (progressive waves) to SWR = 1 (standing waves).   For purposes of velocity measurement, however, the important point is that the *position* of the diffraction order is unaffected either by nonlinearity or by the SWR.   Thus, Eq. (2.2.3) can be used to evaluate the ultrasonic wavelength for any SWR in any transparent medium.

The use of Raman–Nath diffraction for measurements of ultrasonic wave velocity in fluids is very direct and convenient, but the fact that the position of the diffraction order is unaffected by the SWR is especially important to measurement of velocity in transparent solids in which the attenuation is so small that reflected waves are inevitable.   Barnes and Hiedemann[61] set up resonances between parallel faces of glass samples and obtained an improved accuracy in the measurements because the resonance condition increases the number of diffraction orders.   The light source was a slit, which made possible observation of high diffraction orders.

A second type of measurement was realized by setting up volume resonances.   With a point light source, Barnes and Hiedemann[61] observed that the diffraction orders were located on circles around the zero order. One circle corresponded to the longitudinal wave and a (larger) second circle corresponded to the transverse wave arising from mode conversion at the sample surface.   These diffraction patterns previously had been studied by Schaefer and Bergmann,[62] who used them to determine the elastic constants of glasses.[63]   An example of the diffraction pattern produced by volume resonance of a glass sample is shown in Fig. 7.[64]   The inner circle is the locus of first orders produced by longitudinal waves propagating in different directions in the sample.   On the outer circles are found the first orders produced by transverse waves.   The radii of these circles are measures of the corresponding wavelengths in the glass sample.

[59] M. A. Breazeale and E. A. Hiedemann, *J. Acoust. Soc. Am.* **33**, 700–701 (1961).

[60] B. D. Cook and E. A. Hiedemann, *J. Acoust. Soc. Am.* **33**, 945–948 (1961).

[61] J. M. Barnes and E. A. Hiedemann, *J. Acoust. Soc. Am.* **28**, 1218–1221 (1956); **29**, 865 (1957).

[62] Cl. Schaefer and L. Bergmann, *Naturwissenschaften* **23**, 799–800 (1935).

[63] Cl. Schaefer and L. Bergmann, *Ann. Phys.* **3**, 72–81 (1948).

[64] L. Bergmann, "Der Ultraschall," p. 568. Hirzel Verlag, Stuttgart, Germany, 1954; Cl. Schaefer and L. Bergmann, *Naturwissenschaften* **22**, 685–690 (1934).

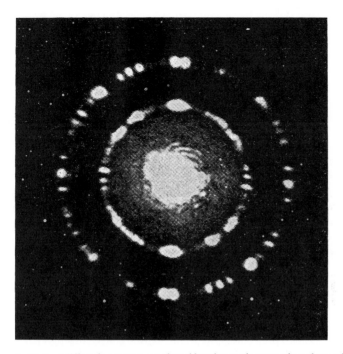

FIG. 7. Optical diffraction pattern produced by ultrasonic waves in a glass cube under volume resonance conditions.[64]

A very interesting extension of this technique was made by Schaefer and Bergmann,[64] who found that the diffraction pattern produced by volume resonance of anisotropic crystals gave a measure of the velocity in various directions and, hence, of the elastic constants of the crystals. An example of the diffraction of light by ultrasonic waves in crystalline quartz is shown in Fig. 8. The similarity between these figures and Laue diagrams obtained by diffracting x rays by crystalline lattices is not coincidental. The same lattice spacing among the atoms that determines the spacing of the diffraction orders in a Laue diagram also leads to anisotropy of the ultrasonic wave velocity. This anisotropy is responsible for the directional variation of the positions of the orders in Fig. 8. In fact, Hiedemann[65] called these figures "Laue diagrams (obtained) with optical waves." The actual diagram observed is dependent on the state of polarization of the incident light since the crystal is birefringent. Figure 8 compares the experimental results of Schaefer and Bergmann[64] with the predictions of the theory of Mueller.[66]

[65] E. A. Hiedemann, "Grundlagen und Ergebnisse der Ultraschallforschung." de Gruyter, Berlin, 1939.
[66] H. Mueller, *Phys. Rev.* **52**, 223–229 (1937).

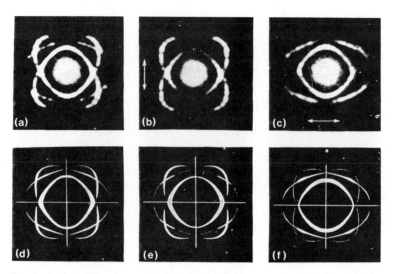

FIG. 8. Optical diffraction patterns produced by ultrasonic waves in quartz.[65] The light is incident parallel to the crystalline y axis: (a) unpolarized light; (b) electric vector parallel to the z axis; (c) electric vector perpendicular to the z axis; (d)–(f) corresponding interference figures calculated by Mueller.[66]

2.2.1.1.2. BRAGG DIFFRACTION. For Bragg diffraction, two conditions must be met (see Fig. 4): (1) the angle of incidence ϕ must be a particular angle ϕ_B, and (2) the parameter Q, defined by Eq. (2.2.1), must satisfy $Q > 9$. If both of these conditions are met, one can rewrite Eq. (2.2.2) in the form

$$n\lambda = 2\lambda^* \sin \phi_B, \qquad (2.2.4)$$

from which the ultrasonic wavelength λ^* can be evaluated. In practical experimental situations one satisfies $Q > 9$ and uses convergent incident light. The ultrasonic beam then selectively diffracts only that portion of the light that is incident at the Bragg angle.

Although the subject has been investigated in considerable detail in other connections, it is only relatively recently that it was realized that Bragg diffraction is ideally suited for measurement of ultrasonic wave velocity at high frequencies. It has proved to be a very effective means of measuring velocity in the frequency range between 100 MHz and a few gigahertz. Krischer[67] measured ultrasonic wave pulses in lanthanum fluoride in the frequency range 200–800 MHz. The estimated absolute accuracy was better than 0.1%. It was limited primarily by the accuracy

[67] C. Krischer, *Appl. Phys. Lett.* **13**, 310 (1968).

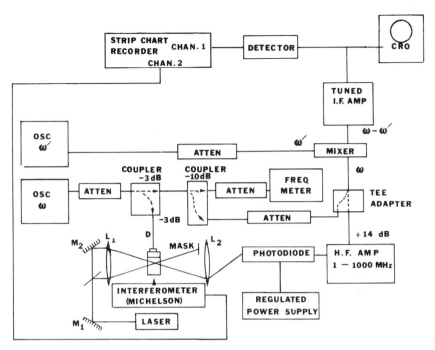

FIG. 9. Experimental arrangement for precision measurement of ultrasonic wave velocity by Bragg diffraction of light.[68]

with which the angle θ_B coulds be measured. Krischer points out that the velocity is measured within a very small volume. (The light beam diameter could be as small as 1 mm.)

Local values of the velocity have been measured with more precision by Simondet et al.[68] The Bragg diffracted light is Doppler shifted by an amount equal to the frequency of the ultrasonic wave. Therefore, superposition of the diffracted light and undiffracted light produces optical heterodyning. The two light beams simultaneously incident upon a photodiode produce an electrical signal whose amplitude depends on the relative phase of the two beams. As the sample is moved in the direction of propagation of the ultrasonic wave, a periodic photodiode output is observed. This period, correlated with the distance of motion of the sample, gives a measure of the wavelength and, hence, the velocity of the ultrasonic wave. Figure 9 is a diagram of a system in which the sample

[68] F. Simondet, F. Michard, and R. Torquet, *Opt. Commun.* **16**, 411–416 (1976); see also F. Michard and B. Perrin, *J. Acoust. Soc. Am.* **64**, 1447–1456 (1978).

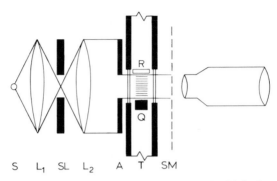

FIG. 10. Experimental arrangement for visibility method. Light from mercury vapor source S passes through lenses L_1 and L_2 and slit SL and is collimated when it passes through aperture A and tank T containing the liquid to be measured. Standing ultrasonic waves generated by quartz transducer Q and reflected from reflector R are imaged in the focal plane of the microscope SM with superimposed micrometer scale.

motion is measured by a Michelson interferometer. The strip chart recorder output is a measure of the ultrasonic wavelength in terms of the light wavelength (6328 Å). Measurements at 300 MHz by Simondet *et al.*[68] in a very homogeneous KBr sample gave a precision of 0.3 m/sec in a measurement of 3035.7 m/sec, or 0.01%. This is significant because it is an accurate measure of the velocity in a very small sample volume. Thus local variations in velocity caused by sample inhomogeneities can be measured.

2.2.1.1.3. ATTENUATION MEASUREMENT. Light diffraction also offers a technique for measuring ultrasonic wave attenuation. For example, Farrow *et al.*[69] have developed an automatic ultrasonic attenuation spectrometer for measurements between 5 and 250 MHz. Either Bragg diffraction or Raman–Nath diffraction can be used. The method depends on satisfying the conditions for linearity between ultrasonic wave intensity and light intensity in the first diffraction order, as does the schlieren technique used by Kannuna.[70] One then scans the length of the ultrasonic beam and measures the attenuation. This method is especially useful at high frequencies and/or high attenuation coefficients.

2.2.1.2. Visibility Method (Fresnel Diffraction Zone). A very convenient and reasonably accurate (between 0.1 and 0.01%) technique for measuring velocity is the visibility method, used extensively by Hiede-

[69] M. M. Farrow, S. L. Olsen, N. Purdie, and E. M. Eyring, *Rev. Sci. Instrum.* **47**, 657 (1976).

[70] M. M. Kannuna, *J. Acoust. Soc. Am.* **27**, 5–8 (1955).

mann.[65,71-76] This method makes use of the diffraction pattern in the optical Fresnel zone, the region along the optical axis immediately behind the ultrasonic wave. Collimated light is incident onto a standing ultrasonic wave. The resulting Fresnel diffraction pattern, which is a series of real images of the ultrasonic wave fronts, is examined by a microscope, as shown in Fig. 10. By interposing a micrometer scale in the focal plane of the microscope, one sees the scale superimposed on the wave field as shown in Fig. 11. This technique has been used by Mayer and Hiedemann[73-76] for measurements of the elastic constants of sapphire. Since the images of the wave fronts are $\lambda/2$ apart, one can quickly obtain the ultrasonic wave velocity in transparent liquids and solids. An even quicker technique is to mount the microscope on a calibrated screw and count fringes as they pass the microscope cross hairs.

It should be pointed out that interference causes the images to change, fade, and reappear as the microscope is moved along the optical axis toward or away from the ultrasonic beam. An image such as the one shown in Fig. 11 appears a short distance behind the ultrasonic beam. Examples of images seen at greater distances are shown in Fig. 12. As this sequence of images is periodic,[77] there is no difficulty in determining the location of the images spaced $\lambda/2$.

The visibility method is one of the most direct methods for measuring ultrasonic wave velocity in transparent samples. Its inherent accuracy is great enough (of the order of 0.01%[78]) that local variations in velocity can be detected.[73]

2.2.2. Pulse Systems

2.2.2.1. Velocity Measurements. 2.2.2.1.1. BASIC PULSE-ECHO METHOD. The basic pulse-echo method for measuring velocity and attenuation has been described by several authors.[79-84] A typical equipment

[71] E. A. Hiedemann and K. H. Hoesch, Z. Phys. 90, 322–326 (1934); 96, 268–272 (1935).
[72] E. A. Hiedemann and K. H. Hoesch, Z. Phys. 107, 463–473 (1937).
[73] W. G. Mayer and E. A. Hiedemann, Acta Crystallogr. 12, 1 (1959).
[74] W. G. Mayer and E. A. Hiedemann, J. Acoust. Soc. Am. 30, 756–760 (1958).
[75] W. G. Mayer and E. A. Hiedemann, J. Acoust. Soc. Am. 32, 1699–1700 (1960).
[76] W. G. Mayer and E. A. Hiedemann, Acta Crystallogr. 14, 323 (1961).
[77] E. A. Hiedemann and M. A. Breazeale, J. Opt. Soc. Am. 49, 372–375 (1959).
[78] E. Schreuer, Akustische Z. 4, 215 (1939).
[79] F. A. Firestone, J. Acoust. Soc. Am. 18, 200 (1946).
[80] J. R. Pellam and J. K. Galt, J. Chem. Phys. 14, 608 (1946).
[81] C. E. Teeter, Jr., J. Acoust. Soc. Am. 18, 488 (1946).
[82] J. M. M. Pinkerton, Nature (London) 160, 128 (1947).
[83] H. B. Huntington, Phys. Rev. 72, 321 (1947).
[84] J. R. Neighbours, F. W. Bratten, and C. S. Smith, J. Appl. Phys. 23, 389 (1952).

FIG. 11. Visibility fringes observed at 4 MHz in water. The scale is in millimeters.

arrangement is shown in Fig. 13. A pulsed rf signal of given frequency is converted by means of a transducer into a pulsed ultrasonic wave of the same frequency. The ultrasonic pulse travels through the sample and is reflected between the sample boundaries until it decays away. Each time

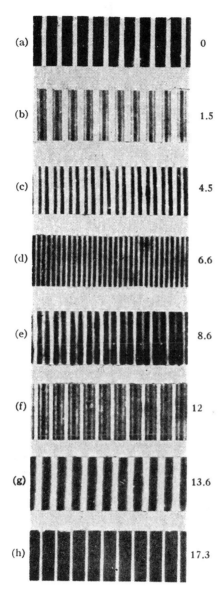

(a)   0

(b)   1.5

(c)   4.5

(d)   6.6

(e)   8.6

(f)   12

(g)   13.6

(h)   17.3

FIG. 12. Images seen at increasing distances (measured in centimeters) from the ultrasonic waves in Fig. 10. The pattern is periodically repeated at a distance $D' = 17.3$ cm with $D'/4 \cong 4.5$, $D'/2 \cong 8.6$, and $3D'/4 \cong 13.6$.[77]

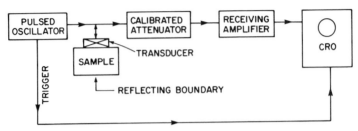

FIG. 13. Basic pulse-echo system.

the ultrasonic pulse strikes the sample end coupled to the transducer, an electrical signal is generated which is amplified and displayed on an oscilloscope. If the pulse length is small compared to a round-trip transit time in the sample, a pulse-echo decay pattern develops as shown in Fig. 14. The velocity of ultrasonic wave propagation is determined by measuring the transit time between the reflected pulses and the corresponding pulse propagation distance in the sample.

A continuously variable time delay may be used to measure the transit time between the pulses that are individually displayed on the oscilloscope by means of an expanded sweep. It is common in such cases to use a detected (i.e., rectified and filtered) signal and to measure the time interval between corresponding reference points of each echo (e.g., the leading edge of the pulse—but see below for refinements). Adjusting the attenuator so that the receiver input signal is constant for each displayed echo minimizes the error due to amplifier nonlinearities and reduces (but does not eliminate) the uncertainty in the signal reference points. Such a

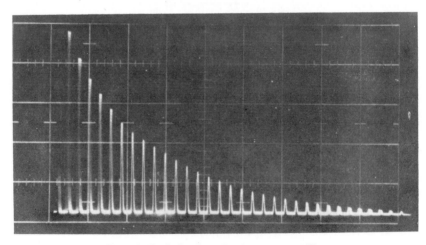

FIG. 14. Typical pulse-echo decay pattern.[135]

procedure also allows one to read directly the ultrasonic attenuation from the calibrated attenuator.

In measurements of solids using contact transducers, Lazarus[85] was among the first to point out that a small transit time error is incurred due to the ultrasonic wave propagating into the transducer and bond. His solution was to measure pulse transit times in samples of different lengths. He plotted the transit times as a function of the sample lengths and assumed that the intercept of the resulting curve was the transit time error. The reciprocal of the slope was assumed to be proportional to the true velocity.

A qualitative explanation of this error was given by Eros and Reitz.[86] They found that partial transmission and reflection of the ultrasonic pulse at the sample–transducer interface changed the shape of the pulse with each successive echo. This means that any attempt to find some characteristic "mark" on the pulses to use as a reference point to measure transit time always leads to limited measurement accuracy. Papadakis[87] pointed out that matching the leading edges of the echoes is an improper procedure for accurate phase velocity measurements and may lead to errors of as much as a few parts in 10^3. McSkimin[88,89] (see Section 2.2.2.1.4) developed a technique that allows cycle-for-cycle matching of the echoes and, hence, eliminates this problem. The related problem of ultrasonic phase shifts upon reflection at the sample–transducer interface is discussed in Section 2.2.2.1.3. In solids, using contact transducers for phase velocity measurements accurate to a few parts in 10^3 and attenuation measurements accurate to about 5 parts in 10^2, the basic pulse-echo method using leading edge echo matching is expedient and in some cases it is adequate.

For samples in which the path length is variable, accuracy may be improved. Pellam and Galt,[80] working with liquids, determined the sound velocity by measuring the distance the transducers had to be moved from the reflector to delay the received echo by a specified increment. With this technique phase distortion due to reflections is less serious, since only the effect of a change in path length is important. They reported velocity measurements accurate to 5 parts in 10^4.

2.2.2.1.2. SING-AROUND METHOD. The sing-around method[90-93]

[85] D. Lazarus, *Phys. Rev.* **76**, 545–553 (1949).
[86] S. Eros and J. R. Reitz, *J. Appl. Phys.* **29**, 683 (1958).
[87] E. P. Papadakis, *J. Acoust. Soc. Am.* **42**, 1045–1051 (1967).
[88] H. J. McSkimin, *J. Acoust. Soc. Am.* **33**, 12–16 (1961).
[89] H. J. McSkimin, *J. Acoust. Soc. Am.* **34**, 404–409 (1962).
[90] R. D. Holbrook, *J. Acoust. Soc. Am.* **20**, 590 (1948).
[91] N. P. Cedrone and D. R. Curran, *J. Acoust. Soc. Am.* **26**, 963 (1954).
[92] M. Greenspan and C. E. Tschiegg, *J. Acoust. Soc. Am.* **28**, 158 (1956).
[93] G. W. Ficken, Jr. and E. A. Hiedemann, *J. Acoust. Soc. Am.* **28**, 921–923 (1956).

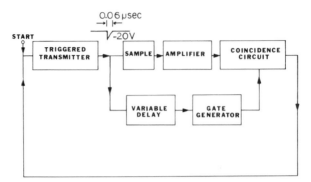

FIG. 15. Block diagram of basic sing-around system.[94]

differs from the basic pulse-echo method by the way in which the pulse timing is done. A block diagram of the basic system is shown in Fig. 15. Usually in this method a second transducer (used as a receiver) is placed at the end of the sample opposite the transmitting transducer. The received signal is used to retrigger the pulse generator thereby generating a continuous succession of pulses. Since the repetition rate of this pulse sequence depends on the travel time (i.e., on the path length and the ultrasonic velocity) in the sample, the ultrasonic velocity may be determined from the measurement of this repetition rate.

An inherent error in the pulse repetition rate (due to time delays in the electronic circuits) limits the accuracy of this method for absolute velocity measurements. Changes in pulse shape resulting from attenuation and the presence of the interface between the contact transducer and the sample also contribute to the error. Accuracies of a few parts in 10^4 for absolute velocity measurements are possible with this method. An improvement to the sing-around method by Forgacs[94] allows relative velocity measurements (i.e., changes in velocity) to be made to one part in 10^7. Millero and Kubinski[95] used this technique to measure relative values of velocity in seawater and pure water.

2.2.2.1.3. GATED DOUBLE-PULSE SUPERPOSITION METHOD. The gated double-pulse superposition method, introduced by Williams and Lamb,[96] may be understood by referring to Fig. 16. A pulsed ultrasonic signal is transmitted into the sample folowed by a second pulsed ultrasonic signal phase locked but delayed in time with respect to the first. The phase locking is obtained by gating a continuously running oscillator as shown in

[94] R. L. Forgacs, *IRE Trans. Instrum.* **9**, 359 (1960).
[95] F. J. Millero and T. Kubinski, *J. Acoust. Soc. Am.* **57**, 312–319 (1975).
[96] J. Williams and J. Lamb, *J. Acoust. Soc. Am.* **30**, 308–313 (1958).

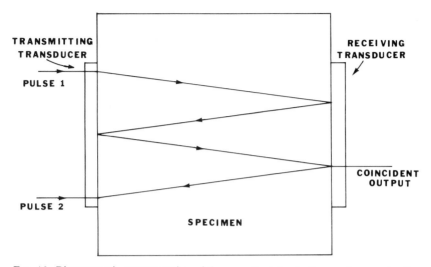

FIG. 16. Diagrammatic representation of the acoustic delay in the specimen and reflections of pulses.[96]

Fig. 17. The two pulsed ultrasonic signals reflect between the sample walls giving rise to two pulse-echo trains. The time delay is adjusted such that superposition of the desired echoes from the two pulse trains is achieved. The resulting signal is received by a transducer at the other

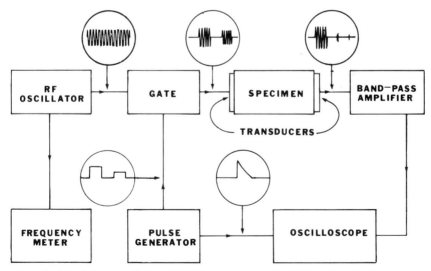

FIG. 17. Block diagram of the gated double-pulse superposition method for velocity measurements in solids.[96]

end of the sample, amplified, and displayed on an oscilloscope as shown in Fig. 18.

If care is taken to make the signals flat topped, a continuous wave analysis may be made with good approximation. Let the signal received from the delayed pulse be

$$y_2 = A \sin \omega t, \qquad (2.2.5)$$

where A is the pulse amplitude and $f = \omega/2\pi$ the ultrasonic frequency. Let τ be the time for a single trip through the sample and γ the phase shift due to reflection at the sample ends. Assuming that the phase shifts at all reflections are equal, we may write, for the initial pulse after one round trip delay,

$$y_1 = A \sin[\omega(t - 2\tau) + 2\gamma]. \qquad (2.2.6)$$

Superposition of the two signals gives the receiving transducer output

$$y_1 + y_2 = 2A \sin[\omega(t - \tau) + \gamma] \cos(\omega\tau - \gamma). \qquad (2.2.7))$$

The superimposed signal may be made zero, independently of time, by adjusting the carrier frequency f such that the condition

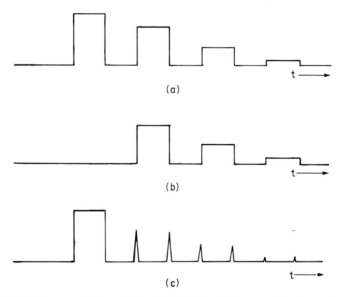

(a)

(b)

(c)

FIG. 18. Oscilloscope displays of demodulated signals demonstrating the method of cancellation in the gated double-pulse method: (a) received signals due to multiple reflection of first pulse; (b) received signals due to multiple reflection of second pulse; (c) cancellation due to addition of parts (a) and (b).[96]

$$\cos(\omega\tau - \tau) = 0 \tag{2.2.8}$$

or

$$\omega\tau - \gamma = (2n + 1)\pi/2, \qquad n = 0, 1, 2, 3, \ldots \tag{2.2.9}$$

is satisfied. The pulse transit time τ can be obtained from Eq. (2.2.9) once n and γ have been evaluated by the following procedures.

Redwood and Lamb,[97] working with solids, published an analytical treatment of the effect that contact transducers have on the reflection of ultrasonic waves. Their analysis showed that the phase angle γ may be expressed by

$$\gamma = \pi - 2 \tan^{-1} a, \tag{2.2.10}$$

where

$$a = \frac{Z_B}{Z_S} \left[\frac{Z_T \tan \theta_T + Z_B \tan \theta_B}{Z_B - Z_T \tan \theta_T \tan \theta_B} \right] \tag{2.2.11}$$

with Z_B, Z_T, and Z_S the mechanical characteristic impedances of the bonding material, transducer, and sample, respectively; $\theta_T = 2\pi f(l_T/v_T)$ and $\theta_B = 2\pi f(l_B/v_B)$, where l_T and v_T are the thickness and velocity of sound in the transducer, respectively, and l_B and v_B are the corresponding parameters in the bonding material.

For zero thickness of the transducer bond, γ may be approximated near the resonance frequency of the transducer f_0 by

$$\gamma = \pi \left[1 - 2 \frac{Z_T}{Z_S} \left(\frac{f - f_0}{f_0} \right) \right]. \tag{2.2.12}$$

Substituting Eq. (2.2.12) into Eq. (2.2.9), we obtain, at each null frequency f_n,

$$2f_n\tau - 1 + 2 \frac{Z_T}{Z_S} \left[\frac{f_n - f_0}{f_0} \right] = n + \frac{1}{2}. \tag{2.2.13}$$

If the bond thickness cannot be neglected, this equation becomes more complicated.

A similar expression for f_{n+1} allows the frequency difference Δf between nulls to be expressed as

$$\Delta f = f_{n+1} - f_n = [2\tau + 2 Z_T/Z_S f_0]^{-1}. \tag{2.2.14}$$

Substituting Eq. (2.2.14) into Eq. (2.2.13) gives

$$f_n/\Delta f - 2 Z_T/Z_S = n + \tfrac{3}{2}. \tag{2.2.15}$$

[97] M. Redwood and J. Lamb, Proc. *Inst. Electr. Eng.* **103B,** 773–780 (1956).

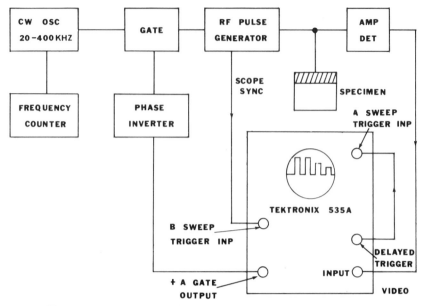

FIG. 19. Block diagram of apparatus for pulse superposition method.[88]

If Z_T/Z_S is known to within ± 0.1, n may be calculated from Eq. (2.2.15) from measurements of the null frequencies f_n and the frequency difference Δf close to the resonant frequency of the transducer f_0. This value of n is then substituted into Eq. (2.2.13) from which the transit time τ is calculated. Some error remains because of the above uncertainty in Z_T/Z_S. However, accuracy can be improved by calculating τ for a series of null frequencies on either side of f_0. By interpolation, a value of τ is obtained for which $f = f_0$ and the factor $2(Z_T/Z_S[(f_n - f_0)/f_0]$ vanishes in Eq. (2.2.13). Accuracies of one part in 10^4 have been reported with the gated double-pulse superposition method using contact transducers.[96] For noncontact transducers, this procedure is unnecessary since $Z_T = 0$ and the factor vanishes regardless of frequency.[98]

2.2.2.1.4. PULSE SUPERPOSITION METHOD. The arrangement of apparatus for the pulse superposition method (developed by McSkimin[99,100]) is shown in Fig. 19. A series of rf pulses from a pulse generator is introduced into the sample. The repetition rate of these pulses, controlled by the frequency of a continuous wave (cw) oscillator, is adjusted to corre-

[98] J. H. Cantrell, Jr. and M. A. Breazeale, *J. Acoust. Soc. Am.* **61**, 403–406 (1977).
[99] H. J. McSkimin, *J. Acoust. Soc. Am.* **37**, 864–871 (1965).
[100] H. J. McSkimin and P. Andreatch, Jr., *J. Acoust. Soc. Am.* **41**, 1052–1057 (1967); **34**, 609–615 (1962).

spond approximately to some multiple p of an acoustic round-trip transit time δ ($= 2\tau$) in the sample. The actually measured time delay T between superimposed "in phase" pulses is the reciprocal of the cw oscillator frequency and may be written

$$T = p\delta - \left(\frac{p\gamma}{360f}\right) + \frac{n}{f}, \qquad (2.2.16)$$

where γ is the phase angle associated with the wave reflection, f the ultrasonic frequency, and n an integer (positive or negative) that indicates the cyclic mismatch between the pulses.

The "in phase" condition is obtained by adjusting T so that the superimposed pulse amplitude is maximized. For $p = 1$, the applied pulse occurs once for every round-trip delay and obscures the echoes if the sequence is not interrupted. In this case the gate and phase inverter are used to interrupt the sequence after it has been on long enough to establish a stable interference pattern. The observed echo pattern during the "interrupt period" is the superimposed sum of all previously applied pulses (see Fig. 20).

For the case $p > 1$, the gate and phase inverter are not necessary. However, the advantages of the $p = 1$ case justify the additional equipment in some situations. In samples in which the attenuation is quite large, the resonance condition can be obtained with greater precision for $p = 1$, and more accurate measurements can be made.

The pulse transit time τ can be obtained from Eq. (2.2.16) if a value of T can be found corresponding to $n = 0$ (i.e., cycle-for-cycle matching of the pulses). This is done by comparing experimental and theoretical values of the quantity ΔT defined by

$$\Delta T = \frac{1}{f_L}\left(n - \frac{p\gamma_L}{360}\right) - \frac{1}{f_H}\left(n - \frac{p\gamma_H}{360}\right), \qquad (2.2.17)$$

where ΔT is the change in T required to maintain the "in phase" condition as the frequency f is changed from some value f_H (usually the transducer resonance frequency) to some other value f_L, 5–10% lower. The phase angles γ_H and γ_L associated with f_H and f_L, respectively, are calculated from Eq. (2.2.10). The experimentally determined T having a ΔT nearest the value calculated from Eq. (2.2.17) for $n = 0$ is the value of T to be used in Eq. (2.2.16).

For measurements in solids involving bonded contact transducers, there is some uncertainty in the values of the γs because the bond thickness is not known. In this case the phase angles are calculated as a function of bond thickness and these values are used to plot ΔT versus bond thickness. A typical example of this is shown in Fig. 21 for different

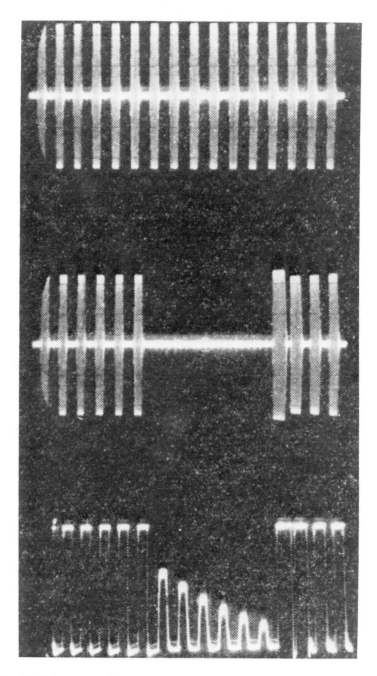

FIG. 20. Pulse superposition oscilloscope patterns for $p = 1$. The top pattern is for repeated rf pulses, the middle pattern for applied rf pulses; and the bottom pattern for envelopes showing received waves.[88]

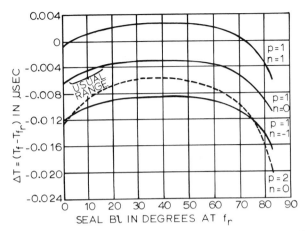

FIG. 21. Change of delay time required to maintain the "in phase" condition with the pulse superposition method plotted as a function of bond thickness.[88]

values of p and n. It is seen that ΔT has a limited range for a given value of p and n. Thus all experimentally measured values of T having a ΔT outside the limits for the $n = 0$ case can be eliminated from consideration. For further details the reader is referred to the work of McSkimin.[88,89]

Accuracies of 2 parts in 10^4 were initially reported for absolute velocity measurements with the pulse superposition method. Refinements in the electronic circuitry[99,100] have led to an increase in the precision of transit time measurements to approximately one part in 10^7 for high Q materials.

2.2.2.1.5. ECHO-OVERLAP METHOD. The echo-overlap method using two transducers was first reported by May.[101] Later a single-transducer modification was used by Papadakis[87,102] to measure the transit time between any pair of echoes in a pulse-echo pattern. A block diagram of the arrangement of the apparatus for use of a single transducer is shown in Fig. 22. As with the pulse superposition method, a series of rf pulses from a pulse generator (4) is introduced into the sample. However, in contrast with that method, the repetition rate of these pulses is low enough that all echoes from a given rf pulse decay away before the next pulse is applied. The pulse repetition rate is controlled by a continuous wave (cw) oscillator (1) (100–1000 KHz) after frequency division (2) by a factor of 10^3 and trigger-shaping by a dc pulse generator (3). The dc pulse generator also is used to trigger a two-channel time delay generator (8) that actuates the Z axis intensity gate on the oscilloscope (7). The hy-

[101] J. E. May, Jr., *IRE Nat. Conv. Rec.* **6,** 134 (1958).
[102] E. P. Papadakis, *J. Appl. Phys.* **35,** 1474 (1964).

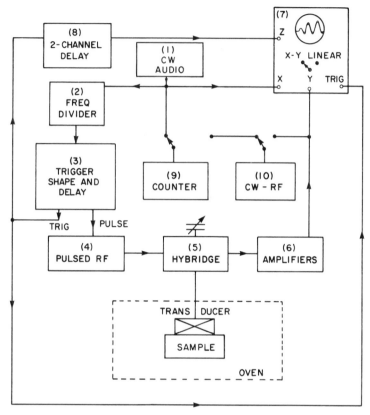

FIG. 22. Block diagram of the circuitry for the ultrasonic pulse-echo-overlap method of velocity measurements.[87]   All the units can be obtained commercially.

bridge (5) allows returning echoes, but not the applied rf pulse, to reach the oscilloscope. The delays (8) are adjusted so that any chosen pair of echoes is intensified.   The oscilloscope is operated with a linear sweep during this adjustment so that many echoes appear on the screen.

During the acoustic transit time measurement the oscilloscope is switched to an $X-Y$ mode of operation in which the cw oscillator (1) provides the sweep.   In this mode the CRT intensity is reduced so that only the two intensified echoes of interest are visible.   The echoes can be made to overlap cycle-for-cycle as shown in Fig. 23 by adjusting the cw oscillator frequency to correspond to the reciprocal of the travel time between the echoes.   An integral multiple m of this frequency also allows this overlap, since an echo then appears for every mth sweep.   The time

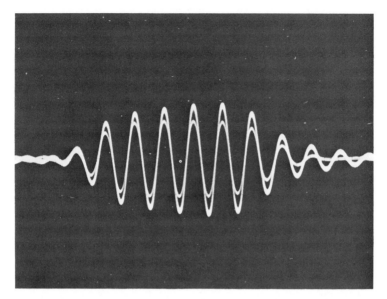

FIG. 23. Cycle-for-cycle overlap of echoes in pulse-echo-overlap method.

delay T actually measured between the echoes is m times the reciprocal of the cw oscillator frequency. The round-trip acoustic transit time δ ($= 2\tau$) in the sample is obtained from Eq. (2.2.16) as in the pulse superposition method. The procedure for obtaining the correct cycle-for-cycle matching of the chosen echo pair (i.e., $n = 0$ condition) and the calculation of the phase shift due to reflection are outlined in Section 2.2.2.1.4.

The counter (9) used to measure the cw oscillator (1) frequency is also used to measure the frequency of the rf oscillator (10), whose output may be superimposed on the ultrasonic echo display to determine the frequency of the ultrasonic pulses.

Absolute velocity measurement accuracies of a few parts in 10^5 have been reported with this method.[87] A critical analysis[103] of the absolute accuracy of the pulse-echo overlap method and the pulse superposition method shows that under favorable circumstances absolute accuracies of a few parts in 10^6 are possible with either method, if phase corrections for diffraction are included in the velocity measurements. A commercial instrument is available.[104]

2.2.2.1.6. PULSE INTERFEROMETER METHODS. A block diagram of a

[103] E. P. Papadakis, *J. Acoust. Soc. Am.* **52**, 843–846, 847–849 (1972).

[104] E. P. Papadakis, *in* "Physical Acoustics" (W. P. Mason and R. N. Thurston, eds.), Vol. XII, pp. 277–374. Academic Press, New York, 1976.

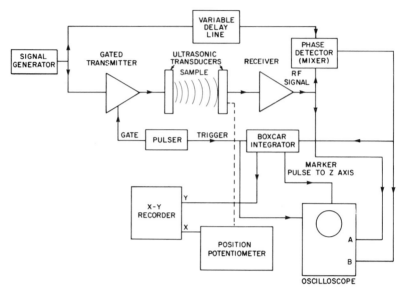

FIG. 24. Block diagram of electronic circuitry of pulse interferometer.[105]

typical pulse interferometer is shown in Fig. 24.   An ultrasonic pulse is transmitted into the sample by gating a cw reference oscillator and the received pulse is combined with the cw reference signal in a phase-sensitive detector, whose output is responsive to variations in acoustic path length and frequency.   As the relative phase between the received acoustic signal and the cw reference frequency is changed by varying either the acoustic path or the frequency, the output of the phase-sensitive detector varies periodically to give a measure of the wavelength.

Several variations of this arrangement have been used for velocity and attenuation measurements.   Pervushin and Filippov[106] report a single-transducer arrangement in which the acoustic path length is fixed and the reference frequency is varied to produce a series of interference maxima and minima that are detected and viewed directly on an oscilloscope. Absolute velocity v is determined from

$$v = 2ls(f_1 - f_2)/n, \qquad (2.2.18)$$

where l is the length of the sample, s the number of acoustic reflections in the sample, and n the number of interference maxima (or minima) corresponding to a change in frequency $(f_1 - f_2)$.   If contact transducers are

[105] R. C. Williamson and D. Eden, *J. Acoust. Soc. Am.* **47**, 1278–1281 (1970).
[106] I. I. Pervushin and L. P. Filippov, *Sov. Phys. Acoust.* **7**, 307–309 (1962).

used, corrections for phase shifts at the appropriate boundaries must be made for accurate measurements.

Chase[107,108] employed a fixed-path, fixed-frequency arrangement using a variable delay line between the reference-signal source and the phase-sensitive detector. Changes in velocity were obtained by measuring the change in delay necessary to maintain a null output from the phase-sensitive detector. Carstensen[109] described a technique that determines the difference in the velocity of sound in two liquids separated by a partitioning membrane. By keeping a fixed distance between transmitter and receiver and moving both relative to the membrane, one can measure the distance the transmitting and receiving transducers must be moved simultaneously to obtain a 360° phase shift in the phase-detected output. This procedure gives a very accurate measure of relative velocity and velocity dispersion. Dispersion of the order of 0.1% of the velocity can be measured to between 3 and 5%. Blume[110] described a fixed acoustic path, phase-locked pulse technique in which the cw reference oscillator was placed under the direct control of a gated automatic frequency control (AFC) oscillator. The AFC oscillator was used to shift the reference frequency to preserve the quadrature condition (i.e., zero of phase detection) between the echo and the reference signals. A fractional change in the transit time of the echo pulse then caused the same fractional change in the reference frequency, which was continuously monitored. Accuracies of a few parts in 10^8 for relative velocity measurements have been reported with this technique.

Williamson and Eden[105] used the variable acoustic path arrangement shown in Fig. 24 at a fixed frequency f. A balanced mixer was used as a phase detector and the mixer output was averaged with a boxcar integrator. The resulting dc signal A is linearly proportional to the amplitude of the received acoustic pulse and is a sinusoidal function of the phase difference ϕ between the reference and received signals as given by

$$A = A_0 e^{-\alpha d} \sin[(2\pi f/v)d + \phi], \qquad (2.2.19)$$

where d is the transducer separation and α the acoustic attenuation. The velocity is obtained by tracking a portion of the received pulse through a series of quadrature conditions corresponding to half-wavelength displacements of the receiving transducer. The amplitude of the pulse at each "fringe" is obtained by adjusting the variable delay line to correspond to a 90° phase shift in the reference signal and recording the re-

[107] C. E. Chase, *J. Phys. Fluids* **1**, 193 (1958).
[108] W. M. Whitney and C. E. Chase, *Phys. Rev. Lett.* **9**, 243 (1962).
[109] E. L. Carstensen, *J. Acoust. Soc. Am.* **26**, 858–861 (1954).
[110] R. J. Blume, *Rev. Sci. Instrum.* **34**, 1400 (1963).

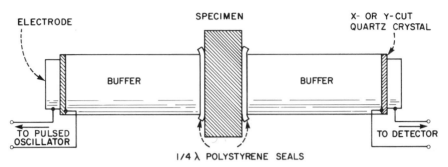

FIG. 25. Experimental arrangement for holding specimen with the long pulse buffer rod method.[111]

sulting output from the boxcar integrator. After each such recording the delay is reset to its original position before proceeding to the next "fringe." Absolute velocity measurements with accuracies of 1 part in 10^4 and attenuation measurements in highly attenuating liquids to 1 part in 10^2 have been reported with this technique.

2.2.2.1.7. LONG PULSE BUFFER ROD METHOD. The long pulse buffer rod method is useful for small, very lossy samples. Several variations of the method have been reported.[104,111-114] Figure 25 shows a two-transducer arrangement for a solid sample. An ultrasonic pulse is transmitted through a buffer rod (typically made of fused silica) and impinges on the sample. The quarter-wavelength seals minimize phase shifts. Partial reflections and transmissions of the pulse between the sample–buffer interfaces give rise to the pulse-echo pattern shown in Fig. 26. Figure 26a shows the case in which the pulse length is short compared to the sample length. The "stepladder" pattern in Fig. 26b results from extending the length of the pulse until the echoes overlap and adjusting the rf frequency until the "in phase" condition is satisfied. (The length of the "stepladder" pattern is shorter than the round-trip travel time in either buffer rod.) The "in phase" condition may be obtained from Eq. (2.2.7) by substituting $\tau = l/v$ for the transit time τ; l is the sample length and v the phase velocity. The time-independent "in phase" condition is

$$\omega_n l/v - \gamma_n = n\pi, \qquad n = 0,1,2,3, \ldots \qquad (2.2.20)$$

or

[111] H. J. McSkimin, *J. Acoust. Soc. Am.* **22**, 413–418 (1950).

[112] H. J. McSkimin, *J. Acoust. Soc. Am.* **29**, 1185–1192 (1957).

[113] H. J. McSkimin, *IRE Trans. Ultrason. Eng.* **5**, 25 (1957).

[114] H. J. McSkimin, *in* "Physical Acoustics" (W. P. Mason, ed.), Vol. 1A, pp. 271–417. Academic Press, New York, 1964.

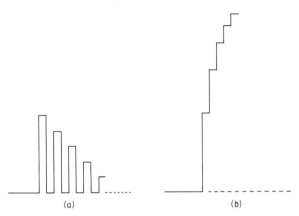

FIG. 26. Long pulse buffer rod method pulse-echo patterns[111]: (a) Pulse length shorter than sample; (b) Pulse length longer than sample so that overlap occurs. Both patterns are for "in phase" frequency adjustment.

$$v = \frac{2lf_n}{n + \gamma_n/\pi},\qquad (2.2.21)$$

where γ_n is the phase change due to reflection at the sample–buffer interface.

If f_m is any other frequency for which the "in phase" condition is satisfied, we may write the analog to Eq. (2.2.20) as

$$(2\pi f_m l/v) - \gamma_m = m\pi.\qquad (2.2.22)$$

From Eqs. (2.2.20) and (2.2.22) we may evaluate n from

$$n = \frac{f_n\,\Delta n}{\Delta f} - \frac{\gamma_n}{\pi} + \frac{f_n}{\pi}\left(\frac{\gamma_m - \gamma_n}{\Delta f}\right),\qquad (2.2.23)$$

where $\Delta f = f_m - f_n$ and $\Delta n = m - n$. The phase shifts may be experimentally evaluated[111] by using the single-transducer arrangement shown in Fig. 27. Phase balance is first made with the buffers and transducers in place as shown. Buffer B and its transducers are then removed and the change in phase necessary to restore the balance condition is the value of γ at the frequency used.

For the case of liquid samples,[112] the phase shifts are negligible, except for samples of extremely high attenuation, and Eqs. (2.2.21) and (2.2.23) may be used with $\gamma_m = \gamma_n = 0$.

A single-transducer arrangement for solids[113] is shown in Fig. 28. The inverted "stepladder" pattern shown on the oscilloscope results from the condition that the waves in the sample are all "in phase" with each other but out of phase with the reflected incident wave. The analysis is analo-

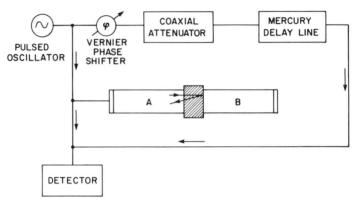

FIG. 27. Circuit for measuring the phase shift at a reflecting interface.[111]

gous to the two-transducer case for solids except that the phase shift correction results from one sample–buffer interface. The γ_n/π term in Eq. (2.2.21) becomes $\gamma_n/2\pi$ in this case.

A detailed analysis[104] for the single-transducer arrangement in which the effect of the reflection coefficient between the specimen and buffer rod is included reveals that, in fact, two characteristic "stepladder" patterns (depending on frequency used) are possible as shown in Fig. 29. Type I [part (a)] results if the overlapping echoes are out of phase; type II

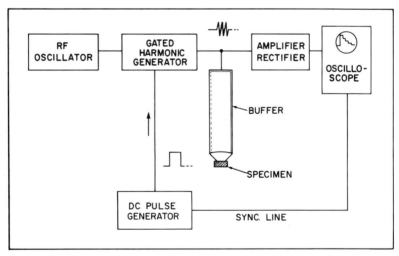

FIG. 28. Block diagram of measuring circuit for single-transducer, long pulse, buffer rod method.[113]

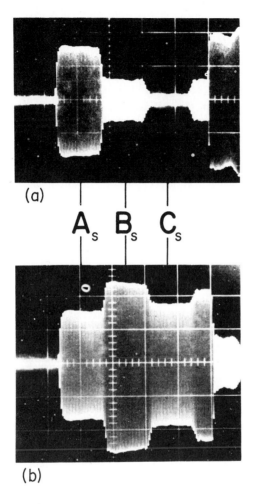

(a)

A_S B_S C_S

(b)

FIG. 29. Echo patterns for ultrasonic pulses much longer than the sample: (a) type I; (b) type II.[115] Multiple reflections of individual echoes interfere to give these two characteristic patterns which differ by π rad of phase in one round trip in the specimen. Velocity measurements are made by noting the phase change as the applied frequency is changed. Attenuation can be computed from the amplitudes A_S, B_S, and C_S of the steps.

[part (b)] results if they are in phase. Velocity measurements good to 1 part in 10^4 have been reported with this method.[114]

2.2.2.1.8. OTHER PULSE VELOCITY METHODS. A number of other useful pulse techniques have been reported in the literature, all of which

[115] E. P. Papadakis, *J. Acoust. Soc. Am.* **44**, 724–734 (1968).

are variations or combinations of the methods already discussed. Any attempt to describe all of them would make this review too cumbersome. Therefore, a representative summary is presented which may prove beneficial to the reader wishing to investigate them further.

McConnell and Mruk[116] describe a pulse technique to measure velocity using a thin, resonant layer of test liquid less than 0.1 cm^3 in volume. Litovitz et al.[117] reported a phase-comparison arrangement using two acoustic paths to measure sonic velocity in highly attenuating liquids. Mason and Bömmel[118] reported a double-balanced modulator circuit to determine changes in velocity to one part in 10^7.

Cunningham and Ivey[119] suggested a buffer technique using two samples of different length bonded to the ends of two fused silica rods to measure velocity and attenuation in rubberlike material. Nolle and Sieck[120] published a method in which the pulse amplitude and transit time delay were measured first in two buffer rods placed end-to-end and then compared with the corresponding measurements made when the sample was sandwiched between the buffer rods. The differences in values were assumed to be due to the additional attenuation and delay time in the sample after correcting for the effects of acoustic impedance mismatch between sample and buffer rods. Nolle and Mowry[121] recommended using a liquid buffer and a reflector in place of the receiving transducer (liquid immersion technique).

Mason et al.[122] and Barlow and Lamb[123] reported a technique in which the shear wave velocity and attenuation in highly viscous liquids are calculated from the measurement of the complex reflection coefficient obtained from waves reflected at a liquid–solid interface. Reissner,[124] Schneider and Burton,[125] and Kono[126] suggested a method using mode conversion to obtain the ratio of shear to longitudinal wave velocities. McSkimin[127] used pulsed torsional waves in a rod to measure the impedance of the surrounding liquid from which the velocity and attenuation may be

[116] R. A. McConnell and W. F. Mruk, J. Acoust. Soc. Am. **27**, 672 (1955).

[117] T. A. Litovitz, T. Lyon, and P. Peselnick, J. Acoust. Soc. Am. **26**, 566 (1954).

[118] W. P. Mason and H. E. Bömmel, J. Acoust. Soc. Am. **28**, 930–943 (1956).

[119] J. R. Cunningham and D. G. Ivey, J. Appl. Phys. **27**, 967 (1956).

[120] A. W. Nolle and P. W. Sieck, J. Appl. Phys. **23**, 888 (1952).

[121] A. W. Nolle and S. C. Mowry, J. Acoust. Soc. Am. **20**, 432 (1948).

[122] W. P. Mason, W. O. Baker, H. J. McSkimin, and J. H. Heiss, Phys. Rev. **75**, 936 (1949).

[123] A. J. Barlow and J. Lamb, Proc. R. Soc. London Ser. A **253**, 52–69 (1959).

[124] H. Reissner, Helv. Phys. Acta **7**, 140 (1938).

[125] W. C. Schneider and C. J. Burton, J. Appl. Phys. **20**, 48 (1949).

[126] R. Kono, J. Phys. Soc. Jpn. **15**, 718 (1960).

[127] H. J. McSkimin, J. Acoust. Soc. Am. **24**, 355–365 (1952).

calculated. McKinney *et al.*[128] developed a system for the direct determination of the dynamic bulk modulus up to approximately 10 kHz. Lacy and Daniel[129] used a digital averaging technique to measure the velocity in solids up to 50 MHz. Asay *et al.*[130] published a modified version of the gated double-pulse superposition technique to measure small changes in velocity in highly attenuating materials to one part in 10^4.

2.2.2.2. **Attenuation Measurements.** When an ultrasonic wave propagates through any medium its amplitude changes because of the interplay of many mechanisms. In addition to the thermodynamic mechanisms that lead to an increase in temperature in the propagating medium (assuming a thermodynamically closed system), there are mechanisms such as diffraction and reflection that in fact may lead to an increase in amplitude as well as a decrease. It is convenient to separate the mechanisms into two classes by using the word "attenuation" to refer to the total change in amplitude resulting from the action of all mechanisms and to reserve the word "absorption" for that loss of amplitude that results in an increase of temperature (however slight) in the propagating medium. The other major contribution to attenuation is scattering, which is treated in detail elsewhere in this book. The measurements described lead to a value of attenuation. The particular situation must be considered before one can determine the magnitude of the absorption or scattering and identify with it a particular physical mechanism.

2.2.2.2.1. BASIC PULSE-ECHO METHODS. In Section 2.2.2.1.1 we pointed out that the basic pulse-echo method can be used for measurement of acoustic attenuation as well as velocity. The acoustic attenuation is obtained from the amount of electrical attenuation introduced into the circuit of Fig. 13 to maintain a constant amplitude of a selected echo into the receiver. In a liquid this is done for a selected echo while moving the reflecting boundary over some measured distance. In a solid the changes in electrical attenuation are recorded individually for a series of echoes in the pulse-echo train, since in this case the sample boundaries remain fixed. A simpler but less accurate technique would be to remove the calibrated attenuator of Fig. 13 and to use the calibrated oscilloscope graticules to measure the pulse height. An advantage of the calibrated attenuator technique is that the receiver nonlinearity does not contribute to the measurement error, whereas in the calibrated oscilloscope technique, it does.

[128] J. E. McKinney, S. Edelman, and R. S. Martin, *J. Appl. Phys.* **27**, 425 (1956).

[129] L. L. Lacy and A. C. Daniel, *J. Acoust. Soc. Am.* **52**, 189–195 (1972).

[130] J. R. Asay, D. L. Lamberson, and A. H. Guenther, *J. Acoust. Soc. Am.* **45**, 566–571 (1969).

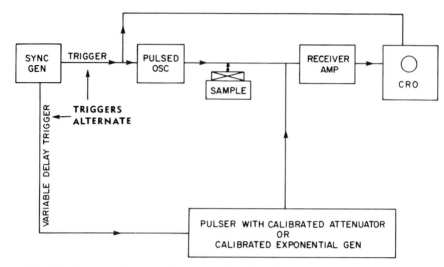

FIG. 30. Apparatus for measuring attenuation using either a calibrated attenuator or a calibrated exponential generator.

An improvement in measurement accuracy can be made with the arrangement shown in Fig. 30. The calibrated attenuator shown in Fig. 13 is removed and either a pulser with calibrated attenuator or a calibrated, continuously variable, exponential generator is added to the circuit as indicated. If the pulser arrangement is used,[131,132] measurements are made by matching the comparison pulse amplitude to the selected echo amplitude as described above. (The comparison pulses and echoes are displayed on alternate sweeps of the oscilloscope.) Any parallax problems associated with the previous methods are now eliminated. Accuracies of the order of 2 parts in 10^2 for measurements in liquids have been reported with this technique.[133] If the exponential generator is used,[134] the generated curve is matched to a selected pair of peaks of the decaying echo train, as shown in Fig. 31. Correction for diffraction losses is made for the echoes selected. For highly attenuating samples the exponential generator technique becomes unreliable but at lower attenuation it is convenient and sensitive.

Automated systems using peak detection of two selected echoes have

[131] R. L. Roderick and R. Truell, *J. Appl. Phys.* **23,** 267–279 (1952).
[132] J. M. M. Pinkerton, *Proc. Phys. Soc. London* **362,** 286 (1949).
[133] J. H. Andreae, R. Bass, E. L. Heasell, and J. Lamb, *Acustica* **8,** 131 (1958).
[134] B. Chick, G. Anderson, and R. Truell, *J. Acoust. Soc. Am.* **32,** 186–193 (1960).

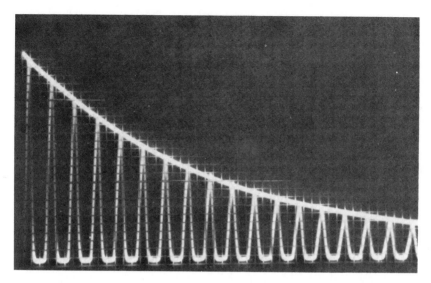

Fig. 31. Pulse-echo decay pattern with superimposed exponential as seen on oscilloscope.[135]

been described by Truell *et al.*[135] and Kamm and Bohm.[136]   They are reported to have achieved sensitivities greater than the manual systems. For measurements of the change in attenuation caused by changes in the sample conditions (e.g., temperature or pressure), monitoring the time-averaged value of a selected echo with a calibrated integrator produces results expeditiously with excellent sensitivity.   Commercial versions of the system are available.

2.2.2.2.2. PULSE SPECTRUM ANALYSIS.   For comparison with theory, one often is interested in measurement of attenuation as a function of frequency.   For making these measurements the use of a broad frequency bandwidth ultrasonic wave and a spectrum analyzer as a receiver has considerable appeal.   The pulse spectrum analysis technique was introduced by Gericke[137] to study the frequency characterization of flaws in materials and was popularized by Adler and Whaley,[138] who first established a quantitative basis for the observed spectrum.   Although the system most

---

[135] R. Truell, C. Elbaum, and B. B. Chick, "Ultrasonic Methods in Solid State Physics." Academic Press, New York, 1969.

[136] G. N. Kamm and H. V. Bohm, *Rev. Sci. Instrum.* **33,** 957 (1962).

[137] O. R. Gericke, *J. Acoust. Soc. Am.* **35,** 364 (1963).

[138] L. Adler and H. L. Whaley, *J. Acoust. Soc. Am.* **51,** 881–887 (1972).

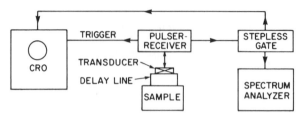

FIG. 32. Typical spectrum analysis equipment arrangement.

often is used for flaw characterization,[139] it is capable of giving attenuation measurements as well. A typical experimental arrangement using a single transducer is shown in Fig. 32. A short, untuned, voltage spike from a pulser shock-excites a highly damped transducer. The highly damped transducer emits a sharp ultrasonic pulse with a relatively large frequency bandwidth which propagates through a delay line (liquid or solid) before striking the sample. The returning echoes from two different regions of the sample are gated singly or together, depending on the properties to be investigated, into a spectrum analyzer. The spectrum analyzer performs a Fourier transform of the time-domain pulses. This information can be used to determine absolute attenuation as well as the velocity of the material under investigation. Papadakis et al.[140] reported a spectrum analysis technique in which both the absolute attenuation and the reflection coefficient are obtained by separately gating into the spectrum analyzer the pulse reflected from the front surface and the first two echoes from the back surface of the sample. Simpson[141] has given the theoretical foundations of pulse spectrum analysis, which would be of great value to the reader interested in using this technique.

An expedient, single-transducer, spectrum analysis technique that is useful for relative attenuation measurements was used by Lizzi et al.[142] for measurement of the attenuation of the lens of the eye. It also has been used for characterization of skin tissue.[143] Let $E_A(\omega)$ and $E_B(\omega)$ denote the frequency spectrum of the returning echoes from surfaces A and B of a sample, respectively. Denoting the magnitude of the ratio $E_B(\omega)/E_A(\omega)$ by $R(\omega)$, they determined the attenuation $\alpha(\omega)$ from the ex-

[139] L. Adler, K. V. Cook, and W. A. Simpson, in "Research Techniques in Nondestructive Testing" (R. S. Sharpe, ed.), Vol. 3, pp. 1–49. Academic Press, New York, 1977.

[140] E. P. Papadakis, K. A. Fowler, and L. C. Lynnworth, J. Acoust. Soc. Am. 53, 1336–1343 (1973).

[141] W. A. Simpson, J. Acoust. Soc. Am. 56, 1776 (1974).

[142] F. Lizzi, L. Katz, L. St. Louis, and D. J. Coleman, Ultrasonics 14, 77 (1976).

[143] J. H. Cantrell, Jr., R. E. Goans, and R. L. Roswell, J. Acoust. Soc. Am. 64, 731–735 (1978).

pression

$$20 \log_{10} R(\omega) = 20 \log_{10}(\rho_B/\rho_A) - 17.4\alpha\tau, \qquad (2.2.24)$$

where ρ_B and ρ_A are the pressure reflection coefficients and τ is the pulse transit time between surfaces A and B. A variation of this method using a two-transducer through transmission arrangement was proposed by Miller et al.[144] in an ultrasonic investigation of myocardial injury.

2.2.2.2.3. OTHER PULSE ATTENUATION METHODS. Variations of the techniques described previously for the measurement of sonic velocity are also adaptable to the measurement of attenuation. The use of the pulse interferometer as suggested by Williamson and Eden[105] for both attenuation and velocity measurements in liquids was described in Section 2.2.2.1.6. Schwan and Carstensen[145] and Carstensen et al.[146] suggested a technique in which transmitting and receiving transducers are mounted on a rigid sliding assembly and immersed in a partitioned test vessel. Half of the vessel is filled with degassed water and the other half filled with the test liquid. The variation of the receiver intensity as a function of assembly displacement is used to calculate the attenuation of the test liquid.

A technique based on the long pulse buffer method (Section 2.2.2.1.7) has been reported[104,115] for attenuation measurements in thin, highly absorptive materials. Hayford et al.[147] published a modification of this technique applicable to highly attenuating specimens of intermediate thickness. McSkimin[112] used two buffer rods separated by a liquid specimen to calculate the attenuation from the measured amplitudes of multiply reflected pulses within the sample.

Fry and Fry[148] have developed a technique using a thermocouple probe capable of measuring absolute acoustic absorption in liquid and biological media.

2.2.3. Continuous Wave Techniques

A third class of techniques for determining the acoustical properties of materials is based on continuous wave (cw) methods. Whereas pulse-echo measurements have a more nearly transient nature, cw measurements are associated with steady-state or near-equilibrium conditions. The infor-

[144] J. G. Miller et al., Proc. IEEE Ultrason. Symp., Annapolis, p. 33 (1976).

[145] H. P. Schwan and E. L. Carstensen, Electronics 216–220 (1952).

[146] E. L. Carstensen, K. Li, and H. P. Schwan, J. Acoust. Soc. Am. 25, 286–288 (1953).

[147] D. T. Hayford, E. G. Hennecke II, and W. W. Stinchcomb, J. Compos. Mater. 11, 429 (1977).

[148] W. J. Fry and R. B. Fry, J. Acoust. Soc. Am. 26, 294–310 (1954); 26, 311–317 (1954).

mation contained in cw measurements, however, is related to that in pulse measurements through a Fourier transform and, theoretically, is identical to it. In essence, the Fourier transform correlates the pulse-echo time base sweep of an oscilloscope with the continuous wave frequency sweep of a frequency-modulated oscillator.

In this section we examine the basic theory of cw analysis for plane propagating waves and describe the experimental apparatus for cw measurements. Attenuation and velocity measurements are discussed in parallel, as they arise together in the discussion of most cw techniques. Variations of the basic cw apparatus have led to successful systems for making measurements. Several of these systems are described and their operating characteristics are discussed.

2.2.3.1. Theory of Continuous Wave Techniques. 2.2.3.1.1. PROPAGATING WAVE MODEL. The basic theory for ultrasonic techniques can be analyzed by use of the one-dimensional propagating plane wave model of Miller and Bolef.[149,150] For simplicity we consider an isolated sample and neglect diffraction effects. The sample has a length $a/2$ with flat and parallel opposing faces where complete internal acoustical reflection occurs.

At time $t = 0$ the $x = 0$ face is caused to vibrate with a particle velocity Re $e^{i\omega t} = \cos \omega t$.* The resulting disturbance propagates with velocity v into the sample as a plane wave of the form $\exp[i(\omega t - kx) - \alpha x]$, where $k = \omega/v$ and α is the acoustic absorption coefficient. At the $x = 0$ face the wave is superimposed on the original driving velocity, resulting in a complex particle velocity $\tilde{U}$, at time $t = a/v$, where $\tilde{U} = \exp(i\omega t)\{1 + \exp[-(ika + \alpha a)]\}$. The wave continues to propagate and reflect at the boundaries, and finally at the $x = 0$ face one finds a particle velocity superposition of the form

$$\tilde{U} = \exp(i\omega t)(1 + \exp[-(ika + \alpha a)] + \exp[-2(ika + \alpha a)] + \cdots).$$

$$(2.2.25)$$

The infinite sum converges for the case of $\alpha > 0$ and may be written

$$\tilde{U} = \exp(i\omega t)/\{1 - \exp[-(ika + \alpha a)]\} \qquad (2.2.26)$$

or, taking the real part,

[149] J. G. Miller and D. I. Bolef, *J. Appl. Phys.* **39**, 4589 (1968).

[150] D. I. Bolef and J. G. Miller, *in* "Physical Acoustics" (W. P. Mason and R. N. Thurston, eds.), Vol. VIII, pp. 96–201. Academic Press, New York, 1971.

---

* All such expressions will be written as $e^{i\omega t}$ with the understanding that only the real part Re $e^{i\omega t} = \cos \omega t$ has physical significance.

$$U = \text{Re } \bar{U} = U_1 \cos \omega t + U_2 \sin \omega t, \qquad (2.2.27)$$

where

$$U_1 = (e^{\alpha a} - \cos ka)/2(\cosh \alpha a - \cos ka) \qquad (2.2.28)$$

and

$$U_2 = \sin ka/2(\cosh \alpha a - \cos ka). \qquad (2.2.29)$$

The amplitude of U is found to be

$$|U| = [U_1{}^2 + U_2{}^2]^{1/2} = e^{\alpha a/2}/[\sqrt{2}(\cosh \alpha a - \cos ka)^{1/2}]. \qquad (2.2.30)$$

The U_1 term is in phase with the driving stimulus, while the U_2 term is in quadrature. These expressions are plotted as a function of frequency in Fig. 33.

The particle velocity response [Eq. (2.2.27)] as a function of frequency consists of a series of standing wave or mechanical resonances whose frequencies correspond to the condition that an integral number of half-

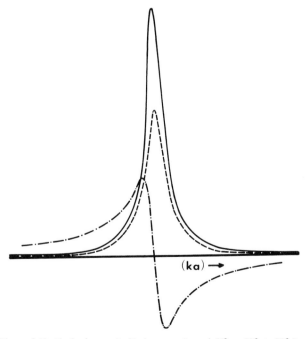

FIG. 33. Plots of U_1 (dashed curve), U_2 (— · — ·), and $U^2 = U_1{}^2 + U_2{}^2$ (solid curve), in arbitrary units, as functions of ka in the vicinity of the mechanical resonance.[149] The vertical scale for U^2 differs from that of U_1 and U_2. The zero of ka is taken at the center of the mechanical resonance.

wavelengths exist in the sample. The resonance condition exists for $ka = 2\alpha m$ ($\alpha a \ll 1$), where m is the harmonic integer. Since $k = \omega/v$, the mth mechanical resonance frequency is written

$$\omega_m = 2\pi m v/a \qquad (2.2.31)$$

and

$$\omega_m - \omega_{m-1} = 2\pi v/a. \qquad (2.2.32)$$

Equation (2.2.32) can be used to determine the ultrasonic velocity from measured resonance frequencies. The acoustic attenuation is determined from the frequency width $\Delta\omega$ of the resonance at half-power by $\alpha = \Delta\omega/2v$. As an example, if we assume a sample length of 2.5 cm with a velocity of 5×10^5 cm sec^{-1} and an absorption of 0.01 cm^{-1}, $(\omega_m - \omega_{m-1})/2\pi = 10^5$ Hz, and $\Delta\omega/2\pi = 10^4/2\pi$ Hz for a Q of 10^3 at $(2\pi) \times 10^7$ Hz.

2.2.3.1.2. SENSITIVITY ENHANCEMENT FACTORS. In many situations measurements of relative acoustic parameters are desired. For example, one may wish to determine the effect of temperature, stress, conductivity, magnetic field, etc., on the ultrasonic velocity or attenuation. The use of resonance conditions with cw measurements provides an increased sensitivity to the results of these changes. Furthermore, since the gain is related to the superposition of acoustic signals (not unlike pulse superposition), the signal-to-noise ratio is improved as well. For such experiments, a change in the measured observable (e.g., $|U|$, U_1, or U_2) must be related to the parameter of interest (v, α, etc.).

The effective gains, called "sensitivity enhancement factors" by Miller and co-workers,[149-151] are derived from Eqs. (2.2.28)–(2.2.30). The variation of the amplitude $|A|$ of any measured observable is

$$\Delta|A| = \frac{\partial|A|}{\partial\alpha} \Delta\alpha + \frac{\partial|A|}{\partial k} \Delta k \qquad (2.2.33)$$

for small $\Delta\alpha$ and Δk. The partial derivatives are the sensitivity enhancement factors for attenuation and dispersion, respectively. If the measured observable is the particle velocity $|U|$, the sensitivity enhancement factors are

$$\frac{\partial|U|}{\partial\alpha} = \frac{a\sqrt{2}\,(e^{-\alpha a/2} - e^{\alpha a/2} \cos ka)}{4(\cosh \alpha a - \cos ka)^{3/2}}, \qquad (2.2.34)$$

$$\frac{\partial|U|}{\partial k} = \frac{a\sqrt{2}\,e^{\alpha a/2} \sin ka}{4(\cosh \alpha a - \cos ka)^{3/2}}. \qquad (2.2.35)$$

[151] J. S. Heyman and J. G. Miller, *J. Appl. Phys.* **44**, 3398–3400 (1973).

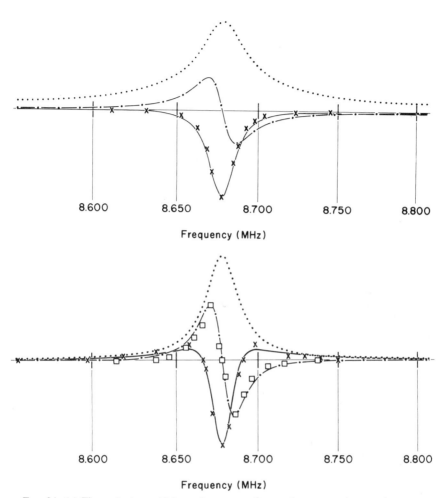

FIG. 34. (a) Theoretical sensitivity enhancement factors for attenuation $(\partial U/\partial \alpha)$ (solid curve) and dispersion $(\partial U/\partial k)$ ($— \cdot — \cdot$) superimposed on a plot of $|U|$ (dotted curve).[151] Experimental values are shown by X. (b) Theoretical line shapes for attenuation $(\partial U_1/\partial \alpha)$ (solid curve) and dispersion $(\partial U_1/\partial k)$ ($— \cdot — \cdot$) superimposed on a plot of U_1 (dotted curve). Pure attenuation signals are obtained for the frequency corresponding to $\partial U_1/\partial k = 0$, while pure dispersion signals are obtained for $\partial U_1/\partial \alpha = 0$.[151] Experimental values are indicated for attenuation (×) and dispersion (□).

These sensitivity enhancement factors are plotted as a functon of frequency in Fig. 34a with $|U|$ for comparison.

Likewise, we may evaluate the partial derivatives of U_1 and U_2 as

$$\frac{\partial U_1}{\partial \alpha} = \frac{-\partial U_2}{\partial k} = \frac{a}{2} \frac{(1 - \cosh \alpha a \cos ka)}{(\cosh \alpha a - \cos ka)^2},\qquad (2.2.36)$$

$$\frac{\partial U_1}{\partial k} = \frac{\partial U_2}{\partial \alpha} = \frac{-a}{2} \frac{\sinh \alpha a \sin ka}{(\cosh \alpha a - \cos ka)^2}.$$ (2.2.37)

These line shapes are shown in Fig. 34b with U_1 for comparison.

Note that at the peak of a mechanical resonance one has, respectively, maximum sensitivity for change in attenuation when measuring $\Delta|U|$ or ΔU_1 with maximum sensitivity for dispersion (changes in velocity) when measuring ΔU_2. Furthermore, both $\partial|U|/\partial k$ and $\partial U_1/\partial k$ vanish at the resonance peak. The presence of zeros in these sensitivity enhancement factors indicates that the experimenter can isolate dispersion from changes in attenuation. In addition, the sensitivity enhancement factor $\partial U_1/\partial \alpha$ passes twice through zero, as shown in Fig. 34b. These zeros occur at frequencies such that the observed amplitude of U_1 is $\sim 50\%$ of its maximum and the dispersion sensitivity enhancement factor is $\sim 75\%$ of its maximum value. Thus, measuring under conditions of slight decrease in signal strength as well as sensitivity may be warranted in order that the output be dispersion data only.

Table I shows the magnitude of many cw expressions for fixed values of a, k, $\Delta\alpha$, and Δk and for two different values of α, assuming a unity input of particle velocity amplitude, with measurements taken at the resonance peak. The large values of these expressions and their sensitivity to small changes result from the wave superposition inherent with cw resonant techniques. For example, we may note that for $\alpha a = 0.001$ and $a = 1.0$ cm, substitution of $\Delta\alpha = 10^{-4}$ results in a $\Delta|U|/|U|$ of nearly 10^{-1} and a sensitivity enhancement factor $\partial|U|/\partial\alpha$ of nearly 10^6. Such a large sensitivity enhancement factor provides obvious experimental advantages.

2.2.3.1.3. EFFECT OF TRANSDUCERS ON MEASUREMENTS. Most practical systems involve the use of a transducer to convert electrical energy into acoustical energy. We therefore briefly examine the effect of transducers on measurement. Transducers of the contacting variety violate the assumptions of the isolated resonator model and produce a composite resonator system.[149,152] The composite system resonances occur at a frequency $\omega_m{}^c$ and require a correction[150,152] $(1 + 2\delta)$ to Eq. (2.2.32). Rearranging for velocity determination by a two-transducer transmission measurement, one can write

$$v_s = (\omega_m{}^c - \omega_{m+1}^c)(1 + 2\delta)a_s/2\pi.$$ (2.2.38)

Here, $\delta = a_T\rho_T/a_s\rho_s$, the sample length is $a_s/2$, the transducer length is $a_T/2$, and the sample and transducer densities are, respectively, ρ_s and ρ_T. The $(1 + 2\delta)$ correction is adequate for small δ (<0.01) and for per-

[152] D. I. Bolef and M. Menes, J. Appl. Phys. 31, 1010–1017 (1960).

TABLE I. Magnitude of cw Expressions $|U|$, U_1, U_2, etc. for Fixed Values of a, k, $\Delta\alpha$, and Δk for Two Different Values of αa[a]

Expression	$\alpha a = 0.01$			$\alpha a = 0.001$		
	$\Delta\alpha = 0$	$\Delta\alpha = 10^{-4}$	$\Delta k = 10^{-4}$	$\Delta\alpha = 0$	$\Delta\alpha = 10^{4}$	$\Delta k = 10^{-4}$
$|U|$	1.01×10^2	9.95×10	1.01×10^2	10^3	9.10×10^2	9.96×10^2
U_1	1.01×10^2	9.95×10	1.01×10^2	10^3	9.10×10^2	9.91×10^2
U_2	0	0	-1.00	0	0	-9.9×10^1
$\Delta|U|/|U|$	—	-9.95×10^{-3}	5.00×10^{-5}	—	9.09×10^{-2}	4.96×10^{-3}
$\Delta U_1/U_1$	—	-9.95×10^{-3}	9.95×10^{-5}	—	9.09×10^{-2}	9.90×10^{-3}
$\Delta U_2/U_2$	—	∞	∞	—	∞	∞
		$a = 1$ cm			$a = 1$ cm	
$\partial|U|/\partial\alpha$	-10^4	-9.8×10^3	-10^4	-10^6	-8.26×10^5	-9.85×10^5
$\partial|U|/\partial k$	0	0	1.01×10^2	0	0	9.86×10^4
$\partial U_1/\partial\alpha$	-10^4	-9.8×10^3	-10^4	-10^6	-8.26×10^5	-9.71×10^5
$\partial U_1/\partial k$	0	0	-2.0×10^2	0	0	-1.96×10^5

[a] To obtain these values it is assumed that the particle velocity amplitude has unit magnitude and that the measurements are taken at the resonance peaks.

cent errors of 10^{-2}. Ringermacher et al.[153] have given an improved correction factor for one-transducer and two-transducer systems. With their more complex correction factors, errors as small as 10^{-7} are possible. For $\delta < 0.005$, uncorrected formula (2.2.32) is accurate to about 1.0% and, therefore, is widely used as an approximation.

2.2.3.2. **Systems for cw Measurements.** The complexity of a cw ultrasonic spectrometer depends on the nature of the experiment or measurement in which it is used. In this and the next few sections, we examine several cw techniques and indicate their attributes and shortcomings in order that the reader can choose an adequate system with the least complexity for the desired measurement.

2.2.3.2.1. THE CW ULTRASONIC INTERFEROMETER. The ultrasonic interferometer provides a straightforward method for measuring attenuation and velocity in liquids and gases. The technique, first reported in 1925 by Pierce,[154] uses a variable path length ultrasonic cell with flat and parallel perfectly reflecting walls perpendicular to the acoustic axis. Several reflection interferometers have been described.[154-160] An example[161] is shown in Fig. 35. By varying only cavity path length and keeping other parameters constant, one can minimize corrections for transducer effects.

Equation (2.2.31) can be used to derive the acoustic velocity from two cavity lengths $a_m/2$ and $a_{m+1}/2$, which correspond to neighboring mechanical resonances of the acoustic path in the fluid. For a fixed-frequency interferometer, the velocity is found to be

$$v = f_m(a_{m+1} - a_m), \tag{2.2.39}$$

where f_m is the frequency.

The absolute attenuation may be determined in a similar fashion by measuring the acoustic signal amplitude at a_m and a_{m+1}. Using the resonance condition $\cos ka = 1$ in Eq. (2.2.30), we find the square of the particle velocity amplitude

[153] H. I. Ringermacher, W. E. Moerner, and J. G. Miller, *J. Appl. Phys.* **45**, 549–552 (1974).

[154] G. W. Pierce, *Proc. Am. Acad. Arts Sci.* **60**, 271 (1925).

[155] J. C. Hubbard, *Phys. Rev.* **38**, 1011 (1931).

[156] J. C. Hubbard, *Phys. Rev.* **41**, 523 (1932).

[157] J. C. Hubbard, *Phys. Rev.* **46**, 525 (1934).

[158] D. R. McMillan, Jr. and R. T. Lagemann, *J. Acoust. Soc. Am.* **19**, 956 (1947).

[159] J. L. Hunter and F. E. Fox, *J. Acoust. Soc. Am.* **22**, 238–242 (1950); J. L. Hunter, *ibid.* **22**, 243–246 (1950).

[160] F. G. Eggers and Th. Funck, *Rev. Sci. Instrum.* **44**, 969–977 (1973).

[161] A. L. Loomis and J. C. Hubbard, *J. Opt. Soc. Am. Rev. Sci. Instrum.* **17**, 295–307 (1928).

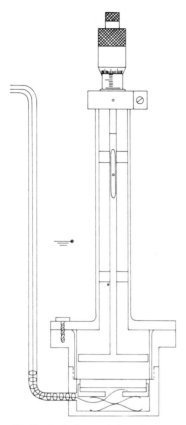

FIG. 35. Typical ultrasonic interferometer.[161]

$$|U|^2(a_m) = G^2 e^{\alpha a_m}/(e^{\alpha a_m} + e^{-\alpha a_m} - 2), \qquad (2.2.40)$$

where G, the electronic gain, is inserted to show the effect of amplification of the signal by electronic means. This gain factor is eliminated by taking the ratio of the particle velocity amplitudes for two cavity lengths. The natural logarithm of this ratio becomes

$$\ln \frac{|U|^2(a_{m+1})}{|U|^2(a_m)} = \alpha(a_{m+1} - a_m) + \ln \left(\frac{e^{\alpha a_m} + e^{-\alpha a_m} - 2}{e^{\alpha a_{m+1}} + e^{-\alpha a_{m+1}} - 2} \right). \quad (2.2.41)$$

Simplifying and keeping terms to third order, we obtain an expression for the attenuation written

$$\alpha = \frac{1}{a_{m+1} - a_m} \left[\ln \left(\frac{|U|^2(a_{m+1})}{|U|^2(a_m)} \right) - 2 \ln \frac{a_m}{a_{m+1}} \right]. \qquad (2.2.42)$$

Other expressions derived from electrical equivalent circuit analysis also may be used.[155-157,159,161-163]

A fixed-length "interferometer" or resonator, described by Eggers and Funck,[160] requires very small sample volumes (about 1 ml). Their analysis results in equations similar to those obtained with the propagating plane wave model. A correction formula is presented for the attenuation determined from resonance width at half-power.

The importance of reflector wall parallelism for accurate interferometric measurements is demonstrated by McMillan and Lagemann.[158] Lack of parallelism leads to inhomogeneous broadening of mechanical resonances along with an uncertainty in the position of the resonance peak.[164] Also, lateral wall losses must be considered if the diameter of the container is not great enough.

Many variations of the basic interferometer have been reported. An automatic device which records velocity and attenuation[165] uses a double-crystal resonant cell.[162,166] Another technique uses an air–liquid reflecting surface[159] to eliminate corrections resulting from transmission of acoustic energy into the reflector itself. With a free liquid surface this system is sensitive to mechanical vibrations; however, a thin mica sheet at the interface[167] produces a very effective reflector.

2.2.3.2.2. cw TRANSMISSION TECHNIQUES. The cw technique can be used for acoustic transmission measurements[168] through flat and parallel homogeneous samples. A block diagram of this technique is shown in Fig. 36, which has been subdivided to show a succession of modifications. Figure 36a is the basic cw apparatus. A stable oscillator is tuned to a mechanical resonance peak ($ka = 1m\pi$), determined by peaking the signal observed on the oscilloscope. The frequency separation between corresponding resonances ($m \to m + 1$) is measured and the ultrasonic velocity calculated from Eq. (2.2.32). The full-frequency width $\Delta\omega$ of each resonance at 50% power (0.707 voltage) is used with the relation $\alpha = \Delta\omega/2v$ to determine the sample attenuation.

Several modifications of the basic cw technique are desirable.[150] Figure 36b shows the addition of a voltage-swept frequency capability along with diode detection so that $|U|$ is electronically plotted as a function of

[162] W. J. Fry, *J. Acoust. Soc.* **21**, 17 (1949).

[163] W. P. Mason, "Piezoelectric Crystals and Their Application to Ultrasonics." Van Nostrand-Reinhold, Princeton New Jersey, 1950.

[164] J. G. Miller and D. I. Bolef, *J. Appl. Phys.* **41**, 2282 (1970).

[165] M. Greenspan and M. C. Thompson, Jr., *J. Acoust. Soc. Am.* **25**, 92 (1953).

[166] M. Greenspan, *J. Acoust. Soc. Am.* **22**, 568 (1950).

[167] E. M. Bains and M. A. Breazeale, *J. Chem. Phys.* **61**, 1238–1243 (1974).

[168] D. I. Bolef, J. de Klerk, and R. B. Gosser, *Rev. Sci. Instrum.* **33**, 631–638 (1962).

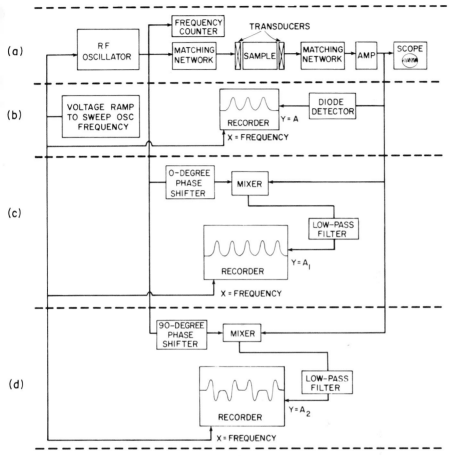

FIG. 36. (a) Black diagram of a simple cw ultrasonic spectrometer. (b) Addition of a voltage-controlled oscillator and a diode detector. (c) Use of a mixer with a 0° phase reference to obtain the A_1 resonator signal. (d) Use of a mixer with a 90° phase reference to obtain the A_2 resonator signal.

frequency.[168,169] By combining a zero-degree phase reference from the oscillator with the output in a double-balanced mixer–modulator[150] shown in Fig. 36c, pure attenuation as well as pure dispersion data may be obtained separately and simultaneously. The mixer and low-pass filter ($\omega_{\text{pass}} < \omega_m{}^c$) give the in-phase response U_1 shown in Fig. 36c. By adding a 90° phase shifter to the mixer–modulator combination, one obtains the quadrature response U_2 as shown in Fig. 36d. Zeros in the corresponding

[169] D. I. Bolef and J. de Klerk, *IEEE Trans. Ultrason. Eng.* **UE-10,** 19 (1963).

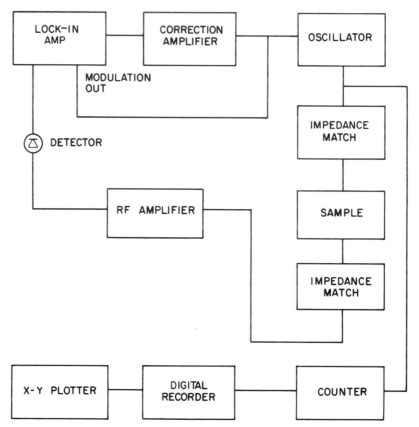

FIG. 37. Block diagram of automatic frequency tracking ultrasonic spectrometer.[170]

sensitivity enhancement factors ensure that small changes in the signal $|U|$, U_1, or U_2 result only from changes in attenuation (dispersion).

A more complex system[170] provides for automatic frequency tracking of a mechanical resonance using an FM technique. A block diagram of the tracking ultrasonic spectrometer is shown in Fig. 37. The frequency modulation leads to an amplitude modulation at the rf detector. The AM signal is in phase with the FM signal for frequencies $\omega \lesssim \omega_m$ and 180° out of phase for frequencies $\omega \gtrsim \omega_m$. The abrupt phase shift at ω_m provides a feedback signal locking the frequency to ω_m (AM phase equals 90°). Relative changes in velocity of a few parts in 10^8 are measured with this technique.[170]

[170] R. G. Leisure and R. W. Moss, *Rev. Sci. Instrum.* **40**, 946–948 (1969).

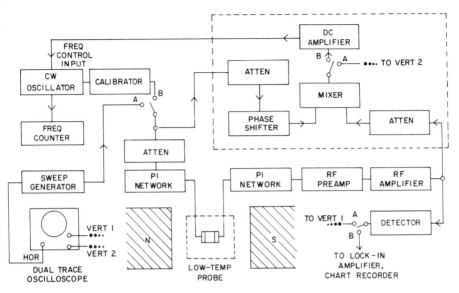

FIG. 38. Block diagram of a cw acoustic transmission spectrometer with the additional components comprising the locking section shown in detail in the box.[171] The spectrometer is adjusted using the sweep generator with switches in position A; then switches are thrown to position B for frequency-locked cw operation.

A second method of frequency locking is possible.[171] The technique utilizes the abrupt 180° phase shift that occurs in the rf signal as the frequency passes through ω_m but does not require frequency modulation of the source. Details of this technique are shown in Fig. 38. As with the FM techniques, the feedback path provides a correction voltage to maintain the spectrometer at the frequency ω_m.

In cw transmission spectroscopy at rf frequencies, sample and transducer construction requires precision and care to prevent direct electromagnetic rf leakage ("cross talk") across the sample.[172] The signal superposition resulting from the presence of leakage leads to anomalous line shapes and can lead to erroneous measurement of amplitudes. Sample probes which minimize cross talk are constructed of well-grounded conductors with any void between the sample and probe packed with indium foil or conductive paint.[172]

2.2.3.2.3. SAMPLED CONTINUOUS WAVE TECHNIQUE. In situations in which electromagnetic cross talk is a problem, one can measure velocity and attenuation by using a sampled continuous wave (scw) spectro-

[171] J. G. Miller and D. I. Bolef, *Rev. Sci. Instrum.* **40**, 361 (1969).
[172] J. G. Miller, *J. Acoust. Soc. Am.* **53**, 710–713 (1973).

meter.[173]   The scw spectrometer eliminates electromagnetic cross talk,[172] provides for both time-domain as well as frequency-domain measurements, and requires transducer contact to only one side of a sample.   A basic scw spectrometer is shown in the block diagram of Fig. 39a.   The rf oscillator output is connected to a sample via the transmitter gate which is turned on (see Fig. 39b) for a sufficient time period to achieve steady-state conditions in the sample.   The receiver gate is turned on after a suitable time delay (~ 200 nsec) following closure of the transmitter gate.   The delay ensures that transmitter switching transients are kept out of the receiver.

Ultrasonic signals typically produced by an scw spectrometer with transducers on both sides of the sample are shown in Fig. 39c.   For this figure the oscilloscope is triggered at T_{off}.   The upper trace shows the T_{on} pulse so that the transmission signal (lower trace) displays the build up and decay in the sample of an acoustic steady-state standing wave mode. The upper trace shows the standard one-transducer (reflection) scw mode decay.   The characteristic steplike response is the result of round-trip times in the sample or individual terms in the infinite sum of Eq. (2.2.25) vanishing with time.   For example, consider the terms in the parentheses: the term "1" vanishes at T_{off}, the term $\exp[-(ika + \alpha a)]$ vanishes at $T_{off} + a/v$, the term $\exp[-2(ika + \alpha a)]$ vanishes at $T_{off} + 2a/v$ . . . , etc.   Thus the acoustic wave decays in a way similar to that by which it is formed—stepping to an equilibrium condition.

In addition to the time-domain decay information, which is useful for determining attenuation, the scw can provide frequency-domain information for determining velocities.   This is accomplished by plotting the signal amplitude while sweeping the oscillator frequency.   An example of such a plot is shown in Figure. 40.   The scw spectrometers use the same fundamental principles as the cw spectrometer, but they often are more complex.   They can be designed to measure U_1 or U_2 or to frequency lock to a mechanical resonance.   Furthermore, pulse-echo data may be obtained with an scw spectrometer by narrowing the transmitter gate width.

2.2.3.2.4. TRANSMISSION OSCILLATOR ULTRASONIC SPECTROMETER. The transmission oscillator ultrasonic spectrometer (TOUS)[174] is a fundamentally different type of cw spectrometer.   The instrument does not use an external oscillator and, in fact, derives its desirable characteristics from its instability.   Basically, the device is an acoustic analog of the marginal oscillator called the "Pound box," used in nuclear magnetic resonance (NMR) techniques.[175]   The first use of marginal oscillators for

[173] J. G. Miller and D. I. Bolef, *Rev. Sci. Instrum.* **40,** 915 (1969).
[174] M. S. Conradi, J. G. Miller, and J. S. Heyman, *Rev. Sci. Instrum.* **45,** 358–360 (1974).
[175] E. M. Purcell, H. C. Torrey, and R. V. Pound, *Phys. Rev.* **69,** 37 (1946).

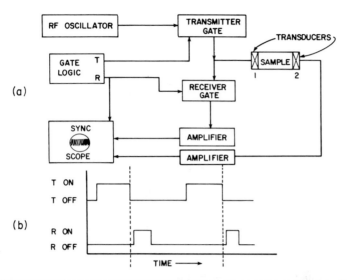

(a)

(b)

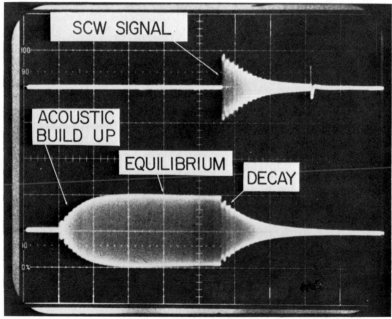

FIG. 39. (a) Block diagram of a basic, sampled, continuous wave, ultrasonic spectrometry (scw). (b) Transmitter and receiver gate logic for the scw. (c) Time-domain decay of an acoustic resonator as measured by a reflection scw instrument (upper trace) and by a transmission scw instrument (lower trace). In the lower trace the receiver gate was left open continuously.

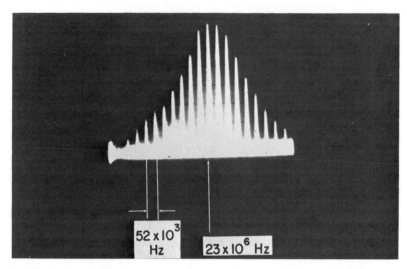

FIG. 40. Frequency-domain response of an acoustic resonator as measured by an scw instrument.

measurements of small changes in acoustic parameters was reported with the discovery of nuclear acoustic resonance (NAR).[176] The device now is called a marginal oscillator ultrasonic spectrometer (MOUS).[177] The MOUS is reported[168,176] to be able to detect changes in acoustic absorption of $10^{-8}-10^{-9}$ cm^{-1} and is considered to exhibit the highest sensitivity to changes in ultrasonic absorption.[114,150] The TOUS appears to have the same high sensitivity characteristics of the MOUS without its inherent complexity and limitations. The TOUS locks to the sample's resonant frequency and therefore utilizes maximum sensitivity enhancement (see Section 2.2.3.1.2) for attenuation measurements. Changes in velocity (or phase) may be determined from changes in the oscillation frequency.

A block diagram of the TOUS is shown in Fig. 41. The circuit oscillates if the closed-loop feedback path has proper gain and phase relationships. Under steady-state conditions, oscillations are stable if the gain factor G and loss factor p product equals 1 and if the phase shift around the loop equals an integral multiple of 2π. If the assumption is made that p is constant, a change in resonance frequency of the sample causes a phase shift around the loop resulting in a change in the oscillation frequency. The frequency change is stabilizing for every second mechan-

[176] D. I. Bolef and M. Menes, *Phys. Rev.* **114**, 1441–1451 (1959).

[177] W. D. Smith, J. G. Miller, D. I. Bolef, and R. K. Sundfors, *J. Appl. Phys.* **40**, 4967 (1969).

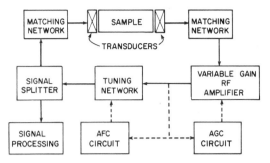

FIG. 41. Block diagram of a transmission oscillator ultrasonic spectrometer (TOUS). Dashed lines indicate connections required to lock the center frequency of the tuning network to the center frequency of the sample's mechanical resonance.

ical resonance since a 180° phase shift exists between succeeding resonances. Therefore, the phase shifts cause the TOUS to track shifts of mechanical resonance frequency, e.g., in response to changes of a controlled environment around the sample.

An analysis of amplitude stability requires a more detailed discussion. A typical rf amplifier has a response curve similar to that in Fig. 42a. For sufficiently small input voltages, the response is nearly linear, and thus the gain $G = V_{out}/V_{in}$ (shown in Fig. 42a as the solid line) is nearly independent of V_{in}. We assume that the TOUS rf amplifier saturates at high input voltages as shown; i.e., $dG/dV_{in} < 0$. Thus, for any nonzero V_{in}, the circuit gain is less than it would be for $V_{in} \to 0$. The parameter $-(dG/dV_{in})^{-1}$ has the general shape shown in Fig. 42b; it approaches infinity as $V_{in} \to 0$. The change in output voltage for a small fractional change in loss $\Delta p/p$ resulting from an increase in acoustic attenuation is given by[174]

$$\Delta V_{out} = -(dG/dV_{in})^{-1}G^2 \, \Delta p/p. \tag{2.2.43}$$

Since the term $-(dG/dV)^{-1}$ becomes large for small V_{in} ("linear" range of amplification), ΔV_{out} also becomes large, even for relatively small changes in $\Delta p/p$.

Using the assumption that the phase is constant, we analyze the circuit for signal amplitudes from the "zero" V_{in} (start-up) condition. Since $G(V_{in} = 0) > G(V_{in} > 0)$, thermal noise in the frequency bandwidth of the tuning network and transducer will increase in amplitude if $G \cdot p > 1$. As V_{in} increases, G decreases until $G \cdot p = 1$ at which point the circuit will be stable. If the acoustic loss increases, V_{in} decreases and G increases until a new $G \cdot p = 1$ condition is met. Conversely, if the acoustic loss decreases, G decreases, so the circuit conditions are always tending toward equilibrium.

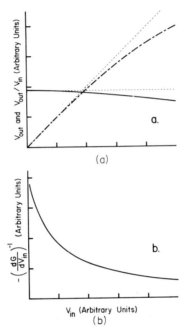

FIG. 42. Response curves of a typical rf amplifier[174]: (a) output voltage V_{out} ($— \cdot — \cdot$) and gain $G = V_{out}/V_{in}$ (solid line) plotted as a function of input voltage V_{in}; (b) the behavior of the parameter $-(dG/dV_{in})^{-1}$ as a function of input voltage V_{in}. Dashed lines represent the response of a truly linear amplifier.

Calibration of marginal oscillator systems involves use of electrical insertion loss methods.[176,178] Use is made of the nearly linear changes in resistance of a small fuse (typically 1/100 A) with current. Other methods use mixer–modulators as variable attenuators[174] to provide changes in insertion loss. A true acoustic calibration is possible with an ultrasonic calibrator,[151] which makes use of the attenuation changes that accompany conductivity changes in piezoelectric semiconductors.[179]

To obtain both attenuation and phase shift data, the TOUS must be operated near the linear range of the amplifier since phase shifts in a saturated amplifier can lead to tracking errors. A wide range of magnitudes of sample attenuation is possible if automatic gain control (AGC) is used with the rf amplifier. Details of such a system have been described[180] in connection with the observation of particulates in flowing liquids. A fur-

[178] W. D. Smith and R. K. Sundfors, *Rev. Sci. Instrum.* **41**, 288 (1970).

[179] A. R. Hutson and D. L. White, *J. Appl. Phys.* **33**, 40 (1962).

[180] J. S. Heyman, D. Dietz, and J. G. Miller, *Proc. IEEE Ultrason. Symp., Los Angeles*, p. 561 (1975).

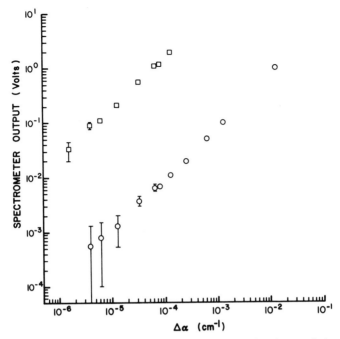

FIG. 43. Output voltages of a TOUS (□) and a more conventional transmission cw spectrometer (○) as functions of the change in ultrasonic attenuation $\Delta\alpha$.[174] The error bars represent the noise present in the spectrometer outputs.

ther refinement of the TOUS uses an FM frequency tracking technique (AFC)[170] to lock the center frequency of the tuning network to that of the sample's mechanical resonance. The locking schemes are shown in Fig. 41 as dashed connections.

Attenuation data obtained with a TOUS[174] are shown in Fig. 43 and are contrasted with data obtained from a simple transmission cw spectrometer (see Fig. 36). For resolving small changes in sample attenuation, the TOUS is shown to have significantly less noise than the cw system which included a high-quality oscillator with which it is compared.

2.2.3.2.5. REFLECTION OSCILLATOR ULTRASONIC SPECTROMETER. An instrumental technique similar to the TOUS is the reflection oscillator ultrasonic spectrometer (ROUS).[181] Unlike the TOUS, the ROUS requires contact to only one side of the sample resonator. It uses the diffracted acoustic beam to close the oscillator feedback path. More complex transducer geometry, as shown in Fig. 44, is required for the ROUS tech-

[181] J. S. Heyman, Proc. IEEE Ultrason. Symp., Annapolis, p. 113 (1976).

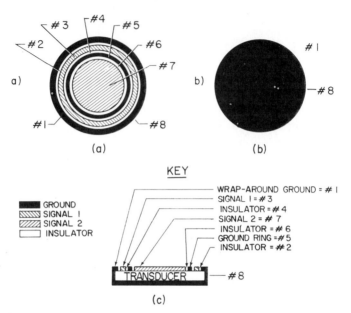

FIG. 44. Reflection oscillator ultrasonic spectrometer (ROUS) transducer electrode configuration[181]: (a) back; (b) front.

nique.   The electrode geometry shown electrically isolates the receiver from the amplifier output.   Without proper isolation, the effects of electromagnetic leakage[172] severely limit the usefulness of this technique. The electrode pattern shown in Fig. 44 was evaporated on a 5-MHz PZT-5 transducer and provided about 60-dB electrical isolation between the transmitter and receiver electrodes when the transducer was mounted in its housing.   For operation with a ratio of acoustic signal-to-leakage of at least 20 dB, the ROUS is limited to samples with $\alpha a < 1$, if a 1% conversion efficiency for the transducer–receiver combination is assumed.   For samples of greater αa, increased total isolation between the receiver and transmitter are necessary.   Higher acoustic isolation may be achieved by milling a slot through ground ring #5 in Fig. 44 or using two separate transducers.

Several applications of the ROUS have been reported, including a technique for measuring changes in bolt preloading[181,182] and changes in attenuation in liquids caused by the passage of particles through the ultrasonic beam.[181]

2.2.3.2.6. OTHER CW METHODS. Another cw technique that is suitable for measuring acoustic absorption is the resonance reverberation

[182] J. S. Heyman, *Exp. Mech.* **17**, 183 (1977).

method.[183]   In the modern forms of this method[184-188] ultrasound is coupled into an isolated spherical cavity (usually a thin-walled glass flask)
filled with a test liquid.   With the excitation turned off, the cavity modes
decay with a time constant related to acoustic absorption in the liquid and
boundary layer losses at the walls of the cavity.   One attempts to minimize wall lossess by selecting a radially symmetric mode when exciting
the cavity.   The time τ for the signal to decay to $1/e$ of its initial amplitude
is related by $\tau = 1/\alpha v$ to total energy absorption by the test liquid, the
viscous and thermal boundary layer at the walls, the glass flask and its
supports and transducers, and the surrounding air, if any.   Relative measurements of τ are obtained with this technique by using calibration liquids.   It is important that the calibration liquids have the same acoustic
velocity as the test liquids to minimize changes in wall effects.[189]
Although an accuracy on the order of 30% was indicated for one measurement set, a possible accuracy of about 5% was reported from minimum
values of many measurements.[187]   Some doubt about the ability to excite
radially symmetric modes has been raised.[190]   Absolute decay times of
1/100 the calculated decay times have been measured for radial resonant
modes of nearly perfectly ground spherical shells of fused quartz filled
with distilled water.   This discrepancy cannot be attributed to cavity
asymmetry resulting from fabrication error; therefore Andreae and Edmonds[190] implicated energy loss associated with suspension of the system
and cavity wall losses associated with residual asymmetric modes.
These discrepancies should be more fully investigated before analysis of
the effects of extraneous energy losses on the reverberation decay time
can be considered complete.
   A different and important resonance technique[160,191-193] makes use of a
sealed cylindrical resonator consisting of two transducers, a thin spacer

[183] C. E. Mulders, *Appl. Sci. Res., Sect. B* **1**(3), 149 (1948), **1**(5), 341 (1950).
[184] W. Kuhn and H. Kuhn, *J. Colloid Sci.* **3**, 11 (1948).
[185] C. J. Moen, *J. Acoust. Soc. Am.* **23**, 62 (1951).
[186] G. Kurtze and K. Tamm, *Acustica* **3**, 33 (1953).
[187] J. Karpovich, *J. Acoust. Soc. Am.* **26**, 819 (1954).
[188] J. Stuehr and E. Yeager, *in* "Physical Acoustics" (W. P. Mason, ed.), Vol. IIA, pp.
351–462. Academic Press, New York, 1965.
[189] T. Ohsawa and Y. Wada, *Jpn. J. Appl. Phys.* **6**, 1351 (1967).
[190] J. H. Andreae and P. D. Edmonds, *Proc. Int. Congr. Acoust., 3rd* (L. Cremer, ed.),
Vol. 1, pp. 556–558 (1959).
[191] F. Eggers and Th. Funck, *J. Acoust. Soc. Am.* **57**, 331–333 (1975); *Naturwissenschaften* **63**, 280–285 (1976).
[192] F. Eggers, Th. Funck, and K. H. Richmann, "Fortschritte der Akustik," p. 577.
DAGA, Braunschweig, 1975, *Rev. Sci. Instrum.* **47**, 361–367 (1976).
[193] A. Labhardt and G. Schwarz, *Ber. Bunsenges. Phys. Chem.* **80**, 83–92 (1976).

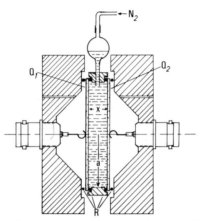

FIG. 45. Cross section of the inner part of a pressurized, 15-mliter ultrasonic, resonator cell with X-cut, 2-MHz quartz transducers (Q_1, Q_2; radius a = 30 mm) mounted between rubber O-rings (R).[191] Transducer separation x. Liquid storage vessel connected to nitrogen bottle (N_2). Themostated, adjustable, holding frame not shown.

with O-rings, and a housing to hold the components and provide electrical coupling, as shown in Fig. 45. Very high Q resonance of the liquid-filled cylindrical cavity is achieved by applying pressure to the liquid under study (~ 1 bar), thus producing a concave bending of both transducers and a net focusing. The result is similar to that achieved with confocal resonators for optical or electromagnetic waves. An improvement factor of nearly 40 in cavity Q was produced by a 1-bar-liquid overpressure. Use of liquid samples as small as 2 ml with a resonance $Q \simeq 3.4 \times 10^4$ at 1.8 MHz has been reported.[192] For a larger cavity (15 ml), a Q of 1.1×10^5 has been achieved at 0.4 MHz.[192] The benefits of this system are high accuracy, small volume, lower frequency range, and fewer effects of gas inclusion. Furthermore, diffraction and side-wall effects are reduced by the focusing geometry.

An ultrasonic technique for observing attenuation with perhaps the oldest antecedants is the acoustic streaming method. Although this effect was recognized by Faraday[194] and Lord Rayleigh,[195] it was not until 1948 that a serious analysis was published.[196] Modern interpretations of streaming[197-199] are based on momentum conservation. If a medium absorbs energy from a propagating acoustic wave, then it must also acquire

[194] M. Faraday, *Proc. R. Soc. London* **3**, 49–51 (1831).
[195] Lord Rayleigh, "Theory of Sound," Vol. II, p. 217. Macmillan, London, 1896.
[196] C. Eckart, *Phys. Rev.* **73**, 68 (1948).
[197] J. J. Markham, *Phys. Rev.* **86**, 497 (1952).
[198] W. L. Nyborg, *J. Acoust. Soc. Am.* **25**, 68 (1953).
[199] J. E. Piercy and J. Lamb, *Proc. R. Soc. London Ser. A* **226**, 43 (1954).

a corresponding amount of momentum. Therefore, absorption in liquids and gases produces force gradients and material transport. An improved method for observing streaming has been reported[199,200] which employs a main and an auxilliary tube containing the test liquid. The main tube is bounded at one end by a transducer and at the other end by a thin diaphragm separating the test liquid from a highly absorbing liquid (acetic acid or methyl cyclohexane). The small, auxiliary sight tube is connected to two points Z_2 and Z_1, between which there is a pressure gradient of the main tube in such a manner that it does not disturb the acoustic beam. Flow of suspended inert particles (e.g., aluminum) in the sight tube is observed through a microscope. The change in pressure from Z_2 and Z_1 can be shown to be[200]

$$\Delta P = E[\exp(-2\alpha Z_1) - \exp(-2\alpha Z_2)], \qquad (2.2.44)$$

where E is the energy density of the acoustic plane wave. The liquid velocity in the sight tube may be determined from Poiseuille's law as

$$V = \Delta P \, R^2/4\eta l, \qquad (2.2.45)$$

where R is the sight tube radius, l its length, and η the shear viscosity of the liquid in question. If the velocity of flow in the sight tube is measured (i.e., by microscopic observation of time of flight of small particles suspended in the liquid), the acoustic attenuation may be calculated by combining Eqs. (2.2.44) and (2.2.45). Accuracies on the order of 6% over the frequency range 0.5–1.5 MHz are reported.[200]

A novel approach to measuring attenuation makes use of an amplitude modulated (AM) radio-frequency source connected to an ultrasonic system.[201,202] Radiation pressure, which is the rate of change of momentum through a unit area,[203] is also modulated at the AM frequency. A low-frequency sensor measures the variation in radiation pressure at the AM frequency with changing path length so that α may be calculated. An improvement of this method uses a vibrating reed nulling electrodynamic balance as the sensor element.[204]

2.3. Conclusion

The three fundamental classes of techniques that have been considered are based on optical, pulse-echo, and continuous wave methods. Optical

[200] D. N. Hall and J. Lamb, *Proc. Phys. Soc. London* **73**, 354 (1959).

[201] A. Barone and M. Nuovo, *Ric. Sci.* **21**, 516 (1951).

[202] F. L. McNamara and R. T. Beyer, *J. Acoust. Soc. Am.* **25**, 259 (1953).

[203] T. F. Hueter and R. H. Bolt, "Sonics," p. 43. Wiley, New York, 1955.

[204] M. Mokhtar and H. Youssef, *J. Acoust. Soc. Am.* **28**, 651 (1956).

techniques have advantages of simplicity and accuracy for velocity measurements, and the visibility method of Hiedemann[205] probably presents the optimum combination of simplicity and accuracy in the megahertz frequency range. Although optical techniques were among the earliest to be investigated, the advantages of the use of Bragg diffraction at high frequencies (of the order of 0.3–2 GHz) and the related subject of Brillouin scattering of light by ultrasonic waves were recognized relatively late.[67,68]

Pulse techniques and continuous wave techniques, in fact, represent a Fourier transform pair of classes and therefore theoretically contain equivalent information. However, in practice, the equivalency may be obscured by lack of experimental precision or by the data processing. Although it was not possible for us to deal with data processing in any detail, it is obvious that it occasionally can introduce errors in the results when the calculations become extensive in the evaluation of a quantity from its measured Fourier transform. This means that the highly sophisticated automated systems must be analyzed with an eye to determining the actual experimental significance of the data output. A good general practice is to use the simplest apparatus capable of achieving the desired result.

With pulse-echo systems it may be necessary to consider the effects of medium nonlinearity if one uses large amplitude drive pulses or to consider the effect of broadband detection and the corresponding increase in background noise. The signal-to-noise ratio may be enhanced in pulse-echo systems by signal averaging, provided the signal sufficiently decays away between pulses.

With continuous wave systems diffraction and nonparallelism of the transducers and/or reflecting surfaces produce complex standing wave mode structures which are awkward to analyze. For liquids this situation has been improved by the approach to a confocal arrangement made by Eggers and Funck.[206] There is no reason a similar arrangement could not be used with solids, although to date the approach has been more in the direction of making use of the diffracted ultrasonic beam to close the oscillator feedback path.[181]

With both pulse and cw techniques to measure solids, the effect of phase shifts in the transducer bond must be considered unless one uses a noncontact transducer.[98] The effect of phase variations in the wave front caused by diffraction can be reduced by using a phase-insensitive trans-

[205] E. Hiedemann, "Grundlagen und Ergebnisse der Ultraschallforschung," p. 80. de Gruyter, Berlin, 1939; L. Bergmann, "Der Ultraschall und Seine Anwendung in Wissenschaft und Technik," p. 310. Hirzel, Stuttgart, 1954.

[206] F. Eggers and Th. Funck, Rev. Sci. Instrum. 44, 969–977 (1973); J. Acoust. Soc. Am. 57, 331–333 (1975).

ducer.[29]   In most cases after correction for diffraction and the effects of transducer bonds, the accuracy of velocity and attenuation measurements is limited primarily by parallelism of sample faces and the accuracies with which the sample length, beam diffraction, and wall losses are determined.   Thus, even though a resolution of a few parts in 10^8 may be available, the most careful length measurement, good to 5 parts in 10^5, becomes the limit to the accuracy of absolute results.   Relative measurements, on the other hand, can approach the limits set by resolution if care is taken to eliminate the effects of equipment instability and temperature and pressure variations.

Acknowledgment

The work, as described in this part, of the first author has been partially supported by the Office of Naval Research.

3. DYNAMIC VISCOSITY MEASUREMENT

By Gilroy Harrison and A. John Barlow

List of Symbols

a	transducer radius
A	constant in the viscosity equation
ΔA	amplitude change of received signal
B	constant in the viscosity equation
ΔB	phase angle change of received signal
c	velocity of shear wave propagation
C	temperature dependence of shear compliance
C_0	transducer equivalent circuit capacitance
ΔD	decay rate change
f	frequency
f_m	$\omega_m/2\pi$, Maxwell relaxation frequency
Δf	frequency change
G	shear modulus
G_r	retardational modulus
$G^*(j\omega)$	complex shear modulus $[=G'(\omega) + jG''(\omega)]$
G_∞	instantaneous shear elastic modulus
h	depth of immersion of rod in liquid
$H(\tau)$	relaxation time spectral distribution function
J_e	equilibrium compliance
J_r	retardational compliance $[=J_1(\omega) - jJ_2(\omega)]$
J_∞	instantaneous shear compliance
$J^*(j\omega)$	complex shear compliance $[=J'(\omega) - jJ''(\omega)]$
K	instrument constant
K_1, K_2	transducer constants
$L(\tau)$	retardation time spectral distribution function
m	ratio of inner to outer radii of tube
n	number of harmonic
r	radius of delay rod
R	magnitude of complex reflection coefficient
R^*	complex shear wave reflection coefficient
R_s	transducer equivalent circuit resistance at resonance
ΔR	change in resistance at resonance
T_g	glass transition temperature
T_0	viscosity equation reference temperature
Z_c	complex mechanical impedance for cylindrical shear waves
Z_L	complex mechanical impedance for plane shear waves $(=R_L + jX_L)$
Z_1, Z_2	shear impedances
β	retardation compliance equation parameter

137

METHODS OF EXPERIMENTAL PHYSICS, VOL. 19

γ shear strain $(=\partial\xi/\partial y)$
Γ propagation constant for shear waves
η viscosity
η_r retardational viscosity
θ phase angle associated with complex reflection coefficient
λ shear wavelength
ξ displacement in shear
ρ density
σ shear stress
τ_m Maxwell relaxation time
τ_r retardation time
ϕ angle of incidence of shear wave at an interface
ψ angle of refraction of shear wave at an interface
ω angular frequency
ω_m Maxwell relaxation angular frequency

3.1. Introduction

This part is concerned with the dynamic behavior of liquids when the existing molecular equilibrium is disturbed by an applied mechanical stress. In general, the stress may be shear or compressional or both, but any compressional component, as occurs, for example, in the propagation of a longitudinal wave at sonic or ultrasonic frequencies, causes a cyclic adiabatic volume change. The resulting local temperature variation will perturb any temperature-sensitive equilibria such as those involving rotational isomers. The present discussion is restricted to the behavior of liquids in shear, in which dynamic volume and temperature variations do not occur. The response of a liquid depends on the relative durations of the applied stress and the time constant or relaxation time associated with the change in equilibrium. If the time over which the shear stress is applied is long compared with the relaxation time, the liquid can respond to the stress and the behavior is then characterized entirely by viscous flow. If the time scale of the deformation is less than the time taken for molecules to diffuse to a new equilibrium position, then the response is elastic and the strain is recoverable. In the region where both elastic and viscous responses to stress are present, the liquid is viscoelastic. A useful guide to the order of magnitude of the time involved is the Maxwell relaxation time $\tau_m = \eta/G_\infty$, where η is the viscosity and G_∞ the instantaneous shear elastic modulus (i.e., as observed at high frequencies or short times). For most liquids, G_∞ is found to be about 10^9 Pa $(= 10^{10}$ dyne cm$^{-2})$, giving $\tau_m \approx 10^{-9}\eta$ sec, where η is the viscosity in Pascal seconds (1 Pa sec = 10 P). In general, the response is characterized not by a single relaxation time but, in phenomenological terms, by a spectrum of relaxation times.
  Conventional methods of generating shear waves are limited to fre-

quencies below about 10^9 Hz; consequently the study of the viscoelastic behavior of liquids is usually restricted to those with viscosities above 1 Pa sec. Most simple organic liquids have viscosities less than this value and, consequently, can only be investigated if the viscosity, and, therefore, the relaxation time, can be increased by cooling or by hydrostatic pressure without crystallization. Materials of very high viscosity, such as high-molecular-mass polymers, may be studied most conveniently by techniques involving direct observation of the stress–strain relationship, either in a stress relaxation experiment in which a step function of strain is applied and the time decay of stress determined or in a creep experiment in which the deformation in response to a step function of stress is measured. These techniques are applicable when the relaxation times range from seconds to weeks. When relaxation times are somewhat less, down to 10^{-3} sec, oscillatory methods are preferred, using mechanical drives in the low-frequency region up to a few hertz and electromechanical transducers up to a few kilohertz. Adequate descriptions of such experimental techniques and of the properties of high-molecular-mass polymers are readily available.[1,2] Methods for investigating the dynamic behavior of liquids in shear at higher frequencies are less widely known, and, accordingly, the majority (Section 3.3) of this part is devoted to a survey of a number of such techniques. The necessary terminology and theoretical background material is presented in Section 3.2 and examples of the results obtained and their interpretation in Section 3.4.

3.2. Phenomenological Theory of Viscoelastic Liquids

3.2.1. Plane Shear Wave Propagation in a Viscoelastic Liquid

Experimental methods for the measurement of dynamic shear properties at high frequencies usually involve the generation and propagation of a shear wave into the liquid. By considering an element $dx\,dy\,dz$ of the liquid, as shown in Fig. 1, with shear stresses σ and $\sigma + ds$ acting on faces normal to the y direction, the equation of motion for a wave propagating in the y direction can be written

$$\rho\,dx\,dy\,dz\,(\partial^2\xi/\partial t^2) = (\partial\sigma/\partial y)\,dx\,dy\,dz, \qquad (3.2.1)$$

where ρ is the density of the liquid and ξ the displacement in the x direction. By defining the shear modulus G as σ/γ, where $\gamma\,(=\partial\xi/\partial y)$ is the shear strain, and substituting in Eq. (3.2.1),

[1] J. D. Ferry, "Viscoelastic Properties of Polymers," 2nd ed. Wiley, New York, 1970.
[2] F. R. Eirich "Rheology," Vols 1–5. Academic Press, New York, 1958–1970.

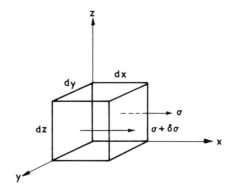

FIG. 1. Shear stresses on an element of material.

$$\partial^2 \xi / \partial t^2 = (G/\rho)\, \partial^2 \xi / \partial y^2, \tag{3.2.2}$$

which is the general equation for shear wave propagation in the y direction. For a sinusoidally varying displacement of angular frequency ω, the solution for a wave propagating in the positive y direction is

$$\xi = \xi_0 \exp[j\omega(t - y/c)] = \xi_0 \exp[j(\omega t - \Gamma y)]. \tag{3.2.3}$$

The velocity of propagation is $c = (G/\rho)^{1/2}$ and the propagation constant $\Gamma = \omega/c$.

For a lossless elastic solid, G is a real quantity, the stress and strain are in phase, and the wave propagates with zero attenuation. For a purely viscous or Newtonian liquid, the stress is 90° out of phase with the strain and the wave is highly attenuated. Newton's law states that $\sigma = \eta \dot{\gamma}$, and since $\dot{\gamma} = d[\gamma_0 \exp(j\omega t)]/dt = j\omega \gamma$, the stress is given by

$$\sigma = \eta \dot{\gamma} = j\omega \eta \gamma$$

and the shear modulus $G = \sigma/\gamma = j\omega \eta$. Generally, however, the shear modulus is complex, is usually denoted by $G^*(j\omega)$, and has frequency-dependent real and imaginary components: $G^*(j\omega) = G'(\omega) + jG''(\omega)$.

In principle, the shear modulus components can be determined from measurements of the velocity and attenuation of the shear wave. However, this is usually impractical except at very low frequencies because of the very high attenuation. For example, for a purely viscous liquid of $\eta = 1$ Pa sec, $\rho = 10^3$ kg m^{-3} and a wave of frequency $f\,(= \omega/2\pi) = 30$ MHz, the distance over which the wave travels before it is attenuated to $1/e$ of its original amplitude is about 6 μm. In practice, recourse is made to determination of the mechanical shear impedance, defined as

$$Z_L = -\sigma/(\partial \xi / \partial t) = R_L + jX_L \tag{3.2.4}$$

by analogy with the specific acoustic impedance for a sound wave (Kinsler and Frey,[3] p. 122). Thus the effect of the liquid on the surface generating the wave is measured rather than the properties of the wave itself. From Eqs. (3.2.3) and (3.2.4).

$$Z_L^2 = \rho G^*(j\omega)$$

or, in terms of modulus and impedance components,

$$G'(\omega) = (R_L^2 - X_L^2)/\rho, \qquad G''(\omega) = 2R_L X_L/\rho. \qquad (3.2.5)$$

For a purely viscous or Newtonian liquid, it follows that $G'(\omega) = 0$, $R_L = X_L$, and $G''(\omega) = \omega\eta$. These relationships also apply in general for a viscoelastic liquid as $\omega \to 0$. Conversely, for an elastic solid or for a viscoelastic liquid, as $\omega \to \infty$, $G'(\omega) = G_\infty = R_L^2/\rho$, $X_L = 0$, and $G''(\omega) = 0$.

The simplest system to exhibit both viscous and elastic properties is the Maxwell model, for which $G^*(j\omega) = G_\infty j\omega\tau_m/(1 + j\omega\tau_m)$. Figure 2 shows the variations of the components of the shear modulus and impedance for the Maxwell model. In mechanical terms, this model can be represented by a spring and a dashpot connected in series, as shown in Fig. 3a.

Alternatively, viscoelastic behavior may be described by the complex compliance $J^*(j\omega)$, defined as the inverse of the complex modulus:

$$J^*(j\omega) = J'(\omega) - jJ''(\omega) = 1/G^*(j\omega). \qquad (3.2.6)$$

In general, as $\omega \to \infty$, $J'(\omega)$ tends to the high-frequency limiting or instantaneous compliance $J_\infty = 1/G_\infty$. For this particular case of the simple Maxwell model, $J'(\omega) = J_\infty$ and is constant (independent of frequency). As mentioned in Chapter 3.1, the behavior of liquids is almost invariably more complicated, and $J'(\omega)$ usually increases with decreasing frequency to reach a limiting value J_e as $\omega \to 0$, where J_e, defined as the equilibrium compliance, is the sum of J_∞ and the retardational compliance J_r (which is zero for the Maxwell model).

3.2.2. The Creep Response

The significance of J_r may be more readily appreciated by considering the time-dependent strain of a liquid in response to a step function of stress. For the Maxwell model, the response is an instantaneous elastic strain determined by G_∞ plus a strain proportional to time which is determined by the viscosity. The creep compliance (ratio of strain to stress is given by

---

[3] L. E. Kinsler and A. R. Frey, "Fundamentals of Acoustics," 2nd ed. Wiley, New York, 1962.

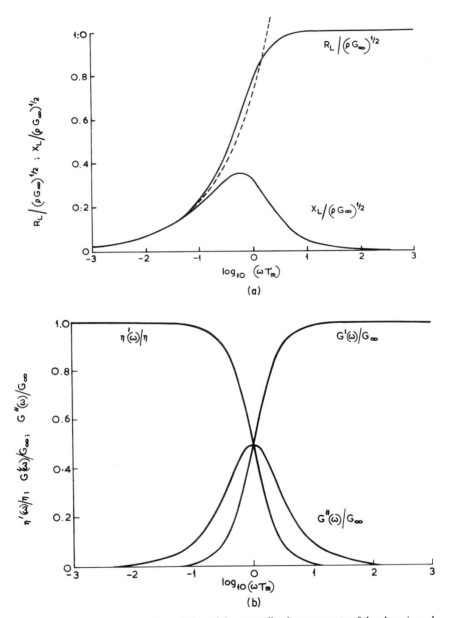

FIG. 2. Maxwell element: (a) variation of the normalized components of the shear impedance as a function of normalized frequency. The dashed curve shows the variation of both
components for a Newtonian liquid ($R_L = X_L$); (b) variation of the normalized components
of the shear modulus and the normalized dynamic viscosity as a function of normalized frequency.

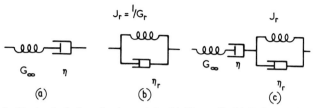

FIG. 3. Mechanical viscoelastic models: (a) Maxwell; (b) Voigt; (c) Burgers.

$$J(t) = \gamma/\sigma_0 = 1/G_\infty + t/\eta. \tag{3.2.7}$$

If the stress is removed, the elastic deformation is immediately recovered, but the viscous strain remains. In the more general case, an additional deformation occurs: the retarded elastic response, which, after a delay described by a retardation time τ_r, attains a magnitude determined by J_r. Again, if the stress is removed, this strain is recovered, but the recovery is delayed. Equation (3.2.7) is therefore generalized to

$$J(t) = \gamma/\sigma_0 = J_\infty + t/\eta + J_r f(t/\tau_r) \tag{3.2.8}$$

with $f(t/\tau_r) \to 1$ as $(t/\tau_r) \to \infty$.

By analogy with the delayed response of solids to constant stress, the retarded strain is termed "creep." (Strictly, only this component is the creep response, but it is common to find the complete response represented by Eq. (3.2.8) described as the creep response.)

The simplest model to embody the general property of viscoelastic liquids, i.e., both instantaneous and retarded elastic responses and steady-state flow, is the Burgers model.[4] This combines the Maxwell and Voigt elements, the latter being defined by the stress–strain relationship

$$\sigma = \eta_r \dot{\gamma} + \gamma/J_r. \tag{3.2.9}$$

The retardation time τ_r is $\eta_r J_r$. Whereas the mechanical model of a Maxwell element consists of a spring and dashpot in series, the Voigt model comprises a spring and dashpot in parallel. These models are illustrated in Fig. 3.

An alternative arrangement with a response identical to that of the Burgers model is a parallel combination of two Maxwell elements.

3.2.3. Relaxation and Retardation Spectra

The creep of the Voigt element in the Burgers model is an exponential function of time proportional to $(1 - e^{-t/\tau_r})]$; most observed creep

[4] M. Reiner "Handbuch der Physik" (S. Flügge, ed.), Vol. VI, p. 472. Springer-Verlag, Berlin and New York, 1958.

responses and, correspondingly, the frequency dependencies of the components of the complex modulus and compliance change more gradually than this simple model allows. In practice, a closer approach to the behavior of real liquids is obtained by extending the Burgers model to include m Voigt elements in series or by using the alternative form of $m + 1$ Maxwell elements in parallel. The choice of extended model is quite arbitrary; the same description of liquid behavior is given by both forms. The first involves m separate retardation times, the second $m + 1$ relaxation times. The interrelations of such canonic forms of model are well understood in terms of the theories of the analogous electrical filter networks, and considerable literature on the subject is available.[5-9] Although it may be mathematically convenient to describe liquid properties in terms of such series or spectra of relaxation and retardation times, the process is quite empirical and the characteristic times obtained are not necessarily related to any physical processes in the liquid. Furthermore, it can be shown that for a given material the retardation and relaxation times are not the same and are not even necessarily similar.

More elegantly, but with an equal lack of physical basis, the discrete spectra may be replaced by distributed spectra. Conventionally, the relaxation spectrum is described by a distribution function $H(\tau)$ and the retardation spectrum by $L(\tau)$, where, taking the latter for example, $L(\tau) \, d\tau$ defines the contribution to the compliance of retardation processes with times in the range τ to $\tau + d\tau$. An extensive analysis of the interrelations among the distribution functions, the complex compliance, and the complex modulus has been given by Gross.[9]

3.2.4. Data Reduction: Time–Temperature/Time–Pressure Superposition

As mentioned in the introduction, the Maxwell relaxation time $\tau_m = \eta/G_\infty = \eta J_\infty$ serves as a guide to the order of magnitude of the time or frequency scale involved in the viscoelastic relaxation process. The variation of J_∞ with temperature is approximately linear and usually small, typically doubling in a few tens of degrees Celsius, while the variation of viscosity is approximately exponential and may change by orders of magni-

[5] S. Whitehead, *J. Sci. Instrum.* **21**, 73–80 (1944).

[6] T. Alfrey, Jr., *Q. Appl. Math.* **3**, 143–150 (1945).

[7] E. A. Guillemin, "Communications Networks," Vol. 2, p. 211. Wiley, New York, 1947.

[8] R. Roscoe, *Brit. J. Appl. Phys.* **1**, 171–173 (1950).

[9] B. Gross, "Mathematical Structure of the Theories of Viscoelasticity." Hermann, Paris, 1953.

tude in a similar range. The variation of τ_m is thus closely linked to the variation of viscosity, and temperature can be used as a variable with which to explore the relaxation region and not simply as a means to bring the region within the frequency or time scale of a particular experimental system. Indeed, the use of temperature as a variable may be unavoidable, if the width of the relaxation region exceeds the limited range of the experimental system or systems available. For example, the range of most high-frequency techniques does not normally exceed one or two decades of frequency, while the relaxation region of comparatively simple organic liquids typically extends over about 2 to 3 decades and that for polymer solutions and melts may be up to 10 or 15 decades wide.[10]

The procedure of reducing data obtained over a temperature range to the equivalent results that would have been obtained at a single (or "reference") temperature over a wide time or frequency range is commonly referred to as the "method of reduced variables" or the use of "time–temperature superposition," "time–temperature reduction," or "thermorheological simplicity." For a Maxwell liquid, this is a reliable procedure, as only one relaxation time is present. Although, ideally, the variation of η and J_∞ with temperature should be known, it is possible to reduce data obtained at various temperatures by the empirical process of shifting the curves until a smooth master curve is obtained. A comprehensive summary of the method is given by Harrison[11] (pp. 56–61). However, in the usual case of results that can only be adequately described by a series of distribution of relaxation processes, the application of time–temperature superposition rests on the assumption that all the physical mechanisms involved have the same temperature dependence. Care must be taken to avoid erroneous results, especially if only a narrow frequency or time range is available experimentally (Ferry,[1] Chapter 11). Conversely, the clear failure of results to reduce to a single curve is a good indication that more than one type of physical process is occurring.

When the variations of η and G_∞ are known, the most convenient method of reducing data is to use normalized variables, each function being divided by the limiting value at the temperature used. For example, the shear modulus components are plotted as $G'(\omega)/G_\infty$ and $G''(\omega)/G_\infty$ against $\omega\tau_m$ ($= \omega\eta/G_\infty$) or $\log(\omega\tau_m)$ as appropriate. Choice of a reference temperature is then unnecessary.

The use of hydrostatic pressure as an alternative or additional variable

[10] A. V. Tobolsky, in "Rheology" (F. R. Eirich, ed.), Vol. II, pp. 63–81. Academic Press, New York, 1958.

[11] G. Harrison, "The Dynamic Properties of Supercooled Liquids." Academic Press, New York, 1976.

to temperature may be treated in the same way, and the foregoing comments apply equally. The pressure dependencies of both η and G_∞ should be determined, so that normalized variables may be used.

3.3. Experimental Techniques

3.3.1. General Considerations

Measurements of the limiting shear elasticity G_∞ are essentially those of $G'(\omega)$ in the region where $G'(\omega) \approx G_\infty$, i.e., at frequencies above and beyond the relaxation region. Even for a Maxwell liquid (with the minimum width of relaxation region), this approximation is only valid for angular frequencies above about $10\omega_m$, where $\omega_m = 1/\tau_m = G_\infty/\eta$. The practical lower limit of usable frequency is set by the glass transition temperature T_g for the liquid. Below T_g the liquid behaves as a glass; that is, the material is not in a state of thermodynamic equilibrium. The value of T_g depends on the time scale of the experiment. By convention, T_g is taken as the temperature at which $\eta \approx 10^{12}$ Pa sec since G_∞ is typically 10^9 Pa and the corresponding value of τ_m is approximately 10^3 sec, which is comparable with the duration of a normal experiment. At T_g or just above, G_∞ may therefore be determined by very low-frequency alternating stress techniques ($\omega \sim 1$ rad sec^{-1}) or by short time ($t \sim 1$ sec) transient methods. At lower temperatures, the time required to attain equilibrium becomes very large.[12] Above T_g, the viscosity decreases by orders of magnitude for a few degrees' rise in temperature. To establish the temperature dependence of G_∞, a minimum range of perhaps 20°C might be required. At $T_g + 20°C$, η may typically have fallen to 10^3 Pa sec, giving $\tau_m = 10^3/10^9$ sec or $f_m = \omega_m/2\pi = 1/2\pi\tau_m = 1.6 \times 10^5$ Hz. Measurements of G_∞ at $T_g + 20°C$ therefore require experimental techniques using shear waves in the megahertz frequency range. Naturally, the higher the frequency available, the greater is the temperature range over which G_∞ may be defined and the more reliable becomes any extrapolation of the variation to higher temperatures. Furthermore, the equivalence of $G'(\omega)$ and G_∞ is only confirmed if measurement of $G'(\omega)$ made at two distinct frequencies at a given temperature are in agreement: a frequency dependence of $G'(\omega)$ indicates that $G'(\omega)$ has not reached its limiting value. An example of the determination of G_∞ in this way is given in Chapter 3.4.

For these reasons, most studies of the viscoelastic behavior of supercooled liquids have been made at frequencies above 1 MHz, and descrip-

---

[12] A. Kovacs, *J. Polym. Sci.* **30**, 131–147 (1958).

tions of suitable experimental techniques are given in Chapter 3.3.2. Measurements at lower frequencies are useful as they extend the available frequency range thereby providing a better check on the validity of time–temperature reduction (Chapters 3.3.3 and 3.3.4). They are also useful in investigating the behavior of polymers and polymer solutions, where particular modes of motion may occur and have long relaxation times, even though the viscosity of the liquid may be comparatively low. Above 1 MHz, most experimental techniques are based on the use of piezoelectric transducers, following the pioneering work of Mason *et al.*[13] and McSkimin.[14] Although measurements have been made up to 3000 MHz,[15] the useful limit of piezoelectric techniques is around 1000 MHz.[16,17] It should be noted that since viscoelastic properties tend to vary logarithmically with frequency rather than linearly, small extensions of the frequency range are not of great utility, particularly if accompanied by loss of accuracy.

Recently, practical and theoretical developments in the study of light scattered from the naturally occurring orientational fluctuations of anisotropic molecules in liquids have complemented and extended the existing frequency range. In several liquids, the depolarized spectrum has been found to consist of a doublet with a spacing of about 1 GHz from the exciting frequency.[18–21] A response of this kind was predicted by Leontovich[22] and Rytov.[23] In principle, measurements yield the frequency of thermally generated shear waves in the liquid and their propagation velocity. Interpretation of the measurements in terms of the Maxwell model is at present limited to an evaluation of a relaxation time and a shear modulus. However, the values obtained are closer to the values for the retardation time and the retardation modulus,[24] and at the present time the light-scattering method cannot be regarded as an established means for the routine investigation of the shear behavior of liquids. It is hoped that

[13] W. P. Mason, W. O. Baker, H. J. McSkimin, and H. J. Heiss, *Phys. Rev.* **75**, 936–946 (1949).

[14] H. J. McSkimin, *J. Acoust. Soc. Am.* **24**, 355–365 (1952).

[15] J. Lamb and H. Seguin, *J. Acoust. Soc. Am.* **39**, 519–526 (1966).

[16] J. Lamb and J. Richter, *Electron. Lett.* **2**, 73–74 (1966).

[17] J. Lamb and J. Richter, *Proc. R. Soc. London Ser. A* **293**, 479–492 (1966).

[18] V. S. Starunov, E. V. Tiganov, and I. L. Fabelinskii, *JETP Lett.* **5**, 260–262 (1967).

[19] G. I. A. Stegeman and B. P. Stoicheff, *Phys. Rev. Lett.* **21**, 202–206 (1968).

[20] G. I. A. Stegeman and B. P. Stoicheff, *Phys. Rev. 3rd Ser. A* **7**, 1160–1177 (1973).

[21] G. D. Enright, G. I. A. Stegeman, and B. P. Stoicheff, *J. Phys. (Paris)* **33**, Cl, 207–213 (1972).

[22] M. A. Leontovich, *J. Phys. USSR* **4**, 499–514 (1941).

[23] S. M. Rytov, *Sov. Phys. JETP* **6**, 401–408, 513–523 (1958).

[24] A. J. Barlow, A. Erginsav, and J. Lamb, *Nature (London) Phys. Sci.* **237**, 87–88 (1972).

further progress may lead to the feasibility of investigating liquids that crystallize (at $\eta < 0.01$ Pa sec) rather than supercool.

3.3.2. Reflection of Plane Shear Waves

The shear impedance of a liquid may be determined from measurements of the complex reflection coefficient for shear waves at a solid–liquid interface. In general, reflection of shear waves incident to an impedance discontinuity gives rise to both longitudinal and shear waves, but by polarizing the shear wave so that the particle motion is parallel to the interface, the conversion to longitudinal is avoided (Thurston,[25] p. 79). Both reflected and refracted waves are then plane shear waves, the angles of incidence and reflection are equal, and the angle of refraction is given by Snell's law. At the interface, two conditions apply: (1) the particle velocity (or displacement) is continuous, and (2) the shear stress is continuous. Consider the simplest case when incident, reflected, and transmitted waves all propagate normal to the interface: using these conditions, the stress amplitude reflection coefficient R^* for normal incidence is given by

$$R^* = (Z_2 - Z_1)/(Z_2 + Z_1), \qquad (3.3.1)$$

where Z_1 and Z_2 are the shear impedances of the two media. In general, R^* is complex, but if the second medium is a vacuum (or air), $Z_2 = 0$ (or very small). If the first medium is a virtually lossless solid, Z_1 is real; R^* is then close to -1. If the second medium is a liquid, $Z_2 = Z_L = R_L + jX_L$ and both Z_2 and R^* are complex. Putting $R^* = R / (\pi - \theta) = -R \cos \theta + jR \sin \theta$.

$$Z_2 = Z_1(1 - R^2 + j2R \sin \theta)/(1 + R^2 + 2R \cos \theta). \qquad (3.3.2)$$

Except for measurements of high accuracy (i.e., errors $< \pm 1\%$) some simplification of Eq. (3.3.2) is possible.

Because G_∞ for a liquid is typically one or two orders of magnitude less than the shear elastic modulus of a solid, Z_L rarely exceeds $0.1Z_1$; therefore R is only a few percent less than -1 and θ is small (usually less than $3°$). It is then valid to assume $\cos \theta = 1$, and Eq. (3.3.2) becomes

$$Z_L = R_L + jX_L = Z_1 \left[\frac{1 - R}{1 + R} \right] + jZ_1 \frac{2R \sin \theta}{(1 + R)^2}. \qquad (3.3.3)$$

Knowing Z_1, one may then determine the value of R_L from measurements

[25] R. N. Thurston, in "Physical Acoustics" (W. P. Mason, ed.), Vol. 1A, pp. 1–110. Academic Press, New York, 1964.

of the magnitude of the reflection coefficient only. Furthermore, as the change in the phase shift θ is so small, its accurate measurement is difficult; this "normal incidence technique" is therefore usually restricted to measurement of R_L. However, McSkimin has shown that phase measurements can be made at frequencies up to 500 MHz using a normal incidence method,[26] and attempts to measure phase at 3000 MHz have been made by Lamb and Seguin.[15] The lack of X_L data is not necessarily a serious disadvantage, as $G'(\omega) = (R_L{}^2 - X_L{}^2)/\rho \cong R_L{}^2/\rho$ in any region where $X_L < 0.2R_L$ and particularly when $G'(\omega)$ approaches G_∞. When R_L and X_L are comparable in magnitude and measurements of both are necessary, the sensitivity of the reflection coefficient to the liquid impedance can be enhanced by directing the incident shear wave at an oblique angle to the interface rather than normally. O'Neill[27] has shown that the liquid impedance is then related to the components of the complex reflection coefficient by

$$Z_L = R_L + jX_L = \frac{\cos \phi}{\cos \psi} \frac{1 - R^2 + 2jR \sin \theta}{1 + R^2 + 2R \cos \theta}, \tag{3.3.4}$$

where ϕ is the angle of incidence and ψ the angle of the refracted shear wave in the liquid. For the usual range of values of Z_L ($Z_L < 0.1Z_1$), ψ is small and, with negligible loss of accuracy, $\cos \psi$ can be taken as unity. The angle ϕ is typically about 75 to 80°, $\cos \phi \cong 0.2$, and the sensitivity is thereby increased five times.

The preceding analysis assumes, for both normal and inclined incidence, that media 1 and 2 are each semi-infinite and that steady-state conditions apply. These conditions are valid for medium 2 if it is a liquid of finite thickness, (e.g., 1 mm), since any wave transmitted into the liquid is highly attenuated and the amplitude of any wave reflected back to the interface is negligible. Medium 1 is a low-loss solid of finite thickness, and a standing wave pattern would be established if continuous excitation were used, making the measurement of the reflection coefficient at the solid–liquid interface difficult. Pulse techniques are therefore employed in shear reflection measurements. The pulse durations are short enough for the successive reflections from the interface to be time separated and identifiable but long enough to contain a sufficient number of cycles so that a single-frequency wave is propagated, and, after the decay of initial transients, steady-state conditions are established.

3.3.2.1. Normal Incidence Method, 1–200 MHz. 3.3.2.1.1. ACOUSTIC SYSTEM. To generate plane shear waves in this frequency range, flat

[26] H. J. McSkimin, *J. Acoust. Soc. Am.* **47**, 163–167 (1970).
[27] H. T. O'Neill, *Phys. Rev.* **75**, 928–935 (1949).

BC- or BT-cut plates of crystal quartz are most commonly used.[28] Other materials such as lithium niobate have higher piezoelectric coupling coefficients but are more expensive and less readily available. The transducer is bonded to one end of a rod or bar, fused quartz being a convenient material for the bar because of its reasonably low-loss, mechanical and chemical stability and because it is easily cut and polished to optical limits of accuracy. The length of the bar is determined by the frequency range. For example, for operation at 10 MHz, steady-state conditions will be valid within a pulse of about 100 cycles or 10-μsec duration. The velocity of shear waves in fused quartz is 3.76×10^3 msec^{-1}, and the minimum length of bar necessary for the separation of transmitted and received pulses is therefore $\frac{1}{2} \times 3.76 \times 10^3 \times 10^{-5}$ m or about 2 cm. In practice, longer bars—typically 5 cm—are used at this frequency; the greater delay allows separation of the shear wave pulses and any spurious longitudinal waves that may be present, and the delay may also simplify the method of determining the amplitudes of the pulses. Bonding is a perennial problem: whereas a liquid bond is adequate for longitudinal waves, a solid or near-solid bond is necessary for shear waves. The bonding material must be of uniform thickness, thin compared to the wavelength, and have a high shear modulus. Good bonds with a wide working temperature range may be made by a cold-welding technique using indium, gold alloys, or other soft metals.[29] Many other materials have been used, including silicones, epoxy cements, grease, wax, phenyl salicylate, and supercooled liquids.[30]

A transducer may be operated at the fundamental resonant frequency and at its odd harmonics, and a given combination of transducer and rod can be used to cover at least a decade of frequency. The upper limit of the range tends to be limited by higher losses in the bond and in the fused quartz. For example, a 5-MHz transducer bonded to a bar 5 cm long may be used at 5, 15, etc., to perhaps 75 MHz. For higher frequencies, either the loss in the fused quartz may be reduced by using a shorter bar, necessitating shorter pulses, or low-loss material such as BT-cut crystal quartz may be employed. The tolerances on the mechanical dimensions of the bar are important but are not particularly severe in this frequency range. A plane wave front launched by the transducer into the bar must return, after reflection at the free surface or solid–liquid interface, as a plane wave front at the same angle; otherwise, variations in the time of arrival across the wave front result in partial cancellation of the received signal.

[28] W. P. Mason, "Piezoelectric Crystals and Their Application to Ultrasonics." Van Nostrand–Reinhold, Princeton, New Jersey, 1950.
[29] A. J. Barlow and S. Subramanian, Brit. J. Appl. Phys. 17, 1201–1214 (1966).
[30] A. J. Matheson, J. Phys. E. Sci. Instrum. 4, 796 (1971).

A difference in angle of 1 minute of arc between launched and returning waves results in a path difference of $\lambda/10$ across a transducer 1.5 cm wide at 75 MHz. For the normal incidence method, where the end faces are parallel, this requirement is easily met since it is not unduly difficult to manufacture bars with optically polished end faces flat to a tenth of a wavelength of light and parallel to a few seconds of arc. The cross-sectional dimensions of the bar and transducer are determined by the divergence of the beam of shear waves. Ideally, the total propagation distance (i.e., twice the length of the bar times the number of reflection echoes) should be within the Fresnel zone and, therefore, less than a^2/λ, where a is the radius of the transducer. In practice, this specification can be relaxed somewhat, and it is usual to make the minimum width or diameter of the active area of the transducer at least 20λ, i.e., 1.5 cm at 5 MHz. To alleviate the effects of beam spreading, the bar is made somewhat wider than the transducer. In order to disperse reflections, the sides are left roughly ground rather than made smooth and polished. Several acoustic systems are described in the literature,[31-33] and examples are shown in Figs. 4a and b. As the transducer is excited by an electric field across the thickness of the plate, electrodes on each side of the plate are necessary. One electrode may be the bond itself if it is metallic or conducting; if not, then a metallic coating is evaporated on to the end of the bar before bonding. For good adhesion, chromium or nichrome are suitable, followed by gold or aluminium for good conduction. The second electrode on the surface of the transducer should also be an evaporated metallic film, but a less-permanent electrode may be formed by conducting paint or even a thin metal foil.

    3.3.2.1.2. ELECTRICAL SYSTEM. A schematic diagram of the simplest electrical system is given in Fig. 4c. The transmitter should be capable of providing a peak-to-peak voltage across the transducer electrodes of a few tens to a few hundred volts, the higher level being necessary at the upper end of the frequency range to compensate for increased acoustic losses. Either vacuum tubes or transistors may be used in a push-pull configuration, with simple coupling networks to match the transmitter output to the predominantly capacitive load presented by the transducer. The transmitter may be a self-excited oscillator or preferably, for better frequency stability, consist of an output stage driven by a separately pulsed oscillator. A crystal-controlled frequency source is not normally required when, as in this case, phase measurements are not attempted.

[31] H. J. McSkimin and P. Andreatch, Jr., *J. Acoust. Soc. Am.* **42,** 248–252 (1967).
[32] H. J. McSkimin and P. Andreatch, Jr., *J. Acoust. Soc. Am.* **41,** 1052–1057 (1967).
[33] A. J. Barlow, G. Harrison, J. Richter, H. Seguin, and J. Lamb, *Lab. Pract.* **10,** 786–801 (1961).

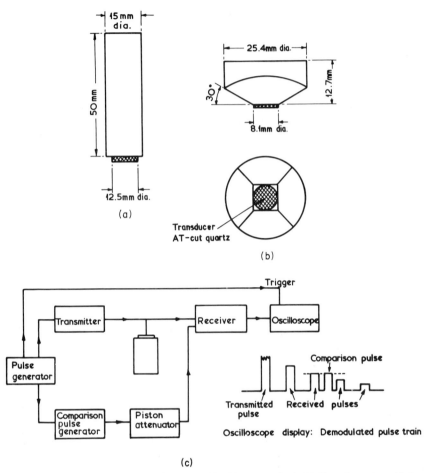

FIG. 4. Two acoustic systems for normal incidence method: (a) Fused quartz cylinder, with BT-cut quartz crystal; (b) AT-cut quartz, with AT-cut quartz crystal. Schematic diagram of normal incidence system is shown in part (c).

The output stage must be operated in Class C, and not Class A, otherwise the quiescent current would generate excessive noise at the receiver input.

The maximum pulse repetition frequency (prf) is set by the time required for the train of successive echoes in the acoustic range to decay to the noise level. For example, using a bar 5 cm long, having a round-trip transit time of 26 μsec, perhaps 100 received echoes may have significant amplitudes, and the maximum prf is then $1/(26 \ \mu\text{sec} \times 100)$ or about 385 Hz. (Although useful measurements may only be feasible of the first ten of these when the bar is loaded by a liquid, accurate readings of these pulses in the initial unloaded state is necessary.) There are advantages in using a prf equal to or locked to the supply frequency of 50 or 60 Hz, mainly as problems arising from hum pickup in the receiver are thereby alleviated.

The comparison pulse generator provides a pulse at the same frequency and repetition rate as produced by the transmitter but at only a few volts or less amplitude. It is also delayed by a variable amount, so that the comparison pulse can be placed adjacent to a received echo on the oscilloscope display. (An alternative technique is to pulse the transmitter and comparison pulse generator alternately, so that the comparison pulse and the received echo train appear on the display to be superimposed.) Amplitude measurements are made by adjusting the attenuator setting to equalize the heights of the comparison pulse and a particular received echo. From the differences in the settings for a number of echoes, taken before and after application of a liquid to the bar, the average attenuation per reflection is readily determined. An accurate and continuously adjustable attenuator is necessary: stepwise attenuators are generally inconvenient and of insufficient precision, and a piston attenuator is almost invariably used. With skill and a clean pulse train, amplitude equalization to ± 0.03 dB or better is possible. Direct radiation from the comparison pulse generator to the receiver input must be avoided, since it would invalidate the amplitude measurements. The received signals from the transducer and the output from the attenuator are combined and amplified in the receiver before being displayed on the oscilloscope. Demodulation of the receiver output is unnecessary if the frequency response of the oscilloscope is sufficient to allow direct display of the rf pulses. A receiver bandwidth of about 1 MHz is required, compatible with pulse rise and fall times of about 1 μsec.

A possible error of ± 0.03 dB in the measurement of the average loss per reflection at the interface due to the application of a liquid corresponds to a possible error of about $\pm 1.5 \times 10^4$ N sec m^{-3} in the determination of R_L. For comparison, the maximum value of R_L (i.e., when $R_L^2/\rho = G_\infty$) is typically 10^6 N sec m^{-3}, so that while the error in R_L may be only $\pm 1.5\%$ near this maximum, the percentage error becomes unacceptably large for values of R_L that are one-tenth or less of the maximum. In this range higher accuracy can be achieved be designing both acoustic and electronic parts of the system for optimum performance at a single frequency and by using the inclined incidence reflection method.[34]

3.3.2.2. Inclined Incidence Method: 5–100 MHz. 3.3.2.2.1.

[34] A. J. Barlow and A. Erginsav, *Proc. R. Soc. London Ser. A* **327**, 175–190 (1972).

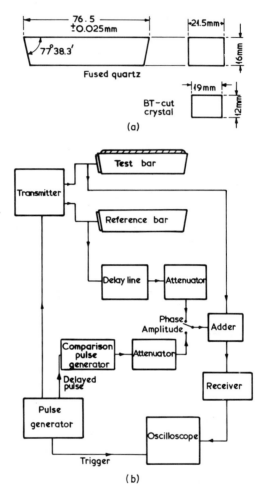

FIG. 5 (a) Acoustic system for inclined incidence technique; (b) schematic diagram of the electrical system for inclined incidence system (after Barlow and Lamb[35]).

ACOUSTIC SYSTEM. As noted earlier, use of the more sensitive inclined incidence method is preferable when X_L is to be determined. The measurement of the phase change at the solid–liquid interface presents both acoustic and electronic difficulties. Figure 5a gives the typical form and dimensions of the acoustic system. The requirement that the wave fronts propagated from and returned to the transducer should be coplanar still applies. However, it is difficult to manufacture bars with

[35] A. J. Barlow and J. Lamb, *Proc. R. Soc. London Ser. A* **253**, 52–69 (1959).

the angle of the end faces defined to significantly better than 1 min and this sets the upper-frequency limit of the method to about 80 or 100 MHz. At higher frequencies recourse has to be made to less-sensitive normal-incidence methods. Temperature stabilization requirements are severe. For a bar in which the distance of a single round trip is 15 cm (Fig. 5a), the first received pulse differs in phase from the transmitted pulse by about 3000 wavelengths at 75 MHz. The velocity of shear waves in fused quartz increases by about 100 ppm/°C temperature rise. For changes in velocity to cause phase changes of less than the errors in phase measurement, i.e., about $\pm 0.3°$, the temperature of the bar and of the liquid to be applied to it must be stabilized to better than $\pm 0.003°C$. Some advantage may accrue from the use of two identical bars, one to which the liquid is applied and the other to provide a phase reference, so that the effects of common temperature changes cancel, but at the cost and difficulty of providing two matched bars, and errors may still arise from differential temperature changes.

3.3.2.2.2. ELECTRICAL SYSTEM. Figure 5b gives a schematic diagram of the electrical system used with two bars. The specifications of the receiver and the components for pulse amplitude measurements are similar to those for the normal incidence method. The transmitter frequency should be stable to 1 part in 10^6 or better during the time required to make a measurement, and a crystal controlled oscillator is necessary. If the drive to a transmitter output stages is obtained by gating the output of a cw oscillator at the required frequency, the gating circuits must have a minimum rejection ratio of 120 dB to prevent residual cw signals interfering with the received signals from the acoustic system. Alternatively, the cw oscillator may be operated at a different frequency and pulsed harmonic multipliers[36] or dividers used to derive the required frequency. For phase measurement, the transducers on each bar are excited simultaneously and the received pulse trains are combined. An attenuator and a variable delay line in the reference channel allows selected pulses from each bar to be made equal in amplitude and opposite in phase, so that cancellation occurs. On applying a liquid to the reflecting surface of the test bar, the resulting phase change is matched by a change in the delay line setting so that cancellation is restored. Because the amplitude of the pulse is reduced, a change in the attenuator setting is also necessary. It is essential that this change not affect the total phase shift in the reference channel; a piston attenuator is therefore used because there is no phase shift along a waveguide operating in an evanescent mode well below cutoff frequency. The change in attenuator setting does not provide a reliable determina-

[36] H. J. McSkimin, J. Acoust. Soc. Am. 34, 404–409 (1962).

tion of the amplitude change caused by the liquid, since small amplitude changes are produced by changes in the variable delay line. For this reason the use of a separate comparison pulse for amplitude measurements is retained. The line is formed from lengths of precision coaxial cable, together with a short continuously variable line. Typically, phase measurements are made with a possible error of $\pm 0.3°$, giving measurements of X_L to within about $\pm 5 \times 10^3$ N sec m^{-3}. A similar uncertainty in R_L arises from a possible error of ± 0.03 dB in the amplitude measurements.

The accuracy can be improved by carefully designing each component of the system for optimum operation at only one frequency and by ensuring adequate thermal and electrical stability. Barlow and Erginsav[34] used a system designed for optimum operation at 30 MHz and showed that the errors in the measured values of R_L and X_L could be reduced to about ± 500 N sec m^{-3} $\pm$ 0.5% for values in the range 10^4–10^5 N sec m^{-3}.

When only a single bar is used, the phase reference signal must be derived from the same crystal-controlled cw source as used for the transmitter, either by direct gating or by using pulsed multipliers or dividers. It is convenient to obtain the amplitude-comparison pulse in this way and to include the variable delay line in this channel, so that the same pulse is used for both amplitude and phase measurements. Amplitude measurements may then be made in the previous manner, with comparison and received pulses adjacent, while phase measurements are made by making the pulses coincident. Details of the method are given by Barlow and Subramanian.[29]

It is common practice to use a single transducer to transmit and receive the pulse of shear waves in a given bar, and consequently there is a direct connection between the transmitter output and the receiver input. The receiver must therefore be designed to withstand the severe overloading on transmission and to recover rapidly so that it is sensitive to the received echoes. With suitable design a recovery time of 1 μsec is readily obtained; the electronic problems involved are less than those that arise from the assembly of an acoustic system with separate transducers for transmission and reception.

3.3.2.3. Normal Incidence Methods: 40–500 MHz. The simple electrical system described in the previous section for use with the inclined-incidence method can also be used for amplitude and phase measurements with a normal-incidence technique: the acoustic system is simpler but the sensitivity and accuracy are lower. Other electrical systems have been devised to allow phase measurements of reasonable accuracy to be made using a normal-incidence acoustic system and to extend the frequency range beyond the useful limit of the inclined-incidence method. For example, instead of measuring the phase change directly, the time delay caused by a liquid at the reflecting interface may be determined. The change in the duration of a round trip for the pulse in the bar is small, and a measurement of the time to 1 part in 10^7 or better is necessary. This method avoids the need for a variable-phase line, but requires a very stable pulse repetition frequency and a precise measurement of this frequency.[31,32] The train of echoes is not allowed to decay to zero between each transmitted pulse. The transmitter pulse occurs at intervals approximately equal to the round-trip time, and the received pulses therefore interfere. The delay T for additive interference between the high-frequency content of selected pulses in successive trains is given by McSkimin[37] as

$$T = p\delta - (p\gamma_0/2\pi f) + nf, \qquad (3.3.5)$$

where p is the number of round trips between the superposed echoes, δ the delay due to the bar alone, γ_0 the phase angle associated with reflection at the ends of the bar, and f the (rf) wave frequency. The integer n is the difference in the cycle number in the superposed pulses and should be made zero. McSkimin[37] and Papadakis[38] discuss the detailed requirements for correct operation. The pulse repetition rate is adjusted so that the superposed odd-number echoes have the same phase, and the resultant amplitude is then a maximum. When a liquid is applied, γ_0 is modified to $\gamma_0 - \theta$, and the repetition frequency for a maximum changes from f_1 to f_2. McSkimin[37] has shown that the phase shift θ is given by

$$\theta = 2\pi f(f_1 - f_2)/pf_1f_2, \qquad (3.3.6)$$

where $p = 2$ for this particular mode of operation. By this method values of R_L and X_L, in the range 10^4–10^5 N sec m^{-3}, have been determined to within $\pm 5\%$ or better.

In an ingenious variation of this technique, McSkimin[26] uses a sequence of two transmitter pulses, the interval between them being chosen so that a selected echo from the first received pulse train is superposed on the first echo of the second train. The interval is kept fixed, and in this interval the frequency of the cw source, from which both transmitter pulses are derived, is temporarily shifted by a small amount. This shift results in a variation of the relative phase between the two superposed received pulses and can be adjusted so that cancellation occurs. Phase measurements with a resolution of 1° have been made at 500 MHz by this method, giving values of X_L to within $\pm 5\%$ at about 10^5 N sec m^{-3}.

[37] H. J. McSkimin, *J. Acoust. Soc. Am.* **33**, 12–16 (1961).
[38] E. P. Papadakis, *J. Acoust. Soc. Am.* **42**, 1045–1051 (1967).

3.3.2.4. Normal Incidence Method: Above 200 MHz. Acoustic systems using bonded transducers operating at high harmonics have been used up to at least 1 GHz. However, the losses in the propagating medium and the bond become increasingly serious above about 200 MHz, and in this region it is difficult to ensure an adequate signal-to-noise ratio at the receiver input. For measurements of useful accuracy, this ratio should be at least 40 dB. The practical limit of the voltage between the transducer electrodes is of the order of a few kilovolts; when the acoustic losses are high, the received-pulse amplitudes may still be insufficient even with such high transmitter voltages. A more efficient acoustic system can be made by eliminating the bond. A thin-film piezoelectric transducer is evaporated or sputtered directly on to the end surface of the bar. A single-crystal film has to be formed, correctly oriented to ensure the generation of shear waves. The delay-rod material need not be piezoelectric, but chosen for low-loss properties. For example, cadmium sulphide, lithium niobate, and zinc oxide transducers have been formed on sapphire, fused and crystal quartz rods and operated at frequencies up to 3 GHz.[39-43] However, a simpler method is to use a bar of piezoelectric material and to generate a shear wave directly at the end surface.[44] The end of the bar is subjected to a high electric field, either by inserting it into a resonant cavity or by the use of specially arranged and phased electrodes, A detailed analysis of this method, using crystal quartz for the bar, has been given by Lamb and Richter.[16,17,45] The orientation of the bar and the direction of the electric field relative to the surface must be carefully chosen if a pure shear mode is to be propagated. Using a BC-cut quartz rod 10 mm in diameter and 15 mm long, about 50 echoes may be observed at 500 MHz. The method of determining the change in pulse amplitudes is the same as that used at lower frequencies, and measurements of R_L have been made in the frequency range 300–200 MHz. The values obtained are estimated to be accurate to within ± 5000 N sec m^{-3}.

As in any receiving system in which repetitive signals occur, various correlation techniques can be used to improve the signal-to-noise ratio. The effect of random noise is reduced by averaging the received pulses

[39] N. F. Foster, G. A. Coquin, G. A. Rozgonyi, and F. A. Vannatta, *IEEE Trans.* **SU-15**, 28–41 (1968).

[40] E. K. Sittig and H. D. Cook, *Proc. IEEE* **56**, 1375–1376 (1968).

[41] N. F. Foster, *J. Appl. Phys.* **40**, 4202–4204 (1969).

[42] W. Duncan, R. H. Hutchins, and P. A. M. Stewart, *J. Vac. Sci. Tech.* **6**, 555–558 (1969).

[43] A. H. Meitzler and E. K. Sittig, *J. Appl. Phys.* **40**, 4341–4352 (1969).

[44] H. E. Bömmel and K. Dransfeld, *Phys. Rev.* **117**, 1245–1252 (1960).

[45] J. Lamb and J. Richter, *J. Acoust. Soc. Am.* **41**, 1043–1051 (1967).

over many echo trains. Improvements of up to about 20 dB are possible without undue complications. Correlation methods may be considered as an alternative or addition to increasing transmitter output, particularly at high frequencies or when the losses in the acoustic system become appreciable.[46-49]

3.3.3. Guided Traveling Waves

The techniques described in the previous section use a free-space wave, where the area of the wave front is determined by the size of the transducer and is independent of the size of the delay rod. The minimum diameter of the rod is controlled by the spreading of the beam in the Fraunhoffer zone and will be inversely proportional to the frequency. Thus the use of free-space-wave techniques at frequencies much below 5 MHz requires delay bars of an inconveniently large size. The techniques described in this section use guided waves in which the boundary of the wave is the surface of the delay medium. This can take the form of a cylindrical rod or tube which will support the propagation of torsional waves or a thin plate in which plane shear waves can be propagated. In both cases the cross section of the waveguide is chosen so that only a single fundamental mode is propagated.

3.3.3.1. Plane Shear Waves: 0.2–2 MHz. A technique using shear waves in a metal-strip delay line has been described by Hunston et al.[50,51] A nondispersive shear wave will propagate in a thin plate provided the thickness of the plate is less than a half-wavelength.[52] In aluminum, the wavelength at a frequency of 2 MHz is 1.6 mm; thus at this frequency the strip must be less than 0.8 mm thick. At 0.2 MHz the corresponding maximum thickness is 8 mm. Under these conditions the displacement is uniform over the thickness of the strip and is at right angles to the direction of propagation (Fig. 6a).

A pulse technique is used, the schematic diagram of which is shown in Fig. 6b. A pulse of shear waves generated by the transducer travels down the strip, is reflected at the end, and returns to the transducer. The resulting train of echoes from successive reflections is displayed on an os-

[46] B. R. Tittman and H. E. Bömmel, *Rev. Sci. Instrum.* **38**, 1491–1496 (1967).

[47] J. H. Simmons and P. B. Macedo, *J. Acoust. Soc. Am.* **43**, 1295–1301 (1968).

[48] H. J. McSkimin and T. B. Bateman, *J. Acoust. Soc. Am.* **45**, 852–858 (1969).

[49] R. B. Hemphill, *Rev. Sci. Instrum.* **40**, 175–176 (1969).

[50] D. L. Hunston, R. R. Myers, and M. B. Palmer, *Trans. Soc. Rheol.* **16**, 33–44 (1972).

[51] D. L. Hunston, C. J. Knauss, M. B. Palmer, and R. R. Myers, *Trans. Soc. Rheol.* **16**, 45–57 (1972).

[52] J. E. May, Jr., *in* "Physical Acoustics," (W. P. Mason, ed.), Vol. 1A, pp. 417–483. Academic Press, New York, 1964.

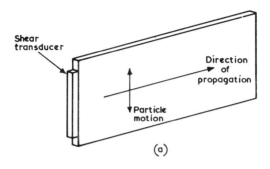

(a)

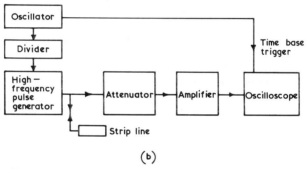

(b)

FIG. 6. (a) Shear waves in strip delay line; (b) schematic diagram of measuring system. (After Hunston et al.[51]).

cilloscope in the usual way, the train of echoes being allowed to decay to zero amplitude before the next pulse is applied. The shear impedance of a liquid is measured by determining the change in the propagation constant of the shear wave when the strip is immersed in the liquid. The impedance is given by the equation

$$Z_{\mathrm{L}} = (\rho c t / 2 l)(\Delta A + j\,\Delta B), \qquad (3.3.7)$$

where ρ is the density of the strip material, c the velocity of shear waves in the strip, t and l the thickness and length of the strip, respectively, ΔA the measured change in the amplitude of the received signal per reflection, and ΔB the measured change in the phase angle. Measurements of these changes in amplitude and phase of the echoes are made using the method of Papadakis.[38] The oscilloscope sweep is triggered at a frequency f_1, which is an integer multiple of the reciprocal of the transit time within the strip, so that a pair of echoes is visually superposed on the oscilloscope display. These echoes are intensified and others blanked out so that only the two chosen echoes are visible. The new repetition frequency f_2 necessary to maintain superposition of the echoes when the strip is immersed in the liquid is used in Eq. (3.3.6) to determine ΔB. The amplitude change ΔA is measured by noting the change in the setting of a series attenuator required to maintain an echo at the same amplitude on the oscilloscope display.

The width of the transducer, and hence, the width of the strip, is determined by the effects of beam spreading and the need to avoid reflections of the wave at the edge of the strip. Such reflections generate longitudinal waves, which lead to spurious echoes. Satisfactory operation was obtained provided the width was more than 2 cm. A delay line of between 5 and 10 cm in length was found to be adequate to obtain time separation of the echoes and to enable steady-state conditions to be reached during the pulse. Hunston *et al.* claim that a delay time of the order of 50 μsec can be measured to an accuracy of 3 nsec corresponding to a change in phase angle of $\pm 2°$. The corresponding accuracy in X_L is approximately ± 1000 N sec m^{-3}. A similar accuracy can be achieved in the measurement of R_L.

3.3.3.2. Torsional Waves: 10–500 kHz.

A technique using torsional waves generated by a quartz crystal and traveling in a metal rod was first proposed by McSkimin.[14] The transducer is a cylinder of crystal quartz which is cut with the crystallographic x axis parallel to the axis of the cylinder[28] (Fig. 7a). Four electrodes on, or near to, the cylindrical surface provide a field in the direction of the y axis. By arranging that the fields on opposite sides of the y axis are of opposite sign, the two resulting shears in the $x-y$ plane cause the crystal to twist about the axis of the cylinder. At the fundamental resonance, the transducer length is equal to a half-wavelength with a node in the center. The resonant frequency is inversely proportional to the length, and a frequency range from 20 to 100 kHz is covered by lengths in the range 10–2 cm. To minimize the change in length of the crystal as it twists and to ensure a pure torsional motion, the diameter of the crystal should be small compared with its length: a ratio of 1:5 has been found to be satisfactory.

The crystal is rigidly attached to the end of a long metal rod having the same diameter as the crystal. Epoxy resin and cyanoacrylate cement have been found to be suitable bonding materials. The measuring system used with this technique is shown in Fig. 7b. A high-stability oscillator or frequency synthesizer is gated to provide pulses containing several cycles of oscillation which are applied to the crystal via a drive amplifier. The gate is a conventional six-diode balanced gate driven from a pulse genera-

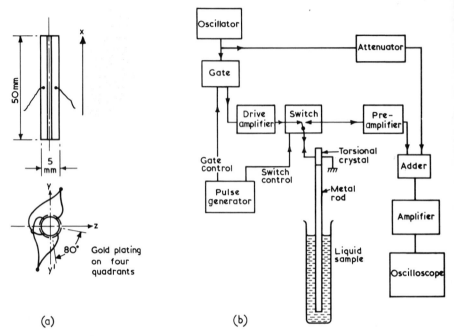

FIG. 7. (a) The 38-kHz torsional quartz crystal; (b) schematic diagram of measuring system for traveling torsional wave technique.

tor. In the equipment currently in use in the authors' laboratory, the torsional crystal is connected to either the drive amplifier or the receiving system by a pair of reed switches acting as a change-over switch. The pulse of torsional waves generated in the rod by the transducer travels the length of the rod, is reflected at the free end, and returns to the transducer. During this time interval the reed switch connects the crystal to the preamplifier, adder, and amplifier, so that the received echoes are displayed on the oscilloscope. The torsional pulse is allowed to travel up and down the rod, generating a train of echoes from successive reflections. The next pulse of torsional waves is generated when the amplitude of the echoes is negligible. This typically results in a pulse repetition rate of between 20 and 50 Hz.

For any selected echo, an amplitude and phase reference is established by comparing the received echo against an attenuated continuous wave from the oscillator. The phase delay, in wavelengths, of the rf signal in the received pulse, relative to the oscillator signal, is equal to the product of the transit time of the echo in the rod and the oscillator frequency. By suitable adjustment of the oscillator frequency and the attenuator setting, the two signals can be made equal in amplitude and opposite in phase, thus giving cancellation for the duration of the pulse. When a liquid is introduced around the rod, the attenuation and velocity of the torsional wave in the rod are changed, and changes in the attenuator and frequency settings are necessary to restore the pulse cancellation. Measurements are made using several pulses, and the average values of the changes in attenuation and frequency are determined. The frequency interval f_c between two successive cancellations in the unloaded rod is determined: this corresponds to a phase change of one wavelength or 2π rad. The phase change introduced by the liquid is then determined from the measured frequency change Δf using the equation

$$\Delta B = 2\pi \, \Delta f/f_c \quad \text{rad.} \tag{3.3.8}$$

The shear impedance Z_c of the liquid is calculated from the measured change in amplitude per reflection ΔA and phase change per reflection ΔB, using the equation[14]

$$Z_c = (\rho cr/2l)(\Delta A + j \, \Delta B). \tag{3.3.9}$$

In this equation, ρ is the density of the rod material, r the radius of the rod, c the velocity of propagation of the torsional wave in the unloaded rod, l the length of rod immersed in the liquid, and Z_c is the impedance of the cylindrical shear waves generated in the liquid at the surface of the rod, from which the plane wave impedance Z_L may be calculated. The two impedances differ significantly only at high values of impedance ($Z_L > 10^4$ N sec m^{-3}).

The length of rod required is determined largely by the need to establish steady conditions during the pulse and to separate the successive reflections in time. The rod material must be resistant to corrosion and be capable of receiving good surface finish, so that surface irregularities are small compared with the wavelength of the shear wave in the liquid. Glass, aluminum, stainless steel, and nickel silver have been used by different workers. The velocity of torsional waves in nickel silver, 2×10^3 m sec^{-1}, is lower than for many other materials, so giving increased sensitivity. A pulse containing 10 cycles at a frequency of 50 kHz would then require a rod at least 40 cm long to ensure time separation of the echoes. The comparatively small number of wavelengths in the rod results in the changes in phase delay caused by the variation of the velocity with temperature being small. A double-wall jacket containing a circulated thermostating liquid controlled to $\pm 0.01°C$ is found to be adequate.

The analysis used in obtaining Eq. (3.3.9) assumes that no energy loss occurs when the torsional wave is reflected at the end of the loaded rod and neglects the effects of reflection of torsional waves at the surface of the liquid. These simplifications lead to negligible error when a substantial part of the rod is immersed in the liquid, but significant errors can be introduced when short lengths of the rod are immersed in liquids of high shear impedance.

When results are averaged over several pulses, it is possible to measure the amplitude change per reflection to an accuracy of ± 0.05 dB and the frequency change to an accuracy of ± 1 Hz. The values of ΔA and ΔB are then accurate to ± 0.006 Np and ± 0.006 rad, respectively, leading to errors in each of the components of Z_L of approximately ± 500 N sec m^{-3} for a typical rod. A modification of this technique has been described by Wada and his co-workers.[53,54] The major feature of their system, which differs from that described above, is that the electrodes for the quartz transducer are not plated on to the quartz but are arranged coaxially around the transducer with a 0.5-mm air gap. This enables the effective length of the electrodes to be varied and leads to an increased conversion efficiency at higher frequencies, up to 500 kHz. Details of the circuits used by these workers are given in the above references.

Torsional traveling waves in a rod can be generated by a magnetostrictive transducer using the Wiedemann effect.[55] A nickel rod or tube is magnetized circumferentially by passing a large current along it for a short time. The transducer consists of a small coil wound round the rod. Current passing through the coil produces a magnetic field along the axis of the rod. This field interacts with the magnetized rod and produces a torsional motion. A pulse of torsional waves can be generated by supplying the coil with a pulse containing several cycles of oscillation of the required frequency. Glover et al.[56] have described a technique using this type of transducer with a hollow nickel rod.

A schematic diagram of the technique is shown in Fig. 8. The transmitter coil is placed over one end of a long tube and a receiver coil placed a short distance below it. The generated pulse of torsional waves travels down the rod and is detected as it passes through the receiving coil. After reflection at the end of the rod, a second signal is received as it passes through the receiving coil again. Further reflections are prevented by coating the upper end of the rod with a high-viscosity liquid or grease; this also absorbs the generated signal which travels up the rod. The amplitude and time delay of the echo which has traveled down the rod and back are determined by transmitting a second pulse which is timed to ar-

[53] H. Nakajima and Y. Wada, *Polym. J.* **1**, 727–735 (1970).
[54] H. Nakajima, H. Okamoto, and Y. Wada, *Polym. J.* **5**, 268–277 (1973).
[55] J. F. W. Bell, B. P. Doyle, and B. S. Smith, *J. Sci. Instrum.* **43**, 28–31 (1966).
[56] G. M. Glover, G. Hall, A. J. Matheson and J. L. Stretton, *J. Phys. E. Sci. Instrum.* **1**, 383–388 (1968).

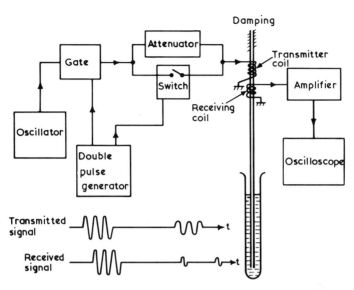

FIG. 8. Traveling torsional wave system using magnetostrictive transducers. (After Glover et al.[56] Copyright the Institute of Physics.)

rive at the receiving coil at the same time. The signal frequency and the amplitude of this pulse are adjusted to make the two signals equal in amplitude and opposite in phase so that the returning echo is cancelled at the receiving transducer. A portion of the tube is then immersed in the test liquid, and the changes in amplitude and frequency necessary to maintain the cancellation of the returning echo are noted. The shear impedance of the liquid is then calculated in the same manner as described above for the system using the piezoelectric transducer, except that Eq. (3.3.9) for a solid rod is modified to give

$$Z_L = (\rho cr/2l)(\Delta A + j\ \Delta B)[(1 - m^4)/(1 + m^3)], \qquad (3.3.10)$$

where m is the ratio of the inner to the outer radii. For a tube of outside diameter 0.125 in. and wall thickness 0.004 in., $m = 0.93$ and the quantity $(1 - m^4)/(1 + m^3)$ has a value of 0.128; thus the sensitivity of the system is increased about eight times. Glover et al. claim an accuracy of ± 150 N sec m^{-3} for their system over a frequency range from 20 to 100 kHz.

A further advantage over the solid rod is that the liquid wets both inner and outer surfaces of the tube, and the plane wave impedance Z_L is determined directly, so that the correction required to obtain Z_L from the cylindrical wave impedance Z_C is avoided. Also, reflection from the air–liquid interface, which is significant with liquids of high shear impedance, can be allowed for by arranging the geometry so that any interfering pulses do not overlap the measuring pulses.

Continuous frequency coverage is possible as the transducer is nonresonant. The upper frequency limit is set by the reduced efficiency of the transducer as the length of the coil becomes comparable to the acoustic wavelength in the rod. With transducers 3 mm in length, an upper limit of 500 kHz is suggested, provided that the electronic system for switching in the extra attenuation for the second transmitter pulse is rapid enough. In all the techniques described in this section, it is essential that the transmitting and receiving systems are broadband, so that the changes in frequency necessary to maintain pulse cancellations are not accompanied by unwanted phase shifts.

3.3.4. Resonance Techniques

This group of techniques involves the measurement of the loading effect on a resonant system when it is immersed in a liquid. A resonant system can be characterized by the frequency of resonance and the width of the resonance curve. The latter property, which depends on the losses in the system, can be expressed in terms of the quality or Q factor, the decay rate of the amplitude of free vibrations, or the effective electrical resistance of the transducer at resonance.

When a transducer is immersed in a liquid, the mechanical motion generates a shear wave in the liquid. The reaction on the transducer results in an increase in the effective inertia, which produces a drop in the resonant frequency, and an increase in the loss which reduces the Q factor and increases the bandwidth of the resonant curve, the decay rate of the oscillations, and the resistance at resonance. Measurement of the changes in the resonant frequency and one of the properties associated with the loss enables the two components of the shear mechanical impedance to be determined.

3.3.4.1. Piezoelectric Techniques: 1–100 kHz. The use of a quartz crystal vibrating in torsion for the measurement of liquid properties was first described by Mason.[57] The cylindrical crystal has been described earlier (Section 3.3.3.2). In this application it is supported at the central nodal region, either by the four wires providing the contact to the electrodes or by four needle points which both support the crystal and make electrical contact with the electrodes. The equivalent circuit of the crystal can be represented as shown in Fig. 9a, where the values of the circuit elements depend on the mechanical and piezoelectric properties of the crystal but are independent of frequency. When the electrical proper-

[57] W. P. Mason, *Trans. Am. Soc. Mech. Eng.* **69**, 359–367 (1947).

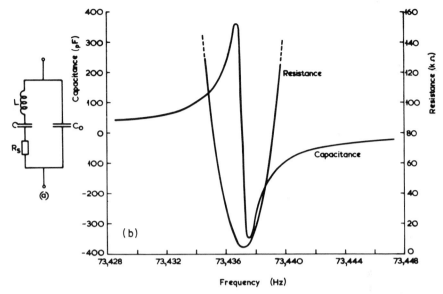

FIG. 9. (a) Equivalent circuit of torsional crystal resonator; (b) variation of parallel eqivalent resistance and capacitance of torsional crystal in vacuum with frequency in the vicinity of resonance.

ties are measured using an alternating-current bridge, it is usual to represent the impedance of the crystal in terms of a parallel combination of resistance (or conductance) and capacitance. These values of resistance and capacitance show a characteristic variation with frequency which is shown in Fig. 9b. The parallel resistance has a minimum value at resonance, equal to R_S, when the parallel capacitance is equal to C_0 which includes stray capacitance associated with the crystal mount. The resonant frequency and the resistance at resonance can be determined by obtaining a bridge balance with the capacitance reading equal to C_0, where the value of C_0 is measured at a frequency remote from the resonant frequency. This technique is simple to carry out unless the bridge has "zero" or calibration settings which vary with frequency. In this situation a rather more complicated measurement procedure has to be employed.[57a] The use of a three-terminal bridge network eliminates the large capacitance of the connecting leads from the measurement and leads to greater accuracy.

Measurements are made of the resonant frequency and the resistance at resonance, initially with the crystal in vacuum and then immersed in the

[57a] G. Harrison, Ph. D. Thesis, Univ. of London (1964).

168   3. DYNAMIC VISCOSITY MEASUREMENT

liquid. The changes in the resonant frequency Δf and the resistance at resonance ΔR are related to the components R_L and X_L of the shear modulus by the equations

$$\Delta R = K_1 R_L, \qquad \Delta f = K_2 X_L. \qquad (3.3.11)$$

The quantities K_1 and K_2 are constants for a particular transducer, except at viscosities of less than 2×10^{-3} Pa sec.[33,58] They can be determined either from measurements on liquids of known viscosity and density or calculated from the dimensions and electrical characteristics of the crystal.

A length-to-diameter ratio of 5 to 1 or greater is desirable in order to ensure a pure torsional mode of vibration. Under these conditions measurements to an accuracy of a few percent are readily obtained on liquids having viscosities of less than about 1 Pa sec. The range of the technique has been extended to about 20 Pa sec by Philippoff[59] using a crystal of length-to-diameter ratio of only 3 to 1 but with some loss of accuracy.

The application of this technique is limited to liquids of low conductivity. Any conduction in the liquid between the electrodes appears as a resistance in parallel with the effective resistance of the crystal. While some allowance can be made for this resistance, considerable reduction in resolution and accuracy occurs especially with liquids of high shear impedance when the resistance of the crystal at resonance may be well above 10^6 Ω.

This difficulty can be overcome by using a more involved technique developed by Robinson and Smedley.[60] They use a composite resonator comprising two matched torsional quartz crystals and a glass rod which has a length of $3\lambda/2$ at the crystal resonance (Fig. 10). The three sections of the resonator are assembled using a cyanoacrylate adhesive. One crystal is used as a driver, the other as a gauge to monitor the amplitude of the motion. The liquid covers only a length $\lambda/4$ at the free end of the glass probe so that the liquid–air interface occurs at a displacement node. A rubber O-ring, or a permanent glass seal, can be made at this point without modifying the resonance of the system. The technique can therefore be used with conducting liquids or for measurements at high temperatures, as only the end of the glass rod is in contact with the test liquid. The measurements consist of determining the resonant frequency and Q factor in air and then with the end section of the probe immersed in liquid. The equations relating the changes in resonant frequency and Q

[58] P. E. Rouse, Jr., E. D. Bailey, and J. A. Minkin, *Proc. Am. Pet. Inst.* **30-III**, 54–78 (1950).
[59] W. Philipoff, *Trans. Soc. Rheol.* **8**, 117–135 (1964).
[60] W. H. Robinson and S. I. Smedley, *J. Appl. Phys.* **49**, 1070–1076 (1978).

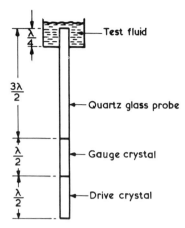

FIG. 10. Composite resonator of Robinson and Smedley.[60]

factor to the liquid properties are complicated, and reference should be made to the original publication. The authors claim a measurement accuracy of better than 0.5% for liquids in the viscosity range 0.0009–2 Pa sec at a temperature of 25°C and a frequency of 40 kHz.

Wada and his co-workers have developed a resonance technique that extends the frequency range down to 2 kHz.[53,54] They use a torsional quartz crystal cemented to a long rod, as in the traveling-wave technique described in Section 3.3.2.2, but operate the system at the resonant frequency of the complete assembly and at harmonics of this frequency. The increased length of the resonator results in a much lower resonant frequency. At this frequency the electrical impedance of the transducer is very large, even at the system resonance, and direct measurement of the impedance is impracticable. Instead, the free decay of the torsional oscillations is observed and measurements are made of the frequency and decay rate with the rod in air and, then, with a known length of the rod immersed in the liquid. The changes in resonant frequency Δf and decay rate ΔD are related to X_L and R_L by the equations

$$\Delta D = K R_L, \qquad \Delta f = (K/2\pi) X_L. \qquad (3.3.12)$$

The quantity K is an instrument constant,

$$K = \frac{2}{\rho r} \left[\frac{h}{l} + \frac{\sin(2n\pi h/l)}{2n\pi} \right], \qquad (3.3.13)$$

where ρ, r, and l are the density, radius, and length of the rod, respectively, h is the length of rod immersed in the liquid, and n is the number of the harmonic at which the system is operated. The decay rate is determined by measuring the amplitude of a reference signal which will cancel the decaying oscillation at a given time after the exciting signal is switched off. In this technique, the height of the liquid surrounding the rod must be chosen carefully so that bending oscillations are not excited. The sensitivity of this technique makes it suitable for the measurement of the properties of dilute solutions of polymers.

3.3.4.2. Magnetostrictive Technique: 4–200 kHz. Waterman[61] has described a technique in which a thin-walled nickel tube is excited into torsional resonance using the Wiedemann effect (Section 3.3.3.2). The tube, typically 40 cm long, is clamped at the center inside a glass tube containing the sample to be measured. Separate driving and receiving coils are used, placed on opposite sides of the central clamp. The resonance curve is determined using a synthesizer, which varies the frequency of the driving signal in small steps, and a digital voltmeter and printer, which monitor the amplitude of the oscillation. The resonant frequency and Q factor of the resonance are determined in air and with the tube immersed in the test liquid. The system is calibrated using a fluid of known properties. By suitable positioning of the drive and pickup coils, higher harmonics up to the 50th can be excited.

3.3.5. High-Pressure Techniques

Some of the techniques described in earlier sections can be modified for use under high hydrostatic pressure. The major difficulties are the restricted space available in a high-pressure vessel, typically 2.5 cm dia by 20 cm long in a vessel for use up to 1.4 GPa, and the restricted number of electrical leads that can be fitted in the vessel closure plug.

The normal-incidence shear wave reflection technique operating at 30 MHz (Section 3.3.2.1) has been used at pressures up to 1.4 GPa.[62] The usual preliminary measurements with a quartz–air interface cannot be made under pressure, and measurements are first made using a liquid of low and known shear impedance as a reference. If the viscosity of the liquid is sufficiently low, then even at the highest pressure of measurement the shear impedance is given by the expression for a Newtonian liquid, i.e., $Z_L = (j\omega\eta\rho)^{1/2}$; isopentane has been found to be a satisfactory reference liquid.

Figure 11 shows a typical arrangement of the acoustic system for use under pressure. An O-ring seal separates the test liquid from the liquid

[61] H. A. Waterman, *Rheol. Acta* **16**, 652 (1977).
[62] A. J. Barlow, G. Harrison, J. B. Irving, M. G. Kim, J. Lamb, and W. C. Pursley, *Proc. R. Soc. London Ser. A* **327**, 403–412 (1972).

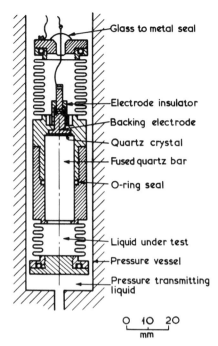

Glass to metal seal

Electrode insulator

Backing electrode

Quartz crystal

Fused quartz bar

O-ring seal

Liquid under test

Pressure vessel

Pressure transmitting liquid

O 10 20
mm

FIG. 11. Arrangement of the acoustic system for measurements of mechanical shear resistance under pressure (Barlow *et al.*[62]).

backing the transducer and metal bellows allow for the compression of the liquids under pressure.

At lower frequencies, the resonant torsional quartz crystal technique (Section 3.3.4.1) can be readily mounted in a pressure vessel and, subject to the maximum viscosity limitation of about 1 Pa sec, provides a simple means of determining the variation of shear impedance with pressure. Careful calibration is necessary, as the transducer constants K_1 and K_2 show significant changes with pressure.

Measurement of the density and viscosity of a liquid as a function of pressure is also necessary if the shear impedance data are to be plotted using reduced variables (Section 3.2.4). Volume changes can be determined by measuring the change in the length of a metal bellows containing the test liquid with a linear voltage differential transformer.[63] The viscosity may be measured by timing the fall of a sinker over a given distance in a tube containing the liquid; coils mounted outside the tube can be used to

[63] G. P. Shakhovskoi, I. A. Lavrov, M. D. Pushinskii, and M. G. Gonikberg, *Prib. Tekh. Eksp.* No. 1, 181–183 *English transl.: Instrum. Exp. Tech.* No. 1, 184–186 (1963)].

sense the position of the sinker.[64] Alternatively a linear voltage differential transformer can be used to monitor the fall of a sinker over a short distance; viscosities up to 10^7 Pa sec can be measured with this technique.[65]

3.4. Analysis and Interpretation of Results

The limited experimental frequency range of most of the techniques described in Chapter 3.3 requires that temperatures (or pressure) be used as an additional variable if the complete relaxation region is to be explored. Data obtained at different temperatures may be reduced to a single temperature, using the time–temperature superposition technique described in Section 3.2.4, and then combined to give a composite, or "master," curve which shows the behavior as a function of frequency at a single temperature. Data obtained as a function of pressure may be similarly reduced to a single pressure, which is usually atmospheric pressure. To carry out this reduction process, the variation of both viscosity η and high-frequency, limiting, shear modulus G_∞ as a function of temperature (or pressure) is required. When shear impedance data are to be reduced, a knowledge of the density variation with temperature is also required in order to calculate the components of the complex elastic modulus using Eq. (3.2.5).

The viscosity variation with temperature is commonly described using either the Arrhenius equation

$$\log(\eta/A) = B/T, \tag{3.4.1}$$

where A and B are constant, or the "free-volume" equation

$$\log(\eta/A) = B/(T - T_0), \tag{3.4.2}$$

where A, B, and T_0 are constant.

The Arrhenius equation is found to apply to low-viscosity liquids at temperatures well above the melting point, but for supercooled liquids, it is necessary to use the free-volume equation to obtain a satisfactory description of the considerable variation of viscosity with temperature. Harrison[11] gives a detailed review of these and other viscosity–temperature and viscosity–pressure equations.

Measurement of the value and the temperature dependence of G_∞ requires the use of techniques using shear waves at frequencies of the order of 1000 MHz in order to reach the region where the liquid is behaving elastically. Figure 12 shows a typical set of measurements of the quantity

[64] J. B. Irving and A. J. Barlow, *J. Phys. E. Sci. Inst.* **4**, 232–234 (1971).
[65] R. J. McLachlan, *J. Phys. E Sci. Instrum.* **9**, 391–394 (1976).

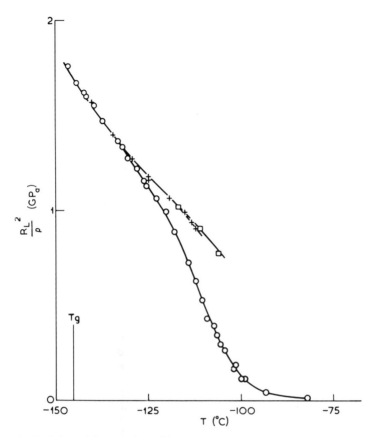

FIG. 12. Variation of the quantity R_L^2/ρ with temperature at frequencies of 30 MHz($\bigcirc$), 450 MHz($+$), and 1000 MHz ($\square$) for sec-butyl benzene.[66]

R_L^2/ρ plotted as a function of temperature. In the lower temperature region of the plot, near T_g, the measured value of R_L^2/ρ becomes independent of frequency. In this region, $R_L \gg X_L$ and from Eq. (3.2.5), $G'(\omega) \simeq R_L^2/\rho$.

When $G'(\omega)$ becomes independent of frequency, it is, by definition, equal to G_∞. Thus, the variation of G_∞ with temperature is given by the extreme-low-temperature region of the curve. To make an extrapolation of G_∞ into the relaxation region at higher temperatures, it is desirable that the variation of G_∞ with temperature be known over as large a temperature range as possible. It is clear from Fig. 12 that a combination of high

[66] A. J. Barlow, J. Lamb, A. J. Matheson, P. R. K. L. Padmini, and J. Richter, *Proc. R. Soc. London Ser. A* **298**, 467–480 (1967).

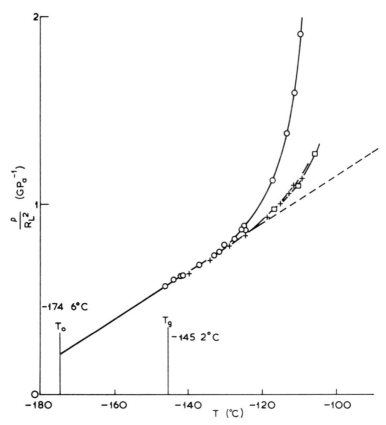

FIG. 13. Variation of the quantity ρ/R_L^2 with temperature for sec-butyl benzene. The dashed line shows the linear variation of $1/G_\infty$ with temperature at frequencies of 30 MHz (O), 450 MHz (+), and 1000 MHz (□).[66]

frequencies and low temperatures is necessary in order to observe the variation of G_∞ with temperature over a sufficiently extensive region.

Results from measurements in many liquids have shown that G_∞ does not decrease linearly with temperature but that the limiting compliance J_∞ ($=1/G_\infty$) can be described by the linear equation

$$J_\infty = 1/G_\infty = J_0 + C(T - T_0),\qquad (3.4.3)$$

as shown in Fig. 13. The extrapolation of J_∞ into the relaxation region is shown by the dashed line in this figure. Harrison[11] has tabulated values of G_∞ and the temperature coefficient C for many simple supercooling liquids. For most liquids, G_∞ has a value of between 0.9 and 5 GPa at the

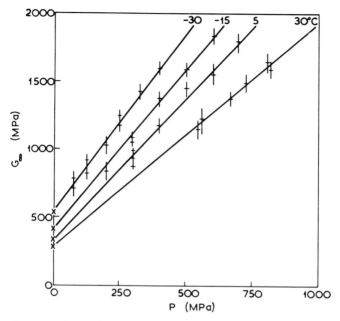

FIG. 14. Variation of G_∞ with pressure for di(2-ethylhexyl)phthalate at several temperatures.[67]

glass transition temperature T_g, and the coefficient C is of the order of $0.01 \, \text{GPa}^{-1} \, \text{K}^{-1}$.

The results of a more restricted number of measurements made under pressure indicate that G_∞ varies linearly with pressure, as shown in Fig. 14.

When the variation of η and G_∞ has been established, the experimental data can be reduced to a single curve; satisfactory reduction to within experimental error is usually an indication that the assumptions inherent in the time–temperature superposition principle have been fulfilled. Figure 15 shows data for a wide range of liquids which all reduced to a single empirical curve.[68]

The behavior shown in Fig. 15, which is found to be typical of many simple liquids, can be described by the empirical equation

$$J^*(j\omega) = \frac{1}{G_\infty} + \frac{1}{j\omega\eta} + 2\left(\frac{1}{j\omega\eta G_\infty}\right)^{1/2}, \qquad (3.4.4)$$

[67] J. F. Hutton and M. C. Phillips. *Nature (London) Phys. Sci.* **238**, 141–142 (1972).
[68] J. Lamb. *Proc. Inst. Mech. Eng.* **182**, Part 3A, 293–310 (1967).

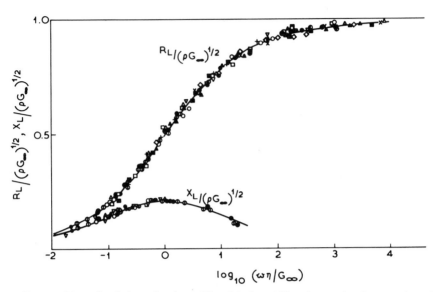

FIG. 15. Normalized plots of $R_L/(\rho G_\infty)^{1/2}$ and $X_L/(\rho G_\infty)^{1/2}$ against $\omega\eta/G_\infty$ for a number of supercooled liquids: squalane ($\triangledown$); squalene ($\oplus$); 6,6,11,11-tetramethyl hexadecane ($\boxtimes$); trichlorethyl phosphate ($\boxdot$); tri(m-tolyl)phosphate ($\odot$); tris(2-ethylhexyl)phosphate ($\times$); tetra(2-ethylhexyl)silicate ($+$); bis(m-(m-phenoxy phenoxy)phenyl)ether ($\triangle$); di(iso-butyl)phthalate ($\oplus$); di(n-butylphthalate ($\bullet$); iso-propyl benzene ($\blacksquare$); n-propyl benzene ($\diamond$); sec-butyl benzene ($\blacktriangle$). (Reproduced by permission of the Council of the Institution of Mechanical Engineers from their Proceedings.)

where there are no arbitrary parameters. Recent work has shown that a more accurate representation is given by the equation

$$J^*(j\omega) = \frac{1}{G_\infty} + \frac{1}{j\omega\eta} + \frac{J_r}{(1 + j\omega\eta\tau_r)^\beta}, \qquad (3.4.5)$$

where J_r is a retardation compliance, τ_r a retardation time, and β a constant lying between 0 and 1. The two equations become effectively equivalent when $\beta = 0.5$, as is often the case, and in the region where $\omega\tau_r \gg 1$. When the data are sufficiently accurate, a plot of the components of the retardational compliance

$$J_r^*(j\omega) = J_r'(j\omega) - J_r''(j\omega) = J^*(j\omega) - \frac{1}{G_\infty} - \frac{1}{j\omega\eta}$$

$$= J_1(\omega) - jJ_2(\omega) \qquad (3.4.6)$$

may be made, to evaluate the values of J_r, τ_r, and β. Typical examples are shown in Fig. 16.[34]

The results obtained from measurements of the type described in this

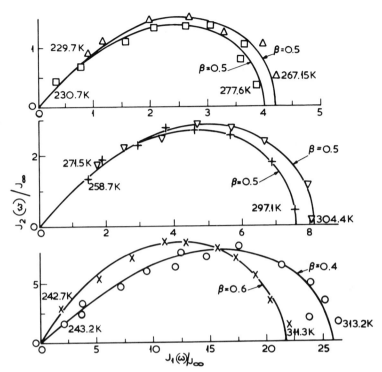

FIG. 16. Components of the retardational compliance at 30 MHz for various liquids: tri(β-chloroethyl)phosphate (○); squalane (×); tri(m-tolyl)phosphate (+); tri(o-tolyl)phosphate (▽); di(n-butyl)phthalate (△); di(isobutyl)phthalate (□) (Barlow and Erginsav[34]).

chapter are of both theoretical and practical importance. When compared with the solid and gas phases, the liquid state is poorly understood, and experimental data are required to test theoretical developments. The measurements made using the techniques described in this chapter are complementary to measurements of dielectric relaxation phenomena. Both types of measurement involve perturbing the equilibrium state of a liquid, and the observed phenomena depend on the resulting molecular motions. In many supercooled liquids the dielectric and shear relaxation processes are similar and have the same temperature dependence which closely follows the viscosity variation. No simple relation can be established between the dielectric relaxation time and the viscoelastic relaxation time, however, although they appear to be related to the same kind of molecular motion.

Attempts to evaluate macroscopic liquid properties from a rigorous analysis of the interactions between molecules have met with some success for the simpler of molecules, such as the rare gases, but yield unrealistic results for liquids containing more complex molecules. In such cases empirical hypotheses, based on Eqs. such as (3.4.4) and (3.4.5) or derived from postulates regarding a particular type of molecular motion such as diffusion, are able to provide a satisfactory description of the observed phenomena. While such equations do not, in general, lead to a further understanding of the fundamental molecular interactions, they have value in characterizing different types of behavior and in describing liquid properties in many practical situations.

For example, in the lubrication of high-speed bearings or gears subjected to high stresses, the hydrostatic pressure in the thin film of lubricant between the elastically deformed surfaces can readily exceed several thousand atmospheres. In such elastohydrodynamic lubrication situations, the transit time of the lubricant through the load-bearing surfaces may be of the same order as the relaxation time which governs the flow of the molecules: the liquid will then exhibit elastic rather than viscous behavior. Similar problems are encountered at much lower shear rates in systems that are exposed to relatively low temperatures. While there are considerable differences of opinion regarding the causes of observed deviations in behavior from the predictions of the existing theoretical analysis, a common feature is the treatment of the lubricant as a viscoelastic material. In recent years, a wide range of new lubricating fluids has been developed, including polymer-additive oils and synthetic lubricants, and it is of interest to compare the properties of these with those of conventional mineral oils. A better understanding of elastohydrodynamic lubrication will require an increased knowledge of liquid properties over wide ranges of time scale, pressure, and temperature. Such information can be obtained using the techniques described in this chapter.

4. ULTRASONIC CHEMICAL RELAXATION SPECTROSCOPY

By Leon J. Slutsky

Remarks on Notation

A unique symbol for each quantity used in this chapter would exhaust the latin and greek alphabets. Thus a number of ambiguities and near ambiguities in notation have been tolerated in circumstance where (it is hoped) multiple use of a given symbol will not lead to real confusion. Specifically, "i" and "k" are used as running indices and also with their usual meanings as $\sqrt{-1}$ and the Boltzmann constant k, respectively. In separate sections "γ" and "γ_i" have been used to denote, respectively, the specific heat ratio and the activity coefficient of the ith chemical species. The symbol "I" has been used for both the acoustic intensity and the ionic strength and "κ" for the thermal conductivity and the inverse Debye length. In Chapter 4.1, "X" has been used as a symbol for a general thermodynamic variable and in Chapter 4.4 to denote the concentration of a specific species.

Variously modified forms of "c" represent both the isobaric heat capacity per unit volume (c_p) or per unit mass ($\bar{c}_p$) and the concentration (c_i) or equilibrium concentration (c_{0i}) of the ith species in cgs units. The symbol "R" has been used for the gas constant and, in inflected forms (R_j, $\bar{R}_j$), for the rates of chemical reactions in terms of concentrations. The symbols "r" and "r_j" are reserved for the often less ambiguous definition of reaction rate in terms of the rate of change of the number of moles of a given component. Short definitions or references to the defining equations for the symbols used in this part are given below.

List of Symbols

a	coefficient of a basis reaction in a normal reaction	A_γ'	Debye–Hückel constant when I is expressed in moles per liter
a_i	activity of the ith chemical species	b	see a
a_{0i}	activity of the ith chemical species at equilibrium	B_i	ith species participating in a given chemical reaction
$\mathscr{A}$	relaxation amplitude [Eq. (4.1.21)]	$[B_i]$	molar concentration of the ith species
A	affinity [Eq. (4.1.12)]		
A_i	affinity for the ith reaction	$[B_i]_0$	molar concentration of the ith species at equilibrium
A_γ	Debye–Hückel constant (cgs)		

179

METHODS OF EXPERIMENTAL PHYSICS, VOL. 19

$\mathscr{B}$ high-frequency limit of α/f^2 [Eq. (4.1.21)]

c_i concentration in moles per milliliter of the ith species

$\bar{c}_i$ concentration in grams per milliliter of the ith species

c_p heat capacity at constant pressure per unit volume

$c_{p\infty}$ "frozen" heat capacity per unit volume

$\bar{c}_p$ heat capacity (at constant pressure) per unit mass

C_P heat capacity at constant pressure

$C_{P\infty}$ "frozen" heat capacity at constant pressure

$\bar{C}_P$ heat capacity per mole (C_P/N_t)

$\bar{C}_{P\infty}$ "frozen" heat capacity per mole

C_v heat capacity at constant volume

d_1 exclusion radius of an iron pair

D_0 dielectric constant of a solvent

$\mathscr{D}$ $(8\pi N_0 e^2/D_0 \mathbf{k} T)^{1/2}$

e charge of the electron

E^* energy of activation

f frequency

f_r relaxation frequency ($= 1/2\pi\tau$)

$\mathbf{F}^0$ initial approximation to secular determinant

F_1, F_2 functions of frequency defined in Eqs. (4.4.6)

$\mathbf{g}, g_{ij}$ defined in Eq. (4.3.17)

G diffusion coefficient

ΔH enthalpy change

$\Delta\mathbf{H}$ a row vector, the components of which are the enthalpy changes of the basis reactions

I acoustic intensity; ionic strength

$\mathbf{I}$ $n \times n$ diagonal unit matrix

$\mathbf{j}$ a vector with one in the jth position, zeros elsewhere

$\mathbf{J}_A$ diffusion current density for species A

k a rate constant

$\bar{k}$ a kinetic coefficient

$\mathbf{k}$ Boltzmann's constant

k_f rate constant for forward reaction

k_b rate constant for reverse reaction

K_i equilibrium constant for the ith reaction

$\mathbf{L}$ transformation that diagonalizes $\mathbf{Rg}$

$\mathbf{L}'$ transformation that diagonalizes $\mathbf{R}^{1/2}\mathbf{g}\mathbf{R}^{1/2}$

$\mathbf{L}^0$ transformation that diagonalizes $\mathbf{F}^0$

L number of independent segments ("Kuhn" segments) in a polymer chain

M_w molecular weight

n_i number of moles of ith species

n_s number of moles of solvent

N_0 Avogadro's number

N_t total number of moles $\left(N_t = \sum_i n_i \right)$

P pressure

q variable specifying the thermodynamic state of a system

$\mathbf{q}$ an eigenvector representing a normal reaction

q_i coefficient of the ith basis reaction in the equation that represents a normal reaction

$\mathscr{Q}_{ij}$ a coefficient in the Bronsted equation [Eq. (4.2.31)]

$\mathscr{Q}$ a particular choice of $\mathscr{Q}_{ij}$

Q_j defined by Eq. (4.3.14)

$\mathbf{r}$ position vector

r rate in moles per second of a chemical reaction

r_d effective radius for reaction

R gas constant

$\mathbf{R}$ a diagonal matrix with elements $\bar{R}_j$

R_j overall rate (in terms of concentrations) of the jth reaction

$\bar{R}_j$ exchange rate of the jth reaction [Eq. (4.3.13)]

R_{fj} rate of the forward reaction for the jth reaction

R_{bj} rate of the reverse reaction for the jth reaction

s condensation ($= \Delta\rho/\rho$)

S entropy

ΔS entropy change

$\bar{S}_i$ partial molar entropy of the ith species

t time

T temperature

v velocity of sound

v_0 low-frequency limit of the velocity of sound

v_∞ high-frequency limit of the velocity of sound

$\mathbf{v}$ velocity of flow

V volume

ΔV	volume change	η_0	shear viscosity of solvent
$\bar{V}_i$	partial molal volume of the ith species	$[\eta]_0$	intrinsic viscosity
$\Delta\mathbf{V}$	a row vector, the components of which are the volume changes for the basis reactions	η_V	volume viscosity
		θ	coefficient of thermal expansion
		θ_∞	"frozen" coefficient of thermal expansion
W_s	molecular weight of the solvent	κ	thermal conductivity; inverse Debye length
X	a thermodynamic variable		
X_i	mole fraction of the ith component	λ	wavelength
$\mathbf{y}$	wave vector	μ_i	chemical potential of the ith species
y	$|\mathbf{y}|$	μ_i^0	chemical potential of the ith species in its standard state
y_0	real part of y		
Y	a thermodynamic variable	ν_i	coefficient of the ith species in a given chemical reaction
z	a Cartesian coordinate		
z_i	charge of the ith species (in units of the electronic charge)	ν_{ij}	coefficient of the ith species in the jth chemical reaction
α	acoustic absorption coefficient (in nepers per centimeter)	ν_{vis}	viscous damping frequency [defined by Eq. (4.4.10)]
α_c	classical acoustic absorption coefficient [Eq. (4.1.20)]	ρ	density
		σ	steric factor in a bimolecular reaction
α_e	excess acoustic absorption coefficient ($= \alpha - \mathscr{B}f^2$)	τ	relaxation time (variously subscripted to indicate which thermodynamic variables are held constant). Unsubscripted τ represents relaxation time at constant pressure and entropy.
β_S	adiabatic compressibility		
$\beta_{S\infty}$	"frozen" adiabatic compressibility		
β_{S0}	static adiabatic compressibility		
γ	C_P/C_V		
γ_i	activity coefficient of the ith species		
Γ	defined by [Eq. (4.1.19)]	τ_d	$(Gy^2)^{-1}$
Γ_x	Γ/N_t	$\psi(\mathbf{r},t)$	a vector specifying the local temperature, pressure, velocity, and chemical composition
Γ_c	Γ/V		
ϵ	advancement [Eq. (4.1.10)] of a chemical reaction in terms of number of moles	ω	circular frequency
		ω_r	circular relaxation frequency ($= \tau^{-1}$)
ζ	advancement in terms of concentrations		
		ω_{vis}	$3/4\beta_S\eta$
η	shear viscosity	Ω	defined in Eq. (4.4.8)

4.1. General and Historical Introduction

4.1.1. Relaxation Spectroscopy, Thermodynamic Preliminaries

Relaxation spectroscopy might reasonably be considered to have its origins in Maxwell's "On the Dynamical Theory of Gases,"[1] wherein the introduction of "a time of relaxation" into the stress–strain relations of a viscoelastic medium is discussed. More generally, any process in which,

[1] J. C. Maxwell, "Scientific Papers," Vol. II, p. 26. Cambridge Univ. Press, London and New York, 1890. or *Philos. Trans. R. Soc. London* **157**, 49 (1867).

at constant values of the external thernodynamic parameters, the rate of change of any quantity q when displaced from its equilibrium value q_0 is

$$dq/dt = -\omega_r(q - q_0) = -(q - q_0)/\tau \qquad (4.1.1)$$

is referred to as a relaxation process with relaxation time τ; the constant ω_r in Eq. (4.1.1), dimensionally a frequency, is often called the "relaxation frequency."

If in a system initially at equilibrium with $q = q_{00}$, a thermodynamic parameter X undergoes a small change at $t = 0$ from X_0 to $X_0 + \delta X$, then for $t > 0$, $q_0 = q_{00} + (dq_0/dX)\,\delta X$ and

$$q = q_{00} + \frac{dq_0}{dX}\,\delta X\,(1 - e^{-\omega_r t}). \qquad (4.1.2)$$

If $X = X_0 + |\delta X|e^{i\omega t}$, then

$$q = q_{00} + \frac{dq_0}{dX}\left\{\frac{1 - i\omega\tau}{1 + \omega^2\tau^2}\right\}|\delta X|e^{i\omega t}. \qquad (4.1.3)$$

The family of techniques for the determination of the rate of energy transfer or structural or chemical change in which the exponential return to equilibrium of an internal parameter such as the chemical composition is observed after equilibrium is perturbed by a fast step in a thermodynamic parameter (temperature, pressure, electric field, etc.) or, alternatively, in which the equilibrium is perturbed periodically and the phase lag between the external field and the internal state is observed indirectly as dispersion or absorption is collectively called "relaxation spectroscopy."

In general, for any function $Y(X, q)$ of the thermodynamic parameters X and the internal coordinate q,

$$\frac{dY}{dX} = \left(\frac{\partial Y}{\partial X}\right)_q + \left(\frac{\partial Y}{\partial q}\right)_X \frac{dq}{dX} = \left(\frac{\partial Y}{\partial X}\right)_q + \left(\frac{\partial Y}{\partial q}\right)_X \left(\frac{dq_0}{dX}\right)\left\{\frac{1 - i\omega\tau}{1 + (\omega\tau)^2}\right\}. \qquad (4.1.4)$$

Derivatives at constant values of q may be interpreted as specifying the variation in Y at frequencies much higher than ω_r when the relaxation process is too slow to admit of significant variation in q within a period of the perturbation and are thus identified with the high-frequency limit of dY/dX and designated by a subscript ∞. Herzfeld and Litovitz[2] refer to derivatives at constant values of the internal parameters as "frozen" compressibilities, heat capacities, coefficients of thermal expansion, etc., and that locution is also used here. Similarly, the low-frequency limit is

[2] K. F. Herzfeld and T. A. Litovitz, "Absorption and Dispersion of Ultrasonic Waves." Academic Press, New York, 1959.

designated by a subscript 0, thus

$$\frac{dY}{dX} = \left(\frac{dY}{dX}\right)_\infty + \left\{\left(\frac{dY}{dX}\right)_0 - \left(\frac{dY}{dX}\right)_\infty\right\}\left\{\frac{1 - i\omega\tau}{1 + (\omega\tau)^2}\right\}. \qquad (4.1.5)$$

Specializing Eq. (4.1.5) to the variation of V with respect to P at constant entropy, the frequency-dependent adiabatic compressibility $\beta_S = V^{-1}$ $(\partial V/\partial P)_S$ is

$$\beta_S = \beta_{S\infty} + (\beta_{S0} - \beta_{S\infty})\frac{1 - i\omega\tau}{1 + (\omega\tau)^2}. \qquad (4.1.6)$$

For a plane acoustic wave propagating in the z direction, we may write

$$\delta P = |\delta P|e^{i\omega t - iyz},$$

where ω is real and $y = y_0 - i\alpha$. For small-amplitude adiabatic propagation in an isotropic fluid with negligible viscosity,

$$y^2/\omega^2 = \beta_S\rho = v^{-1}, \qquad (4.1.7)$$

where v is the velocity of sound.

In the usual circumstance, when the "relaxing compressibility" is small compared to β_∞, the acoustic dispersion and the absorption coefficient α due to a single process with relaxation time τ as a function of the circular frequency ω are,[2] respectively,

$$\left(\frac{v_0}{v}\right)^2 = 1 - \frac{v_\infty^2 - v_0^2}{v_\infty^2}\frac{(\omega\tau)^2}{1 + (\omega\tau)^2},$$

and $\qquad\qquad\qquad\qquad\qquad\qquad\qquad\qquad\qquad\qquad\qquad (4.1.8a)$

$$\alpha = \frac{(v_\infty^2 - v_0^2)}{2v_\infty^2 v}\frac{\omega^2\tau}{1 + (\omega\tau)^2},$$

where v_0 is the low-frequency limit of the velocity of sound and v_∞ the velocity at frequencies much higher than the relaxation frequency where the internal equilibrium does not respond to the temperature and pressure variations of the acoustic wave. When the dispersion is not small, the absorption coefficient is given by

$$\alpha\left(\frac{v_0}{v}\right)^2 = \frac{(v_\infty^2 - v_0^2)}{2v_\infty^2 v}\frac{\omega^2\tau}{1 + (\omega\tau)^2}. \qquad (4.1.8b)$$

Explicitly, in the case of chemical reaction, we let B_i designate the ith chemical species and write the equation representing the reaction as $|\nu_1|B_1 + |\nu_2|B_2 + \cdots \rightarrow |\nu_n|B_n + |\nu_{n+1}|B_{n+1} + \cdots$ or, if the stoichiometric coefficients ν_i are taken to be positive for products and negative for reactants, as

$$\sum_i \nu_i B_i = 0. \qquad (4.1.9)$$

Then, if n_i designates the number of moles of the ith species in an element of volume V, with dimensions small compared to the wavelength of sound, within which the temperature, pressure, and composition may be taken to be uniform, and if the change in local composition is the result of the progress of a single chemical reaction or structural transformation, then a single parameter ϵ (the "advancement") defined by

$$dn_i = \nu_i \, d\epsilon \qquad (4.1.10)$$

suffices to specify the variation in composition and plays the role of the relaxing quantity q in Eq. (4.1.1). The rate of the reaction r is then unambiguously defined in terms of the number of moles of any participating species by

$$r = (1/\nu_i) \, dn_i/dt = d\epsilon/dt. \qquad (4.1.11)$$

In Section 4.4.1 we shall attempt to make plausible the assumption implicit in Eqs. (4.1.10) and (4.1.11) that for even the shortest ultrasonic wavelengths, diffusion does not contribute significantly to the variation in local composition.

The equations that determine the time evolution of the chemical composition in a system far from equilibrium are not, in general, linear. However, it will be shown in Section 4.2.1 that the propagation of an ultrasonic wave of moderate intensity in a system initially at equilibrium induces only very small displacement from the equilibrium chemical composition. Hence, the rate equations can be linearized; the relaxation times being functions of both the equilibrium concentrations and rate constants. Explicit examples will be considered in Section 4.3.1.

In general, it is required for chemical equilibrium that $\Sigma_i \nu_i \mu_i = 0$, where μ_i is the chemical potential of the ith species or, defining the affinity A by $A = -\Sigma_i \nu_i \mu_i$, the requirement for equilibrium is $A = 0$. At equilibrium the rate is zero and near equilibrium the rate is proportional[3] to A. If both the rate and the affinity are expanded about ϵ_0, the equilibrium value of ϵ, retaining only linear terms

$$d\epsilon/dt = -\bar{k} \, (\partial A/\partial \epsilon) \, (\epsilon - \epsilon_0), \qquad (4.1.12)$$

where the relaxation time τ is $[\bar{k}(\partial A/\partial \epsilon)]^{-1}$.

In a classical kinetic study the rate of a reaction is observed, insofar as

[3] I. Prigogine, "Thermodynamics of Irreversible Processes," 3rd ed., p. 55. Wiley (Interscience), New York, 1967.

is possible, at constant temperature and pressure; relaxation processes may well proceed isoentropically or at constant volume. The relationships between relaxation times observed under various conditions are implicit in Eq. (4.1.12). For example, if A is considered to be a function of T, P, and the chemical composition, then

$$dA = (\partial A/\partial T)_{P,\epsilon} \, dT + (\partial A/\partial P)_{T,\epsilon} \, dP + (\partial A/\partial \epsilon)_{T,P} \, d\epsilon. \quad (4.1.13)$$

Since $(\partial A/\partial T)_{P,\epsilon} = -\Sigma_i \, \nu_i \, (\partial \mu_i/\partial T)_{P,\epsilon} = \Sigma_i \, \nu_i \overline{S}_i = \Delta S$, where $\overline{S}_i$ is the partial molal entropy of the ith species and ΔS the entropy change for the process in question, and similarly, since $(\partial A/\partial P)_{T,\epsilon} = -\Sigma_i \, \nu_i \, (\partial \mu_i/\partial P)_{T,\epsilon} = -\Sigma_i \, \nu_i \overline{V}_i = -\Delta V$, Eq. (4.1.13) may be rewritten

$$dA = \Delta S \, dT - \Delta V \, dP + (\partial A/\partial \epsilon)_{T,P} \, d\epsilon \quad (4.1.14)$$

and

$$dA/d\epsilon = \Delta S \, dT/d\epsilon - \Delta V \, dP/d\epsilon + (\partial A/\partial \epsilon)_{T,P}. \quad (4.1.15)$$

In general,

$$(\partial y/\partial x)_z = -(\partial z/\partial x)_y \, (\partial y/\partial z)_x \quad (4.1.16)$$

and, in particular, $(\partial T/\partial \epsilon)_{S,P} = -(\partial S/\partial \epsilon)_{T,P} \, (\partial T/\partial S)_{\epsilon,P} = -T \, \Delta S/C_{p\infty}$, where $C_{p\infty}$ is the heat capacity at constant pressure and chemical composition.

Thus, if Eq. (4.1.15) is specialized to constant pressure and entropy, then

$$(\partial A/\partial \epsilon)_{S,P} = (\partial A/\partial \epsilon)_{T,P} - (T/C_{p\infty})(\Delta S)^2. \quad (4.1.17)$$

From Eq. (4.1.12),

$$\tau_{T,P}/\tau_{S,P} = 1 - (T/C_{p\infty})(\Delta S)^2/(\partial A/\partial \epsilon)_{T,P} = 1 + \Gamma(\Delta S)^2/RC_{p\infty}, \quad (4.1.18)$$

where Γ is defined by the relation

$$\Gamma^{-1} = -(1/RT)(\partial A/\partial \epsilon)_{T,P}. \quad (4.1.19)$$

By appropriate specialization of Eq. (4.1.15) and use of Eq. (4.1.16), relations between relaxation times with various constraints on the external thermodynamic parameters are easily derived. The relaxation time in Eqs. (4.1.6) and (4.1.8) is appropriately $\tau_{S,P}$. For acoustic relaxation in a relatively incompressible system $\tau_{S,V} = (\beta_{S\infty}/\beta_{S0})\tau_{S,P} = (v_0/v_\infty)^2 \tau_{S,P}$ is also of particular interest.

An estimate of the magnitude of the difference between isoentropic and isothermal relaxation times must await the explicit evaluation of Γ in terms of the equilibrium thermodynamic properties in Section 4.2.1. It will be found that in many cases of interest the distinction may be ignored.

4.1.2. Acoustic Relaxation

Lord Rayleigh[4,5] conjectured in 1899 that slow energy exchange between translational and internal degrees of freedom might be a major source of the absorption of sound in polyatomic gases. In 1920 Einstein[6,7] nominated chemical reaction as a relaxing internal degree of freedom, briefly discussed frequency-dependent acoustic absorption, derived an expression for the dispersion due to the perturbation of a vapor-phase dissociation equilibrium, and suggested that acoustic dispersion might be a suitable technique for the determination of the rates of fast gas-phase reactions. Experimental work addressed to this end was initiated in Nernst's laboratory[8] and by Gruneisen and Goens.[9]

Much of the theory relevant to the propagation of sound in a chemically reacting fluid is equally applicable to the effect of slow energy exchange between translational and rotational or vibrational states, and in this later context, Rice and Herzfeld[10] developed explicit expressions for acoustic absorption as well as dispersion at low frequencies in a relaxing medium. From Eq. (4.1.8) it is clear that measurement of the frequency dependence of either the absorption or the velocity will permit the determination of τ. Dispersion has proven useful in the investigation of energy-transfer rates in gases. It has, however, been the absorption that has played the central role in the applications of physical acoustics to chemical kinetics.

In the absence of any relaxation processes, viscosity[11] and the deviation from perfectly adiabatic propagation due to conduction of heat from compressed to rarefied regions[12] will produce absorption. At frequencies low compared to $\omega_{vis} \simeq (3/4\beta_S\eta)$, this "classical" absorption α_c in terms of the viscosity η, thermal conductivity κ, ratio of the heat capacity at constant pressure to heat capacity at constant volume γ, and the specific heat (per gram) of the medium $\tilde{c}_p$ is[2]

$$\alpha_c = \frac{w\omega^2}{3v^3\rho_0}\left\{\eta + \frac{3}{4}(\gamma - 1)\frac{\kappa}{\tilde{c}_p}\right\} \equiv \mathscr{B}_c f^2, \qquad (4.1.20)$$

[4] Lord Rayleigh, *Philos. Mag.* **47**, 308 (1899).
[5] R. B. Lindsay (ed.), "Acoustics, Historical and Philosophical Development." Benchmark Papers in Acoustics, Dowden, Hutchinson & Ross, Stroudsburg, Pennsylvania, 1973.
[6] A. Einstein, *Sitzberichte Preuss. Akad. Wiss. Berlin* **24**, 380 (1920).
[7] R. B. Lindsay (ed.), "Physical Acoustics." Benchmark Papers in Acoustics, Dowden, Hutchinson & Ross , Stroudsburg, Pennsylvania, 1974.
[8] F. Keutel, Dissertation, Berlin, 1910, cited in Einstein.[6]
[9] E. Gruneisen and E. Goens, *Ann. Phys. (Leipzig)* **72**, 193 (1923).
[10] K. F. Herzfeld and F. O. Rice, *Phys. Rev.* **31**, 691 (1928).
[11] G. Stokes, *Trans. Cambridge Philos. Soc.* **8**, 287 (1845).
[12] G. Kirchhoff, *Poggendorf's Ann. Phys.* **184**, 177 (1868).

where ρ_0 is the unperturbed density of the medium. For water at 4°C, $\omega_{vis} \simeq 6 \times 10^{11}$ sec^{-1}; for air at 300°K and 1 atm, $\omega_{vis} \simeq 10^{10}$ sec^{-1}. In polymer solution it may be necessary to introduce a frequency-dependent shear viscosity, otherwise in simple solvents losses due to viscosity and thermal conductivity are proportional to ω^2 over the range of frequencies of interest in chemical kinetics. For relaxation times sufficiently short so that $\omega\tau \ll 1$, Eq. (4.1.8) also predicts an acoustic absorption proportional to ω^2. The classical absorption and the absorption due to any processes much faster than the process of immediate interest are thus often combined and Eq. (4.1.8) is rewritten in terms of the frequency $f(f = \omega/2\pi)$ as

$$\frac{\alpha}{f^2} = \frac{2\pi^2(v_\infty^2 - v_0^2)}{v_\infty^2 v} \frac{\tau}{1 + (\omega\tau)^2} + \mathcal{B} = \frac{\mathcal{A}\tau}{1 + (f/f_r)^2} + \mathcal{B} \quad (4.1.21)$$

or in terms of the excess absorption ($\alpha_e = \alpha - \mathcal{B}f^2$) per wavelength ($\lambda$),

$$\alpha_e\lambda = \frac{\pi(v_\infty^2 - v_0^2)}{v_\infty^2} \frac{(2\pi f\tau)}{1 + (2\pi f\tau)^2} = \frac{\pi(v_\infty^2 - v_0^2)}{v_\infty^2} \frac{(f/f_r)}{1 + (f/f_r)^2}, \quad (4.1.22)$$

where the relaxation frequency $f_r = \omega_r/2\pi = 1/2\pi\tau$.

In the solvents most commonly employed in studies of chemical relaxation in solution, $\mathcal{B}$ is about three times α_c as calculated from Eq. (4.1.20).[2] Moreover, in the moderately concentrated solutions (0.1–1 molar) often employed in such studies, $\mathcal{B}$ may vary significantly from its value in the pure solvent. Hence, Eqs. (4.1.21) and (4.1.22) involve three experimentally determined parameters and a really satisfactory deduction of τ will, in general, require measurement of α from roughly a decade below to a decade above f_r. Data at low or high frequencies determine $\mathcal{A}\tau + \mathcal{B}$ or $\mathcal{A}/\tau$ and $\mathcal{B}$, only. It will be seen in Section 4.2.2 that $\mathcal{A}$ in Eq. (4.1.21) can be calculated from the volume change, enthalpy change, and equilibrium constant for the reaction and the thermal expansion, heat capacity, and density of the medium. Thus, given sufficient knowledge of the thermodynamics of a process, it will in favorable cases be possible to obtain kinetic information from data which does not span a large range on either side of the relaxation frequency.

"Conventional" ultrasonic equipment (cylindrical resonators[13] and pulse-echo spectrometers operating on overtones of quartz transducers) span roughly the range $\omega = 10^6 - 3 \times 10^9$ sec^{-1}. Large spherical resonators and sputtered thin-film transducers can add something more than a decade to this range. Hence, processes with relaxation times between

[13] F. Eggers and Th. Funck, *J. Acoust. Soc. Am.* **57**, 331 (1975); F. Eggers, Th. Funck, and K. H. Richman, *Rev. Sci. Instrum.* **47**, 361 (1976).

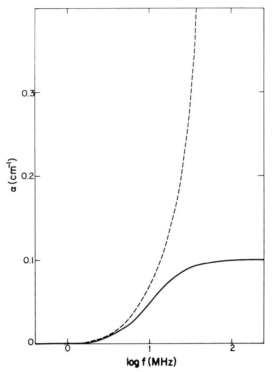

FIG. 1. Total acoustic absorption (dashed curve) and excess acoustic absorption (solid curve) as a function of frequency for a single relaxation with amplitude $\mathscr{A}\tau = 100 \times 10^{-17}$ sec^2/cm and a relaxation frequency of 10 MHz ($\tau = 1.6 \times 10^{-8}$ sec) in a medium with $\mathscr{B} = 20 \times 10^{-17}$ sec^2/cm.

10^{-5} and 10^{-10} sec are potentially accessible to study by acoustic techniques.

By way of illustration, we show in Fig. 1 the total absorption and the excess absorption associated with a single process with a relaxation frequency of 10 MHz ($\tau \simeq 1.6 \times 10^{-8}$ sec) and $\mathscr{A}\tau = 100 \times 10^{-17}$ sec^2/cm in aqueous solution near room temperature ($\beta \simeq 20 \times 10^{-17}$ sec^2/cm). In Fig. 2, α/f^2 and $\alpha_e\lambda$ are displayed. For this relaxation ($v_\infty - v_0)/v_\infty \simeq 2 \times 10^{-4}$ and it has proved to be easier to measure the excess absorption coefficient even in circumstances where α_e is a relatively small fraction of the total absorption (in this case 5% at 100 MHz) than to measure the velocity with the accuracy necessary to determine τ from the dispersion. (However, with the development[14] of accurate "au-

---

[14] D. Eden and J. G. Elias, *J. Acoust. Soc. Am.* **65**, S48 (1979); *Rev. Sci. Instrum.* **50**, 1299 (1979).

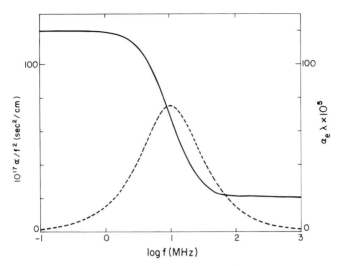

FIG. 2. α/f^2 (solid curve) and $\alpha_e\lambda$ (dashed curve) versus frequency for a single relaxation with $\mathcal{A}\tau = 100 \times 10^{-17}$ sec²/cm and $f_r = 10$ MHz in a medium with $\mathcal{B} = 20 \times 10^{-17}$ sec²/cm.

tomated'' systems for the determination of the velocity of sound, dispersion may well be a more attractive alternative.) It will be obvious from an inspection of Fig. 1 that, particularly for fast processes, there is a considerable incentive to work in media with low B—water, methanol, and acetone being common choices. Table I briefly lists the physical properties of these solvents.

Representative experimental results for a 0.01-molar solution of the peptide antibiotic bacitracin in 0.1-molar phosphate buffer at pH 7 and 4°C are displayed in Fig. 3 where filled circles represent data obtained in a 1.5-ml capacity cylindrical resonator, open circles measurements in a 15-ml cylindrical resonator, and crosses results in a variable path length apparatus of conventional design. The relaxation parameters that define the continuous curves are $\mathcal{B} = 13 \times 10^{-17}$ sec²/cm, $\mathcal{A}\tau = 2300 \times 10^{-17}$ sec²/cm, and $f_r = 1.2$ MHz. The value of $\mathcal{B}$ used to construct the plot of $\alpha_e\lambda$ versus $\ln f$ is deduced from measurements in the frequency range 30–150 MHz (not plotted). In circumstances where a reasonably accurate high-frequency limit of α/f^2 can be determined, a plot of $\alpha_e\lambda$ versus f or $\ln f$ is probably the most convenient representation of experimental results. However, it should be noted that even in this rather dilute solution, $\mathcal{B}$ is about 25% higher than in the pure buffer, and there is presumably a second relaxation process with f_r higher than 150 MHz. Otherwise, with conventional equipment the relaxation parameters are in this

TABLE I. Physical Properties of Three Important Solvents

Solvent	Liquid range (°C) at 1 atm	Density (g/cc) at 25°C	$V = V_0\{1 + a10^{-3}t + b10^{-5}t^2 + c10^{-8}t^3\}$ (t in °C)			Velocity of sound $\times 10^{-5}$ (cm/sec) at 25°C	Temperature derivative of velocity of sound (cm/sec °K) at 25°C	$\bar{C}_\mathrm{p}$ (J/°K)		$\mathscr{B} \times 10^{17}$ (sec²/cm) at 25°	$\dfrac{d \ln \mathscr{B}}{dT}$
			a	b	c			at -23°C	at 25°C		
Methanol	-98–65	0.7868	1.1324	1.3635	0.8741	1.104	325	94.48	81.6	33	-0.011
Acetone	-94–56	0.7850	1.324	3.809	0.8798	1.177	430	119	125	29	—
Water	0–100	0.99707	-0.6427	8.5053	-6.79	1.40273	$+5.03358 \times 10^{-3}t$ $-5.79506 \times 10^{-5}t^2$ $+3.31636 \times 10^{-7}t^3$ $+1.45262 \times 10^{-9}t^4$ $+3.0449 \times 10^{-12}t^5$	—	75.15	22	-0.036

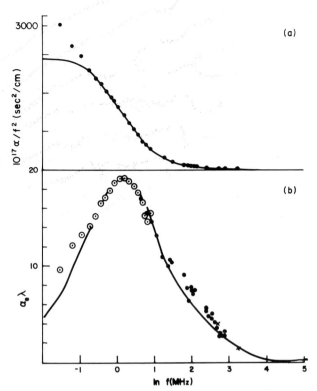

Fig. 3. The excess acoustic absorption in a 0.01-molar solution of the peptide antibiotic bacitracin in 0.1-molar phosphate buffer at pH 7 and 4°C as a function of frequency. In part (a), α/f^2 for the solution $-\alpha/f^2$ for the pure buffer is plotted. In part (b), wherein $\alpha_e\lambda$ is plotted as a function of frequency, the open circles represent measurements in a 15-ml capacity cylindrical resonator, the filled circles represent data taken in a 1.5-ml capacity cylindrical resonator,[14] and the crosses show results in a variable path length pulse-echo apparatus.

case rather easily determined on rather small sample volumes in rather dilute solution. What remains problematic is an unambiguous identification of the microscopic process responsible for the observed relaxation. This problem will be considered in Section 4.5.

Equation (4.1.21) or (4.1.22) has been the basis of the applications of acoustics to chemical kinetics and our principal concern in the next two sections will be an explicit expression for the relaxation amplitude $\mathscr{A}$ in terms of the equilibrium thermodynamic properties of the medium and for τ in terms of the parameters of the kinetic model of the chemical reaction or structural change under consideration. However, the absorption due to chemical relaxation is not strictly independent of and additive to the effects of diffusion, viscosity, thermal conduction, and electrical transport, and although the coupling between reaction and transport has not in fact proven to be important[15-17] in the ultrasonic determination of reaction rates, it might legitimately be judged more satisfying logically to develop these topics together. Meixner and his co-workers[15] have made such a development in terms of the phenomenological coupling coefficients of irreversible thermodynamics, concluding that when, as is almost universally the case, $(\alpha\lambda)_e$ is small, Eq. (4.1.21) is satisfactory. Schurr[16,17] has developed a more detailed hydrodynamic theory of the coupling between chemical reaction and transport in a context (dynamical light scattering) where such coupling may well be important.

There is at present a substantial literature[2,18,19] on the determination of characteristic times for energy transfer in gases by studies of acoustic propagation as well as by such mixed techniques as the "spectrophone"[20] or the "acoustic calorimeter,"[21] wherein the phase lag or time delay between a periodic or pulsed optical perturbation and an acoustic response is determined. However, although interest in gas-phase chemical kinetics provided the impetus both for Einstein's theoretical work and for the early experiments, acoustic methods have not yet proven to be of great importance in this discipline. The theoretical situation[2,18,19] and what experimental results are available[18] with respect to the effect of small perturbations of chemical equilibria on acoustic propagation in gases have been thoroughly reviewed by Kneser,[18] Bauer,[19] and Herzfeld and Litovitz.[2]

More recently there has been some theoretical interest[22-24] and at least one experimental investigation[25] of the propagation of sound in a vapor with a chemical reaction far from equilibrium. Interesting phenomena such as spontaneous acoustic oscillation and acoustic gain have been pre-

[15] J. Meixner, *Acustica* **2**, 101 (1952); S. M. T. de la Selva, L. S. Garcia-Colin, and J. Meixner, *Adv. Mol. Relaxation Processes* **11**, 73 (1977).

[16] J. M. Schurr, *J. Phys. Chem.* **73**, 2820 (1969).

[17] J. M. Schurr *Crit. Rev. Biochem.* **4**, 371 (1977).

[18] H. O. Kneser, Relaxation processes in gases, in "Physical Acoustics" (W. P. Mason, ed.) Vol IIA, p. 133. Academic Press, New York, 1965.

[19] H. J. Bauer, Phenomenological theory of the relaxation phenomena in gases, *in* "Physical Acoustics" (W. P. Mason, ed.), Vol. IIA, p. 48. Academic Press, New York, 1965.

[20] H. J. Bauer, *J. Chem. Phys.* **57**, 3130 (1977).

[21] J. Callis, W. W. Parson, and M. Gouterman, *Biochim. Biophys. Acta* **267**, 348 (1972); J. Callis, *J. Res. Natl. Bur. Stand.* **80A**, 413 (1976).

[22] R. Gilbert, P. Ortoleva, and J. Ross, *J. Chem. Phys.* **58**, 3625 (1973).

[23] R. G. Gilbert, H. S. Hahn, P. Ortoleva, and J. Ross, *J. Chem. Phys.* **57**, 2672 (1972).

[24] C. A. Garris, T. Y. Toong, and J. P. Patureau, *Acta Astronaut.* **2**, 981 (1975).

[25] J. P. Patureau, T. Y. Toong, and C. A. Garris, *Proc. Int. Symp. Combust.* (1976).

dicted and perhaps observed.[25]   These phenomena are considered, also rather briefly, in Section 4.4.

Ultrasonic investigations of fast chemical and conformational equilibria in condensed phases are more recent and have proven somewhat more profitable than studies in the gas phase.   In 1949 Leonard[26] determined that the excess acoustic absorption in sea water was primarily due to the presence of $MgSO_4$, and Liebermann[27] concluded that the observed acoustic relaxation in sea water was the result of the perturbation of what he denominated as the "dissociation" equilibrium of $MgSO_4$ but which in more modern parlance[28] would be regarded as the transition between an $Mg^{2+}-SO_4^=$ ion pair and a solvent-separated ion pair.   In 1953 Freedman[29] interpreted the acoustic relaxation observed by Lamb and Pinkerton[30] in liquid acetic acid as the result of the perturbation of an equilibrium between monomers and cyclic hydrogen-bonded dimers and at about the same time Kurtze and Tamm initiated their work on proton-transfer reactions[31] and relaxation in aqueous solutions of divalent electrolytes.[32]   Since these early studies, measurement of the frequency dependence of acoustic attenuation in liquids and solutions has developed into a specialized but occasionally very useful technique for the determination of the kinetic and thermodynamic parameters of fast chemical and structural change; processes with characteristic times as short as 10^{-10} sec and as long as 10^{-5} sec having been successfully investigated.

We shall be concerned with the theory, utility, limitations, interpretative problems, and techniques of acoustic methods in chemical kinetics, principally solution kinetics.   Only a few representative recent results will be cited by way of illustration.   Reviews by Lamb[33] and by Stuehr and Yeager[34] report an extensive body of experimental results of ultrasonic studies in, respectively, nonelectrolytes and electrolyte solution. The merit of an acoustic technique for any specific problem is of course appropriately judged in comparison with the other methods for determination of the rates of fast processes.   Bernasconi[35] has given a readable gen-

[26] R. Leonard, P. Combs, and L. Skidmore, *J. Acoust. Soc. Am.* **21**, 63A (1949).

[27] L. Liebermann, *Phys. Rev.* **76**, 1520 (1949).

[28] H. Diebler and M. Eigen, *Z. Phys. Chem. (Frankfurt)* **20**, 299 (1959).

[29] E. Freedman, *J. Chem. Phys.* **21**, 1784 (1953).

[30] J. Lamb and J. Pinkerton, *Proc. R. Soc. London Ser. A* **199**, 114 (1949).

[31] K. Tamm, G. Kurtze, and H. Kaiser, *Acustica* **4**, 380 (1954).

[32] G. Kurtze and K. Tamm, *Acustica* **3**, 33 (1953).

[33] J. Lamb, in "Physical Acoustics" (W. P. Mason, ed.) Vol. IIA, p. 203. Academic Press, New York, 1965.

[34] J. Stuehr and E. Yeager, in "Physical Acoustics" (W. P. Mason, ed.), Vol. IIA, p. 351. Academic Press, New York, 1965.

[35] C. Bernasconi, "Relaxation Kinetics." Academic Press, New York, 1976.

eral introduction to relaxation spectroscopy and a very useful survey of the recent literature. A more extensive discussion of the theory, methods, and results of all forms of relaxation spectroscopy has appeared under the general editorship of Hammes.[36] The 1963 review by Eigen and DeMaeyer[37] is still in many respects the standard work in this field.

To summarize briefly, techniques based on perturbation by rapid joule heating[35,36,38] or on rapid pressure jump[35,36,39] with conductiometric or spectrophotometric detection of the change in composition have generally been most successful for times greater than 10^{-6} sec. Perturbation of an equilibrium by a pulsed electrical field or studies of high-field dielectric relaxation have been used in media of low conductivity to determine relaxation times of chemical reactions as fast as 10^{-8} sec.[35] In ultrasonics the identification of the process or processes responsible for the observed acoustic absorption is often difficult. In favorable cases, fast pulse or step methods with optical sensing (spectrophotometric, fluorimetric, optical rotation, light scattering) offer the possibility of monitoring the concentration of a given species or class of species with a concommitant simplification of interpretative problems.

Relaxation times as short as 3×10^{-8} sec have been observed in "cable temperature jump"[40] systems and recently[41] a temperature jump of 1°K with a rise time of 10^{-9} sec has been achieved in a sample volume of 0.1 ml of aqueous solution with a pulsed iodine laser operating at 13,150 Å. In the future, refinement of the laser temperature jump method may well offer an attractive alternative to ultrasonic methods for some fast processes.

As we proceed with the development of the basic equations for ultrasonic relaxation times and relaxation amplitudes, the comparative advantages and limitations of acoustic methods will, we believe, become apparent. In general, determination of acoustic absorption, particularly in the range $\omega = 10^7 - 10^9$ sec^{-1}, is not technically difficult in comparison to the fast-pulsed methods and there is a substantial area of research in chemical kinetics in which ultrasonics will be found to be the most suitable technique.

[36] G. Hammes (ed.), "Techniques of Chemistry," 3rd ed., Vol. 6, Part 2. Wiley (Interscience), New York, 1974.

[37] S. L. Friess, E. S. Lewis, and A. Weissburger (eds.), "Techniques of Organic Chemistry," Vol. 8, No. 2. Wiley (Interscience), New York, 1963.

[38] G. H. Czerlinski and M. Eigen, Z. Electrochem. 63, 652 (1959).

[39] S. Ljunggren and O. Lamm, Acta Chem. Scand. 12, 1834 (1958).

[40] D. Porschke, Biopolymers 17, 315 (1978); G. W. Hoffman, Rev. Sci. Instrum. 42, 1643 (1971).

[41] J. F. Holsworth, A. Schmitt, H. Wolf, and R. Volk, J. Phys. Chem. 81, 2300 (1977).

4.2. Relaxation Amplitudes and the Magnitude of the Chemical Contribution to the Equation of State

In Section 4.2.1 the magnitude of the deviation from the local equilibrium chemical composition induced by an acoustic wave is calculated and a justification is presented for the linerization of the equations that describe the reaction kinetics. In Section 4.2.2 the size of the chemical contribution to the heat capacity, coefficient of thermal expansion, and adiabatic compressibility is explored and the basic equation [Eq. (4.2.19)] for the relaxation amplitude in terms of the thermodynamic properties of an ideal mixture is derived. In Section 4.2.3 the appropriate modification of Eq. (4.2.19) in a real solution is discussed and in Table II representative volume changes, which might enter into a consideration of the feasibility or the analysis of the results of an ultrasonic investigation of reaction kinetics, are tabulated.

4.2.1. The Variation of Equilibrium Chemical Composition with Temperature and Pressure

In a medium with a density ρ in which the velocity of sound is v, the pressure amplitude $|\delta P|$ associated with an acoustic wave of intensity I is[42] given by $|\delta P| = (\rho v I)^{1/2}$. In water at room temperature $\rho = 1$ gm/ml and $v = 1.5 \times 10^5$ cm/sec. Thus, for a "typical" intensity of 1 mW/cm², the pressure amplitude is about 4×10^4 dyn/cm² or 0.04 atm. In representative organic solvents such as acetone or benzene, the pressure variation associated with a 1-mW/cm² acoustic intensity is about 0.03 atm.

Assuming that the propagation of the sound wave is very nearly adiabatic, the temperature variation is readily calculated from

$$(\partial T/\partial P)_S = TV\theta/C_p = T\theta/c_p, \qquad (4.2.1)$$

where V is the volume, θ the coefficient of thermal expansion, C_p the heat capacity at constant pressure and c_p the isobaric heat capacity per unit volume. For water at room temperature, $c_p \simeq 1$cal/°K ml $\simeq 4.2 \times 10^7$ ergs/°K ml and $\theta = 0.2 \times 10^{-3}$ °K. Hence, $(\partial T/\partial P)_S = 1.5 \times 10^{-9}$ °K cm²/dyn $= 1.5 \times 10^{-3}$ °K/atm and the temperature variation ΔT associated with an ultrasonic intensity of 1 mW/cm² is approximately 6×10^{-5} °K.

Near room temperature the coefficients of thermal expansion of most

---

[42] P. M. Morse and K. O. Ingard, "Theoretical Acoustics." McGraw Hill, New York, 1968.

organic liquids are appreciably greater than that of water (i.e., $1.24 \times 10^{-3}/°K$ for benzene, $1.49 \times 10^{-3}/°K$ for acetone at $298°K$)^2 and c_p is significantly smaller (i.e., at $298°K$, $c_p = 1.56 \times 10^7$ ergs/$°K$ ml for acetone, 1.72×10^7 erg/$°K$ ml for benzene). Thus the temperature variation associated with an adiabatic pressure fluctuation in a typical organic liquid is larger than in water (approximately $0.001°K$ for 1 mW/cm^2), but $\Delta T/T$ nonetheless remains small.

The heat capacity, compressibility, and coefficient of volume expansion of a concentrated solution may, to be sure, be rather different from those of the solvent and the contribution of a chemical or structural equilibrium to these properties may in certain circumstances be large. These effects will be dealt with explicitly in succeeding sections. However, the conclusions of this introductory section as to the order of magnitude of the temperature and pressure variation and the consequent variations in equilibrium chemical composition associated with moderate acoustic intensities will not be greatly altered.

If the local temperature and pressure are varied subject to the condition that chemical equilibrium is maintained, then $dA = 0$ or, from Eq. (4.1.14),

$$(\partial A/\partial \epsilon)_{T,P} \, d\epsilon = -\Delta S \, dT + \Delta V \, dP. \qquad (4.2.2)$$

Introducing Γ as defined by Eq. (4.1.19) and making use of the requirement that $\Delta G = \Delta H - T \, \Delta S = 0$ (since we suppose chemical equilibrium to be maintained), one obtains

$$d\epsilon = (\Gamma/RT)\{(\Delta H/T) \, dT - \Delta V \, dP\}. \qquad (4.2.3)$$

The chemical potential of any species in a complex mixture may be expressed in terms of μ_i^0, the chemical potential of that species in its standard state, and a_i the activity of the species as

$$\mu_i = \mu_i^0 + RT \ln a_i. \qquad (4.2.4)$$

In mixtures of nonelectrolytes, μ_i^0 is taken to be the chemical potential of the pure ith species in its stable state of aggregation at the temperature and pressure in question. An ideal solution is defined by the relation $\mu_i = \mu_i^0 + RT \ln X_i$, where X_i (the mole fraction of the ith component) is defined by $X_1 = n_i/\Sigma_i \, n_i = n_i/N_t$ and, since μ_i^0 is by definition independent of the composition,

$$\Gamma_{\text{ideal}}^{-1} = \frac{\partial}{\partial \epsilon} \left(\sum_i \nu_i \ln X_i \right) = \frac{\partial}{\partial \epsilon} \left(\sum_i \nu_i \left[\ln n_i - \ln \sum_i n_i \right] \right)$$

$$= \sum_i \frac{\nu_i}{n_i} \frac{dn_i}{d\epsilon} - \frac{1}{N_t} \sum_i \nu_i \left(\sum_i \frac{dn_i}{d\epsilon} \right)$$

(continues)

$$= \sum_i \frac{\nu_i^2}{n_i} - \frac{1}{N_t} \left(\sum_i \nu_i \right)^2$$

$$= \frac{1}{N_t} \left\{ \sum_i \frac{\nu_i^2}{X_i} - \left(\sum_i \nu_i \right)^2 \right\}. \tag{4.2.5}$$

In aqueous solutions of electrolytes and, more generally, in dilute solution, the standard state is often taken to be a (hypothetical) 1-molar or 1-molal ideal solution. The relative activity in an "ideal dilute" solution is then the molar concentration ($[B_i] = 1000n_i/V$) or the molality ($m_i = n_i 1000/n_s W_s$, where n_s and W_s are, respectively, the number of moles and the molecular weight of the solvent). Equation (4.2.5) is, of course, not altered by an alternate choice of the standard state.

Γ, as defined by Eq. (4.1.19) and approximated by Eq. (4.2.5), is an extensive quantity. It is often useful to define an intensive analog $\Gamma_x = \Gamma/N_t$ or $\Gamma_c = \Gamma/V$. Then

$$\Gamma_c^{-1} = V\Gamma^{-1} = \sum_i \frac{\nu_i^2}{c_i} - \frac{V}{N_t} \left(\sum_i \nu_i \right)^2, \tag{4.2.6}$$

where $c_i = n_i/V$. Usually the volume per mole (~ 18 ml in an aqueous solution) is small compared to the reciprocal concentrations of the least abundant reactants and Eq. (4.2.6) may be approximated by

$$\Gamma_c^{-1} = \sum_i (\nu_i^2/c_i). \tag{4.2.7}$$

The calculation of Γ for a real solution will be discussed in the next section; however, we are now in a position to estimate the order of magnitude of the deviations from the equilibrium chemical composition which may be induced by the passage of an ultrasonic wave. From Eqs. (4.1.10) and (4.2.3), the relative variation in the equilibrium composition is

$$\frac{dn_i}{n_i} = \frac{\nu_i}{n_i} d\epsilon = \frac{\nu_i \Gamma}{n_i RT} \left(\Delta H \frac{dT}{T} - \Delta V \, dP \right). \tag{4.2.8}$$

Equation (4.2.8) also specifies the maximum deviation of the local composition from the equilibrium composition; the maximum deviation obviously corresponding to a circumstance where the reaction is too slow to permit significant variation in composition in a period of the pressure oscillation.

From Eq. (4.2.5) it is easy to show that the maximum value of Γ/n_i is ν_i^{-2}. Hence,

$$\frac{dn_i}{n_i} < \frac{1}{\nu_i} \left\{ \frac{\Delta H}{RT} \frac{dT}{T} - \frac{\Delta V \, dP}{RT} \right\}.$$

If Eq. (4.1.9) is written in such a way that the coefficients are one or small integers, ΔV in a condensed phase will rarely exceed 100 ml/mole. At room temperature RT is 2.5×10^{10} ergs/mole. Thus $\Delta V/RT$ is rarely larger than 4×10^{-9} cm^2/dyn and for a 4×10^4 dyn/cm^2 pressure amplitude, the second term on the right-hand side of Eq. (4.2.8) is not usually larger than 2×10^{-4}. Similarly, if ΔH is as large as 100 kcal, at room temperature $\Delta H/RT \cong 170$. For a typical organic liquid at room temperature $dT/T = 3 \times 10^{-6}$ when the acoustic intensity is 1 mW/cm^2; the first term in Eq. (4.2.8) is then about 5×10^{-4}. Thus, even for rather large volume and enthalpy changes in condensed phases, one may assume that the deviations from equilibrium composition induced by moderate acoustic intensities are small and that the kinetic equations that describe the time evolution of the composition may be linearized without significant error. If one were to seek for exceptions, one might perhaps imagine a highly "cooperative" process, more nearly a phase transition than a chemical reaction, of the form $lA = A_l$, where, for sufficiently large l, dn/n for A_l would not necessarily be small at moderate acoustic intensities.

4.2.2. Chemical Contributions to the Heat Capacity, Coefficient of Thermal Expansion, and Adiabatic Compressibility

With the aid of Eq. (4.1.10), the general expression for the variation in the volume

$$dV = \left(\frac{\partial V}{\partial T}\right)_{P,n_j} dT + \left(\frac{\partial V}{\partial P}\right)_{T,n_j} dP + \sum_i \left(\frac{\partial V}{\partial n_i}\right)_{T,P,n_j} dn_i$$

$$= \left(\frac{\partial V}{\partial T}\right)_{P,n_j} dT + \left(\frac{\partial V}{\partial P}\right)_{T,n_j} dp + \sum_i \overline{V}_i \, dn_i$$

(where $\overline{V}_i$ is the partial molar volume of the ith species) may be rewritten

$$dV = \left(\frac{\partial V}{\partial T}\right)_{P,\epsilon} dT + \left(\frac{\partial V}{\partial P}\right)_{T,\epsilon} dP + \sum_i \nu_i \overline{V}_i \, d\epsilon$$

$$= \left(\frac{\partial V}{\partial T}\right)_{P,\epsilon} dT + \left(\frac{\partial V}{\partial P}\right)_{T,\epsilon} dP + \Delta V \, d\epsilon, \qquad (4.2.9)$$

where ΔV is the volume change for the chemical reaction under the actual experimetal (as opposed to standard) conditions. Thus, from Eq. (4.2.3),

$$\theta = \frac{1}{V}\left(\frac{\partial V}{\partial T}\right)_P = \frac{1}{V}\left(\frac{\partial V}{\partial T}\right)_{P,\epsilon} + \frac{\Delta V}{V}\left(\frac{\partial \epsilon}{\partial T}\right)_P = \theta_\infty + \frac{\Gamma \, \Delta H \, \Delta V}{RT^2 V}, \qquad (4.2.10)$$

$$C_p = \left(\frac{\partial H}{\partial T}\right)_P = \left(\frac{\partial H}{\partial T}\right)_{P,\epsilon} + \Delta H \left(\frac{\partial \epsilon}{\partial T}\right)_P = C_{p\infty} + \frac{\Gamma \, \Delta H^2}{RT^2} \qquad (4.2.11)$$

or for the heat capacity per unit volume c_p and the heat capacity per mole $\overline{C}_p = C_p/N_t$,

$$c_p = c_{p\infty} + \Gamma_c \,\Delta H^2/RT^2, \qquad \overline{C}_p = \overline{C}_{p\infty} + \Gamma_x \,\Delta H^2/RT^2.$$

With the requirement that at equilibrium $\Delta H = T\,\Delta S$, Eq. (4.2.11) permits the simplification of Eq. (4.1.18) to

$$\tau_{T,P}/\tau_{S,P} = C_p/C_{p\infty}. \qquad (4.2.12)$$

The chemical contribution to the heat capacity is not usually large. For example, for a two-state equilibrium in a pure substance, $\Gamma_x = K/(1 + K)^2$, where K is the equilibrium constant for the process in question. For processes such as equilibrium between two rotational isomers, for which ΔS° is presumably small, $K = \exp(-\Delta H^\circ/RT)$ and the chemical contribution to the heat capacity reaches a maximum value of $(\overline{C}_p - \overline{C}_{p\infty})/R = 0.44$ when $\Delta H^\circ/RT = 2.4$. In the absence of any large contribution from conformational or chemical equilibrium, $\overline{C}_p/R$ for "typical" liquids at room temperature is approximately 10.

For processes in solution, even with an optimal equilibrium constant, Γ is smaller than in a pure liquid. Thus, $\Gamma(\Delta H/RT)^2$ is usually considerably less than $C_{p\infty}/R$ and in most cases it is possible to ignore the distinction between adiabatic and isothermal relaxation times.[43]

To obtain the chemical contribution to the equilibrium adiabatic compressibility, one may specialize Eq. (4.2.9) to adiabatic conditions:

$$\left(\frac{\partial V}{\partial P}\right)_S = \left(\frac{\partial V}{\partial T}\right)_{P,\epsilon}\left(\frac{\partial T}{\partial P}\right)_S + \left(\frac{\partial V}{\partial P}\right)_{T,\epsilon} + \Delta V\left(\frac{\partial \epsilon}{\partial P}\right)_S, \quad (4.2.13)$$

$$\left(\frac{\partial \epsilon}{\partial P}\right)_S = \frac{\Gamma}{RT}\left\{\frac{\Delta H}{T}\left(\frac{\partial T}{\partial P}\right)_S - \Delta V\right\} \qquad (4.2.14)$$

or, substituting Eq. (4.2.14) in (4.2.13) and dividing both sides by V,

$$-\beta_{S0} = \frac{1}{V}\left(\frac{\partial V}{\partial P}\right)_S$$

$$= \frac{1}{V}\left\{\left(\frac{\partial V}{\partial T}\right)_{P,\epsilon} + \frac{\Gamma\,\Delta V\,\Delta H}{RT^2}\right\}\left(\frac{\partial T}{\partial P}\right)_S + \frac{1}{V}\left(\frac{\partial V}{\partial P}\right)_{T,\epsilon} - \frac{\Gamma(\Delta V)^2}{RTV}$$

and using Eq. (4.2.1) and (4.2.10) one obtains

$$-\beta_{S0} = \frac{\theta^2 TV}{C_p} + \frac{1}{V}\left(\frac{\partial V}{\partial P}\right)_{T,\epsilon} - \frac{\Gamma(\Delta V)^2}{VRT}.$$

Since the adiabatic compressibility at infinite frequency is

$$-V^{-1}\,(\partial V/\partial P)_{T,\epsilon} - TV\theta_\infty^2/C_{p\infty},$$

[43] J. Meixner, *Kolloid Z.* **134**, 3 (1953).

then

$$\beta_{S\infty} - \beta_{S0} = TV \left\{\frac{\theta^2}{C_p} - \frac{\theta_\infty{}^2}{C_{p\infty}}\right\} - \frac{\Gamma(\Delta V)^2}{RTV}. \qquad (4.2.15)$$

By appropriate substitution of Eqs. (4.2.10) and (4.2.11) and some algebraic rearrangement, Eq. (4.2.15) may be rewritten

$$\beta_{S0} - \beta_{S\infty} = \frac{\Gamma V}{RT}\frac{C_{p\infty}}{C_p}\left\{\frac{\theta_\infty \Delta H}{C_{p\infty}} - \frac{\Delta V}{V}\right\}^2 = \frac{\Gamma_c}{RT}\frac{C_{p\infty}}{C_p}\left\{\frac{\theta_\infty \Delta H}{c_{p\infty}} - \Delta V\right\}^2$$

$$= \frac{\Gamma_c}{RT}\frac{C_p}{C_{p\infty}}\left\{\frac{\theta \Delta H}{c_p} - \Delta V\right\}^2 \qquad (4.2.16)$$

or, if the relaxing heat capacity is small,

$$\beta_{S0} - \beta_{S\infty} = \frac{\Gamma_c}{RT}\left\{\frac{\theta \Delta H}{c_p} - \Delta V\right\}^2. \qquad (4.2.17)$$

From Eq. (4.1.7) the velocity of sound $v = (\rho\beta_S)^{-1/2}$; hence

$$\frac{v_\infty{}^2 - v_0{}^2}{v_\infty{}^2} = \frac{\beta_{S0} - \beta_{S\infty}}{\beta_{S0}} = \rho v_0{}^2 \frac{\Gamma_c}{RT}\left\{\frac{\theta \Delta H}{c_p} - \Delta V\right\}^2 \frac{C_p}{C_{p\infty}} \quad (4.2.18)$$

and the constant $\mathscr{A}$ in Eq. (4.1.21) is

$$\mathscr{A} = 2\pi^2\rho \frac{v_0{}^2}{v}\frac{C_p}{C_{p\infty}}\frac{\Gamma_c}{RT}\left\{\frac{\theta \Delta H}{c_p} - \Delta V\right\}^2$$

$$= 2\pi^2\rho \frac{v_0{}^2}{v}\frac{C_{p\infty}}{C_p}\frac{\Gamma_c}{RT}\left\{\frac{\theta_\infty \Delta H}{c_{p\infty}} - \Delta V\right\}^2 \qquad (4.2.19a)$$

$$\cong 2\pi^2\rho v_0 \frac{\Gamma_c}{RT}\left\{\frac{\theta \Delta H}{c_p} - \Delta V\right\}^2. \qquad (4.2.19b)$$

It is Eq. (4.2.19b), satisfactory when the dispersion, relaxing heat capacity, and relaxing thermal expansion are small, that has been the usual basis for the estimation and analysis of relaxation amplitudes.

It is possible to devise examples, principally reactions with large $\Delta H°$ and large compensatory entropy changes of the same sign, in which the use of (4.2.19b) is not altogether appropriate. For example, for a reaction of the form $A \rightleftharpoons B$ in 1 molar aqueous solution at 298°K with $\Delta H° = \Delta H = 10,000$ cal, $\Delta V = 10$ ml and $\Delta S° = 33.6$ cal/°K, the equilibrium constant is one and $c_p - c_{p\infty} = 0.28$ cal/°K ml, $C_p/C_{p\infty} = 1.29$, $\theta - \theta_\infty = -2.8 \times 10^{-4}/°K$, and $\theta_\infty \Delta H/c_{p\infty} = 2$ ml. The value of $(\beta_{S0} - \beta_{S\infty})/\beta_{S0}$ calculated from Eq. (4.2.16) is 0.05 and $\mathscr{A} = 6.7 \times 10^{-6}$ sec/cm. An error in $\mathscr{A}$ of 28% is made by the use of Eq. (4.2.19b), and there is a similar error in the assumption that the adiabatic and isothermal relaxation times are equivalent. The adiabatic relaxation time at constant volume $\tau_{SV} = (\beta_{S\infty}/\beta_{S0})\tau_{SP}$ is about 5% less than the value at constant pressure.

Thus on occasion one may encounter a reaction with simultaneously a large ΔH and a large Γ, in which case Eq. (4.2.19a) is the appropriate form for the amplitude and the distinctions between the various relaxation times implicit in Eq. (4.1.12) are of some importance. In any case, when the amplitude is large, Eq. (4.1.8b) is the more appropriate form for α. More usually, however, Eqs. (4.1.8a) and (4.2.19b) suffice.

For a reaction to be amenable to kinetic study by ultrasonic techniques, there must be an appreciable ΔV or, in nonaqueous systems, an appreciable ΔH and all the reactants and products must be present at equilibrium at appreciable concentrations. The calorimetric data necessary for the evaluation of ΔH and the equilibrium composition are extensively and systematically tabulated.[44] Volume changes are perhaps not always so conveniently available. However, as discussed previously, in aqueous solution it is usually the volume change that provides the important coupling between an acoustic wave and a chemical equilibrium. We have, therefore, provided a short table of volume changes for some representative chemical reactions and a number of references to the literature. Of the classes of reactions enumerated in Table II, proton-transfer reactions, association reactions between small ions or small molecules, and conformational changes in small molecules have been extensively and successfully investigated by ultrasonic techniques. Conformational changes in large molecules and micelle formation have likewise been widely studied, but these systems present greater interpretative difficulties and the results of such studies are correspondingly less conclusive.

4.2.3 The Evaluation of Γ in a Nonideal System

The activity of any component of a real solution is conventionally specified in terms of the mole fraction and an activity coefficient γ_i by $a_i = \gamma_i X_i$. The activity coefficient will, in general, depend on the composition as well as on the temperature and pressure. Thus, from Eq. (4.1.19),

$$\Gamma^{-1} = \frac{\partial}{\partial \epsilon} \left(\sum_i \nu_i \ln \gamma_i X_i \right)$$

$$= \frac{1}{N_t} \left\{ \sum_i \frac{\nu_i^2}{X_i} - \left(\sum_i \nu_i \right)^2 + \sum_i \nu_i \left(\frac{\partial \ln \gamma_i}{\partial \epsilon} \right)_{T,P} \right\}$$

$$= \frac{1}{N_t} \left\{ \sum_i \frac{\nu_i^2}{X_i} - \left(\sum_i \nu_i \right)^2 + \sum_{ij} \left(\frac{\partial \ln \gamma_i}{\partial X_j} \right)_{T,P} \left[\nu_i \nu_j - \nu_i X_j \sum_k \nu_k \right] \right\}.$$

$$(4.2.20)$$

[44] U.S. National Bureau of Standards Circ. 500. US Govt. Printing Office, Washington, D.C.; "Handbook of Chemistry and Physics." CRC Press, Cleveland, Ohio; Thermal Properties of Aqueous Uni-Univalent Electrolytes, NSRDS Circ. 2. US Govt. Printing Office, Washington, D.C., 1965.

TABLE II. Volume Changes for Representative Reactions[a]

Reaction	Formula	Value (ml)	Ref.
Acidic and basic ionization reactions of small molecules[45-52]	$CH_3COOH \longrightarrow CH_3COO^- + H^+$	-10.3	45
	$HOOC—COOH \longrightarrow HOOC—COO^- + H^+$	-10.5	47
	$HOOC—COO^- \longrightarrow C_2O_4^= + H^+$	-15.0	47
	$H_2O \longrightarrow OH^- + H^+$	-21.0	45
	$H_2PO_4^- \longrightarrow HPO_4^= + H^+$	-21.7	51
	$CO_2 + H_2O \longrightarrow HCO_3^- + H^+$	-27.6[b]	53
		-88[c]	
Interconversion between classical and zwitterionic forms[46,55,56]	(valine)$^+NH_3—CH(C_3H_7)—COOH \longrightarrow {}^+NH_3—CH(C_3H_7)—COO^- + H^+$	-8	49
	$CN^- + H_2O \longrightarrow HCN + OH^-$	-12	54
	$^-OOC—CH_2—NH_2 + H_2O \longrightarrow {}^-OOC—CH_2—NH_3^+ + OH^-$	-26	48
	$CH_3NH_2 + H_2O \longrightarrow CH_3NH_3^+ + OH^-$	-27	45
	$NH_2—CH_2—COOH \longrightarrow {}^+NH_3—CH_2—COO^-$	-13.5	46
Ion Pairing[58]	$Be^{2+} + SO_4^= \longrightarrow Be^{2+}—H_2O—SO_4^{2-}$ (free ions) → (solvent-separated ion pair)	10	57
	$Be^{2+}—H_2O—SO_4^= \longrightarrow Be^{2+}SO_4^= + H_2O$ (solvent-separated ion pair) → ("inner sphere" ion pair)	3	57
	$Li^+ + Cl^- \longrightarrow Li^+Cl^-$ (in 2-propanol)		
Other association reactions	Na^+ + dinactin (in methanol)	21.2	59
	Dimerization of adenosine-5'-phosphate	-16.5	60
	Stacking of 9-methyl purine	±7.8	61
	Stacking of $^6N, {}^9N$ dimethyl adenine	-4	62
Binding of small molecules to proteins[66]	myoglobin + F^-	-7	63
	myoglobin + imidazole	-3.3	64
	myoglobin + $HCOO^-$	0	64
	ribonuclease + cytidine 2'3' monophosphate	7.5	64
	"riboflavin binding protein" + flavin mononucleotide	23	65,66
		-3.3	67

Class	Reaction	Value	Reference
Conformational change[33]	Isomerization of 1,1,2 trichloroethane (in n-heptane)	~1	68
	cis-trans isomerization of ethyl acetate	0	69
	syn-anti conversion in adenosine	0.5	70,71
	Helix-coil transition in poly-L-glutamic acid	0.5–1.0	72
	Helix-coil transition in poly-L-lysine	1.0–1.5	73
	Internal complexation between ring systems in flavinyl trypotophan peptides	−1.8–4.3	74
Denaturation of proteins[75-77]	Chymotrypsinogen	−94	75
Macromolecular associations[66]	Dimerization of lyzozyme (pH 6.7)	0	76
	Polymerization of myosin	6.4×10^{-4}/gm	77
	2 hexamer $\longrightarrow$ dodecamer, lobster haemocyanin	$<0^d$ 390^e 120^f	78
	Melting of DNA (c. perfringens)	0–8.6×10^{-3}/gm	79
Micelle formation[g]	Na$^+$CH$_3$(CH$_2$)$_7$SO$_4^-$	4.9	80
	Na$^+$CH$_3$(CH$_2$)$_9$SO$_4^-$	7.9	80
	Na$^+$CH$_3$(CH$_2$)$_{11}$SO$_4^-$	10.0	80
	Na$^+$CH$_3$(CH$_2$)$_{13}$SO$_4^-$	11.3	80

[a] Results are in water at 25°C unless otherwise noted. General references for classes of reactions are given when appropriate, as well as references for data cited.

[b] Value computed at 25°C.

[c] Value computed at 250°C.

[d] Value computed at pH = 8.

[e] Value computed at pH = 8.46.

[f] Value computed at pH = 9.6.

[g] Volumes changes per mole of monomer.

In electrolyte solution the activity coefficient is expressed most naturally in terms of the concentration. Hence

$$\frac{\partial}{\partial \epsilon} \sum_i \nu_i \ln \gamma_i = \sum_{ij} \nu_i \left(\frac{\partial \ln \gamma_i}{\partial c_j} \right)_{T,P} \frac{\partial c_j}{\partial \epsilon}$$

$$= \frac{1}{V} \sum_{ij} \left(\frac{\partial \ln \gamma_i}{\partial c_j} \right)_{T,P} [\nu_i \nu_j - \nu_i c_j \, \Delta V] \qquad (4.2.21)$$

[45] H. H. Weber, *Biochem. Z.* **218**, 1 (1930).

[46] E. J. Cohn and J. T. Edsall (eds.), "Proteins, Amino Acids, and Peptides as Ions and Dipolar Ions," p. 159. Van Nostrand-Reinhold, Princeton, New Jersey, 1943.

[47] W. Kauzmann, A. Bodansky, and J. Rasper, *J. Am. Chem. Soc.* **84**, 1777 (1962); W. Kauzmann and J. Rasper, *ibid.* **84**, 1771 (1962).

[48] K. R. Applegate, L. J. Slutsky, and R. C. Parker, *J. Am. Chem. Soc.* **90**, 6909 (1968).

[49] S. Brun, J. Rassing, and E. Wyn-Jones, *Adv. Mol. Relaxation Processes* **5**, 313 (1973).

[50] H. Inoue, *J. Sci. Hiroshima Univ.* **34**, 17 (1970).

[51] F. J. Millero, *in* "Water and Aqueous Solutions" (R. A. Horne, ed.), Chapter 16. Wiley, New York, 1972.

[52] T. Asano and W. J. LeNoble, *Chem. Rev.* **78**, 407 (1978).

[53] A. J. Read, *J. Solution Chem.* **4**, 53 (1975).

[54] J. Stuehr, E. Yeager, T. Sachs, and F. Hvorka, *J. Chem. Phys.* **38**, 587 (1963).

[55] E. J. Cohn, T. L. McMeekin, J. T. Edsall, and M. H. Blanchard, *J. Am. Chem. Soc.* **56**, 784 (1934).

[56] R. D. White and L. J. Slutsky, *J. Phys. Chem.* **75**, 161 (1971); **76**, 1327 (1972).

[57] W. Knoche, *in* "Chemical and Biological Applications of Relaxation Spectrometry" (E. Wyn-Jones, ed.), p. 265. Reidel, Boston, Massachusetts, 1975.

[58] P. Hemmes, *J. Phys. Chem.* **76**, 895 (1977).

[59] T. Noveske, J. Stuehr, and D. F. Evans, *J. Solution Chem.* **1**, 93 (1972).

[60] P. Chock, F. Eggers, M. Eigen, and R. Winkler, *Biophys. Chem.* **6**, 239 (1977).

[61] J. Lang, S. Yiv, and R. Zana, *Stud. Biophys.* **57**, 13 (1976).

[62] U. Gaarz and H. D. Lüdemann, *Ber. Bunsenges. Phys. Chem.* **80**, 607 (1976).

[63] D. Porschke and F. Eggers, *Eur. J. Biochem.* **26**, 490 (1972).

[64] G. B. Ogunmola, W. Kauzmann, and A. Zipp, *Proc. Natl. Acad. Sci. U.S.* **73**, 4271 (1976).

[65] W. F. Harrington and G. Kegeles, *in* "Methods in Enzymol." (C. H. W. Hirs and S. N. Timasheff, eds.), Vol. XXVII, Part D, p. 324. Academic Press, New York, 1973.

[66] G. D. Fasman (ed.), "Handbook of Biochemistry and Molecular Biology," 3rd ed., Proteins, Vol. II, p. 593 CRC Press, Cleveland, Ohio, 1976.

[67] T. M. Li, J. W. Hook, II, H. G. Drickamer, and G. Weber, *Biochemistry* **15**, 3205 (1976).

[68] K. H. Crook and E. Wyn-Jones, *J. Chem. Phys.* **50**, 3445 (1969).

[69] W. M. Slie and T. A. Litovitz, *J. Chem. Phys.* **39**, 1538 (1963).

[70] L. M. Rhodes and P. R. Schimmel, *Biochemistry* **10**, 4426 (1971).

[71] P. R. Hemmes, L. Oppenheimer, and F. Jordan, *J. Am. Chem. Soc.* **96**, 6023 (1974).

[72] H. Noguchi and T. Yang, *Biopolymers* **1**, 359 (1963).

[73] H. Noguchi, *Biopolymers* **4**, 1105 (1966).

[74] A. J. W. G. Visser, T. M. Li, H. G. Drickamer, and G. Weber, *Biochemistry* **16**, 4883 (1977).

[75] S. A. Hawley and R. M. Mitchell, *Biochemistry* **14**, 3257 (1975).

[76] J. F. Brandts, R. J. Oliverira, and C. Westort, *Biochemistry* **2**, 1038 (1970).

and

$$\Gamma^{-1} = \frac{1}{V} \left\{ \sum_i \frac{\nu_i^2}{c_i} - \frac{V}{N_t} \left(\sum_i \nu_i \right)^2 \right.$$

$$\left. + \sum_{ij} \left(\frac{\partial \ln \gamma_i}{\partial c_j} \right)_{T,P} [\nu_i \nu_j - \nu_i c_j \, \Delta V] \right\} \quad (4.2.22)$$

If the activity coefficients depend on the composition only through the ionic strength $I = \frac{1}{2} \sum_i c_i z_i^2$, where z_i is the charge in units of the electronic charge e, then Eq. (4.2.22) simplifies to

$$\Gamma^{-1} = \frac{1}{V} \left\{ \sum_i \frac{\nu_i^2}{c_i} - \frac{V}{N_t} \left(\sum_i \nu_i \right)^2 \right.$$

$$\left. + \frac{1}{2} \sum (\nu_i z_i^2 - c_i z_i^2 \, \Delta V) \sum_i \nu_i \left(\frac{\partial \ln \gamma_i}{\partial I} \right)_{T,P} \right\} \quad (4.2.23)$$

In dilute ionic solution the activity coefficient of a species with charge $z_i e$ in a medium of dielectric constant D_0 may be approximated by the Debye–Hückel limiting law

$$\ln \gamma_i = -(z_i^2 e^2 / 2kTD_0)\kappa. \quad (4.2.24)$$

where, if N_0 is Avogadro's number, the inverse Debye length κ is

$$\kappa = (8\pi N_0 e^2 I / D_0 kT)^{1/2}. \quad (4.2.25)$$

When the Debye–Hückel Law is applicable, the last term of Eq. (4.2.23) becomes

$$\frac{\partial}{\partial \epsilon} \sum_i \nu_i \ln \gamma_i = -\frac{e^2}{2kTD_0} \left(\sum_i \nu_i z_i^2 \right) \left(\frac{\partial \kappa}{\partial I} \right) \left(\frac{\partial I}{\partial \epsilon} \right)$$

$$= -\frac{e^2 \kappa}{4kTD_0 I} \left(\sum_i \nu_i z_i^2 \right) \left(\sum_i z_i^2 \left(\frac{\partial c_i}{\partial \epsilon} \right) \right)$$

$$= -\frac{e^2 \kappa}{8kTVD_0 I} \left(\sum_i \nu_i z_i^2 \right) \left(\sum_i \nu_i z_i^2 - 2I \, \Delta V \right)$$

$$\approx -\frac{e^2 \kappa}{8kTID_0 V} \left(\sum_i \nu_i z_i^2 \right)^2 \quad (4.2.26)$$

[77] A. Zipp and W. Kauzmann, *Biochemistry* **12**, 4217 (1973).
[78] V. P. Saxena, G. Kegeles, and R. Kikas, *Biophys. Chem.* **5**, 161 (1976).
[79] S. A. Hawley and R. M. Macleod, *Biopolymers* **13**, 1417 (1974).
[80] S. Kaneshima, N. Tankaka, T. Tomida, and R. Matuura, *J. Colloid Interface Sci.* **48**, 450 (1974).

$$\Gamma^{-1} = \frac{1}{V} \left\{ \sum_i \frac{\nu_i^2}{c_i} - \frac{V}{N_t} \left(\sum_i \nu_i \right)^2 \right.$$

$$\left. - \frac{e^2\kappa}{8kTD_0I} \left(\sum_i \nu_i z_i^2 \right) \left(\sum_i \nu_i z_i^2 - 2I\,\Delta V \right) \right\} \qquad (4.2.27)$$

and usually with small error,

$$\Gamma^{-1} = \frac{1}{V} \left\{ \sum_i \frac{\nu_i^2}{c_i} - \frac{e^2\kappa}{8kTD_0I} \left(\sum_i \nu_i z_i^2 \right)^2 \right\}. \qquad (4.2.28)$$

If the ionic strength is expressed in moles per liter, then in water at room temperature $\kappa \approx 0.3 \times 10^8 \sqrt{I}$ cm^{-1} and $e^2\kappa/8kTD_0 \approx 0.3\sqrt{I}$. Even in the absence of supporting electrolyte, the ionic strength is of course always larger than the concentration of any ionic species and thus, unless $\sum_i \nu_i z_i^2$ is quite large, the contribution of the variation of the activity coefficients to Γ^{-1} is relatively small.

In solutions that are too concentrated to reasonably expect the Debye-Huckel limiting law to apply, there are a number of theoretical and semi-empirical expressions for the single ion and mean activity coefficients of electrolytes. Among the most useful are the Debye–Hückel equation

$$\ln \gamma_i = -z_i^2 e^2 \kappa / 2kTD_0(1 + d_i\kappa) = -z_i^2 A_\gamma \sqrt{I}/1 + 2d_i\sqrt{I}, \qquad (4.2.29)$$

where $A_\gamma = (2\pi N_0)^{1/2}(e^2/D_0kT)^{3/2}$, $\mathcal{Q} = (8\pi N_0 e^2/D_0kT)^{1/2}$, and d_i is the exclusion radius of the ith ion, and the simplification, suggested by Guggenheim,[81,82] is

$$\ln \gamma_i = -A_\gamma' z_i^2 \sqrt{I}/(1 + \sqrt{I}), \qquad (4.2.30)$$

where I is expressed in moles per liter and $A_\gamma' = A_\gamma(\rho/1000)^{1/2}$ (where ρ is the density of the solvent and a standard value of 3 Å is assumed for d_i). In more concentrated solution the Bronsted equation,[81,82] written

$$\ln \gamma_i = A_\gamma' z_i^2 \frac{\sqrt{I}}{1 + \sqrt{I}} + \sum_j \mathcal{D}_{ij}c_j, \qquad (4.2.31)$$

where j is restricted to ions of charge opposite in sign to that of the ith ion, provides a useful semiempirical form for the activity coefficient. Extensive tables of $\mathcal{D}_{ij}$ for various ion pairs as a function of the temperature are available[82] and these may serve as a basis for the estimation, by analogy, of $\mathcal{D}_{ij}$ in solutions where the equilibrium thermodynamic properties are

[81] E. A. Guggenheim, *Philos. Mag.* **19**, 588 (1935).
[82] J. N. Bronsted, *J. Am. Chem. Soc.* **44**, 1938 (1922); G. N. Lewis, M. Randall, K. S. Pitzer, and L. Brewer, "Thermodynamics," 2nd ed. McGraw-Hill, New York, 1961.

not well known.   Alternatively, Davies has suggested[83] a standard choice of $\mathcal{D}_{ij}$ for all ions.   With this approximation

$$\ln \gamma_i = -A'_\gamma z_i^2 [(I)^{1/2}(1 + I^{1/2})^{-1} + \mathcal{D}I]$$

and   (4.2.32)

$$\Gamma^{-1} = \frac{1}{V} \left\{ \sum \frac{\nu_i^2}{c_i} - A'_\gamma \left(\sum_i \nu_i z_i^2 \right)^2 \left[\frac{1}{\sqrt{I}(1 + \sqrt{I})^2} - \mathcal{D} \right] \right\}$$

Scatchard[84,85] has given empirical expressions for the activity coefficients of dipolar ions in ionic solution that may be used in conjunction with Eq. (4.2.21) to compute Γ.   (Note, however, that the equations in Scatchard and Prentiss[84] and Scatchard[85] are in terms of weight molalities and must be transformed[86] to moles per milliliter).   Hussey and Edmonds[87] have discussed explicitly the effect of various approximate forms on the calculated value of Γ for aqueous solutions of glycine.

4.3. Linearized Rate Equations

4.3.1. Rate Laws, Elementary Steps, Reaction Mechanisms

The term "rate law" is us used here to designate an expression that relates the rate (in terms of molar concentrations) of a chemical reaction proceeding isothermally to the concentrations of the reactants; that is, $R_j = V^{-1} d\epsilon_j/dt = V^{-1}r_j = d\zeta_j/dt = f([B_1], [B_2], \ldots, [B_i], \ldots)$. Often even a reaction of rather simple stoichiometry will occur *via* a sequence of elementary steps, a "reaction mechanism," which may involve the formation of intermediate species or the participation of substances not appearing in the equation which represents the overall process, and it is not, in general, possible to deduce the rate law from the stoichiometric equation for the reaction.   However, the rates of each of the elementary steps should be proportional to the number of encounters among the participating species.   Thus, in an ideal system, for each elementary step, although not for the overall reaction,

$$R_f = k_f \prod_i [B_i]^{|\nu_i|}, \qquad R_b = k_b \prod_i [B_i]^{|\nu_i|}, \qquad (4.3.1)$$

[83] C. W. Davies, "Ion Association." Butterworths, London, 1962.

[84] G. Scatchard and S. S. Prentiss, *J. Am. Chem. Soc.* **86**, 2314 (1934).

[85] G. Scatchard, "Proteins, Amino Acids and Peptides as Ions and Dipolar Ions" (E. J. Cohn and J. T. Edsall, eds.).   Van Nostrand-Reinhold, Princeton, New Jersey, 1943.

[86] H. Harned and B. Owen, "The Physical Chemistry of Electrolytes." Van Nostrand-Reinhold, Princeton, New Jersey, 1958.

[87] M. Hussey and P. D. Edmonds, *J. Acoust. Soc. Am.* **49**, 1907 (1971).

where R_f and R_b and k_f and k_b are, respectively, the rates and rate constants for the forward and reverse reactions, and the continued product extends only over reactants for the forward reaction and only over products for the reverse reaction.

In a nonideal mixture, Eq. (4.3.1) becomes[88]

$$R_f = k_f' \left(\prod_i \gamma_i^{|\nu_i|} \right) \left(\prod_i [B_i]^{|\nu_i|} \right), \qquad R_b = k_b' \left(\prod_i \gamma_i^{|\nu_i|} \right) \left(\prod_i [B_i]^{|\nu_i|} \right),$$
(4.3.2)

where γ_i is the activity coefficient of the ith species. The rate constants k_f' and k_b' are functions of the temperature and pressure, as are k_f and k_b, but the activity coefficients will, in general, depend on the concentration; thus in a nonideal solution, even for an elementary step, deviations from the simple mass action expression for the rate [Eq. (4.3.1)] may be encountered. However, in many systems Eq. (4.3.1) does in fact give an adequate description of the rate of change of the composition and in any case the use of Eq. (4.3.2) introduces considerable algebraic and numerical complication but no conceptual difficulties. Thus, we shall initially discuss reaction kinetics in an ideal system and then extend the argument to include nonideal mixtures.

4.3.2. A Simple Example

If the reaction $HA \rightleftharpoons H^+ + A^-$ is displaced from equilibrium, the molar concentrations of the various reactants $[B_i]$ may be expressed in terms of the advancement and the equilibrium concentrations $[B_i]_0$ by $[HA] = [HA]_0 - \zeta$, $[H^+] = [H^+]_0 + \zeta$, and $[A^-] = [A^-]_0 + \zeta$, or in general $[B_i] = [B_i]_0 + \nu_i \zeta$. If the simple mass action expression for the rate of the reaction is valid, then

$$d[HA]/dt = -d\zeta/dt = k_f[HA] - k_b[H^+][A^-]$$
$$= k_f([HA]_0 - \zeta) - k_b([H^+]_0 + \zeta)([A^-]_0 + \zeta)$$
(4.3.3)

Neglecting terms in ζ^2 and making use of the requirement that the net rate at equilibrium $(k_f[HA]_0 - k_b[H^+]_0[A^-]_0)$ is zero, one obtains

$$d\zeta/dt = -\{k_f + k_b([H^+]_0 + [A^-]_0)\}\zeta = -k_b\{K + [H^+]_0 + [A^-]_0\}\zeta,$$
(4.3.4)

where $K = k_f/k_b$ is the equilibrium constant for the dissociation reaction. The solution to Eq. (4.3.4) is then $\zeta = \zeta_0 \exp(-\omega_r t)$ or $\zeta = \zeta_0 \exp(-t/\tau)$,

---

[88] I. Amdur and G. Hammes, "Chemical Kinetics." McGraw-Hill, New York, 1966.

where

$$\omega_r = 1/\tau = k_f + k_b([H^+]_0 + [A^-]_0). \tag{4.3.5}$$

The relaxation time τ or relaxation frequency ω_r corresponding to any presumed rate law is easily obtained, as above, as a function of rate constants and equilibrium concentration of the reactants by substitution of $[B_i] = [B_i]_0 + \nu_i \zeta$ into the rate law and the linearization and solution of the resulting equation subject to the requirement that the rate at equilibrium is zero. Bernasconi,[35] Castellan,[89] and Eigen and DeMaeyer[37] have given tabulations of formulas for various rate laws.

4.3.3. Coupled Reactions

Consider the dissociation of an unsymmetrical dibasic acid

$$HABH \underset{k_b}{\overset{k_f}{\rightleftharpoons}} HAB^- + H^+, \tag{4.3.6a}$$

$$HABH \underset{k_b'}{\overset{k_f'}{\rightleftharpoons}} ABH^- + H^+. \tag{4.3.6b}$$

If the advancements of reactions (4.3.6a) and (4.3.6b) are, respectively, ζ_1 and ζ_2, then $[HABH] = [HABH]_0 - \zeta_1 - \zeta_2$, $[HAB^-] = [HAB^-]_0 + \zeta_1$, $[ABH^-] = [ABH^-]_0 + \zeta_2$, and $[H^+] = [H^+]_0 + \zeta_1 + \zeta_2$. More generally, if ν_{ij} is the coefficient of the ith species in the equation that represents the jth reaction, $[B_i] = [B_i]_0 + \Sigma_j \nu_{ij}\zeta_j$, and the linearized kinetic equations corresponding to Eq. (4.3.6) are

$$\begin{aligned}
d[HAB^-]/dt = d\zeta_1/dt &= k_f[HABH] - k_b[H^+][HAB^-] \\
&= -\{k_f + k_b([HAB^-]_0 + [H^+]_0)\}\zeta_1 \\
&\quad - \{k_f + k_b[HAB^-]_0\}\zeta_2, \tag{4.3.7}
\end{aligned}$$

$$\begin{aligned}
d[ABH^-]/dt = d\zeta_2/dt &= -\{k_f' + k_b'[ABH^-]_0\}\zeta_1 \\
&\quad - \{k_f' + k_b'([ABH^-]_0 + [H^+]_0)\}\zeta_2.
\end{aligned}$$

Substitution of $\zeta_1 = q_1 \exp(-\omega t)$, $\zeta_2 = q_2 \exp(-\omega t)$, or, in general,

$$\zeta_i = q_i e^{-\omega t} \tag{4.3.8}$$

into Eq. (4.3.7) gives a pair of linear homogeneous equations in the amplitudes q_1 and q_2:

$$\{k_f + k_b([HAB^-]_0 + [H^+]_0) - \omega\}q_1 + \{k_f + k_b[HAB^-]_0\}q_2 = 0,$$

$$\{k_f' + k_b'[ABH^-]_0\}q_1 + \{k_f' + k_b'([ABH^-]_0 + [H^+]_0) - \omega\}q_2 = 0,$$

[89] G. W. Castellan, *Ber. Bunsenges. Phys. Chem.* **67**, 898 (1963).

which have either the trivial solution $q_1 = q_2 = 0$ or

$$\begin{vmatrix} k_f + k_b([HAB^-]_0 + [H^+]_0) - \omega & k_f + k_b[HAB^-]_0 \\ k_f' + k_b'[ABH^-]_0 & k_f' + k_b'([ABH^-]_0 + [H^+]) - \omega \end{vmatrix} = 0.$$

For a system of n reactions linearization of the rate equations and substitution of Eq. (4.3.8) gives an $n \times n$ secular equation analogous to Eq. (4.3.9). The linear response of a system with n reactions to a step or periodic perturbation is a superposition of the n independent normal modes of relaxation defined by the eigenvectors of the secular determinant with characteristic times given by the corresponding eigenvalues.

In the case of two coupled reactions, the quadratic secular equation that determines the relaxation frequencies and the relative magnitudes of q_1 and q_2 is easily solved algebraically. For more complex schemes numerical or approximate methods are necessary and these shall be briefly discussed in Section 4.3.5. By way of illustration, we consider here the simple case in which $k_f = k_f'$ and $k_b = k_b'$. With this additional symmetry, the eigenvectors are $q_1 = 1$, $q_2 = 1$ and $q_1 = 1$, $q_2 = -1$, the corresponding relaxation frequencies and normal reactions being

$$\mathbf{q} = \begin{pmatrix} 1 \\ 1 \end{pmatrix}, \qquad \mathbf{q} = \begin{pmatrix} 1 \\ -1 \end{pmatrix},$$

$$\omega_r = 2k_f + k_b([ABH^-]_0 + [HAB^-]_0 + 2[H^+]_0) \qquad \omega_r = k_b[H^+]_0,$$

$$= 2k_b(K + [ABH^-]_0 + [H^+]_0),$$

$$2HABH \rightleftharpoons HAB^- + ABH^- + 2H^+, \qquad HAB^- \rightleftharpoons ABH^-,$$

where $K = k_f/k_b = k_f'/k_b'$ and, for the simple case here, $[ABH^-]_0 = [HAB^-]_0$.

The normal modes of the system of two dissociation reactions represented by Eq. (4.3.6) (the "basis reactions") are, for the symmetrical example chosen here, overall dissociation and intramolecular proton transfer (not independent dissociation of $-AH$ and $-BH$). The volume and enthalpy changes for the normal reactions are simply the sum and difference of the volume and enthalpy changes of the basis reactions. More generally, if the volume and enthalpy changes for the basis reactions are represented as row vectors [i.e., $\Delta\mathbf{V} = (\Delta V_1, \Delta V_2, \ldots)$ and $\Delta\mathbf{H} = (\Delta H_1, \Delta H_2, \ldots)$] and $\mathbf{q}_k$ is the eigenvector corresponding to the kth normal mode, then the volume and enthalpy changes for the kth normal reaction (ΔV_k and ΔH_k) are simply

$$\Delta V_k = \mathbf{q}_k \cdot \Delta\mathbf{V}, \qquad \Delta H_k = \mathbf{q}_k \cdot \Delta\mathbf{H}. \qquad (4.3.10)$$

As may be seen by inspection of Eq. (4.3.9), the secular determinant is not, in general, symmetric, and it is the transpose of the matrix generated by the elementary methods of this section that should be used for the explicit computation of the eigenvectors.

4.3.4. Redundant Reactions

Intramolecular proton transfer based on Eq. (4.3.6) proceeds by ionization at one site and bimolecular recombination at the other. However, there might well be competing unimolecular modes of internal proton exchange either direct or through water bridges as represented schematically in Eq. (4.3.11).

$$
\begin{pmatrix} B^- \\ \quad_{A}\diagdown^{H} \end{pmatrix} \rightleftarrows \begin{pmatrix} B\diagdown^H \\ \quad_{A^-} \end{pmatrix} \quad \text{or} \quad \begin{pmatrix} B^- \quad H\underset{O}{}H \\ \quad_{A}\diagdown^{H} \end{pmatrix} \rightleftarrows \begin{pmatrix} B\diagdown^{H} \quad H\underset{O}{}H \\ \quad_{A^-} \end{pmatrix} \tag{4.3.11}
$$

or generally

$$
\text{HAB}^- \underset{k_{\text{b}}''}{\overset{k_{\text{f}}''}{\rightleftharpoons}} \text{ABH}^-.
$$

The addition of Eq. (4.3.11) to the basis set adds nothing to the thermodynamic constraints on the equilibrium composition of the system since Eq. (4.3.11) is a linear combination of equations already in the basis set [i.e., Eq. (4.3.6b) − Eq. (4.3.6a)]. However, Eq. (4.3.11) does represent a new path for reaction and if direct transfers proceed with appreciable rate, Eq. (4.3.11) must be added to the kinetic basis set.

For the basis set of Eqs. (4.3.6) and (4.3.11), Eq. (4.3.7) becomes

$$[\text{HABH}] = [\text{HABH}]_0 - \zeta_1 - \zeta_2, \qquad [\text{HAB}^-] = [\text{HAB}^-]_0 + \zeta_1 - \zeta_3,$$

$$[\text{ABH}^-] = [\text{ABH}^-]_0 + \zeta_2 + \zeta_3, \qquad [\text{H}^+] = [\text{H}^+]_0 + \zeta_1 + \zeta_2,$$

and the linearized kinetic equations

$$d\zeta_1/dt = -\{k_{\text{f}} + k_{\text{b}}([\text{H}^+]_0 + [\text{HAB}^-]_0)\}\zeta_1 - \{k_{\text{f}} + k_{\text{b}}[\text{HAB}^-]_0\}\zeta_2$$
$$+ k_{\text{b}}[\text{H}^+]_0\zeta_3,$$

$$d\zeta_2/dt = -\{k_{\text{f}}' + k_{\text{b}}'[\text{ABH}^-]_0\}\zeta_1 - \{k_{\text{f}}' + k_{\text{b}}'([\text{ABH}^-]_0 + [\text{H}^+]_0)\}\zeta_2$$
$$- k_{\text{b}}'[\text{H}^+]_0\zeta_3,$$

$$d\zeta_3/dt = k_{\text{f}}''\zeta_1 - k_{\text{b}}''\zeta_2 - (k_{\text{f}}'' + k_{\text{b}}'')\zeta_3.$$

The secular equation is

$$\begin{vmatrix} k_f + k_b([H^+]_0 + [HAB^-]_0) - \omega & k_f + k_b[HAB^-]_0 & -k_b[H^+]_0 \\ k'_f + k'_b[ABH^-]_0 & k'_f + k'_b([ABH^-]_0 + [H^+]_0) - \omega & k'_b[H^+]_0 \\ -k''_f & k''_b & k''_f + k''_b - \omega \end{vmatrix}$$

equals 0, where $\omega = 0$ is an eigenvalue independent of any assumptions about the rate constants. The corresponding normal reaction is Eq. (4.3.6a) − Eq. (4.3.6b) + Eq. (4.3.11) or $(ABH^- \rightarrow HAB^-) + (HAB^- \rightarrow ABH^-)$ or no reaction. In general, the introduction of p linear dependences into the basis set will introduce p zero eigenvalues.

If, as previously, we assume $k_f = k'_f$ and $k_b = k'_b$ and, hence, $k''_f = k''_b$, then $q_1 = q_2 = 1$, $q_3 = 0$ is an eigenvector; neither the eigenvector nor the eigenvalue corresponding to overall dissociation is affected by the introduction of Eq. (4.3.11). The third eigenvector and eigenvalue are

$$\mathbf{q} = \begin{pmatrix} 1 \\ -1 \\ -2k''_f/k_b[H^+]_0 \end{pmatrix}; \qquad \omega_r + k_b[H^+]_0 + 2k''_f.$$

Intramolecular proton exchange remains a normal mode, and the eigenvector $\mathbf{q}$ specifies the relative importance of the alternate dissociative and direct pathways.

4.3.5. A More General Formulation

From Eq. (4.3.2), the rate of the jth elementary step is

$$R_j = d\zeta_j/dt = k_{fj} \prod_i a_i^{|\nu_{ij}|} - k_{bj} \prod_i a_i^{|\nu_{ij}|}, \qquad (4.3.12)$$

where in the first term the continued product extends over reactants in the second term over products. At equilibrium the rates of the forward and reverse reactions are equal, so, defining the exchange rate[89] $\overline{R}_j$ in terms of the equilibrium activities a_{0i}, one obtains

$$\overline{R}_j = k_{fj} \prod_i a_{0i}^{|\nu_{ij}|} = k_{bj} \prod_i a_{0i}^{|\nu_{ij}|},$$

$$(4.3.13)$$

$$R_j = \overline{R}_j \left\{ \prod_i (a_i/a_{0i})^{|\nu_{ij}|} - \prod_i (a_i/a_{0i})^{|\nu_{ij}|} \right\}.$$

Again the continued product extends over negative ν_{ij} (i.e., reactants) in the first term and positive ν_{ij} in the second term.

The equilibrium constant for the jth reaction is $K_j = \prod_i a_{0i}^{\nu}$, where the continued product here and in all subsequent relations extends over all i.

If Q_j is defined by $Q_j = \Pi_i \, a_i^{\nu_{ij}}$, then Eq. (4.3.13) may be rewritten

$$R_j = \overline{R}_j \prod_i (a_i/a_{0i})^{|\nu_{ij}|}\{1 - Q_j/K_j\}. \qquad (4.3.14)$$

The affinity of the jth reaction is expressed by $A_j = -\Sigma_i \, \nu_{ij}\mu_i = -\Sigma_i \, \nu_{ij}(\mu_i^0 + RT \ln a_i)$. More generally, in terms of the quantities defined here, $A_j = RT(\ln K_j - \ln Q_j)$ or $Q_j/K_j = \exp(-A_j/RT)$. Thus, when A_j/RT is small, $1 - Q_j/K_j \cong A_j/RT$. For small departure from equilibrium, $a_i/a_{0i} \approx 1$ and $A_j/RT \ll 1$. Hence, Eq. (4.3.14) becomes

$$R_j = \overline{R}_j A_j/RT. \qquad (4.3.15)$$

Near equilibrium at constant temperature and pressure

$$
\begin{aligned}
A_j &= \sum_{j'} \left(\frac{\partial A_j}{\partial \zeta_{j'}}\right)_{T,P} \zeta_{j'} = -RT \sum_{ij'} \nu_{ij} \left(\frac{\partial \ln a_i}{\partial \zeta_{j'}}\right)_{T,P} \zeta_{j'} \\
&= -RT \sum_{ikj'} \nu_{ij} \left(\frac{\partial \ln a_i}{\partial c_k}\right)\left(\frac{\partial c_k}{\partial \zeta_{j'}}\right) \zeta_j \\
&= -RT \sum_{ikj'} \nu_{ij}\nu_{kj'} \left(\frac{\partial \ln a_i}{\partial c_k}\right)_{T,P} \zeta_{j'} \\
&= -RT \sum_{j'} \left\{\sum_{ik} \nu_{kj'} \left(\frac{\partial(\ln c_i + \ln \gamma_i)}{\partial c_k}\right)_{T,P}\right\} \zeta_{j'} \\
&= -RT \sum_{j'} \left\{\sum_i \frac{\nu_{ij}\nu_{ij'}}{c_i} + \sum_{ik} \nu_{ij}\nu_{kj'} \left(\frac{\partial \ln \gamma_i}{\partial c_k}\right)_{T,P}\right\} \zeta_{j'} \\
&= -RT \sum_{j'} g_{jj'} \zeta_{j'}, \qquad (4.3.16)
\end{aligned}
$$

where

$$g_{jj'} = \sum_i \frac{\nu_{ij}\nu_{ij'}}{c_i} + \sum_{ik} \nu_{ij}\nu_{kj'} \left(\frac{\partial \ln \gamma_i}{\partial c_k}\right)_{T,P}. \qquad (4.3.17)$$

Substitution of Eq. (4.3.16) into (4.3.15) gives

$$R_j = \frac{d\zeta_j}{dt} = -\sum_{j'} \overline{R}_j g_{jj'} \zeta_{j'}. \qquad (4.3.18)$$

If for a system of n reactions, $\mathbf{R}$ is an $n \times n$ diagonal matrix with diagonal elements $\overline{R}_j$ as defined by Eq. (4.3.13), $\mathbf{g}$ is an $n \times n$ matrix with elements defined by Eq. (4.3.17), and $\boldsymbol{\zeta}$ is an n-dimensional column vector with components $\zeta_{j'}$, then Eq. (4.3.18) may be written in matrix notation as

$$d\boldsymbol{\zeta}/dt = -\mathbf{R}\mathbf{g}\boldsymbol{\zeta}, \qquad (4.3.19)$$

and on substitution of Eq. (4.3.8), the secular equation

$$|\mathbf{Rg} - \mathbf{I}\omega| = 0, \qquad (4.3.20)$$

which determines the normal modes and relaxation frequencies, is obtained. Here, $\mathbf{I}$ is the $n \times n$ diagonal unit matrix.

Although $\mathbf{R}$ and $\mathbf{g}$ are both symmetric, the product $\mathbf{Rg}$ in general is not. It is often convenient to rewrite Eq. (4.3.20) in a symmetrical form. If $\mathbf{R}^{1/2}$ and its inverse $\mathbf{R}^{-1/2}$ are diagonal matrices with elements, respectively, equal to $R_{jj}^{1/2}$ and $R_{jj}^{-1/2}$ and if new reaction coordinates z defined by $\mathbf{z} = (\mathbf{R}^{-1/2})\boldsymbol{\zeta}$ are introduced, then Eq. (4.3.19) becomes

$$d\mathbf{z}/dt = -\mathbf{R}^{1/2}\mathbf{g}\mathbf{R}^{1/2}\mathbf{z}. \qquad (4.3.21)$$

The symmetrized secular equation

$$|\mathbf{R}^{1/2}\mathbf{g}\mathbf{R}^{1/2} - \mathbf{I}\omega| = 0 \qquad (4.3.22)$$

is better adapted to many computational routines than Eq. (4.3.20). Normal reactions $\mathbf{q}$ in terms of the original basis set $\boldsymbol{\zeta}$ are easily recovered from $\mathbf{q}'$ the eigenvectors of Eq. (4.3.22) by the transformation $\mathbf{q} = \mathbf{R}^{1/2}\mathbf{q}'$. Similarly, the transformation $\mathbf{L}$ which diagonalizes $\mathbf{Rg}$ is given in terms of the matrix $\mathbf{L}'$ which diagonalizes $\mathbf{R}^{1/2}\mathbf{g}\mathbf{R}^{1/2}$ by $\mathbf{L} = \mathbf{R}^{1/2}\mathbf{L}'$. Since the eigenvectors contain an arbitrary multiplicative constant, $\mathbf{R}^{1/2}$ and $\mathbf{R}^{-1/2}$ may be scaled as convenient in these operations. The volume change, the enthalpy change, and Γ^{-1} for any normal reaction will individually depend on the "normalization." The quantity C in Eq. (4.2.19) is independent of the choice of multiplicative constant.

To illustrate the utility of Eq. (4.3.22), we may consider the reactions $A \underset{k_b}{\overset{k_f}{\rightleftharpoons}} B$ and $A \underset{k_{b'}}{\overset{k_{f'}}{\rightleftharpoons}} C$ in ideal solution. In this case

$$\mathbf{R} = \begin{pmatrix} k_f[A]_0 & 0 \\ 0 & k_f'[A]_0 \end{pmatrix},$$

$$\mathbf{g} = \begin{pmatrix} \dfrac{1}{[A]_0} + \dfrac{1}{[B]_0} & \dfrac{1}{[A]_0} \\ \dfrac{1}{[A]_0} & \dfrac{1}{[A]_0} + \dfrac{1}{[C]_0} \end{pmatrix},$$

$$\mathbf{Rg} = \begin{pmatrix} k_f \dfrac{(K+1)}{K} & k_f \\ k_f' & k_f' \dfrac{(K'+1)}{K'} \end{pmatrix},$$

$$\mathbf{R}^{1/2}\mathbf{g}\mathbf{R}^{1/2} = \begin{pmatrix} k_f \dfrac{(K+1)}{K} & (k_f k_f')^{1/2} \\ (k_f k_f')^{1/2} & k_f' \dfrac{(K'+1)}{K'} \end{pmatrix},$$

where K and K' are the equilibrium constants for A $\rightleftharpoons$ B and A $\rightleftharpoons$ C and $[A]_0$ denotes the equilibrium constant of species A, etc. If, for example, $K = 1$, $K' = 2$, $k_f = 10^6$ sec^{-1}, and $k_f' = (4/3) \times 10^6$ sec^{-1}, then the roots of Eq. (4.3.20) or (4.3.22) are $\omega = 3.2 \times 10^6$ sec^{-1} and 0.8×10^6 sec^{-1}. The eigenvectors of (4.3.20) are (1, 1) and 1, -1). The normal reactions in terms of the original basis set are then

$$\begin{pmatrix} 1 & 0 \\ 0 & \dfrac{2}{\sqrt{3}} \end{pmatrix}\begin{pmatrix} 1 \\ -1 \end{pmatrix} = \begin{pmatrix} 1 \\ -\dfrac{2}{\sqrt{3}} \end{pmatrix} \quad \text{or} \quad 1.15C \longrightarrow B + 0.15A,$$

$$\begin{pmatrix} 1 & 0 \\ 0 & \dfrac{2}{\sqrt{3}} \end{pmatrix}\begin{pmatrix} 1 \\ 1 \end{pmatrix} = \begin{pmatrix} 1 \\ \dfrac{2}{\sqrt{3}} \end{pmatrix} \quad \text{or} \quad 2.15A \longrightarrow B + 1.15C,$$

and

$$L = \begin{pmatrix} 1 & 1 \\ \dfrac{2}{\sqrt{3}} & -\dfrac{2}{\sqrt{3}} \end{pmatrix}.$$

The columns of L directly give the coefficients of the basis reactions in the normal reactions. If one chooses to represent the volume changes and enthalpy changes of the basis reactions as column vectors, then Eq. (4.3.10) of for ΔH and Γ^{-1} of the jth normal reaction can be rewritten in matrix notation as

$$\Delta V_j = \mathbf{j} \cdot \mathbf{L}\dagger \, \Delta\mathbf{V}, \qquad \Delta H_j = \mathbf{j} \cdot \mathbf{L}\dagger \, \Delta\mathbf{H}, \qquad \Gamma_j^{-1} = \mathbf{j}\mathbf{L}\dagger\mathbf{g}\mathbf{L}\mathbf{j}\dagger,$$

where $\mathbf{j}$ is a row vector with 1 in the jth position, zero elsewhere, and $\mathbf{j}\dagger$ is the corresponding column vector.

The development here rather closely parallels that of Castellan.[89] White has given a similar derivation[90] in terms of a somewhat more compact notation. Castellan proceeds to show how, in the case of a set of basis reaction with linear dependences, one can reduce the order of Eq. (4.3.20) through the use of the coefficients of the equations that explicitly express the dependent reactions in terms of the subset of reactions chosen to be the independent basis set. If such a procedure is adopted, the eigenvectors of the reduced system of equations are no longer susceptible of easy physical interpretation. Equations (4.3.17) and (4.3.18), which, in the case of reactions in ideal systems, allow the secular equation to be written by inspection from the stoichiometric equations representing the basis reactions and which permit simple intuitive interpretation of the eigenvectors, will generally prove to be a satisfactory basis for calculation.

[90] R. D. White, Ph.D. Thesis, Univ. of Washington, Seattle, Washington (1969).

The numerical solution of secular equations is of course routine, and the determination of the relaxation times and normal reactions of even a large system of basis reactions for a given set of rate constants and equilibrium concentrations present no difficulty. The inverse problem of determining the rate constants from observed relaxation times in the absence of an explicit algebraic solution of the secular equation is not quite so routine. Formally this problem is equivalent to the determination of force constants from molecular vibrational frequencies and the methods[91-94] used by spectroscopists may occasionally be of value.

In order to perform a least-squares adjustment of the rate constants, it is necessary to know the derivatives of the calculated relaxation frequencies with respect to the rate constants. In the absence of an exact or approximate algebraic solution of the secular equation, one may guess "reasonable" trial values of the rate constants and so generate an approximation ($\mathbf{F}^0$) to $\mathbf{Rg}$ or $\mathbf{R}^{1/2}\mathbf{gR}^{1/2}$. The secular equation

$$|\mathbf{F}^0 - \mathbf{I}\omega| = 0$$

so generated may be solved computationally for the eigenvalues $\omega_j{}^0$, the eigenvectors, and hence for the transformation $\mathbf{L}^0$ that diagonalizes $\mathbf{F}^0$. The variation of the eigenvalues when the matrix elements are varied is then approximately

$$\Delta\omega_m = \sum_l (L^0_{lm})^2 \, \Delta F_{ll} + 2 \sum_{p>l,\, lp} L^0_{lm} L^0_{pm} \, \Delta F_{lp}. \qquad (4.3.23)$$

Since the matrix elements are simple linear functions of the rate constants, the variation of the relaxation frequency with variation in each rate constant is easily calculated from Eq. (4.3.23). If $\Delta\omega_m$ is identified with the difference between the observed frequency and $\omega_m{}^0$ and if a number of relaxation frequencies equal to or greater than the number of independent rate constants is observed or, alternatively and more probably, if a single relaxation frequency has been observed at a sufficient number of different equilibrium compositions, then application of linear least squares to Eq. (4.3.23) permits the computation of corrections to the trial rate constants. A new $\mathbf{F}^0$ and a new $\mathbf{L}^0$ may then be calculated and the process repeated until convergence. As in determination of force constants, the identification of a given experimental frequency with a calculated eigenvalue is

[91] D. E. Mann, T. Shimanouchi, J. H. Meal, and L. Fano, *J. Chem. Phys.* **27**, 43, 51 (1957).

[92] W. T. King, I. M. Mills, and B. L. Crawford, *J. Chem. Phys.* **27**, 455 (1957).

[93] J. H. Schachtschneider and R. G. Snyder, *Spectrochem. Acta* **19**, 117 (1963).

[94] J. Overend and J. R. Scherer, *J. Chem. Phys.* **32**, 1289, 1296, 1720 (1960); **33**, 446 (1960).

prerequisite to the numerical procedure and multiple solutions always exist.

4.4. Coupling with Transport and Irreversible Reactions

In this section we are concerned primarily with providing a plausible foundation, if not a genuine theoretical justification, for the assumptions that underlie Eqs. (4.1.10), and (4.1.21), and (4.1.22) and with a brief introduction to the theoretical literature on interaction of sound with a reaction far from equilibrium. With the exception of Patureau et al.,[25] the results of this section have not, in fact, been important in the analysis of experimental data. Section 4.4.1 is offered only as a justification of the conventional analysis and Section 4.4.2 as introducing an (at least to this reviewer) interesting body of predictions not yet extensively exploited experimentally.

4.4.1. Coupling between Reaction and Diffusion

Apart from any conceptual difficulties, a general consideration of the coupling between transport and chemical reaction is inevitably rather involved algebraically. A greatly simplified example may serve to illustrate the general approach of Schurr,[16] to justify partially the assumption made uniformly throughout this part that chemical reaction is the important source of any variation in the local composition induced by an acoustic wave and to lend some credence to the proposition that the contributions of chemical reaction to acoustic absorption and dispersion are to a good approximation additive to the contributions of transport.

Consider the reaction $A \rightleftharpoons B$ in an aqueous solution sufficiently dilute that the solution may be taken to be ideal and at a temperature sufficiently close to the density maximum so that acoustic propagation may be assumed to be isothermal. Some further simplification in notation can be obtained by taking the equilibrium constant at zero applied pressure to be 1 so that, in the absence of external pressure, $k_f = k_b = k$ and the equilibrium concentrations of the reactants are equal or $c_{0A} = c_{0B} = X$.

The partial molal volumes $\overline{V}_A$ and $\overline{V}_B$ of the reactants, the standard volume change for the reaction ΔV^0, and the diffusion coefficient (which will be assumed to be equal to G for both species) are assumed to be independent of the pressure and local composition. We shall consider only coupling between reaction and diffusion and thus the viscous terms in the equation of motion of the fluid shall be neglected.

With these simplifications, the rates of change of the local concentrations of the two reactants are

$$\frac{d\delta c_A}{dt} = -k \left(\delta c_A - \delta c_B - \frac{\Delta V^0 \, \delta PX}{RT} \right) - \nabla \cdot \mathbf{J}_A,$$

$$\frac{d\delta c_B}{dt} = k \left(\delta c_A - \delta c_B - \frac{\Delta V^0 \, \delta PX}{RT} \right) - \nabla \cdot \mathbf{J}_B,$$

$$(4.4.1)$$

where δc_i represents the displacement of the concentration of the ith species from its equilibrium value at zero applied pressure.

The diffusion current density for the ith species $\mathbf{J}_i$ is proportional to the diffusion coefficient, to the local concentration, and to the gradient in the chemical potential. Since the solution is ideal and the temperature uniform, $\nabla \mu_i = \nabla(\mu_i^0 + RT \ln c_i) = \nabla \mu_i^0 + (RT/c_i)\nabla \, \delta c_i$. Taking the divergence of $\mathbf{J}_i$ and retaining only first-order terms in small quantities, Eq. (4.4.1) becomes

$$\frac{d\delta c_A}{dt} = -k \left(\delta c_A - \delta c_B - \frac{X \, \Delta V^0 \, \delta P}{RT} \right) + G \left\{ \frac{\overline{V}_A X \nabla^2 \, \delta P}{RT} + \nabla^2 \, \delta c_A \right\},$$

$$(4.4.2)$$

$$\frac{d\delta c_B}{dt} = k \left(\delta c_A - \delta c_B - \frac{X \, \Delta V^0 \, \delta P}{RT} \right) + G \{ \overline{V}_B X \nabla^2 \, \delta c_B + \nabla^2 \delta c_B \}.$$

To obtain a dispersion relation, the equations of motion and continuity for the fluid

$$\rho \, d\delta \mathbf{v}/dt = -\nabla P,$$

$$\partial \, \delta s / \partial t = -\nabla \cdot \delta \mathbf{v}$$

$$(4.4.3)$$

must be added to the set of relations represented by Eqs. (4.4.2). The velocity of the fluid is $\mathbf{v}$, the density is ρ, and the condensation δs is defined by $\delta s = \delta\rho/\rho$. In terms of the variations in concentration and pressure, δs may be expressed as

$$\delta s = \{M_W/\rho_0 - \overline{V}_A\} \, \delta c_A + \{M_W/\rho_0 - \overline{V}_B\} \, \delta c_B + \beta \, \delta P, \quad (4.4.4)$$

where M_W is the molecular weight of A and B, ρ_0 the equilibrium density, and β the compressibility at constant composition. Solutions of the form

$$\begin{Bmatrix} \delta P \\ \delta c_A \\ \delta c_B \\ \delta \mathbf{v} \\ \delta s \end{Bmatrix} = \begin{Bmatrix} \delta P^0 \\ \delta c_A{}^0 \\ \delta c_B{}^0 \\ \delta \mathbf{v}^0 \\ \delta s^0 \end{Bmatrix} e^{i\mathbf{y} \cdot \mathbf{r}} e^{i\omega t} \qquad (4.4.5)$$

are sought.   The substitution of Eq. (4.4.5) into Eqs. (4.4.2)–(4.4.4) gives

$$(i\omega + k + Gy^2)\,\delta c_A{}^0 - k\,\delta c_B{}^0 = \frac{X}{RT}\,(k\,\Delta V^0 - Gy^2\overline{V}_A)\,\delta P^0,$$

$$-k\,\delta c_A{}^0 + (i\omega + k + Gy^2)\,\delta c_B{}^0 = -\frac{X}{RT}\,(k\,\Delta V^0 + \overline{V}_B Gy^2)\,\delta P^0,$$

$$\left(\frac{y^2}{\omega^2} - \rho\beta\right)\delta P^0 = \frac{M_W}{\rho_0}\,(\delta c_A{}^0 + \delta c_B{}^0)$$

$$- (\overline{V}_A\,\delta c_A{}^0 + \overline{V}_B\,\delta c_B{}^0). \qquad (4.4.6)$$

Introducing the definitions

$$F_1 = \frac{2Gy^2}{i\omega + Gy^2}, \qquad F_2 = \frac{2k + Gy^2}{i\omega + 2k + Gy^2},$$

one obtains the solutions of Eq. (4.4.6)

$$\delta c_A{}^0 = \frac{X}{2RT}\,[\Delta V^0\,F_2 - (\overline{V}_A + \overline{V}_B)F_1]\,\delta P^0,$$

$$\delta c_B{}^0 = -\frac{X}{2RT}\,[\Delta V^0\,F_2 + (\overline{V}_A + \overline{V}_B)F_1]\,\delta P^0,$$

$$\frac{y^2}{\omega^2} = \beta\rho_0 + \frac{X}{RT}\left\{\left[-\frac{M_W}{\rho_0} + \frac{1}{2}(\overline{V}_A + \overline{V}_B)\right](\overline{V}_A + \overline{V}_B)F_1 + \Delta V^{0^2}F_2\right\}.$$

Defining $\tau_d = 1/Gy^2$, one may express F_1 as $F_1 = 2(1 - i\omega\tau_d)/[1 + (\omega\tau_d)^2]$. Since $y \cong \omega/v_\infty$, then $\tau_d \cong v_\infty{}^2/G\omega^2$ and $\omega\tau_d \cong v_\infty{}^2/G\omega$. For a rapidly diffusing species in aqueous solution $G \cong 10^{-5}$ cm²/sec. Thus, even at the highest ultrasonic frequencies ($\omega \cong 10^{11}$ sec), $\tau_d \cong 10^{-6}$, $\omega\tau_d \cong 10^5$ and both the real and imaginary parts of F_1 are negligibly small.   Hence, even at the shortest ultrasonic wavelength, concentration variations due to diffusion can be neglected as can the contribution of diffusion to acoustic absorption and dispersion.

In a viscous medium, the equation of motion (4.4.3) becomes

$$\rho\,\partial\delta v/\partial t = -\nabla\,\delta P + [\tfrac{1}{3}\eta + \eta_V][\nabla\,(\nabla\cdot\delta v) + \nabla^2\,\delta v],$$

where η and η_V are the shear and volume viscosities, respectively.   For a longitudinal wave $\nabla\cdot\mathbf{v} = i|\mathbf{y}||\mathbf{v}|$ and $\nabla^2\mathbf{v} = -y^2\mathbf{v}$; thus for Eq. (4.4.3) one obtains

$$\rho\,\partial\delta v/\partial t = -\{\nabla P + (\tfrac{4}{3}\eta + \eta_V)y^2\,\delta v$$

or

$$\rho i \omega \, \delta v^0 = -iy \, \delta P^0 - (\tfrac{4}{3}\eta + \eta_V) y^2 \, \delta v^0,$$

$$\omega \, \delta s^0 = -y|\delta \mathbf{v}^0|. \tag{4.4.7}$$

The simultaneous solution of Eqs. (4.4.7), (4.4.4), and the linearized chemical rate equation (4.4.2) involves only algebraic complication. Markham et al.,[95] give a general expression for the complex dispersion relation in a viscous medium with a single relaxation that reduces to Eq. (4.1.21) when the viscous and chemical absorptions (per wavelength) are both small. In aqueous solution at room temperature, the *total* classical absorption per wavelength is less than 0.04 at 1000 MHz, the dispersion due to chemical reaction is rarely larger than 1%, and neglect of the coupling between reactive volume change and viscous processes is almost always justifiable.

4.4.2. Reactions Far from Equilibrium

Gilbert et al.[22] have discussed the propagation of acoustic waves in a system with a single chemical reaction sufficiently far from equilibrium that the contribution of the reverse reaction to the time dependence of the chemical composition may be ignored. The local state as characterized by the pressure, velocity, temperature, and chemical composition, all functions of position $\mathbf{r}$ and time t, is designated by a vector $\boldsymbol{\psi}(P, \mathbf{v}, T, X) = \boldsymbol{\psi}(\mathbf{r}, t)$ that is decomposed into a spatially homogeneous part $\boldsymbol{\psi}^0(t)$, which specifies the development in time of the properties of the system averaged over the volume, and a small deviation $\delta\boldsymbol{\psi}(\mathbf{r}, t)$ described in terms of its Fourier components $\delta\boldsymbol{\psi}(\mathbf{y}, t)$.

Formally the equation of motion for $\delta\boldsymbol{\psi}(\mathbf{y}, t)$ is

$$\partial\delta\boldsymbol{\psi}(\mathbf{y}, t)/\partial t = \boldsymbol{\Omega} \, \delta\boldsymbol{\psi}(\mathbf{y}, t). \tag{4.4.8}$$

In Gilbert et al.,[22] the operator $\boldsymbol{\Omega}$ is constructed from the usual linearized hydrodynamic equations, neglecting thermal conductivity and diffusion, and the kinetic equations for the reaction in question, presuming that the rate constant may depend on both the temperature and the pressure. The viscosity and heat capacity are assumed not to vary as the reaction proceeds.

The general result of Gilbert et al.[22] for the pressure amplitude of a

---

[95] J. J. Markham, R. T. Beyer, and R. B. Lindsay, *Rev. Mod. Phys.* **23**, 353 (1951).

standing wave with inverse wavelength $|y|$ is

$$\frac{\delta P(t)}{\delta P(0)} = \left(\frac{\beta_S(0)}{\beta_S(t)}\right)^{1/4} \exp\left[-\nu_{\text{vis}}t + \frac{1}{2}\int_0^t (\Omega_{\text{PP}} + \frac{\theta T}{\rho \bar{c}_p}\Omega_{\text{PT}})\, dt'\right]$$

(4.4.9)

where ρ is the density, $\bar{c}_p$ the heat capacity per unit mass, and

$$\nu_{\text{vis}} = \tfrac{1}{2}\{\tfrac{4}{3}\eta + \eta_V\}y^2/\rho.$$ (4.4.10)

The same result is applicable to the amplitude $\delta P(r, t)$ at the middle $\bar{r}$ of a wave packet composed of a narrow band of Fourier components centered about y and initially centered (in space) at $r = 0$ if $\delta P(t)/\delta P(0)$ is replaced by $\delta P(r, t)/\delta P(0, 0)$, where r is calculated from the (time-dependent) group velocity $(\rho\beta_S)^{-1/2}$ by

$$r = \int_0^t (\rho\beta_S)^{-1/2}\, dt'.$$ (4.4.11)

Gilbert *et al.*[22] evaluate Ω explicitly for reactions of the form A → B and A → B + C in ideal gases with Δc_v for the reaction equal to zero. Their result for A → B, further specialized to the case where the rate coefficient is independent of the pressure, is

$$\frac{\delta P(t)}{\delta P(0)} = \left(\frac{\beta_S(0)}{\beta_S(t)}\right) \exp\left[-\nu_{\text{vis}}t - \int_0^t \frac{X_A}{2\bar{c}_p}\frac{k\,\Delta H}{T}\left\{1 + \frac{P}{\rho\bar{c}_v}\frac{\partial \ln k}{\partial T}\right\} dt'\right].$$

(4.4.12)

If the usual Arrhenius temperature dependence of the rate is assumed, then $(\partial \ln k/\partial T) = E^*/RT^2$, where E^* is the energy of activation. It is clear that for an exothermic reaction sufficiently far from equilibrium the integral in Eq. (4.4.12) is negative and may, for long waves, be greater than $\nu_{\text{vis}}t$. Thus, amplification of a propagating wave packet and instability of the homogeneously evolving state with respect to long-wave acoustic oscillation are predicted.

Patureau *et al.*[25] have studied the propagation of 300-Hz pulses in a mixture of hydrogen and chlorine diluted with argon before, during, and after the interval in which the photochemically initiated reaction $H_2 + Cl_2 \rightarrow 2HCl$ runs its course. They find, if not overall amplification, a decreased attenuation due to the irreversible reaction.

At present it is not unfair to say that the topic of acoustic propagation in irreversibly reacting systems has been rather more popular with theoreticians than experimentalists. The phenomena associated with coupling between the irreversible reaction and a sound wave seem at this time to be more interesting for their own sake than as experimental methods in chemical kinetics.

4.5. Interpretative Problems

The identification of the particular chemical or structural equilibria perturbed in a given acoustic experiment is the principal interpretative problem encountered in the applications of acoustics to chemical kinetics. The specific expressions for the dependence of the relaxation amplitudes and relaxation times on the equilibrium chemical composition, the volume and enthalpy changes for the reaction, and the properties of the medium that are developed in Chapters 4.2 and 4.3 are the usual basis for the attribution of an observed relaxation to a given microscopic process. In Sections 4.5.1 and 4.5.2, examples, drawn from the literature of proton-transfer kinetics, where we believe a reasonably convincing attribution is possible, will be offered by way of illustration. In Section 4.5.3 some possible sources of absorption in polymer and polyelectrolyte solution will be discussed as examples of an area of endeavor where interpretative problems rather than experimental difficulties predominate.

4.5.1. A Straightforward Case

Ionizations of glutamic acid in basic solution of the form

$$R—NH_2 + H_2O \underset{k_b}{\overset{k_f}{\rightleftharpoons}} R—NH_3^+ + OH^- \qquad (4.5.1)$$

are more than usually amenable to study by ultrasonic techniques. The volume changes are large (20–30 ml/mole) and the equilibrium constants are easily determined by classical techniques or estimated by analogy from rather extensive tabulations.[96]

If we denote the total concentration of amine ($[R—NH_2]$ + $[R—NH_3^+]$) by C_0 and the equilibrium constant for the reverse reaction by K, then

$$\Gamma_{ideal} = \frac{KC_0[OH^-]}{KC_0 + (1 + K[OH^-])^2},$$

$$\omega_r = \frac{1}{\tau} = k_b([R—NH_3^+] + [OH^-]) + k_f \qquad (4.5.2)$$

$$= \frac{k_b\{KC_0 + (1 + K[OH^-])\}}{K(1 + K[OH^-])},$$

[96] M. Kotake (ed.), "Constants of Organic Compounds." Asakura Pub., Tokyo, 1963.

where Γ has a maximum value at $[OH^-] = (1 + KC_0)^{1/2}/K$ and τ a maximum at $[OH^-] = (KC_0)^{1/2} - 1$. Unless $R—NH_2$ is a rather strong base and the solution is rather dilute, $KC_0 \gg 1$. Hence the relaxation amplitude $\mathscr{A}\tau$ as a function of pH will be a steep-sided bell-shaped curve with a maximum at

$$pH_{max} = \tfrac{1}{2}(pK_w + pK_a + \log C_0),$$

where pK_w and pK_a are the negative logarithms of, respectively, the dissociation constant of water and the acid ionization constant of $R—NH_3^+$.

In any case, from Eqs. (4.5.2) it is possible to calculate the form of the variation of the amplitude with pH. From a single kinetic parameter k_b determined at any pH, it is possible to predict the form of the variation of α_e with pH. From k_b and a known or estimated volume change, one may deduce the magnitude of α_e due to the process represented by Eq. (4.5.1) at any frequency and pH. Thus it is not at all difficult to decide whether the acoustic absorption which might be observed in basic solutions of an amine is attributable to an ionization reaction. Indeed, if the form of the variation of α_e with pH is such as to justify such an attribution, the equilibrium constant and volume change for the reaction, if not known, are easily deducible from Eqs. (4.5.2) and (4.2.19).

In the example chosen here, 0.5 M L-glutamic acid $[R = —OOC—CH(CH_2)_2COO^-]$ at pH 11.73, α was measured from 3 to 100 MHz and fitted to the form of Eqs. (4.1.21)–(4.1.22) with $\tau = 1.3 \times 10^{-8}$ sec, $\mathscr{A}\tau = 140 \times 10^{-17}$ sec^2/cm, and $\mathscr{B} = 31.8 \times 10^{-17}$ sec^2/cm. The data are presented in the form of a plot of $\alpha_e\lambda$ versus frequency in Fig. 4.

From the known K (2×10^4 liters/mole), k_b and ΔV are deduced from the experimentally determined $\mathscr{A}$ and τ and, assuming ideal-solution behavior, the absorption at 5.26 MHz is calculated as a function of pH and compared with experiment in Fig. 5. The agreement is sufficiently good to nominate the process represented by Eq. (4.5.1), proceeding by the simple kinetic scheme implied by Eq. (4.5.1), as the important source of absorption. Only the fact that $\mathscr{B}$ is significantly higher than the absorption in pure water suggests the existence of a second process (conformational change? association? modification of structural relaxation in water by the solute? perturbation of solvation equilibria?) with a relaxation frequency much higher than the highest measuring frequency.

There are to be sure other proton-transfer processes possible in glutamic acid solution. However, Γ is less than c/ν^2 for the least abundant species and, thus, at the high pH at which these measurements were performed, Γ for any process which involves protonation or deprotonation of the carboxyl groups ($pK_a = 2.115$ and 4.346) are neglible. Similarly, at low pH, proton transfers at the carboxyl groups can be investigated[49,87]

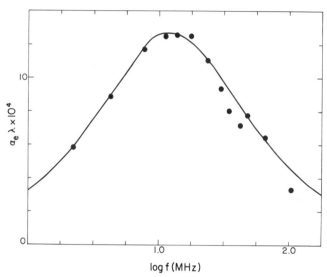

FIG. 4. Plot of $\alpha_e \lambda$ versus frequency for 0.5 M aqueous glutamic acid at 25°C and pH 11.73.

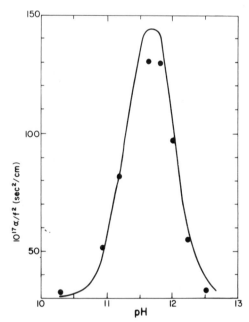

FIG. 5. Plot of α/f^2 at 5.26 MHz as a function of pH for 0.5 M aqueous L-glutamic acid at 25°C; the solid curve represents a theoretical fit using the known dissociation constant and a rate constant and volume change derived from the data in Fig. 4.

uncomplicated by reaction at the amine group.   It will be seen that it is not always possible to choose conditions so that only a single process contributes significantly to the relaxation of proton-transfer equilibria.

Simple reactions of the form of Eq. (4.5.1) or its counterpart in acid solution (HA $\rightleftharpoons$ H$^+$ + A$^-$) have been extensively investigated by ultrasonic techniques.   For weak acids with reasonable exceptions (i.e., sterically hindered or internally hydrogen-bonded systems), the rates of the bimolecular step have been successfully rationalized in terms of the Debye–Smoluchowski theory with r_d, the effective radius for reaction, taken to be of the order of an O–H—O or N–H—O hydrogen bond distance (2.7 Å) and the effective dielectric constant D being equal to the bulk dielectric constant of water.   The Debye–Smoluchowski result is[97]

$$k_b = 4\pi\sigma N_0 z_A z_B e_0^2 (G_A + G_B)/DkT[\exp(z_A z_B e_0^2/Dr_d kT) - 1],$$
(4.5.3)

where N_0 is Avogadro's number, z_A and z_B the algebraic charges of species A and B, e_0 the electronic charge, T the temperature in degrees Kelvin, r_d the reaction radius, G_A and G_B the diffusion coefficients of species A and B, k the Boltzmann constant, and σ the steric factor.   The diffusion coefficient of the hydroxide or hydronium ion is almost always much greater than that of the other acid or base; hence, except for systems with exceptionally small steric factors, k_b is roughly independent of the chemical nature of the acid or base.   The approximate transferrability of bimolecular rate constants as deduced from observations on simple systems will be of considerable assistance in interpreting results in more complex cases.

4.5.2. Ionization Reactions of p-Aminobenzoic Acid: Kinetic Models

In contrast to glutamic acid, in the neutral form of p-aminobenzoic acid, both the classical (HAB) and zwitterionic (ABH) forms are present in appreciable concentrations.   The kinetics of the interconversion of the classical and zwitterionic forms

(4.5.4)

[97] P. Debye, *Trans. Electrochem. Soc.* **82**, 265 (1942).

are not easily accessible to other techniques and the equilibrium constant, 6.8, and volume change, -11.4 ml/mole, are such as to suggest that even with the limited solubility (~ 0.03 M) of aminobenzoic acid in water an ultrasonic method might be appropriate. In slightly acidic solutions of p-aminobenzoic acid, the species AB^-, H^+, and $HABH^+$ are present in appreciable concentration and a reasonable choice of linearly independent basis reactions might be as summarized in the accompanying tabulation.

Reaction	K (M/liter)	ΔV (ml)	No.
$HAB \underset{k_{b1}}{\overset{k_{f1}}{\rightleftharpoons}} H^+ + AB^-$	1.4×10^{-5}	-9.4	(R1)
$ABH \underset{k_{b2}}{\overset{k_{f2}}{\rightleftharpoons}} H^+ + AB^-$	9.3×10^{-5}	2.0	(R2)
$HABH^+ \underset{k_{b3}}{\overset{k_{f3}}{\rightleftharpoons}} HAB + H^+$	6.2×10^{-4}	-9.4	(R3)

The intramolecular proton-exchange process is contained in this set [(R1)–R2)]. The circumstances under which Eq. (4.5.4) corresponds to a kinetic process distinct from ionization and recombination is a question one might hope to answer in an experimental study.

There are, however, a larger number of thermodynamically redundant but kinetically distinct reactions that might also contribute to the relaxation of internal and overall ionization equilibria.

$$HABH^+ \underset{k_{b4}}{\overset{k_{f4}}{\rightleftharpoons}} ABH + H^+, \qquad (R4)$$

$$HABH^+ + {}^-AB \underset{k_{b5}}{\overset{k_{f5}}{\rightleftharpoons}} 2HAB, \qquad (R5)$$

$${}^-AB + HABH^+ \underset{k_{b6}}{\overset{k_{f6}}{\rightleftharpoons}} 2ABH, \qquad (R6)$$

$$HABH^+ + ABH \underset{k_{b7}}{\overset{k_{f7}}{\rightleftharpoons}} HAB + HABH^+, \qquad (R7)$$

$$ABH + {}^-AB \underset{k_{b8}}{\overset{k_{f8}}{\rightleftharpoons}} {}^-AB + HAB. \qquad (R8)$$

The number of independent rate constants in such a scheme is too large to permit progress from an observed relaxation time to rate constants for the individual steps. Indeed, the limited solubility of p-aminobenzoic acid in water makes it rather difficult to determine α at low frequencies in a pulse-echo apparatus with sufficient accuracy to determine uniquely a relaxation time or times (Fig. 6). However, the reverse process of pro-

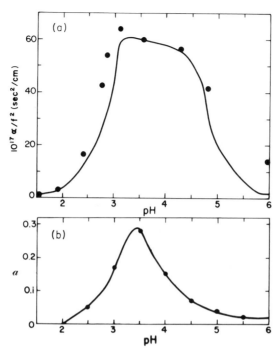

FIG. 6. (a) Measured and calculated (solid curve) values of α/f^2 in 0.025 molar p-aminobenzoic acid at 5.4 MHz and 25°C as a function of pH. (b) The quantity a which specifies the deviation of the important normal mode from pure interconversion of classical and zwitterionic forms is plotted as a function of pH.

ceeding from known equilibrium constants and volume changes and from assumed rate constants, that is, from a kinetic model, to normal reactions, relaxation times, and amplitudes and then to acoustic absorption as a function of pH and frequency, remains relatively easy. Thus, if there is sufficient knowledge of the equilibrium thermodynamics, the adequacy of a complex kinetic scheme may on occasion be tested in circumstances where resolution of the observed acoustic absorption into distinct relaxation times and amplitudes is difficult. The system of equations (R1)–(R8) are offered as an illustration of this approach.

As discussed previously, the rate constants for the bimolecular recombination reaction between H^+ and the carboxylate group [(R1) and (R4)] or the primary amine group [(R2) and (R3)] do not depend greatly on the chemical nature of the acid or amine. Average values for (R1) and (R4) are 4×10^{10} moles/liter sec; 1.5×10^{10} moles/liter sec is typical for (R2) and (R3). Bimolecular proton transfer processes with equilibrium con-

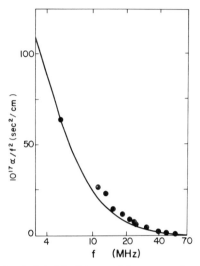

FIG. 7. Measured and calculated (solid curve) frequency dependence of α/f^2 in 0.025 molar p-aminobenzoic acid at pH 3.56 and 25°C.

stants greater than one [(R5–R8)] are in general diffusion controlled. From the known diffusion coefficient of p-aminobenzoic acid (0.843×10^{-5} mole/cm^2 sec) and for a reaction radius of 2.7 Å, Eq. (4.5.3) predicts in water at 25°C a rate constant $k_f = 0.98 \times 10^{10}$ mole/sec when the transfer is between a negatively charged carboxylate group and a positive $-NH_3^+$ as in (R5) and $k_f = 0.34 \times 10^{10}$ mole/sec when the transfer is between neutral groups (protonated carboxyl and $-NH_2$) as in (R6). Reactions (R5)–(R8) all require simultaneous orientation of one reactive site in each molecule along the line joining the two molecules; hence, $\frac{1}{4}$ is a reasonable rough estimate for σ.

The equilibrium constants and volume changes for reactions (R4)–(R8) are deducible from those for the basis reactions (R1)–(R3). Thus, all the information necessary to compute the frequency dependence and pH dependence of the absorption is available. The predictions from the kinetic model shown in Figs. 6 and 7 are in fairly good agreement with experiment and this agreement could easily be improved by chemically reasonable *ad hoc* adjustment of the kinetic parameters.

At pH greater than 2.5, a single mode accounts for more than 90% of the calculated absorption. With a redundant set of reactions there is always latitude in the representation of a normal mode in terms of independent basis reactions. If the important mode is expressed as ABH $\rightarrow$ HAB + a($^-$AB + HABH$^+$ $\rightarrow$ 2HAB) + b(HABH$^+$ $\rightarrow$ HAB + H$^+$), then b is large (3.7) at pH 2, significant at pH 2.5 (.27), and negligible above pH 3. The quantity a is plotted as a function of pH in Fig. 6b. It will be seen that in this analysis that the acoustic absorption is primarily due to perturbation of the equilibrium between the classical and zwitterionic forms of p-aminobenzoic acid, although over part of the pH range there is a significant admixture of the reaction between cationic and anionic forms to give the neutral acid. Moreover, the intramolecular proton transfer proceeds by a series of biomolecular diffusion-controlled steps. There is no reason to adduce a direct intramolecular transfer mechanism.

It should also be noted that the mixture of an ionization reaction into a normal mode, which is principally internal proton transfer, means that there is a second Wien effect and a conductivity change associated with that mode, and one should not immediately reject the possibility that such a process might contribute to the relaxation spectrum observed in, for example, a pulsed E field experiment.[98]

As one proceeds to yet more complicated systems (i.e., biological macromolecules), one must expect to find many cases where it is not possible to resolve the observed acoustic absorption into a set of discrete relaxations, each one identifiable by the methods of Section 4.5.1 with a single elementary process chosen as one of the basis reactions. The alternate analysis illustrated here, of constructing a model consonant with what is known about the physical chemistry of these systems and then using the acoustic absorption to refine the parameters of the model, may on occasion be preferable.

4.5.3. Polymers and Other Complex Systems

Many of the relaxation phenomena observed in polymer solution are of course in no sense peculiar to macromolecules. For example, such phenomena as relaxation of side-chain ionization equilibria in polypeptides or proteins are interpreted by the methods appropriate to small molecules, somewhat complicated by the distribution of effective ionization constants inevitable in a polyelectrolyte. In polymer solution, as discussed previously, Eqs. (4.1.20) and (4.1.21) should be modified by the introduction of a frequency-dependent shear viscosity. Fortunately, there is a simple and fairly satisfactory theory[99] that allows an estimate of the relaxation times and amplitudes associated with shear viscous relaxation.

Explicitly, for the dynamic shear viscosity of a dilute solution of free-draining Gaussian-coiled macromolecules, Zimm and Rouse obtain[99]

$$\eta = \eta_0 + \frac{\tilde{c}}{M_{\rm w}} RT \sum_l \frac{\tau_l}{1 + i\omega\tau_l}, \qquad (4.5.5)$$

[98] M. Eigen and E. Eyring, *J. Am. Chem. Soc.* **84**, 3254 (1962).

[99] B. H. Zimm, *J. Chem. Phys.* **24**, 269 (1956); P. E. Rouse, *ibid.* **21**, 1272 (1953).

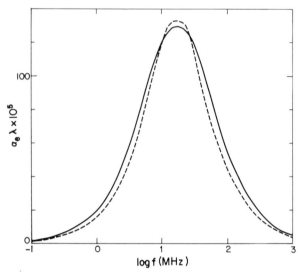

FIG. 8. The solid curve represents $\alpha_e \lambda$ versus frequency for a system with two relaxation process, both with $\mathscr{A} = 6.3 \times 10^{-8}$ sec/cm. For one process $f_r = 10$ MHz ($\mathscr{A}\tau = 100 \times 10^{-17}$ sec^2/cm); for the second process $f_r = 30$ MHz ($\mathscr{A}\tau = 33 \times 10^{-17}$ sec^2/cm). The dashed curve represents $\alpha_e \lambda$ for a single relaxation with $f_r = 17$ MHz and $\mathscr{A} = 1.1 \times 10^{-7}$ sec/cm.

where η_0 is the shear viscosity of the solvent, $\bar{c}$ the concentration in grams of polymer per milliliter, M_w the molecular weight of the polymer, and the relaxation times τ_l are given by

$$\tau_l = 6M_w \eta_0 [\eta]_0 / \pi^2 RT l^2, \qquad (4.5.6)$$

where $[\eta]_0$ is the intrinsic viscosity of the solution at zero frequency. The substitution of Eqs. (4.5.5) and (4.5.6) into Eq. (4.1.20) gives for the acoustic absorption

$$\frac{\alpha}{f^2} = \frac{8\pi^2 RT\bar{c}}{3M_w \rho v^3} \sum_l \frac{\tau_l}{1 + (2\pi f \tau_l)^2}. \qquad (4.5.7)$$

The sum over l in Eqs. (4.5.5) and (4.5.7) extends from 1 to L, the total number of independent segments in the polymer chain. A precise definition of L is not without its difficulties; fortunately, the acoustic absorption is principally associated with the lower l modes and is not very sensitive to the choice of the length of an independent segment of the coil. Equation (4.5.6) has been modified,[99] with only slight complication, to take hydrodynamic interactions into account. Unfortunately, there is not a comparably simple and satisfactory theory of dynamic volume viscosity.

It is natural to seek the simplest explanation of any observed acoustic relaxation, and often when a single relaxation apparently gives a satisfactory account of the data, viscous relaxation is ignored. Figure 8 constitutes an argument in support of the proposition that with only a modest experimental error it is as easy to arrive at an erroneously oversimplified interpretation as it is to err by overinterpretation. The absorption predicted by Eq. (4.5.7) is usually not large. For example, for a 10% by weight aqueous solution of polyethylene oxide of molecular weight 20,000, $\tau_1 \cong 1.6 \times 10^{-7}$ sec $1/2\pi\tau_1 \cong 1$ MHz, and $\mathscr{A}\tau_1 \cong 12 \times 10^{-17}$ sec^2/cm. Shear viscous relaxation is not usually the sole, and often not the principal, acoustic relaxation phenomenon in polymer solution. It is, however, generally present and its contribution is approximately calculable from easily measured quantities.

The results in Markham et al.[95] or Chapter 4.4 imply that the effect of viscous relaxation is, to a good approximation, independent of and additive to the effect of chemical relaxation. A valid interpretation[100] of excess absorption in dilute solutions of flexible macromolecules probably begins with a deduction of the contribution of shear relaxation as obtained from Eq. (4.5.7) or from the somewhat more sophisticated theories which include hydrodynamic interactions. It has been suggested[101,102] that if the contribution of the Zimm–Rouse modes to the volume viscosity were properly evaluated, viscous relaxation might fully account for the observed absorption in such systems as aqueous polyethylene oxide and Dextran. Theoretical progress in this matter would be of considerable utility.

In the absence of a satisfactory theory of the contribution of structural relaxation, it is a distinct temptation to attribute any absorption with concentration dependence appropriate to a unimolecular process to perturbation of solvation equilibria,[103–105] where "solvation" is not necessarily to be understood simply as association between a solvent molecule and a segment of the polymer but includes the induction of altered "structures" of solvent in the neighborhood of the polymer. In the case of aqueous solution in particular, by assuming a suitable degree of cooperativity, one may make plausible whatever volume change one wills and the anticipated characteristic times can be slowed from the very fast relaxations associated with molecular reorientation and structural relaxation in liquid water to a desired value in the ultrasonic range.

[100] M. A. Cochran, J. H. Dunbar, A. M. North, and R. A. Pethrick, J. Chem. Soc. Faraday II 70, 215 (1973).
[101] S. A. Hawley and F. Dunn, J. Chem. Phys. 50, 3523 (1969).
[102] L. W. Kessler, W. D. O'Brien, Jr., and F. Dunn, J. Phys. Chem. 74, 4096 (1970).
[103] G. Hammes and T. B. Lewis, J. Phys. Chem. 70, 1610.
[104] G. Hammes and P. Schimmel, J. Am. Chem. Soc. 89, 442 (1967).
[105] G. Hammes, and P. B. Roberts, J. Am. Chem. Soc. 90, 7119 (1968); J. Chem. Phys. 52, 5496 (1970).

It is to be hoped that more sophisticated computation of solvent–macromolecule[106] potentials, large-scale Monte Carlo and molecular dynamics calculations which may make the kinetic and structural[107] consequences of a potential explicit, and the increased ability of crystallographers to locate the water molecules associated with macromolecules may shortly place some theoretical and experimental constraints on the relaxation parameters associated with solvation, but with the present state of knowledge as it is, it is indeed possible to postulate that perturbation of the translational or orientational distribution of solvent molecules in the neighborhood of a polymer molecule is the principal source of acoustic absorption in a given system or to conclude, in this particular case, as have others more generally[108] that the concept of structural relaxation of solvated water is so ill-defined as to be without substantial utility.

The case of aqueous polyethylene oxide may serve to illustrate some difficulties and possible interpretative pitfalls. Hammes and Schimmel[104] and Hammes and Lewis[103] have made what is, on the basis of the internal ultrasonic evidence, a plausible case for perturbation of a cooperative hydration equilibrium as the important source of ultrasonic absorption. They find that their results can be described by a single relaxation time which increases with increasing molecular weight up to $M_w = 3400$ (presumably the approximate size of the cooperative unit) and is independent of molecular weight thereafter and also that τ at high molecular weight exhibits a rather sharp transition from higher to lower values as the concentration of the "structure-breaking" solute urea is increased. The presumably less cooperative relaxation at low molecular weight is relatively insensitive to the presence of urea.

On the other hand, Kessler et al.[102] working over a somewhat greater range of frequencies found that more than one relaxation was required to describe their data and that the contribution of shear viscous relaxation as calculated from Eq. (4.5.7) was a significant fraction of the observed absorption. Hawley and Dunn[101] argued plausibly that if both shear relaxation and a second process contribute to the absorption, then as the molecular weight increases and the frequency of the first Zimm–Rouse mode drops, f_r for the best single-relaxation approximation to the data will decrease until the viscous relaxation frequency falls well below the lowest measuring frequency at which point τ becomes independent of molecular weight as observed in Hammes and Lewis[103] and Hammes and

[106] L. Carazzo, G. Corongiu, C. Petronglio, and E. Clementi, J. Chem. Phys. 68, 787 (1978).
[107] A. Rahman and F. Stillinger, J. Am. Chem. Soc. 95, 7943 (1973); J. Chem. Phys. 60, 1545 (1974).
[108] A. Holtzer and M. F. Emerson, J. Phys. Chem. 73, 26 (1969).

Schimmel.[104]   Moreover, Jones and Stockmayer[109] found that the spin–spin and spin–lattice relaxation times as well as the chemical shifts of the polymer protons showed only a small and gradual transition with urea concentration arguing strongly against a highly cooperative transition in water structure.   Thus, even in a rather simple polymer in which no chemical relaxation is possible, interpretation of the acoustic observations is difficult.

The helix-coil transition in uniform polypeptides is perhaps the best characterized of the reversible conformational transitions in macromolecular systems.   In particular, the conveniently synthesized, water-soluble, polyamino acids, poly-L-lysine,[110] poly-L-ornithine,[111] and poly-L-glutamic,[112–115] have been fairly extensively studied by ultrasonic and other relaxation techniques.   The number of processes possible in a fairly concentrated solution of flexible polyelectrolytes is large and it is easier to determine a relaxation time and relaxation amplitude than it is to arrive at a unique attribution of the observed relaxation to a given microscopic process.   In the case of the helix-coil transition, perturbation of side-chain ionization or intramolecular proton-transfer equilibria[116] remain reasonable alternative explanations of the acoustic and other relaxation spectra which are observed in the regime of temperature, pH, and ionic strength in which the helix-coil transition occurs and neither studies by pulsed electrical field,[117,118] T jump,[115] or ultrasonic techniques have yet produced an unequivocal result for the characteristic time for the addition of a helical segment to a preexisting sequence of helical segments.   Proteins and other nonuniform polyelectrolytes present yet greater problems.[119–122]

With the recent development of instruments capable of accurate measurements on small sample volumes[13,122] and sensitive automated pulse

[109] A. A. Jones and W. H. Stockmayer, *J. Phys. Chem.* **78**, 1528 (1974).

[110] R. C. Parker, L. J. Slutsky, and K. R. Applegate, *J. Phys. Chem.* **72**, 3177 (1968).

[111] G. Hammes and P. B. Roberts, *J. Am. Chem. Soc.* **91**, 1812 (1969).

[112] J. Burke, G. Hammes, and T. Lewis, *J. Chem. Phys.* **42**, 3520 (1965).

[113] A. Barksdale and J. Stuehr, *J. Am. Chem. Soc.* **94**, 3334 (1972).

[114] R. Zana, *J. Am. Chem. Soc.* **94**, 3646 (1972).

[115] F. Eggers and Th. Funck, *Stud. Biophys.* **57**, 101 (1976).

[116] L. Madsen and L. J. Slutsky, *J. Phys. Chem.* **81**, 2264 (1977).

[117] A. L. Cummings and E. M. Eyring, "Chemical and Biological Applications of Relaxation Spectrometry" (E. Wyn-Jones, ed.).   Reidel, Boston, Massachusetts, 1975.

[118] T. Yasunaga, Y. Tsuji, T. Sano, and H. Takenaka, *J. Am. Chem. Soc.* **98**, 813 (1976).

[119] E. L. Carstensen and H. F. Schwann, *J. Acoust. Soc. Am.* **31**, 305 (1959).

[120] R. D. White and L. J. Slutsky, *Biopolymers* **11**, 1973 (1972).

[121] P. D. Edmonds, *Biochem. Biophys. Acta* **63**, 216 (1962).

[122] M. A. Breazeale, J. H. Cantrell, Jr., and J. S. Heyman, Part 2, this volume, pp. 67–135 (1981).

systems[14] suitable for the determination of differential acoustic absorption in very dilute solution, ultrasonics has become a kinetic technique of considerable generality applicable to most fast equilibria in which, at some set of experimentally accessible conditions, all reactants and products are present at appreciable concentrations. However, when there is not a satisfactory theory or body of experimental knowledge for the relaxation amplitudes associated with the process in question, and with plausible alternative or concomitant processes, the generality of the technique is a source of ambiguity. The case of reversible conformation transitions in macromolecules is, we believe, representative of the difficulties that can be encountered in interpreting results in systems with many chemical and structural degrees of freedom. It should be noted, however, that others[123] have taken a more optimistic view of the current state of knowledge, at least for polymers with nonionic side chains.

Greater success has been achieved in systems of lower molecular weight but greater chemical complexity. Kinetic studies of the conformational changes coupled to the ion complexation reactions of small peptide antibiotics[124] and synthetic analogs of ionophore antibiotics[125,126] have yielded information only difficultly accessible by other techniques. In addition to the established utility of ultrasonics in fast association reactions, proton-transfer reactions, and conformational changes in small molecules, there seems every prospect that the technique will continue to prove useful in the study of fast processes in chemically complex systems where it is possible either from an adequate experimental or theoretical knowledge of the equilibrium thermodynamics to make an unequivocal identification of the chemical processes which contribute to the relaxing volume or heat capacity.

We have been concerned almost exclusively with the utility of acoustic relaxation as a method for the study of very fast chemical and structural transformations. However, pulse-echo techniques do lend themselves naturally to time-resolved measurements of ultrasonic velocity and absorption. Thus either velocity or absorption can be used as an analytical technique to monitor changing chemical composition as a reaction runs its course whether or not there is an adequate microscopic understanding of

[123] B. Gruenewald, C. U. Nicola, A. Lustig, and G. Schwarz, *Biophys. Chem.* **9**, 137 (1979).

[124] F. Eggers and Th. Funck, *Die Naturwiss.* **63**, 280 (1976); E. Grell and Th. Funck, *J. Supramol. Struct.* **1**, 307 (1973).

[125] L. J. Rodriquez, G. W. Liesegang, R. D. White, M. M. Farrow, N. Purdie, and E. M. Eyring, *J. Phys. Chem.* **81**, 2118 (1977).

[126] L. J. Rodriquez, G. W. Liesegang, M. M. Farrow, N. Purdie, and E. M. Eyring, *J. Phys. Chem.* **82**, 647 (1978).

the differences in the acoustic properties of reactants and products. Ultrasonic studies of the kinetics of protein denaturation with a sampling interval of 5 msec[127] have been reported and, although the technique has not been extensively exploited, acoustic methods may occasionally offer a useful alternative to the traditional optical or conductiometric methods of following moderately fast reactions. Similarly, a technique in which equilibrium is perturbed optically and the response sensed by condensor microphone has been used to determine the kinetics of the release of protons from membrane-bound bacteriorhodopsin into the surrounding buffered aqueous medium.[128] Although the characteristic time for this process (~ 400 μsec) is in a range accessible to a variety of relaxation methods, the "hybrid" acoustic technique, which places no restriction on the concentration of the buffer and which requires no indicator, has proven convenient in this rather complex system.

[127] K. Yamanaka, H. Nakajima, and Y. Wada, *Biopolymers* **17**, 2159 (1978).
[128] D. R. Ort and W. W. Parson, *J. Biol. Chem.* **253**, 6158 (1978).

5. SCATTERING IN POLYCRYSTALLINE MEDIA

By Emmanuel P. Papadakis

5.1. Introduction

5.1.1. General Comments

5.1.1.1. Interest in Engineering and Science. Polycrystalline media present particular problems and afford special opportunities for the scientist or engineer working with ultrasound. Both the problems and the opportunities can be traced directly to the polycrystalline constitution of the medium because the grains of the polycrystal affect ultrasonic waves very strongly. For one scientist the grains of the polycrystal obscure the properties of the crystal structure he would study, while for another scientist the grains permit the performance of scattering experiments in classical physics. For one engineer the grains make the penetration inadequate for testing for flaws in forgings, while for another engineer the grains permit testing for the improper fabrication and heat treatment of parts. Other specialists have their unique problems; for instance, communications personnel working with ultrasonic delay lines are concerned with the interactions of elastic waves and grains because they are interested in propagating ultrasonic waves with no loss of information. In general, persons working with ultrasound are interested in the rate of change of energy in an ultrasonic wave for one of two purposes: to find what these parameters can tell them about a material and to find the way the parameters characterize the material as a propagation medium. The outlooks and approaches are different: persons of the first group use ultrasonic waves to investigate materials, whereas those of the second group use materials to support the propagation of ultrasonic waves. Both groups need to know how to handle the propagation of ultrasonic waves in polycrystalline media. This part is concerned with scattering that abstracts energy from the wave, resulting in attenuation.

5.1.1.2. Description of Polycrystallinity. A polycrystalline medium consists of grains of the constituent material.[1,2] In general, these grains

---

[1] C. S. Barrett, "Structure of Metals," 2nd ed. McGraw-Hill, New York, 1952.

[2] R. M. Brick and A. Phillips, "Structure and Properties of Alloys," 2nd ed. McGraw-Hill, New York, 1949.

are of various shapes and sizes, filling all space within the boundaries of the medium. The grain boundaries themselves are fairly flat curved surfaces. The grains may form by crystallization from the melt or by recrystallization during heat treatment, as in a metal, or they may be brought together by pressure and sintering, as in a ceramic. Other forms of growth, such as pyrolysis or devitrification, can produce polycrystalline media. Partial devitrification can result in the presence of grains within an amorphous matrix filling the rest of space in the sample—an interesting situation. If the grains have voids or inclusions at their boundaries, then the situation is more complicated than if the boundaries are between the constituent material of the grains themselves. A single grain may be a single crystal of the constituent material or it may have two or more phases breaking up the grain. Not all the grains in one piece of metal need have the same proportion of chemical constituents, the same lattice constant, or even the same crystal structure. A two-phase alloy, for instance, may have some grains of one phase and some of the other. Sometimes one phase can be grown or precipitated within grains of another phase. At any rate, each grain can be assigned a set of axes corresponding to the crystal axes of its major constituent or of its contents as they were before the multiple phases subdivided it. These grain axes are oriented differently from grain to grain. If any orientation with respect to the symmetry of the sample is more probable than others, then there is preferred orientation. The geometry of the grains as a whole may be elongated, flattened, or fairly spherical. Grains of a more or less spherical shape are termed *equiaxed*. If a treatment has made all the grains of a sample fairly spherical, then one says that the sample has been equiaxed. The simplest polycrystalline sample is an equiaxed, homogeneous, single-phase material with no preferred orientation, no inclusions, and no voids. Many metals and alloys can be processed to yield such samples, although preferred orientation is difficult to remove. Much theoretical and experimental work has been done on the propagation of elastic waves in such media. Modifying any of the restrictions mentioned above, however, makes the analysis more complex and the experiments more interesting.

   5.1.1.3. Wave Propagation. To discuss the propagation of ultrasonic waves[3] let us first consider a plane wave passing through a medium that can abstract energy from the wave. Then the wave is represented by

$$A = A_0 e^{-\alpha L^z} \sin(\beta z - \omega t), \qquad (5.1.1)$$

where A is the magnitude of some quantity such as stress, strain, pres-

[3] R. B. Lindsay, "Mechanical Radiation." McGraw-Hill, New York, 1960.

sure, particle displacement, particle velocity, or particle acceleration and A_0 that magnitude at the origin. The wave of frequency $f = \omega/2\pi$ has a propagation constant $\beta = 2\pi f/v$, v being the phase velocity. The wave decays with a rate constant of attenuation α_L per unit length of travel in the medium. Often the attenuation α_T is measured in nepers per unit time on an oscilloscope. Then $\alpha_T = v\alpha_L$. For traveling waves, α and v are the measurable quantities. In this part the attenuation will be the quantity studied.

5.1.1.4. Principal Effects of Polycrystallinity on Measurable Quantities. How does the structure of a polycrystalline material influence attenuation and velocity? Principally, in two ways. The attenuation of a polycrystalline material in most cases is determined almost entirely by grain scattering, which disperses the energy in the traveling wave. The velocity is determined by the elastic moduli and the preferred orientation of the grains, the latter property making the medium as a whole elastically anisotropic, so that the velocity is a function of the direction of propagation. Every parameter and condition mentioned in the description of polycrystallinity affects the grain scattering strongly and the velocity to some degree. Anisotropy in the velocity affects the beam spreading corrections to be applied to attenuation measurements.

5.1.2. Scope of the Part

This part will cover attenuation caused by grain scattering and will touch on the subject of velocity anisotropy caused by preferred orientation as it affects beam spreading. Both theory and experiment will be treated, with the emphasis on results. The theoretical approach will be outlined, and the results presented with enough detail to make them useful to most workers. Similarly, the experimental methods and equipment will be sketched out sufficiently to permit most readers to grasp the essentials of an experiment without being burdened with details. Emphasis will be placed on the way the results of the experiment verify a particular theory or illustrate a certain point of interest. For the worker desiring a complete knowledge of a theory, experiment, or technique there will be copious references to the literature. Material already summarized in textbooks and reference works will be treated briefly; brevity does not indicate disfavor or lack of value. Some of the older work may be superseded by more modern treatments, of course. Where this is so, it will be noted. Where effects other than those under study appear in experiments to obscure the results, methods of correcting for them will be given. The emphasis, as has been stated, will be on attenuation caused by grain scattering. First, consideration will be given to theory and experiment on

scattering by single-crystal grains, by grains exhibiting preferred orientation, by grains broken up into several phases or disoriented regions, and
by nonequiaxed grains. Carbon steel will be emphasized as a case of
phase transformations.

5.2. Attenuation Caused by Grain Scattering

5.2.1. Theory of Grain Scattering

5.2.1.1. General Considerations. A wave impinging upon an inhomogeneity in a medium will be scattered. The disturbance within the inhomogeneity will differ from the incident wave; the difference will give
rise to other waves outside the inhomogeneity. The scattered waves will
depend on the propagation characteristics within the inhomogeneity and
outside it, on the mode of the incident wave, and on the boundary conditions at the surface of the inhomogeneity. The boundary conditions ior
elastic waves are the continuity of stress and displacement across the
boundary. After the scattered waves are found, the power they carry off
may be computed as the outgoing flux density integrated over a sphere in
the radiation region far from the inhomogeneity. The scattered power
will be a fraction of the incident power; for a unit volume containing N inhomogeneities scattering independently (no multiple scattering), the fraction will be N times as large and will determine the rate of attenuation of
the incident wave.

Scattering by single inhomogeneities, particularly spheres and cylinders, has been studied extensively.[4-10] In general, an inhomogeneity will
scatter elastic waves if it differs in modulus or density from the surrounding medium. The scattering depends also on the wavelength λ relative to the size of the inhomogeneity. For scattering by single inhomogeneities of diameter D this scattered power is proportional to D^4, D
being the particle diameter, when $\lambda \approx D$, and to D^6 when $\lambda \gg D$.

A polycrystalline medium is a space totally filled with inhomogeneities.
One grain is different from all those around it since it is elastically anisotropic and since its crystallographic axes are misoriented with respect to

[4] Lord Rayleigh, "The Theory of Sound," pp. 149–152. Macmillan, London, 1894,
and Dover, New York (first Am. ed., 1945).
[5] P. M. Morse, "Vibration and Sound," 2nd ed., pp. 346–357. McGraw-Hill, New York,
1948.
[6] C. F. Ying and R. Truell, J. Appl. Phys. 27, 1086–1097 (1956).
[7] R. M. White, J. Acoust. Soc. Am. 30, 771–785 (1958).
[8] N. G. Einspruch and R. Truell, J. Acoust. Soc. Am. 32, 214–220 (1960).
[9] N. G. Einspruch, E. J. Witterholdt, and R. Truell, J. Appl. Phys. 31, 806–818 (1960).
[10] Y. H. Pao, and C. C. Mow, J. Appl. Phys. 34, 493–499 (1963).

the axes of its neighbors. The latter condition makes the modulus different from grain to grain. The grains may also differ in density; a two-phase alloy is a possible example. However, the elastic anisotropy of a grain, resulting from its crystalline nature, usually has the larger effect. The preferred orientation, if any, has a large effect, since it modifies the change in modulus from grain to grain by aligning the crystallographic axes. It is a secondary aspect, however, and will be considered later.

From what has been said, it will be seen that the treatment of the elastic anisotropy of a grain is of central importance in the grain scattering problem. Another question, of equal importance, is the grain-size distribution in the polycrystalline medium, for the scattering depends strongly on the particle size. Obviously, averaging over a grain-size distribution will affect powers of D considerably. Similarly, the number of scatterers per unit volume (which is taken as the reciprocal of the average grain volume) will be affected by a grain-size distribution.

In the next two sections the twin problems of the elastic anisotropy and the grain-size distribution as they contribute to grain scattering will be treated. The current theories will be summarized and the results explained in useful forms. After that there will be two short sections on the effects of microstructure and preferred orientation. Theories on these are only qualitative so far; the analytical treatment being restricted to certain simplified two-dimensional models. However, they represent progress toward understanding the experiments on scattering in polycrystalline media with microstructure or preferred orientation. Various experimental work on scattering will be presented in Section 5.2.3.

5.2.1.2. Grain Scattering Formulas for Anisotropic Grains. The early work on grain scattering by Mason and McSkimin[11,12] was an attempt to adapt the formula of Rayleigh[4] to scattering by grains in a metal. Rayleigh's work originally applied to an inhomogeneity in a fluid for the case $\lambda \gg D$. Two concepts were introduced by Mason and McSkimin: the idea that the number of scatterers was inversely proportional to the average grain volume and the idea that the mean-square fractional variation of the modulus of a grain with azimuth and declination could express the elastic difference between the grain as an inhomogeneity and all the other grains as a medium in which the one scatters. The first concept is still used, but the second has been modified considerably in the more exact formulations of the grain scattering theory made more recently. An acknowledged shortcoming of the early theory was its failure to account for mode conversion. More recent theoretical and experimental work has shown that most (80%) of the scattered energy from an incident longi-

---

[11] W. P. Mason and H. J. McSkimin, *J. Acoust. Soc. Am.* **19**, 464–473 (1947).

[12] W. P. Mason and H. J. McSkimin, *J. Appl. Phys.* **19**, 940–946 (1948).

tudinal wave is carried off by shear waves in the case $\lambda \gg D$, so the more exact formulation is necessary. Other investigators[13-15] have suggested mechanisms for grain scattering for the cases $\lambda \simeq D$ and $\lambda \ll D$.

The grain scattering problem was solved in the presently accepted manner by Lifshits and Parkhomovskii.[16] The restrictions on the validity of their solution are as follows:

(1)   The anisotropy of the grains must be small; that is, the variation in modulus with azimuth and declination must be much smaller than the average value of the modulus.

(2)   Preferred orientation must be absent.

(3)   The grains must be equiaxed.

(4)   The material must have single-crystal grains of a single phase with no voids or inclusions.

Solutions emerge for two ratios of λ to $\overline{D}$, namely, $\lambda > 2\pi\overline{D}$ and $\lambda < 2\pi\overline{D}$, where $\overline{D}$ is the average grain diameter. Merkulov[17] specialized the general solution to the common cases of cubic and hexagonal metals. His expressions are summarized here. The following material will be referred to as the LPM theory (Lifshits–Parkhomovskii–Merkulov theory).

Case I: Rayleigh scattering with $\lambda > 2\pi\overline{D}$.

Cubic Crystallites:

$$\alpha_l = \frac{8\pi^3 \mu^2 T f^4}{375 \rho^2 v_l^3} \left\{ \frac{2}{v_l^5} + \frac{3}{v_t^5} \right\}, \tag{5.2.1}$$

$$\alpha_t = \frac{2\pi^3 \mu^2 T f^4}{125 \rho^2 v_t^3} \left\{ \frac{2}{v_l^5} + \frac{3}{v_t^5} \right\}, \tag{5.2.2}$$

with

$$\mu = c_{11} - c_{12} - 2c_{44}. \tag{5.2.3}$$

Hexagonal Crystallites:

$$\alpha_l = \frac{4\pi^3 T f^4}{450 \rho^2 v_l^3} \left\{ \frac{a_1}{v_l^5} + \frac{b_1}{v_t^5} \right\}, \tag{5.2.4}$$

$$\alpha_t = \frac{4\pi^3 T f^4}{450 \rho^2 v_t^3} \left\{ \frac{a_2}{v_l^5} + \frac{b_2}{v_t^5} \right\}, \tag{5.2.5}$$

[13] H. B. Huntington, *J. Acoust. Soc. Am.* **22**, 362–364 (1950).

[14] C. L. Pekeris, *Phys. Rev.* **71**, 268–269 (1947).

[15] W. Roth, *J. Appl. Phys.* **19**, 901–910 (1948).

[16] E. M. Lifshitz and G. D. Parkhomovskii, *Zh. Eksp. Teor. Fiz.* **20**, 175–182 (1950).

[17] L. G. Merkulov, *Sov. Phys. Tech. Phys.* (*English transl.*) **1**, 59–69 (1956); *Zh. Tekh. Fiz.* **26**, 64–75 (1956).

with

$$a_1 = \tfrac{88}{15}\gamma^2 + 40\chi^2 + 96\eta^2 + \tfrac{80}{3}\chi\gamma + \tfrac{128}{3}\gamma\eta + \tfrac{320}{3}\chi\eta, \qquad (5.2.6)$$

$$b_1 = \tfrac{82}{15}\gamma^2 + 30\chi^2 + \tfrac{272}{3}\eta^2 + 30\chi\gamma + \tfrac{112}{3}\gamma\eta + 80\chi\eta, \qquad (5.2.7)$$

$$a_2 = \tfrac{41}{15}\gamma^2 + 15\chi^2 + \tfrac{136}{3}\eta^2 + 10\chi\gamma + \tfrac{56}{3}\gamma\eta + 40\chi\eta, \qquad (5.2.8)$$

$$b_2 = \tfrac{8}{5}\gamma^2 + 28\eta^2 + 8\gamma\eta, \qquad (5.2.9)$$

where

$$\gamma = c_{11} + c_{33} - 2(c_{13} + 2c_{44}), \qquad (5.2.10)$$

$$\chi = c_{13} - c_{12}, \qquad (5.2.11)$$

$$\eta = c_{44} + (c_{12} - c_{11})/2. \qquad (5.2.12)$$

Case II: Intermediate scattering with $\lambda < 2\pi\overline{D}$.
Cubic Crystallites:

$$\alpha_l = \frac{16\pi^2\mu^2\overline{D}f^2}{525v_l^6\rho^2}, \qquad (5.2.13)$$

$$\alpha_t = \frac{4\pi^2\mu^2\overline{D}f^2}{210v_t^6\rho^2}, \qquad (5.2.14)$$

with μ as given in (5.2.3).
Hexagonal Crystallites:

$$\alpha_l = \frac{16\pi^2\overline{D}f^2 a_3}{1575v_l^6\rho^2}. \qquad (5.2.15)$$

with

$$a_3 = 7\gamma^2 + 35\chi^2 + 140\eta^2 + 140\chi\eta + 30\chi\gamma + 60\eta\gamma, \qquad (5.2.16)$$

where γ, χ, and η as given in Eqs. (5.2.10)–(5.2.12).   Unfortunately, α_t has not been given.

In these formulas the attenuation α is in nepers per unit length.   The subscripts l and t refer to longitudinal and transverse incident waves, respectively, the elastic moduli of the crystalline material in the grains are designated c_{ij}, and the anisotropy is expressed by μ, γ, χ, and η.   The density is ρ, the elastic wave velocity is v, and the ultrasonic frequency is f.   In Rayleigh scattering, T is a measure of the grain size with the dimensions of volume, and in the intermediate scattering range $\overline{D}$ is the average grain diameter.

An equivalent theoretical analysis was performed by Bhatia and Moore,[18] based on Bhatia,[19] for Rayleigh scattering in materials of ortho-

---

[18] A. B. Bhatia and R. A. Moore, *J. Acoust. Soc. Am.* **31**, 1140–1142 (1959).

[19] A. B. Bhatia, *J. Acoust. Soc. Am.* **31**, 16–23 (1959).

rhombic grain symmetry. By equating certain pairs of moduli this symmetry can be reduced to a hexagonal or cubic one. The attenuation formulas derived by Bhatia and Moore[18] do indeed reduce to those presented by Merkulov[17] when this reduction is performed. The orthorhombic Rayleigh formulas are presented here.

Case 1: Rayleigh scattering with $\lambda > 2\pi\overline{D}$.
Orthorhombic Crystallites:

$$\alpha_l = \frac{4\pi^3 Tf^4}{5\rho^2 v_l^3} \left\{ \frac{B_1}{2v_l^5} + \frac{A_1}{v_t^5} \right\}, \tag{5.2.17}$$

$$\alpha_t = \frac{4\pi^2 Tf^4}{5\rho^2 v_t^3} \left\{ \frac{B_6}{2v_l^5} + \frac{A_6}{v_t^5} \right\}, \tag{5.2.18}$$

with

$$A_1 = B_6 = \frac{2}{225} P^2 + \frac{1}{135} (24a + 7b + 13c + d), \tag{5.2.19}$$

$$B_1 = \frac{8}{675} P^2 + \frac{8}{135} (4a + 2b + 3c + d), \tag{5.2.20}$$

$$A_6 = \frac{1}{150} P^2 + \frac{1}{90} (12a + b + 4c - 2d), \tag{5.2.21}$$

where

$$P = (c_{11} + c_{22} + c_{33}) - (c_{12} + c_{13} + c_{23}) - 2(c_{44} + c_{55} + c_{66}), \tag{5.2.22}$$
$$a = (c_{44} + c_{55} + c_{66})^2 - 3(c_{44}c_{55} + c_{55}c_{66} + c_{44}c_{66}), \tag{5.2.23}$$
$$b = (c_{11} + c_{22} + c_{33})^2 - 3(c_{11}c_{22} + c_{22}c_{33} + c_{11}c_{33}), \tag{5.2.24}$$
$$c = (c_{12} + c_{13} + c_{23})^2 - 3(c_{12}c_{13} + c_{13}c_{23} + c_{12}c_{23}), \tag{5.2.25}$$
$$d = c_{11}(c_{12} + c_{13} - 2c_{23}) + (c_{12} + c_{23} - 2c_{13}) + c_{33}(c_{13} + c_{23} - 2c_{12}). \tag{5.2.26}$$

The definitions of the quantities in these formulas are the same as those given above with Merkulov's expressions. Bhatia and Moore[18] give more general formulas for the As and Bs for use in further studies of other grain symmetries.

5.2.1.3. Grain Scattering Tables. Equations (5.2.1), (5.2.2), (5.2.4), and (5.2.5) can be written in the form

$$\alpha = Tf^4 S \tag{5.2.27}$$

to separate the dependence of the attenuation into a constant coefficient S and two experimental variables: f, the ultrasonic frequency, and T, the

TABLE I. Scattering Coefficients and Elastic-Wave Velocities

Material	S (dB/cm (MHz)^4 cm^3)		Σ (dB/cm (MHz)^2 cm)		v (10^5 cm/sec)	
	S_l	S_t	Σ_l	Σ_t	v_l	v_t
Aluminum	40.2	284	0.059	3.27	6.42	3.04
Chromium[a]	49.3	164	0.271	3.33	6.77	4.12
Copper	3065	24,600	2.19	158	5.01	2.27
Cu$_3$Au[a]	5850	48,100	2.73	205	4.11	1.85
Gold	16,300	241,000	1.80	438	3.24	1.20
Iron	700	3260	1.73	42.0	5.95	3.24
Iron[a]	977	4290	2.04	57.7	5.83	3.09
Fe–30%Ni	4210	14,000	10.4	128	4.55	2.77
Lead	54,500	938,000	1.71	562	1.96	0.69
Lead[a]	147,000	1,960,000	8.96	1760	2.21	0.85
Nickel	896	5480	1.46	61.1	6.04	3.00
Niobium[a]	801	6990	0.541	45.8	5.19	2.29
Palladium[a]	4240	43,600	1.70	199	4.57	1.91
Silver	9810	85,700	3.27	277	3.65	1.61
Tantalum[a]	1080	7530	0.676	35.9	4.11	1.96
Thorium[a]	33,700	163,000	17.7	46.3	2.83	1.52
Tungsten	0.023	0.151	2.8×10^{-15}	1.3×10^{-13}	5.41	2.64
Vanadium[a]	63.1	445	0.082	4.56	6.06	2.87

[a] Voigt velocities used and listed.[20]

appropriate average grain volume, where S is a parameter of the material and will vary little from sample to sample since p, v_l, v_t, and the c_{ij} vary little among equiaxed samples of the same nominal composition. Similarly, Eqs. (5.2.13)–(5.2.15) can be written

$$\alpha = \overline{D}f^2\Sigma \qquad (5.2.28)$$

with a different coefficient Σ. The scattering coefficients S and Σ have been tabulated for a number of elements and compounds by Papadakis.[20] This tabulation is different from earlier tabulations in the literature (Mason,[21] p. 208) because the earlier ones used the formulas of the semi-quantitative theory of Mason and McSkimin.[11,12] Some of the scattering coefficients from the 1965 reference are listed in Table I. These calculations used Merkulov's expressions[17] and, where available, elastic-wave velocities found experimentally in polycrystalline samples (Mason,[21] pp. 17–20); in other cases, Voigt velocities[22] found from single-crystal moduli

[20] E. P. Papadakis, *J. Acoust. Soc. Am.* **37**, 703–710 (1965).
[21] W. P. Mason, "Physical Acoustics and the Properties of Solids." Van Nostrand-Reinhold, Princeton, New Jersey, 1958.
[22] R. F. S. Hearmon, "Applied Anisotropic Elasticity," pp. 41–44. Oxford Univ. Press, London and New York, 1961.

(Mason,[21] pp. 355–373) were used. The velocities v_l and v_t are also listed in Table I. Where they are Voigt velocities, this fact is indicated.

In two cases in Table I, iron and lead, calculations were made for both measured velocities and Voigt velocities. The scattering factors differ appreciably. It is not known whether the discrepancies are caused by errors in the measured velocities (possibly due to preferred orientation) or by the bias in the Voigt averaging of the moduli. Hearmon[22] mentions that the Voigt and Reuss velocities differ by up to 30%. As is well known, Voigt's averaging method assumed continuity of strain across boundaries, whereas Reuss's method was based on continuity of stress. Discrepancies from this cause are to be expected.

In Table I no hexagonal materials were included, since formulas (5.2.6)–(5.2.12) permit negative values of the scattering factors because of the cross terms $\gamma\chi$, $\chi\eta$, and $\eta\gamma$. This anomaly was pointed out earlier.[20] Some hexagonal materials actually yield negative values.

5.2.1.4. Graphing of Attenuation Data. Values from Table I will be used in later sections, in which theory and experiment will be compared. Generally the method is as follows. The condition $\lambda_B = 2\pi\overline{D}$ is the dividing line, or boundary, between the Rayleigh scattering region and the intermediate region. Corresponding to λ_B there is a frequency f_B, called the boundary frequency. At $f_R = 0.1f_B$ the attenuation is computed as $\alpha_R = Tf_R{}^4S$ for the Rayleigh region and at $f_I = 10f_B$ it is found from $\alpha_I = \overline{D}f_I{}^2\Sigma$ for the intermediate region. On logarithmic graph paper a pair of lines is drawn through these points with slope 4 at f_R and 2 at f_I. The data are compared with a curve for which these lines are asymptotic.

At even higher frequencies, where $\lambda \ll \overline{D}$, the scattering is purely a reflection problem,[12] the grain boundaries being considered partially reflecting mirrors. In this region the attenuation due to scattering is independent of frequency and inversely proportional to the grain diameter. However, elastic hysteresis again becomes important, and thermoelastic damping enters once more.[23] In the frequency-dependent scattering region these two effects are swamped by the scattering in most cases. They produce attenuation linear and quadratic with frequency, respectively. Merkulov[23] has seen evidence of hysteresis and thermoelastic damping when $\lambda \ll \overline{D}$ in aluminum. The functional dependence of the attenuation upon frequency and grain size is summarized in Table II. This summary was graphed conveniently by Smith and Stephens.[24]

[23] L. G. Merkulov, Sov. Phys. Tech. Phys. (English transl.) 2, 953–957 (1957).
[24] R. T. Smith and R. W. B. Stephens, In "Progress in Applied Materials Research" (E. G. Stanford, J. H. Fearon, and W. J. McGonnagle, eds.), Vol. 5, pp. 41–64. Gordon and Breach, New York, 1964.

TABLE II.   Functional Dependence
of Attenuation[36]

Range	Dependence[a,b]
$\lambda > 2\pi\overline{D}$	$B_1 f + A_4 D^3 f^4$
$\lambda < 2\pi\overline{D}$	$A_2 \overline{D} f^2$
$\lambda \ll D_{min}$	$B_1 f + B_2 f^2 + A_0/\overline{D}$

[a] $B_1 f$ is elastic hysteresis loss and $B_2 f^2$ is thermoelastic loss.
[b] Coefficients A and B are *not* related to those given in Eqs. (5.1.1), (5.2.17)–(5.2.21), (5.2.35c)–(5.2.38), or (5.2.52)–(5.2.55).

5.2.1.5. Grain-Size Distribution. After the derivation of the scattering formulas in Section 5.2.1.2, the last remaining difficulty in comparing theory and experiment lay in finding the proper grain size to insert into the formulas. Outlined here is a statistical solution, proposed by the author,[25,26] which is appropriate to the problem for approximately spherical grains.

Several writers[27–32] have pointed out that a polycrystal (particularly a metal prepared by ordinary mechanical working and heat-treatment methods) has a distribution of grain sizes. In the ordinary metal the distribution is approximately lognormal. Let $N_V(R)\, dR$ be the number of grains per unit volume of radius between R and $R + dR$. The average of a power of R over the distribution is

$$\langle R^n \rangle_{av} = \int_0^\infty R^n N_V(R)\, dR / \int_0^\infty N_V(R)\, dR. \qquad (5.2.29)$$

Since the scattering power of a single grain in Rayleigh scattering is proportional to the square of the volume $T^2 = (4\pi/3)^2 R^6$, one needs $\langle R^6 \rangle_{av}$ over the distribution to find $\langle T^2 \rangle_{av}$. The assumption that the number of grains per unit volume is inversely proportional to the average grain volume means that one needs $\langle R^3 \rangle_{av}$ to find $\langle T \rangle_{av}$. The quantity T appearing in the equations of Section 5.2.1.2 is not $\langle T \rangle_{av}$ but, in a strict sense, is $\langle T^2 \rangle_{av}/\langle T \rangle_{av}$, which is larger than $\langle T \rangle_{av}$. Hence it should be understood

[25] E. P. Papadakis, *J. Acoust. Soc. Am.* **33**, 1616–1621 (1961).
[26] E. P. Papadakis, *J. Appl. Phys.* **35**, 1586–1594 (1964).
[27] S. D. Wicksell, *Biometrika* **17**, 85–89 (1925).
[28] E. Scheil, *Z. Metallkd.* **27**, 199–209 (1935).
[29] E. Scheil and H. Wurst, *Z. Metallkd.* **28**, 340–343 (1936).
[30] H. J. Seemann and W. Bentz, *Z. Metallkd.* **45**, 663–669 (1954).
[31] P. Feltham, *Acta Metall.* **5**, 97–105 (1957).
[32] R. T. DeHoff, *Trans. AIME* **233**, 25–29 (1965).

that everywhere T appears in Section 5.2.1.2, the following substitutions are to be made:

$$T \longrightarrow \frac{4\pi}{3} \frac{\langle R^6 \rangle_{av}}{\langle R^3 \rangle_{av}}, \tag{5.2.30}$$

$$\overline{D} \longrightarrow 2\langle R \rangle_{av}. \tag{5.2.31}$$

How is $N_V(R)$ to be found so that the averaging may be performed? The value of $N_V(R)$, a distribution of three-dimensional objects in three-dimensional space, must be found from data in two dimensions on a plane surface. The best datum is a good photomicrograph showing many grain images. The images are slices through grains and show diameters smaller than the true grain diameters. Let r be the radius of a grain image and let $n_A(r) \, dr$ be the number of grain images per unit area of radius between r and $r + dr$. Also let $N_A(r)$ be the number of grain images per unit area smaller than or equal to r. For spherical grains randomly distributed in space, it can be shown[25] that $N_A(r)$ is

$$N_A(r) = 2 \int_0^\infty R N_V(R) \, dR - 2 \int_r^\infty N_V(R)(R^2 - r^2)^{1/2} \, dR. \tag{5.2.32}$$

Differentiation of this formula[26] produces an equation

$$n_A(r) \, dr = \int_r^\infty \frac{2r N_V(R) \, dR}{(R^2 - r^2)^{1/2}} \, dr, \tag{5.2.33}$$

which can be changed to a summation

$$n_A(r_i) \, \Delta r_i = \sum_{j=1}^m \frac{2r_i N_V(R_j) \, \Delta R_j \, \Delta r_i}{(R_j^2 - r_i^2)^{1/2}} \tag{5.2.34}$$

for numerical analysis. Now R terminates at R_{max} instead of infinity. With the definitions

$$n_i \equiv n_A(r_i) \, \Delta r_i, \tag{5.2.35a}$$

$$N_j \equiv N_V(R_j), \tag{5.2.35b}$$

$$A_{ij} \equiv \frac{2r_i \, \Delta R_j \, \Delta r_i}{(R_j^2 - r_i^2)^{1/2}}, \tag{5.2.35c}$$

Eq. (5.2.34) becomes

$$n_i = \sum_{j=1}^m A_{ij} N_j. \tag{5.2.36}$$

The grain-size distribution N_i can be found by inverting the A matrix and postmultiplying by the image distribution to give

$$N_i = \sum_{j=1}^{m} A_{ij}^{-1} n_j. \tag{5.2.37}$$

For computational purposes, one must choose the magnitudes of R_j and r_i. In this choice the difference $R_j - r_j$ is defined; this quantity is crucial because $(R_j^2 - r_j^2)^{1/2}$ appears in the denominators of the diagonal elements in the matrix (A_{ij}). If equal intervals $\Delta R = \Delta r$ are taken and $R_j - r_j$ is made proportional to ΔR or $R_j - r_j = c \, \Delta R$, the matrix elements become

$$A_{ij} = \begin{cases} \dfrac{(2R_j \, \Delta R)^{1/2} \, \Delta R \, [1 - (\Delta R/R_j)(j - i + c)]}{(j - i + c)^{1/2}[1 - (\Delta R/2R_j)(j - i + c)]^{1/2}} & \text{for } j \geq i \\ 0 & \text{for } j < i. \end{cases} \tag{5.2.38}$$

In the work by Papadakis[26] a computer program was written to compute the matrix elements, invert the matrix, and multiply the inverse by the area-distribution column to find the grain-size distribution within the polycrystalline medium. The computer program also calculated $\langle R \rangle_{av}$, $\langle R^3 \rangle_{av}$, $\langle R^6 \rangle_{av}$, and $(4\pi/3)\langle R^6 \rangle_{av}/\langle R^3 \rangle_{av}$ for the ultrasonic scattering. The computer program was tested on hypothetical grain-size distributions operated on first by Eq. (5.2.23) to produce grain-image area distributions n_j to feed back into Eq. (5.2.37). The program produced the best replicas of the original distributions when c was set at $c = 0.25$. A 10×10 matrix yielded adequate definition of the distributions. An inverse matrix normalized to unit maximum grain radius is presented in Table III for the case $m = 10$, $c = 0.25$, $R_1 = 0.1$, $\Delta R = 0.1$, and $R_{10} = R_{max} = 1.0$. The grain-size distribution N_i can be found by postmultiplying by the grain-image area distribution n_j as a ten-element column matrix made up of the number of grain images of radii between r_j and r_{j-1}. Intervals are taken at $r_j = R_j$ and $\Delta r = \Delta R$. (The fact that $c = 0.25$ means that one should write $r_j = R_j - 0.25 \, \Delta R$ did not seem to make a measurable difference in the results, and r_j was taken as R_j with adequate accuracy.) From the grain-size distribution the necessary averages are calculated.

In practice one makes an enlargement of a photomicrograph containing 200 or more images of equiaxed grains. Then R_{10} is chosen as slightly larger than the radius of the largest image and $\Delta R = R_{10}/10$. Overlays, either circles or octagons for convenience, are cut from an opaque material and attached to handles for ease in manipulation. Their diameters are $2R_9$, $2R_8$, $2R_7$, . . . , $2R_1$. With the aid of the overlays grain images larger than $2R_9$ (that is, from $2R_9$ to $2R_{10}$) are checked off and counted as

TABLE III.[26]   Inverse 10 × 10 Matrix[a]

44.095855	-9.8692758	-1.8736601	-0.6836014	-0.32656580	-0.18166478	-0.11157226	-0.073481978	-0.050982716	-0.036830361
0	27.664167	-8.6621376	-1.9520020	-0.79488678	-0.41004926	-0.24144197	-0.15491877	-0.10563838	-0.075395154
0	0	21.799234	-7.6604916	-1.8635616	-0.80286200	-0.43253311	-0.26361767	-0.17396461	-0.12142337
0	0	0	18.559214	-6.9168850	-1.7575630	-0.78370619	-0.43419285	-0.27084567	-0.18227003
0	0	0	0	16.434205	-6.3475218	-1.6589414	-0.75711851	-0.42774978	-0.27132484
0	0	0	0	0	14.903598	-5.8961218	-1.5716352	-0.72939104	-0.41808648
0	0	0	0	0	0	13.733701	-5.5277571	-1.4950712	-0.70269606
0	0	0	0	0	0	0	12.802023	-5.2201025	-1.4277798
0	0	0	0	0	0	0	0	12.037357	-4.9582967
0	0	0	0	0	0	0	0	0	11.395121

[a] Computed for the case where $c = 0.25$, $R_1 = 0.1$, and $\Delta R = 0.1$.

n_{10}, then the remaining ones larger than $2R_8$ become n_9, and so on, giving the n_js.

Papadakis[26] showed that the average grain diameter $\overline{D}$ was larger than the median grain-image diameter $d_{50} = 2r_{50}$ by a factor of 1.45 (half the images are larger in diameter than d_{50}, half are smaller), so

$$\overline{D} = 2\langle R \rangle_{av} = 1.45d_{50}. \qquad (5.2.39)$$

Similarly, it was shown that the scattering volume T of Eq. (5.2.30) could be approximated by

$$T \approx 1.0d_{95}^3, \qquad (5.2.40)$$

where d_{95} is the grain-image diameter in the 95th percentile where only 5% of the images are larger. These largest images cover about 25% of the photomicrograph, so d_{95} can be found quickly. Much better estimates of $\overline{D}$ and T are found by going through the entire counting procedure, doing the matrix multiplication, and then performing the required averages. Examples of the entire procedure will be given in Section 5.2.3. It should be noted in passing that the largest grains in a distribution may be too large to contribute to Rayleigh scattering in some samples, so the average $\langle R^6 \rangle_{av}/\langle R^3 \rangle_{av}$ may be excessive.

5.2.1.6. Effect of Preferred Orientation. How may preferred orientation change the grain scattering of elastic waves? The answer has not been worked out quantitatively, but we may make some qualitative remarks. First, the mechanism will be an interaction of the elastic moduli among the anisotropic grains. If the orienting operation increases the probability of high-modulus directions in some grains coinciding with low-modulus directions in adjacent grains, then the scattering should increase. Metals with more than one texture contributing to their final state (Barrett,[1] Chapters 18, 19, 21) might very well exhibit this property. On the other hand, if the orienting operation increases the probability of the alignment of high-modulus directions in adjacent grains, then the scattering should decrease. In an ideal case, such as that of cube-texture material (Barrett,[1] pp. 478–480, 494–502), in which the crystallographic axes of all the grains are parallel to one another, the scattering should disappear. Examples of such structures are cube-textured copper and oriented electrical steel.

A semiquantitative, two-dimensional analysis by Papadakis[33] for slightly anisotropic grains of hexagonal symmetry aligned with a small degree of preferred orientation along one direction is summarized here. For the coordinates, see Fig. 1. A grain with axes $a-c$ is imbedded in an

[33] E. P. Papadakis, J. Appl. Phys. 36, 1738–1740 (1965).

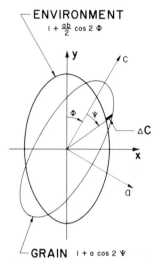

ENVIRONMENT

$1 + \frac{ab}{2} \cos 2\,\Phi$

GRAIN $1 + a \cos 2\,\Psi$

FIG. 1. Two-dimensional model of the elastic modulus of a grain in an environment provided by other grains with preferred orientation. The rms value of the difference in modulus between a grain and its environment is reduced by the worked texture. (Papadakis,[33] by permission of American Institute of Physics.)

environment with axes $x-y$. The normalized elastic modulus of the grain is

$$C = 1 + a \cos 2\psi, \qquad (5.2.41)$$

where ψ is measured from axis c. The orientation of axis c is at an angle Φ from axis y. The preferred orientation of axis c with respect to axis y is given by a weight $W(\Phi)$ expressing the probability that a grain will be found with axis c at angle Φ. The weight is

$$W(\Phi) = 1 + b \cos 2\Phi. \qquad (5.2.42)$$

The average modulus of the environment $\overline{C}$ is found as a weighted average of the grain modulus, Eq. (5.2.41), over all orientations. This treatment is analogous to the analysis of Mason and McSkimin.[11,12] Thus,

$$\overline{C} = 1 + (ab/2) \cos 2\Phi. \qquad (5.2.43)$$

The mean-square average $\langle (\Delta C / \overline{C})^2 \rangle_{av}$ is

$$\langle (\Delta C / \overline{C})^2 \rangle_{av} = (a^2/2) - (a^2 b^2/8) + O(a^4 b^2 \text{ and } a^4 b^4). \qquad (5.2.44)$$

Here a expresses the grain anisotropy and b the degree of preferred orientation. By the assumptions of small anisotropy and small preferred orientation, $a^2 \ll 1$ and $b^2 \ll 1$. The scattering factor for a random material is $a^2/2$; with the introduction of the preferred orientation, the scattering is reduced by $a^2b^2/8$. It seems that a three-dimensional analysis along this line would be of value.

5.2.1.7. Effect of Polyphase or Multiparticle Structure within Grains. How may structure within grains change the scattering? Here, as in the case of the effect of preferred orientation, the problem has not been worked out in detail. Qualitative statements can be made, however, concerning microstructure and scattering. As will be seen in Section 5.2.3, several experimentalists have found corroborating evidence. With the formation of a polyphase, or multiparticle structure, within a grain, the grain is broken up into a number of smaller sections—often a very large number—which differ in crystallographic orientation and sometimes in composition from each other and from the original grain (Brick and Phillips,[2] Chapters 3, 5–12, 14–16). In a diffusionless transformation such as the martensitic transformation, in which the rearrangement of atoms comes about by shearing strains, there is a difference in orientation (but not composition) among the small sections of the microstructure. In a diffusion-controlled transformation the regions differ in both composition and crystal lattice structure. Examples are the growth of pearlite in carbon steel, the precipitation of the β' phase in zinc-rich brass, and the age-hardening of β-titanium alloy. The effect of this structure within the original grain boundaries is twofold: each region of the structure may scatter individually and the original grain may scatter as a unit, its properties being determined by a summation over its structure, as suggested by Kamigaki.[34] The attenuation is then the sum of these two contributions.

The first of these is an obvious corollary of the general grain scattering theory. A region of the structure may be a large fraction of the original grain, as in the case of a region of pearlite (parallel layers of iron and iron carbide) in a slowly cooled carbon steel, or it may be less than one one-thousandth of the grain in the hardened-steel structure of martensite. Thus, depending on the size and anisotropy of the regions of intragranular structure and on their relative densities, the scattering by individual regions might be either larger or smaller than the scattering by the original grain. The effect of preferred orientation may have to be introduced, of course, because the crystallographic axes of the microstructure may form in only a few directions with respect to the axes of the host grain. An example is martensite in carbon steel, in which the [110] planes and [1$\bar{1}$1] directions of the body-centered-tetragonal martensite lie parallel to the [111] planes and the [$\bar{1}$10] directions in the prior face-centered-cubic austenite grain.[35] In carbon-free iron alloys the martensite is body-

[34] K. Kamigaki, *Sci. Rep. Tohoku Univ. First Ser.* A-9, 48–77 (1957).
[35] G. Kurdjumov and G. Sachs, *Z. Phys.* **64,** 325–343 (1930).

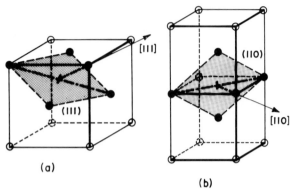

FIG. 2. The shear transformation from (a) austenite (fcc) to (b) martensite (bcc or bct). The pyramid of five atoms changes shape slightly during the transformation, such that the angles of the rhombic base plane (shaded) change and the angle between the base plane and the line from the fifth atom to the center of the base plane also changes. (Adapted from Brick and Phillips,[2] pp. 269–271. Used by permission of McGraw-Hill.)

centered-cubic. The total number of possible orientations is 24. The formation of martensite from austenite by a shear transformation (Brick and Phillips,[2] pp. 269–271) is illustrated in Fig. 2.

The second effect of microstructure is not so obvious but becomes clear after a consideration of the "preferred orientation" of the interior structure of the grain relative to the original grain axes. If the structure forms in only a finite number of directions within each grain, then the grain volume will remain anisotropic and will continue to be a scattering center.[34] The anisotropy factor (μ for cubic grains and combinations of γ, χ, and η for hexagonal grains; Section 5.2.1.2) will be reduced. The scattering factor will change in proportion to the change in $\langle (\Delta C/\overline{C})^2 \rangle_{av}$. In this case C will be a sum over all the orientations of the moduli weighted by the prevalence of the orientation and $\Delta C = C - \overline{C}$.

The following is a brief analysis of a model of cubic martensite in two dimensions. Figure 3 defines the axes of the model. The [111] and [$\bar{1}$11] directions in the face-centered-cubic austenite are rotated about the y axis into the x–y plane to form the two-dimensional model of austenite. These directions are at 35° and 145° to the positive x axis and are the directions of the [110] type of axis of the two-dimensional model of martensite formed from the austenite. For the austenite with its modulus written

$$C = C_0(1 + a \cos 4\theta), \tag{5.2.45}$$

the average modulus $\overline{C}$ is C_0, and the value of $\langle (\Delta C/\overline{C})^2 \rangle_{av}$ is

$$\langle (\Delta C/\overline{C})^2 \rangle_{av|A} = a^2/2. \tag{5.2.46}$$

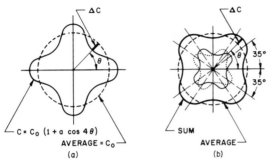

FIG. 3. Elastic anisotropy of grains: (a) single-crystal grain; (b) martensite at $\pm 35°$. Coordinates for the analysis of the two-dimensional martensite model. The directions at $35°$ and $145°$ to the x axis correspond to the [111] direction in the two-dimensional austenite model and to the [110] type of direction in the two-dimensional martensite model. The rosettes represent the martensite modulus, and the flattened rosette is the sum of the moduli of the two possible orientations. The average $\langle (\Delta C/C)^2 \rangle_{av}$ of the sum is smaller than the anisotropy of the single martensite platelet and may be smaller than the anisotropy of the austenite.[36]

Here, the A subscript indicates the austenite average. On the other hand, the martensite modulus in Fig. 3 is

$$C' = C_1[1 + b \cos 4(\theta + 35°)] \qquad (5.2.47)$$

for half the regions and

$$C' = C_1[1 + b \cos 4(\theta + 145°)] \qquad (5.2.48)$$

for the other half, so the effective modulus is

$$C'' = C_1\{1 + (b/2)[\cos 4(\theta + 35°) + \cos 4(\theta + 145°)]\}. \qquad (5.2.49)$$

Performing the averaging to find $\langle (\Delta C/\overline{C})^2 \rangle_{av}$, one obtains

$$\langle (\Delta C/\overline{C})^2 \rangle_{av|M} = 0.413b^2/2. \qquad (5.2.50)$$

Here, M indicates the martensite average. The ratio of Eq. (5.2.50) to Eq. (5.2.46) is

$$\langle (\Delta C/\overline{C})^2 \rangle_{av|M} / \langle (\Delta C/\overline{C})^2 \rangle_{av|A} = 0.413(b/a)^2. \qquad (5.2.51)$$

The equation expresses the change in the anisotropy factor in the two-dimensional model of the transformation from austenite to martensite. Besides the constant factor 0.413, the ratio depends only on the ratio b/a

[36] E. P. Papadakis, in "Physical Acoustics," (W. P. Mason, ed.), Vol. 4, Part B, pp. 269–328. Academic Press, New York, 1968.

TABLE IV.[36] Anisotropy for Fe–30%Ni, Fe, and Ni

Metal	Lattice	Value
Fe–30%Ni	a	-15.5×10^{11} dyn/cm^2
Fe	b	-13.9×10^{11} dyn/cm^2
Ni		-14.7×10^{11} dyn/cm^2
	ratio b/a: 0.897	

of anisotropies of the martensite lattice b and the austenite lattice a. If $b/a < \sqrt{2}$, the scattering will decrease appreciably in a hypothetical martensitic transformation in two dimensions. In three-dimensional martensite the coefficient multiplying $(b/a)^2$ probably will be smaller than 0.4 because of the larger number of orientations of the martensite axes. In other types of transformation, too, the coefficient is likely to be smaller than unity. Thus it is likely that a phase transformation producing several particles within the original grain volume would reduce the attenuation of ultrasound by lowering the scattering. Only if b/a were considerably larger than unity could scattering increase. Kamigaki[34] has suggested $b/a \gg 1$ in pearlite versus austenite. What is a likely range of b/a in comparing martensite to austenite? The best values will be hardly more than a guess, because one of the two phases is usually inaccessible to elastic-modulus measurements, either because it exists only at elevated temperatures or because it exists only in microscopic particles. For one guess, the difference between fcc Fe–30%Ni alloy[37] and bcc iron (Mason,[21] pp. 355–373) can be examined. Computing $\mu = c_{11} - c_{12} - 2c_{44}$ for each and calling them a and b, respectively, one finds the values in Table IV for a, b, and their ratio. It is near 0.9. Nickel by itself (Mason,[21] pp. 355–373) is included for comparison.

Investigations must cover two types of transformation: the diffusionless and the diffusion controlled. In the first category one has the martensitic transformation in carbon steel on one hand, and in various alloys (not necessarily containing iron) on the other. In the second category there is the growth of pearlite and bainite in carbon steel and the growth of secondary phases in other metastable alloys through annealing. Other interesting examples could be found.

5.2.1.8. Nonequiaxed Grains. If the grains of a metal are elongated or flattened, then the scattering is changed. Besides the preferred orientation, which is likely to be present, the grain shape will introduce a factor affecting not only the magnitude but also the functional dependence of the scattering. For instance, scattering by severely elongated grains will re-

[37] G. A. Alers, J. R. Neighbours, and H. Sato, *Phys. Chem. Solids* 13, 40–55 (1960).

duce to scattering by cylinders.   In Rayleigh scattering by cylinders,[5] the scattering power of a single cylinder depends on the square of the cross-sectional area and the cube of the frequency.   Other shapes will have other shape factors.   Another problem arises owing to the possibility that the long dimension or dimensions of a grain are in the size category $\lambda < 2\pi\overline{D}_2$, while the short dimension or dimensions fulfill the condition $\lambda > 2\pi\overline{D}_1$, in this case $\overline{D}_1$ being the average short dimension and $\overline{D}_2$ the average long dimension.   Then the scattering must obey different laws in different directions.   This is analogous to the possibility that the largest grains in an equiaxed distribution fall under a different scattering law from that of the rest of the grains.   Nothing quantitative has been worked out so far on scattering by nonequiaxed grain samples.   For a rough estimate one could use the formulas of Morse[5] for the scattering powers of rigid spheres and the cylinders in a fluid.   Taking objects of the same diameter D, one finds that a cylinder scatters more strongly by a factor of λ/D than a group of spheres which, if placed on a line and touching, would equal the cylinder in length.

5.2.2. Some Experimental Methods

5.2.2.1. Scope. This section on experimentation is subdivided to correspond to various experimental techniques that can be applied successfully to Section 5.2.1 on theory.   A short subsection concerns the pulse-echo method of ultrasonic measurement, since all the experiments discussed depend on it.   Both rf bursts and short pulses with spectrum analysis are treated.

5.2.2.2. Pulse-Echo Method with Direct Bonding. The pulse-echo method of measuring ultrasonic attenuation and velocity is widely used in the megahertz and gigahertz regions of frequency.   The work reported in this part lies below 100 MHz, at which the ultrasonic wave is usually generated by a piezoelectric transducer bonded to one of two plane-parallel faces of the sample.   The method is explained here with reference to the block diagram in Fig. 4.   Various authors describe it more fully.[38-40]   The transducer of radius a is excited by a burst of rf voltage from a pulsed or gated source.   The trigger initiating the burst has also initiated the horizontal sweep of an oscilloscope.   The burst passes through a hybrid circuit (or a gate) arranged to allow the burst to reach the transducer but not the amplifiers.   The radio frequency is set at the frequency at which the transducer is half a wavelength thick, or $(2n + 1) \lambda'/2$ for odd harmonics, for the wavelength λ' of the elastic wave in a bulk sample of the trans-

[38] R. L. Roderick and R. Truell, *J. Acoust. Soc. Am.* **23**, 267–279 (1952).
[39] B. Chick, G. Anderson, and R. Truell, *J. Acoust. Soc. Am.* **32**, 186–193 (1960).
[40] E. P. Papadakis, *IEEE Trans. Sonics Ultrason.* **SU-11**, 19–29 (1964).

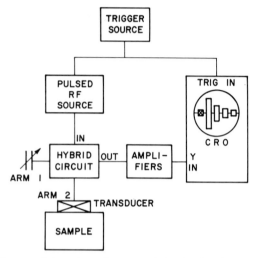

FIG. 4. Block diagram of the pulse-echo system in its simplest form. A repetitive trigger activates the pulsed oscillator and the oscilloscope sweep. The rf pulse from the oscillator arrives at the transducer by way of a hybrid circuit or a gate, which prevents the rf pulse from overloading the amplifier. An ultrasonic wave train is generated by the transducer; the waves echo back and forth within the sample bonded to the transducer, which picks up the echoes again. They are amplified and displayed on the oscilloscope. The applied frequency must be exactly proper to operate the transducer plate at an odd harmonic.[36]

ducer material. The transducer is many wavelengths in diameter; that is, $a \gg \lambda'$. The rf burst excites the piezoelectric transducer, which in turn radiates elastic waves into the sample. The burst is shorter than the travel time in the sample, so, as the elastic wave echoes back and forth between the plane-parallel faces, the echoes are separate. In the sample the wavelength is λ. The piezoelectric transducer picks up a signal proportional to the amplitude of each echo while abstracting only a small portion of the wave energy. The signal is permitted to pass into the amplifiers and be displayed on the oscilloscope. The attenuation is found from the ratio of echo amplitudes and the velocity from the travel time between echoes, as described in Section 5.1.1.3. Various calibration and comparison methods are used for improving the accuracy of the measurements. See Roderick and Truell.[38] May,[41] Chick et al.,[39] Forgacs,[42] McSkimin,[43] McSkimin and Andreatch,[44] Bolef and deKlerk,[45] and Papadakis.[40]

[41] J. E. May, Jr., IRE Nat. Conv. Record 6, Pt. 2, 134–142 (1958).
[42] R. L. Forgacs, IRE Trans. Instrum. I-9, 359–367 (1960).
[43] H. J. McSkimin, J. Acoust. Soc. Am. 33, 12–16 (1961).
[44] H. J. McSkimin and P. Andreatch, J. Acoust. Soc. Am. 34, 609–615 (1962).
[45] D. I. Bolef and J. de Klerk, IEEE Trans. Ultrason. Eng. UE-10, 19–26 (1963).

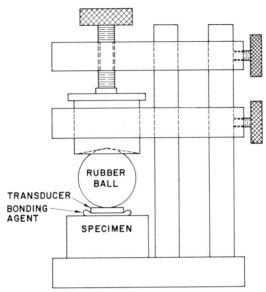

FIG. 5. Nonuniform pressure bonding jig. A rubber sphere or ball presses the bonding agent (e.g., epoxy) into a thin layer because the pressure distribution under the ball is a hemiellipse with a radial gradient at all points.

5.2.2.3. Bonding of Transducer Plates.
It is necessary to produce thin flat bonds between transducer plates and the specimens of interest. The specimens must be flat and parallel.   As a rule of thumb, finely ground faces with a parallelism of 10^{-4} are adequate up to 100 MHz. Transducer plates polished for overtone operation should be used.   These should be ground, lapped, and polished to the desired thickness and then plated with a durable coating of minimum thickness compatible with high conductivity, not "plated-to-frequency" as crystal manufacturers are wont to do.

Bonding layers must be dirt free and thin.   For permanent bonds, epoxy pressed out by the nonuniform pressure method[46] is advised.   A sketch of a bonding fixture is shown in Fig. 5.   The epoxy must be filtered (2 μm or better) and outgassed; it is handled in a clean atmosphere, preferably in a laminar-flow hood.

For temporary bonds in the vicinity of room temperature, phenyl salicylate (salol) is adequate.   It is a chemical melting at 43°C and recrystallizing by seeding after supercooling to room temperature.   The best technique for its application is to place a few small crystals next to the trans-

[46] E. P. Papadakis, *J. Adhes.* **3**, 181–194 (1971).

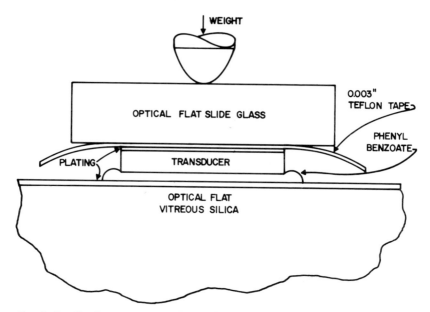

FIG. 6. Bonding fixture adequate for use with salol. The pressure is distributed over thin transducer plates with a slide glass or other flat material. (Papadakis,[47] by permission of the Institute of Electrical and Electronics Engineers.)

ducer plate which is precleaned and laid in position on the cleaned specimen, heat the group with a hot plate or heat lamp, let the salol melt and flow under the transducer plate, apply a pressure to the top of the transducer plate, let the system cool, and seed. Again, the nonuniform pressure bonding method may be used with a rubber sphere pressure applicator. Another adequate method is to use a jig such as in Fig. 6 which distributes a point force over the transducer through a flat plate such as a piece cut from a microscope slide. Telfon tape keeps the salol from bonding the slide to the transducer. Cleansing of all surfaces can be accomplished with an artist's brush and reagent grade acetone.

5.2.2.4. Beam Spreading. Inasmuch as one is propagating a wave from a finite transducer, the effect of beam spreading upon his loss and travel-time measurements must be considered. The beam-spreading problem is one of diffraction from a single aperture. It has been solved by Seki et al.[48] and by Tjadens[49] for longitudinal waves from a circular transducer of radius a, radiating waves of wavelength λ into fluids along

[47] E. P. Papadakis, IEEE Trans. Sonics Ultrason. SU-16, 210–218 (1969).
[48] H. Seki, A. Granato, and R. Truell, J. Acoust. Soc. Am. 28, 230–238 (1956).
[49] K. Tjadens, Acustica 11, 127–136 (1964).

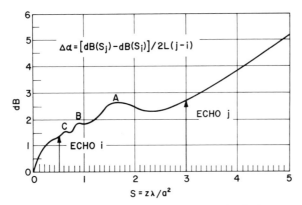

FIG. 7. Loss versus normalized distance for circular, longitudinal, wave transducers on isotropic media. This plot shows the nonmonotonic character of the loss in the Fresnel region of the radiation pattern of the transducer. The normalized distance $S = z\lambda/a^2$ allows all distances z, wavelengths λ, and transducer radii a to be represented by one curve. The diffraction loss can change the attenuation measured between echoes; diffraction corrections[52] must be applied to attenuation measurements[48] 1956). (Papadakis,[53] by permission of American Institute of Physics.)

the coordinate z. These solutions work well for isotropic solids.[50,51] The curve of loss versus normalized distance $S = z\lambda/a^2$, given by Seki *et al.*,[48] is presented in Fig. 7; the loss is not a monotonic function. This curve is used for making diffraction corrections for the loss of longitudinal waves supported by isotropic media, as outlined by Papadakis.[52] The problem has not been solved for shear waves. It has, however, been solved by Papadakis[53,54] for longitudinal waves in anisotropic media along axes of threefold, fourfold, and sixfold symmetry. The solution is of some importance in materials with preferred orientation, since the diffraction loss will depend on the degree and symmetry of the preferred orientation. For instance, circular bar stock has sixfold symmetry along its axis and is amenable to treatment by the existing diffraction theory. Care should always be taken to propagate only one pure mode at a time in anisotropic samples. For a complete treatment of diffraction corrections, see the work of Papadakis.[55]

[50] E. P. Papadakis, *J. Acoust. Soc. Am.* **35**, 490–494 (1963).

[51] E. P. Papadakis, *J. Appl. Phys.* **34**, 265–269 (1963).

[52] E. P. Papadakis, *J. Acoust. Soc. Am.* **31**, 150–152 (1959).

[53] E. P. Papadakis, *J. Acoust. Soc. Am.* **40**, 863–876 (1966).

[54] E. P. Papadakis, *J. Acoust. Soc. Am.* **36**, 414–422 (1964).

[55] E. P. Papadakis, *in* "Physical Acoustics" (W. P. Mason and R. N. Thurston, eds.), Vol. 11, pp. 151–211. Academic Press, New York, 1975.

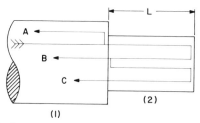

FIG. 8. Definition of echoes A, B, and C from a specimen at the end of a buffer, where $A = R$, $B = (1 - R^2)e^{-2\alpha L}$, and $C = -R(1 - R^2)e^{-4\alpha L}$. (From Papadakis et al.,[56] by permission of the American Institute of Physics.)

5.2.2.5. Use of Buffer Rods.

The curve shown in Fig. 7 of loss versus normalized distance (dB versus $S = z\lambda/a^2$) can be used in the case of buffer rods or liquid columns which convey the ultrasonic waves from the transducer to the specimen and back as in Fig. 8. The theory for diffraction corrections in buffer–specimen systems was presented earlier.[56]

In a buffer–specimen system, one needs the amplitudes of three separate echoes, A, B, and C (or A', A, and B, where echo A' is echo A before the specimen is attached in Fig. 8), for the calculation of the attenuation and reflection coefficients. In bulk specimens, these three echoes are affected to different degrees by diffraction (beam spreading). To be specific, each echo is smaller than it would have been in the absence of diffraction. The principle invoked in diffraction corrections in this case is the correction of each echo amplitude to its undiffracted value and the subsequent calculation of R and α from the corrected echo amplitudes.

First, S is calculated for each echo, since S is the abscissa of the loss versus distance curve in Fig. 7. For anisotropic materials, use the appropriate curves.[53,55] Then, look up the corresponding losses on the ordinate. After that, correct the amplitudes A', A, B, and C of the echoes to eliminate the above loss, getting A_0', A_0, B_0, and C_0. The formulas for R and α are then used:

Echoes A, B, and C:

$$R = [\tilde{A}_0\tilde{C}_0/(\tilde{A}_0\tilde{C}_0 - 1)]^{1/2} \qquad (5.2.52)$$

and

$$\alpha = [\ln(-R/\tilde{C}_0)]/2L, \qquad (5.2.53)$$

where $\tilde{A}_0 \equiv A_0/B_0$, $\tilde{C}_0/B_0$, and L is specimen length.

[56] E. P. Papadakis, K. A. Fowler, and L. C. Lynnworth, *J. Acoust. Soc. Am.* **53**, 1336–1343 (1973).

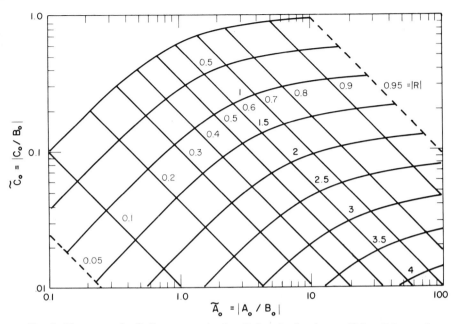

FIG. 9. Nomogram for finding attenuation loss $2\alpha L$ and reflection coefficient R from echo amplitudes A, B, and C of separate echoes in a buffer system as in Eqs. (5.2.52) and (5.2.53). The subscript 0 indicates that diffraction corrections have been applied if needed (for bulk and plate waves, not for wire waves). (From Papadakis et al.,[56] by permission of American Institute of Physics.)

Echoes A, A', and B

$$R = A_0/A_0'$$ (5.2.54)

and

$$\alpha = \{\ln[A_0'(1 - R^2)/B_0]\}/2L.$$ (5.2.55)

In the present formulation, the relative signs of A', A, B, and C *must* be used. The experimenter must note whether A', B, or C are inverted with respect to A using unrectified echoes.

The sign of R may be determined from the equation

$$R = (Z_B - Z_S)/(Z_B + Z_S),$$ (5.2.56)

where Z_B and Z_S are the specific acoustic impedances of the buffer and the specimen, respectively. Equation (5.2.56) is not adequate for determining the exact numerical value of R because of phase shifts in the bond layer between the buffer and the specimen. Best results are obtained for $Z_B < Z_S$.

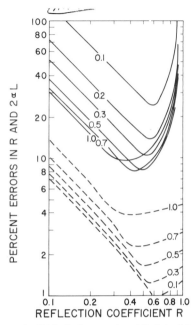

FIG. 10. The probable error in R (dashed curves) and $2\alpha L$ (solid curves) as a function of the magnitude of R for various values of $2\alpha L$ when the probable error in the magnitudes A, B, and C of separate echoes in a buffer system is 1% of the largest of these magnitudes. The measurement of α is best when R is between 0.4 and 0.6. (From Papadakis,[58] by permission of American Institute of Physics.)

The relative signs of A, B, and C must be recorded in order for the equations for R and α to apply. Because of the reflections at the buffer–specimen interface, echoes A and C are always of opposite sign; B may be of either sign with respect to A. As a result, $\tilde{A}_0$ and $\tilde{C}_0$ are of opposite sign. If $\tilde{C}_0$ is positive, R must be negative to satisfy Eq. (5.2.53) and vice versa. It will be found that the sign of R agrees with the definition in Eq. (5.2.56) for the reflection coefficient.

A nomogram[56,57] is presented in Fig. 9 for determining R and $2\alpha L$ from $\tilde{A}_0$ and $\tilde{C}_0$. From the curved isolines on logarithmic graph paper it can be seen that the accuracy must be a function of $\tilde{A}_0$ and $\tilde{C}_0$. A graph is presented in Fig. 10 for estimating the errors to be expected in R and $2\alpha L$ if the error in measuring each of A, B, and C is $\pm 1\%$ of the largest of the three amplitudes. It can be seen that the errors are lowest if R is made about 0.6. The buffer method is useful in wires where there is no beam spreading. It is indispensable in the momentary-contact method with

[57] L. C. Lynnworth, *Ultrasonics* **12**, 72–75 (1974).
[58] E. P. Papadakis, *J. Acoust. Soc. Am.* **44**, 1437–1441 (1968).

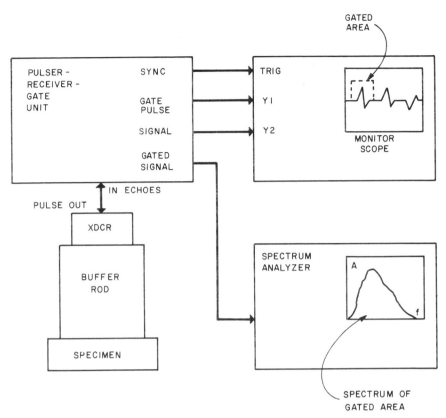

FIG. 11. Block diagram of the electrical and ultrasonic system for spectrum analysis of separate echoes in a buffer–specimen system. The gate transfers one echo at a time to the spectrum analyzer, which gives amplitude as a function of frequency. Equations (5.2.52) and (5.2.53) or the nomogram in Fig. 9 can then be used to find R and α. (From Papadakis *et al.*,[56] with permission of The Institute of Electrical and Electronics Engineers, and American Institute of Physics, respectively.)

pressure coupling[59] for measuring hot specimens where the transducer must be mounted on the end of a buffer rod to protect it from the high temperatures. It is also useful when commercial nondestructive testing (NDT) transducers are to be used. One must evaluate commercial transducers to assure that they act as piston sources [60] before using them for accurate measurements.

5.2.2.6. Spectrum Analysis of Broadband Echoes. Earlier work on

[59] E. P. Papadakis, L. C. Lynnworth, K. A. Fowler, and E. H. Carnevale, *J. Acoust. Soc. Am.* **52,** 850–857 (1972).

[60] E. P. Papadakis, *Proc. IEEE Ultrason. Symp., Phoenix, Arizona,* October 26–28, Paper DD-1, pp. 104–112 (1977).

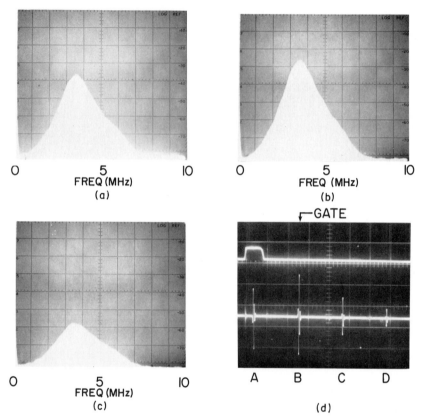

O   5   IO   O   5   IO
    FREQ (MHz)   FREQ (MHz)
        (a)   (b)

                                    ┌─GATE

O   5   IO   A   B   C   D
    FREQ (MHz)
        (c)   (d)

FIG. 12. Spectra of echoes A [part(a)], B[part(b)], and C[part(c)] in the buffer–specimen system shown in Fig. 11. In part (d) is shown the video view of the echoes and the gate synchronized with echo A. (From Papadakis[60b] © 1972 IEEE.)

spectrum analysis of ultrasonic pulses[60a] has been refined[56] to permit quantitative measurement of ultrasonic attenuation. A buffer rod is interposed between a transducer plate or a commercial broadband transducer (such as those used in nondestructive testing applications) and the specimen to eliminate the transducer's effect upon multiple echoes. Each echo and reverberation A, B, and C is gated separately into a spectrum analyzer by the system shown in Fig. 11. The resulting spectra shown in Fig. 12 are $A(f)$, $B(f)$, and $C(f)$. Use of Eqs. (5.2.52) and (5.2.53), or of the nomogram in Fig. 9, yields the attenuation $\alpha(f)$ as a function of fre-

[60a] O. R. Gericke, *Mater. Res. Stand.* **5**, 23–30 (1965).
[60b] E. P. Papadakis, *Proc. IEEE Ultrason. Symp., Boston, Massachusetts*, October 4–7, pp. 81–86 (1972).

quency. In this way, one rugged transducer or one permanently bonded transducer plate can do the work of ten fragile removable quartz plates in less time with no need for inconvenient details like tuning. It is necessary to perform diffraction corrections properly[56] at each frequency utilized from within the spectrum.

5.2.3. Experiments on Grain Scattering

5.2.3.1. **Early Experiments.** In the early experiments[11-13,16] there was some uncertainty about whether the highest power of frequency visible in the attenuation was f, f^2, or f^4. Mason's contention that f^4 was present for $\lambda > 3\bar{D}$ was accepted at least as a working hypothesis for experimentalists. In the Rayleigh region the attenuation[12] was written

$$\alpha = a_1 f + a_4 f^4, \tag{5.2.57}$$

where the linear term accounted for elastic hysteresis found by Wegel and Walther.[61] At higher frequencies the attenuation varied as $\bar{D}f^2$, and at still higher frequencies it became inversely proportional to $\bar{D}$ and seemingly dependent on f^0 or f^1. Hirone and Kamigaki[62,63] Merkulov,[17] and Kamigaki[34] used Eq. (5.2.57) to express the attenuation in the Rayleigh region in later work. Merkulov found experimentally that: (1) there is Rayleigh scattering proportional to $\bar{D}^3 f^4$ when $\lambda > 10\bar{D}$; (2) the dependence of the scattering goes over to $\bar{D}f^2$ in the range $4 < \lambda/\bar{D} < 10$; (3) the dependence becomes $1/\bar{D}$ at higher frequencies ($\lambda/\bar{D} < 4$); and (4) the LPM theory (Section 5.2.1.2) underestimates the attenuation by a factor of 3 to 5 but gives the correct ratio for shear to longitudinal wave attenuation. One suspects that a large part of the discrepancy in magnitude lay in the grain-size determination. Hirone and Kamigaki[62] found the attenuation in aluminum to be proportional to $1/\bar{D}$ when $\lambda < \bar{D}$. In other work Hirone and Kamigaki[63] and Kamigaki[34] found that the coefficient a_4 for Rayleigh scattering in Eq. (5.2.57) was proportional to the austenitic grain volume in carbon steel, stainless steel, and cast iron. They inferred that the prior austenite grain was a scattering center with an anisotropy dependent upon its interior structure. They found that the anisotropy of a grain transformed to lamellar pearlite was 10,000 times higher than that of an aluminum grain and, hence, higher than that of iron. Merkulov[64] reported, however, that the grain anisotropy of pearlite was lower than that of iron.

[61] R. L. Wegel and H. Walther, *Physics* **6**, 141–157 (1935).
[62] T. Hirone and K. Kamigaki, *Sci. Rep. Tohoku Univ. First Ser.* **A-7**, 455–464 (1955).
[63] T. Hirone and K. Kamigaki, *Sci. Rep. Tohoku Univ. First Ser.* **A-10**, 276–282 (1958).
[64] L. G. Merkulov, *Sov. Phys. Tech. Phys. (English transl.)* **2**, 1282–1286 (1957).

TABLE V.[36]   Treatment and Properties of the Metal Specimens Studied
in Section 5.2.3.2

	Metal	
Property	Nickel	Fe–30%NiAU1
Annealing temp. (°C)	700	815
Annealing time (hours)	3	1
Quench	Air	Air
Velocity^a (10^5 cm/sec)		
Polarization (a)	5.730	—
Polarization (b)	5.732	4.65
Polarization (c)	2.988	—
Polarization (d)	2.980	2.69
Polarization (e)	2.989	—
$\langle R \rangle_{av}$ (cm)	5.54×10^{-3}	2.80×10^{-3}
T (cm³)	1.10×10^{-5}	5.76×10^{-7}

^{a} Values are for polarization (a)–(e) in Fig. 13.

These experiments have shown that the basic elements of elastic-wave-scattering theory are essentially correct and have opened up new avenues of research, especially on microstructure. Quantitative verification of the LPM scattering theory will be presented in the next section.

5.2.3.2. Quantitative Agreement with Scattering Theory. Ultrasonic

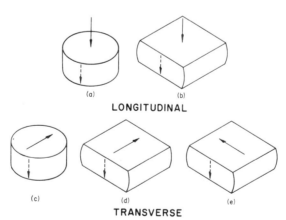

FIG. 13. Polarization directions in specimens cut from bar stock. The dotted arrows indicate the propagation direction and the solid arrows show the direction of particle motion. Longitudinal [parts (a) and (b)] and transverse [parts (c)–(e)] waves were generated, respectively, by X-cut and Y-cut circular quartz transducers 1.27 cm in diameter. Specimens were about 1.78 cm thick and 3 cm in diameter, except the Fe–30%Ni specimens which were about 2 cm in diameter but had rough side-walls (Papadakis,[65] by permission of American Institute of Physics.)

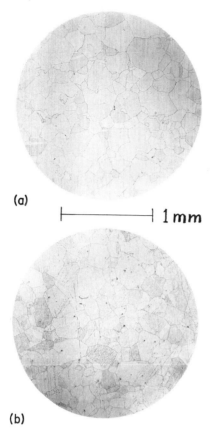

(a)

⊢――――――⊣ 1 mm

(b)

FIG. 14. Photomicrographs of recrystallized nickel specimen: (a) axial view; (b) radial view. The grains are equiaxed single crystals with no preferred orientation (the latter being inferred from the isotropy in the velocity given in Table V). (Papadakis,[65] by permission of American Institute of Physics.)

attenuation measurements of four polycrystalline metals with well-determined grain-size distributions have been presented.[65] The data were analyzed by the methods presented in Sections 5.2.1.2–5.2.1.5, 5.2.2.5, and 5.2.2.6. One of these metals, nickel, will be treated as an example here. Another alloy, 30%Ni in Fe, will also be treated. Data on this alloy were taken in conjunction with a study of microstructure, made by Papadakis and Reed,[66] and have since been analyzed.[26] Two austenitic stainless steels also were studied,[65] but they will not be treated here, because the single-crystal moduli are not known.

[65] E. P. Papadakis, J. Acoust. Soc. Am. **37**, 711–717 (1965).
[66] E. P. Papadakis and E. L. Reed, J. Appl. Phys. **32**, 682–687 (1961).

1 mm

FIG. 15. Photomicrograph of equiaxed and quenched Fe–30%Ni. The quench into liquid nitrogen transformed the alloy 90% to a martensitic microstructure, but the austenitic (fcc) grain size is still clearly visible. The ultrasonic measurements discussed in this section were performed on the equiaxed material in the 100% fcc condition at room temperature before quenching. (Papadakis and Reed,[66] by permission of American Institute of Physics.)

The moduli of nickel are well known (Mason,[21] pp. 355–373). Alers *et al.*[37] measured the moduli of the Fe–30%Ni alloy. The longitudinal and shear velocities were measured in the course of the attenuation work.[65,66] Data on the annealing treatments, ultrasonic velocities, and moments of

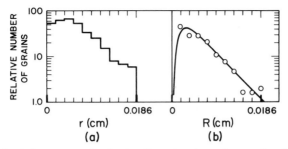

FIG. 16. (a) Grain-image area distribution from the photomicrographs of recrystallized nickel in Fig. 14; (b) grain-size distribution within the nickel, computed from the image distribution. The radii r and R refer to images and grains, respectively (Papadakis,[65] by permission of American Institute of Physics.)

the grain-size distributions, $\langle R \rangle_{av}$ and $T = (4\pi/3)\langle R^6 \rangle_{av}/\langle R^3 \rangle_{av}$, are summarized in Table V. Polarizations 1–5 in the table are defined in Fig. 13. The metal specimens in question were homogeneous and essentially isotropic with equiaxed single-crystal grains showing no other character-istics in their microstructure.

Micrographs of the nickel and Fe–30%Ni alloy appear in Figs. 14 and 15, respectively. Grain-image area histograms were made from these, and grain-size distribution curves were then computed. These pairs of graphs are shown in Figs. 16 and 17. From the grain-size distributions the moments $\langle R \rangle_{av}$ and T were computed. This procedure followed Section 5.2.1.5 exactly. The moments and other data in Table V were then used in the attenuation formulas given in Section 5.2.1.2, for constructing graphs such as those described in Section 5.2.1.4.

Figures 18 and 19 are the logarithmic graphs of attenuation versus fre-

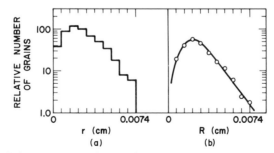

FIG. 17. (a) Grain-image area distribution from the photomicrograph of recrystallized Fe–30%Ni alloy in Fig. 15, with the austenite grains measured; (b) grain-size distribution in this alloy (Papadkis,[26] by permission of American Institute of Physics.)

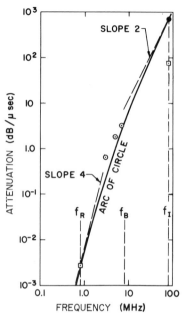

FIG. 18. Attenuation versus frequency of longitudinal waves in polycrystalline nickel. The data points ($\odot$) are to be compared with the theoretical curve approximated by the arc of a circle drawn through the Rayleigh point ($\square$) at frequency f_R with a slope 4 and through the intermediate frequency f_I with a slope 2; f_B is the boundary frequency. The theoretical intermediate point ($\square$) for longitudinal waves seems too low in this case (Papadakis,[65] by permission of American Institute of Physics.)

quency for longitudinal and shear waves, respectively, in the nickel specimens. Note that there is a discrepancy in the longitudinal theory. One cannot draw a curve of monotonically decreasing slope through the theoretical points and make the slope 4 at the Rayleigh scattering frequency and 2 at the intermediate frequency. Either the slope is less than 2 at $f = f_I$ or one of the two points is of the wrong magnitude. Figure 18 illustrates one useful ad hoc approximation, if one assumes the theory predicts an intermediate frequency attenuation too small in magnitude. The approximation consists of an arc of a circle passing through the Rayleigh scattering point with a slope of 4 and having a slope of 2 at the intermediate frequency. The data points fall fairly close to this curve. Figure 19, on the other hand, contains a curve drawn through the two theoretical points with the proper slope at both of them. The shear wave data points lie above it by factors of from 2 to 3. The ratio of shear to longitudinal attenuation at 5 MHz is about 4, close to the proper value, according to the LPM theory. According to Table I, the ratio α_t/α_l, with α in decibels per

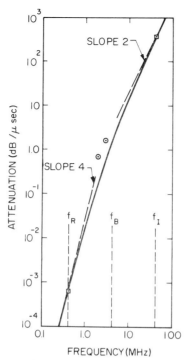

FIG. 19. Attenuation versus frequency of shear waves in polycrystalline nickel. The data points (⊙) are to be compared with the theoretical curve drawn through the Rayleigh point (▫) at frequency f_R with a slope 4 and through the intermediate frequency f_i with a slope 2 (▫). These shear wave data points are excessive in attenuation because no diffraction correction was available. The circular arc approximation is not needed for shear waves (Papadakis,[65] by permission of American Institute of Physics.)

microsecond, should be 3 in the Rayleigh region and higher above it. This ratio is simply $(3/4)(v_l/v_t)^2$ for attenuation per unit time; see Eqs. (5.2.1) and (5.2.2) with $\alpha_T = v\alpha_L$ as in Section 5.1.1.3.

Figure 20 is a graph on logarithmic scales of attenuation versus frequency for longitudinal waves in the equiaxed Fe–30%Ni specimen. In this case the data points lie on the properly sloping curve connecting the theoretical points; the theory for $\lambda < 2\pi\overline{D}$ is adequate.

Thus, in some cases the theory and data agree in this range, while in other cases the theory seems low. Between f_R and f_B the data agree with the properly sloping curve through the theoretical points if $\alpha(f_I)$ is large enough for this curve to be drawn; if not, then the data agree with the arc of a circle drawn through the Rayleigh point as described. Agreement is within a factor of 2 in all cases and sometimes is much better. Pre-

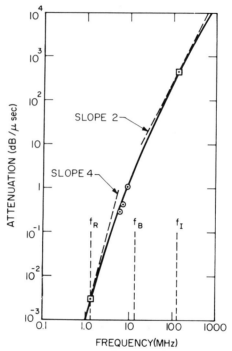

FIG. 20. Attenuation versus frequency in an equiaxed Fe–30%Ni specimen. The data points (○) fall on the curve drawn between the theoretical points (□) with the proper slopes at those points. In this case the scattering theory for the attenuation of longitudinal waves is adequate over the whole frequency range (Papadakis,[26] by permission of American Institute of Physics.)

viously,[17] discrepancies of factors of 5 were common. The accuracies of the LPM theory for Rayleigh scattering and of the grain-counting procedure are attested to by the agreement of the attenuation with curves of monotonic decreasing slope beginning with slope 4 at the Rayleigh point. The LPM theory also gives the ratio α_t/α_l correctly between f_R and f_B.

Note, however, that in both the metals studied by the author and in all the metals studied by Merkulov[17] the data for comparing the ratios α_t/α_l of theory and of experiment were obtained in the range $7 < (\lambda/\overline{D}) < 10$, which is not far from the value $\lambda_B = 2\pi\overline{D}$ defining f_b. This means that a really good comparison deep in the Rayleigh region has not yet been made.

The buffer method with spectrum analysis[56] as set forth in Sections 5.2.2.5 and 5.2.2.6 was applied to the same nickel specimen reported on above. Agreement between this method and the direct bonding method in Section 5.2.2.2 was exact. The data are presented in Fig. 21. Only

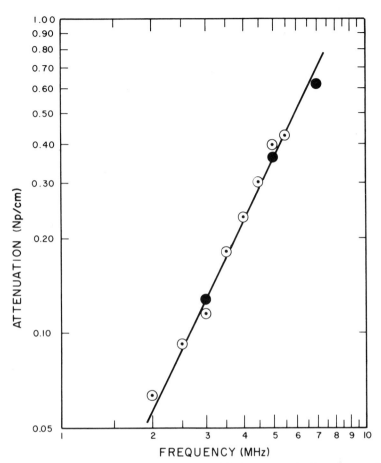

FIG. 21. Attenuation in a specimen of grade A nickel on a water buffer column. Agreement between rf bursts from a directly bonded quartz transducer (●) and the spectrum analysis of broadband pulses using the buffer (○) is essentially perfect. Diffraction corrections were performed by the appropriate methods[55] for the rf bursts and for the spectrum-analyzed echoes in the buffer column. (From Papadakis et al.,[56] by permission of American Institute of Physics.)

longitudinal waves were used, as water was used for the buffer column. Experimental parameters are given in Table VI.

5.2.3.3. Effect of Preferred Orientation on Scattering. Although a great deal has been done on elastic anisotropy arising from preferred orientation, not much work has been reported on the effect of preferred orientation on ultrasonic attenuation. Wilson[67] discovered a maximum in

[67] R. Wilson, Sheet Metal Ind. **30**, 146–160 (1953).

TABLE VI.   Data on Water Buffer, Nickel Specimen, and Transducer[a]

Datum	Transducer	Buffer	Specimen
Mode	Longitudinal	—	—
Crystal	5 MHz	—	—
Diameter (cm)	1.27	1.53	3.78
Length (cm)	—	1.335	1.775
Velocity (cm/μsec)	—	0.150	0.573

[a] From Papadakis et al.,[56] by permission of the American Institute of Physics.

the attenuation of shear waves in rolled mild steel when the propagation direction coincided with the direction transverse to rolling.   Frederick[68] suggested preferred orientation as one possible cause of anomalous attenuation in uranium.   Truell[69] found that the orientation of striations in a Cr–Ni–Mo steel could be correlated with the attenuation of longitudinal waves; the loss was highest for propagation parallel to the striations and lowest for propagation normal to them.

An experiment[33] on material with preferred orientation has shown anisotropic attenuation explicable in terms of the scattering theory presented in Section 5.2.1.6.   A drawn zinc bar was cut into slabs at various angles to its axis, as in Fig. 22.   Longitudinal ultrasonic waves were propagated through the slabs; velocity and attenuation were measured. The velocities are listed in Table VII.

The velocity in the 0° specimen is only 8% lower than the velocity along the a axis of a zinc crystal (Mason,[21] pp. 355–373), whereas the velocity in the 90° specimen is 26% lower than the velocity along the a axis of a zinc crystal and 30% higher than the velocity along its c axis.   Thus, along the bar axis there are crystalline a axes almost exclusively, whereas along the bar diameter there are crystalline a and c axes almost equally.   This type of orientation is expected in drawn zinc (Barrett,[1] pp. 444–445) and other hexagonal metals.

The effect upon the ultrasonic attenuation was striking; see Fig. 23. At 3 MHz the attenuation in the radial direction was six times that in the axial direction and amounted to an extra 10 dB or more between echoes. Since anisotropy would change the diffraction corrections 2 dB at most[53,54] and since dislocation damping[70] is probably small, the major portion of the attenuation must come from scattering.   The explanation is

[68] C. L. Frederick, REP. W-31-109-Eng-52.   Hanford Works, Hanford, Washington (1960).

[69] R. Truell, Mech. Eng. 77, 585–587 (1955).

[70] A. Granato and K. Lucke, J. Appl. Phys. 27, 583–593, 789–805 (1956).

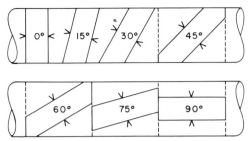

FIG. 22. Specimens cut from zinc bar stock having a axes of grains aligned strongly with the bar axis. Longitudinal ultrasonic waves were propagated normal to the finished faces. Both the velocity and the attenuation were found to be very anisotropic (Papadakis,[33] by permission of American Institute of Physics.)

this: As the longitudinal wave propagates along the bar axis, it traverses grains with their a axes along that direction. From grain to grain there is no change in modulus, so the wave solution in any grain is identical with the solution outside it. Thus the boundary conditions are matched exactly without recourse to a scattered wave. Hence, almost no scattering! In the radial direction, however, the wave traverses grains rotated in any degree about one a axis parallel to the bar axis. In one grain the propagation vector may be along an a axis; in the next, along a c axis. This is a condition of maximal scattering and is found to be such experimentally.

5.2.3.4. Effects of Microstructure on Grain Scattering. 5.2.3.4.1. TRIPARTITE APPROACH TO THIS SECTION. In the earlier discussion of theory it was pointed out that the scattering could arise both from the regions of the microstructure acting as scattering centers with their own anisotropy and from the original grains with modified anisotropy. The earliest experiments on grains with interior structure were carried out with carbon steel, a very complicated system. Chronological order will be reversed in this presentation, in order to present simple systems first.

TABLE VII.[33]   Longitudinal Wave Velocities in
Zinc Bar Stock

Specimen orientation (deg)	Velocity (10^5cm/sec)
0	4.40
15	4.29
30	3.98
45	3.86
60	3.76
75	3.60
90	3.54

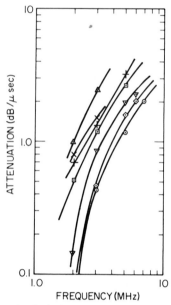

FIG. 23. Attenuation of longitudinal ultrasonic waves in zinc bar stock cut as in Fig. 22 at the following angles: 0° (○), 15° (◇), 30° (△), 45° (□), 60° (+), 75° (×), and 90° (△). The waves propagating axially and encountering only a axes of zinc grains were attenuated much less than waves propagating radially and encountering both a and c axes (Papadakis,[33] by permission of American Institute of Physics.)

Two simple systems will be discussed. They have the following features in common:

(1) No parameter of the original grain shape or grain-size distribution changes during the procedure introducing complex interior structure into the grains.

(2) Small changes in the velocity and density occur but are not be large enough to account for the changes in attenuation.

(3) The attenuation decreases, showing that the change in anisotropy of the original grains is enough to offset any new scattering introduced by the regions of the interior structure.

Carbon steel can never fulfill the first condition, since to arrive at such a state as to be ready to transform into some product having a microstructure it must completely recrystallize to an fcc structure (austenitize) at 850°C or higher. The microstructure develops immediately upon the cooling of the steel and depends on the rate of cooling for its form. There is no intermediate condition of single-crystal grains at room temperature.

The alloys considered in the next two subsections can be treated to pre-

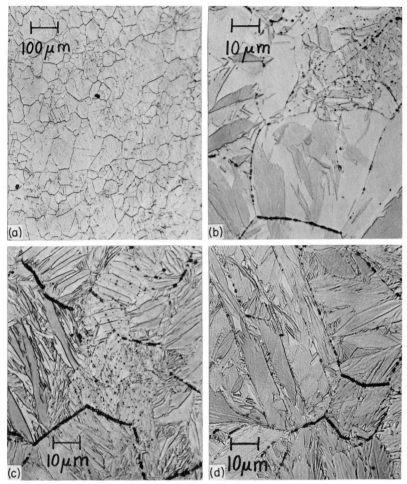

FIG. 24. Microstructure of Fe–30%Ni during heat treatment and quenching. Photomicrographs of block AU1 in the following conditions: (a) equiaxed, 100% austenite; (b) quenched in dry ice and acetone, 60% austenite and 40% martensite; (c) quenched in liquid nitrogen, 10% austenite and 90% martensite; (d) annealed at 150°C, same composition as (c) (Papadakis and Reed,[66] by permission of American Institute of Physics.)

serve single-crystal grain structure at room temperature. The transformation to the polyphase intragranular structure can be effected with no grain growth. These alloys fall into two classes: those with diffusionless transformations and those with diffusion-controlled transformations. Examples of both are given. In a third subsection, carbon steel with low to moderate alloy content is treated at some length.

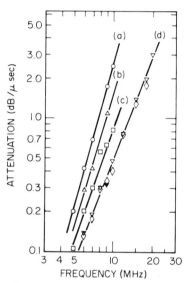

FREQUENCY (MHz)

FIG. 25. Attenuation in Fe–30%Ni specimen AU1 for longitudinal ultrasonic waves. Comparison of the attenuation in one block in five successive conditions: (a) as received, cold-rolled; (b) equiaxed; (c) quenched in dry ice and acetone; (d) quenched in liquid nitrogen (∇) and annealed at 150°C (◇). Both the attenuation and its slope decreased on quenching, but stress-relieving changed nothing. The lowering of the attenuation on quenching is ascribed to the lowering of the scattering when the grains are broken up by the microstructure (Papadakis and Reed,[66] by permission of American Institute of Physics.)

5.2.3.4.2. DIFFUSIONLESS TRANSFORMATIONS. Diffusionless transformations are characterized by the motion of large aggregates of atoms in unison in shearing motion. Relative motions of adjacent atoms are only a small fraction of a lattice spacing. [See the discussions in Section 5.2.1.7, and in Brick and Phillips,[2] Chapters 3,5–12, and 14–16).] Because the M_s temperature at which martensite begins to form in some alloys may be below room temperature and because the temperature at which the reverse transformation begins may be well above room temperature,[71] these alloys are very convenient for the ultrasonic study of microstructure; one which has been investigated is iron with a 30% nickel content.[66] In the experiment several specimens of equal size from a bar cold-rolled into a flat section were austenitized and equiaxed at 815°C for one hour, returned to room temperature still in the fcc structure, and measured for attenuation. Then one was held as a control, while the rest were subjected to grain growth at higher temperatures. The grain growth changed

[71] P. F. Fopiano, WADD TR60-273. Manufacturing Lab. Inc., Cambridge, Massachusetts (1960).

the grain-size distribution, as demonstrated previously,[26] yielding anomalous results when comparisons were made among the various specimens. Hence, only specimen AU1, the control, is considered here.
After the equiaxing it was quenched in dry ice and acetone ($-77°C$) and
later in liquid nitrogen ($-195°C$). The first quench produced 40% bcc
martensite, and the second quench increased it to 90%. Subsequently,
the specimen was stress relieved at 150°C for 2 hours. (Micrographs appear in Fig. 24.) Ultrasonic attenuation data were taken before and after
equiaxing, after each quench, and after annealing. A composite graph of
attenuation versus frequency for these steps of the treatment is given in
Fig. 25. After the equiaxing each quench lowered the attenuation and
also its slope on the logarithmic plot. This type of decrease indicates a
lowering of the part of the attenuation proportional to a high power of the
frequency. The stress-relieving treatment did not change the attenuation. The attenuation was analyzed in terms of Eq. (5.2.57) to give the
changes in coefficients a_1 and a_4. These are listed in Table VIII.

As shown in Fig. 20, the analysis in terms of $a_4 f^4$ is not quite correct,
because the frequency dependence of the scattering portion of the attenuation is below f^4 at frequencies approaching f_B, the boundary frequency,
where $\lambda_B = 2\pi D$. Curve (b) of Fig. 25 has a slope of 3.2, indicating that
the attenuation in the material arose principally from scattering. This
experiment showed conclusively that the martensitic transformation
lowers the scattering contribution to the attenuation in polycrystalline
metals. The fact that the grain-size distribution remained constant in the
stepwise transformation contributed to the unequivocal character of the
results.

5.2.3.4.3. DIFFUSION-CONTROLLED TRANSFORMATIONS. These transformations are characterized by the migration of atoms over many lattice
distances to form regions of a stable structure in a previously metastable
matrix. Typically, a metal can be heated to a certain temperature and
held there long enough to achieve an equilibrium distribution of atoms in
single-crystal grains. Rapid quenching can cause this composition and

TABLE VIII.[66]   Elastic Hysteresis and Rayleigh Scattering Coefficients[a] for the Equation
$\alpha = a_1 f + a_4 f^4$ Applied to Fe–30%Ni Specimen AU1

Treatment	$a_1 (\times 10^{-2})$	$a_4 (\times 10^{-5})$
Equiaxing	1.4	14.5
Dry ice and acetone quench	2.5	6.0
Liquid nitrogen quence	3.1	1.5
Stress-relieving	3.2	1.3

[a] Frequency in megahertz.

TABLE IX.[36]   Heat Treatment of
β-Titanium Alloy

Treatment	β-titanium
Solution treating	
Temperature	790°C
Time	1 hr
Quench	Water
Aging	
Temperature	595°C
Time	64 hr
Quench	Furnace

structure to be retained in a metastable form at room temperature, if the room temperature is low enough to inhibit diffusion. Reheating to a temperature lower than the original treatment temperature but high enough to permit diffusion produces precipitates of another phase within the metastable phase. These precipitates break up the grains and change the scattering.

One example is presented here; it is β-titanium, a titanium alloy containing 13% vanadium, 11% chromium, and 3% aluminum.[36]   The titanium alloy is all β phase (bcc) after solution-treating (a high-temperature annealing) and quenching but forms a precipitate of a second phase after aging at a lower annealing temperature.

Specimens of bar stock were procured, treated to produce the single-phase metastable product, measured for the attenuation of longitudinal ultrasonic waves, aged to produce the more complex microstructure, and measured again. The heat treatments are summarized in Table IX.

The solution treatments equiaxed the β-titanium grains. Aging produced no change in size or shape. Photomicrographs of the grains before and after aging are shown in Fig. 26. Ultrasonic attenuation before and after aging is shown in Fig. 27. The plot presents data for longitudinal ultrasonic waves propagating along the bar diameter as in Fig. 13b. Other directions of propagation yielded essentially the same curves. The data were taken by the pulse-echo method and were corrected for diffraction as described in Sections 5.2.2.2–5.2.2.4.

The attenuation was high initially and showed a slope of 3.0 on the logarithmic plot. The slope is quite reasonable for scattering by grains, particularly in view of the fact that the grain size puts the experiment in the region between the Rayleigh and the intermediate scattering ranges.

The aging treatment lowered the attenuation in both metals by a factor of 2. The decrease may be attributed to a decrease in ultrasonic scattering, and this in turn may be ascribed to the introduction of the complex

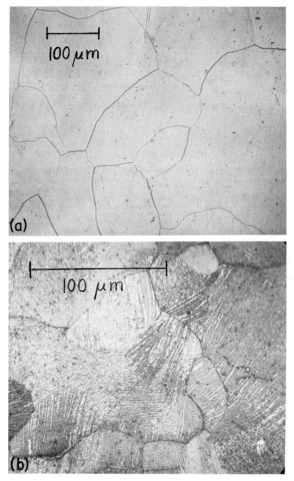

FIG. 26. Grain structure of β-titanium alloy (13% vanadium, 11% chromium, 3% aluminum): (a) solution-treated specimen held 1 hr at 788°C and water quenched to retain the β-phase grain structure. (b) specimen subsequently aged 64 hr at 493°C and cooled slowly to precipitate the second phase. The precipitate seems finely divided but arranged in clusters of elongated regions, seen sometimes sideways and sometimes end-on.[36]

intragranular structure, which breaks up the grains and makes them more elastically isotropic. Indeed, in the titanium alloy the difference curve (Fig. 27) starts with a slope of 2.6, showing that the change upon aging is a decrease in scattering.

For a check on whether the change was due to stress relief in the low-temperature annealing (aging), specimens of commercially pure, hot-rolled α-titanium were given the aging treatment. Since there are no

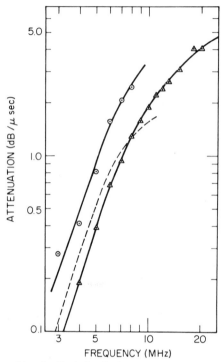

FIG. 27. Attenuation of longitudinal ultrasonic waves in β-titanium alloy (○). Propagation was along the bar diameter, as in Fig. 13b. The attenuation decreased by a factor of 2 during the aging (△) because of the breakup of the grains into a substructure. The presence of the substructure lowered the ultrasonic grain scattering.[36] The dashed curve represents the difference between the two attenuation curves.

alloying elements in the α-titanium to produce metastable phases, the aging procedure did not cause the appearance of a second phase, and the attenuation did not change upon aging. Thus, stress relief was ruled out as a cause of the change of attenuation in the alloys undergoing precipitation of a second phase. The decrease in attenuation must be ascribed to a decrease in scattering, which occurs because of the breakup of the grains as scattering centers. These two experiments on titanium have shown that a diffusion-controlled reaction resulting in a complex structure within the grains of a metal lowers the ultrasonic grain scattering in the metal.

5.2.3.4.4. CARBON STEEL. Carbon steels can be forced to transform by either process mentioned above.[1,2] When quenched rapidly enough, steel transforms from fcc austenite to martensite, a body-centered-tetragonal lattice with interstitial carbon (trapped, not dissolved) by a diffusionless process. Many tiny martensite platelets fill the volume of the prior austenite grain. When cooled very slowly, the steel transforms by diffusion of the carbon into pearlite, a layered structure of iron and iron carbide, and another product. Hypereutectoid steel (above about 0.8% C) precipitates cementite (Fe_3C) at the grain boundaries and elsewhere, whereas hypoeutectoid steel separates into pearlite and ferrite (pure bcc iron) in regions that are a fairly large fraction of the prior austenite grain in size. When steels are cooled to an intermediate temperature and held there for an isothermal transformation, another product, known as bainite, forms. It is a dendritic structure involving iron and iron carbide and may be considered midway between pearlite and martensite in fineness.

Several workers have found very large differences in attenuation among the various microstructures. For a given prior austenite grain size Kamigaki[34] found that pearlite may scatter 100 times as strongly as martensite. He also found that the scattering in pearlitic carbon steel was directly proportional to the prior austenitic grain volume. Comparing pearlitic carbon steel, iron, and aluminum, he found that the pearlite scattered 1000 times more strongly than aluminum and 30 times more strongly than iron. From these results he concluded that pearlite is much more elastically anisotropic than iron and most other metals. To explain this he hypothesized that the lamellar pearlite had an anisotropic modulus dominated by the more compliant layers in a direction normal to the layers and by the less compliant layers in a direction tangent to the layers. Thus, Kamigaki views pearlite as a structure having hexagonal symmetry or perhaps a highly exaggerated tetragonal symmetry and showing greater elastic anistropy than iron.

Merkulov,[64] on the other hand, found pearlite to be less anisotropic than iron. In his experiment he studied attenuation as a function of carbon content in steel slowly cooled from the austenitic state. Merkulov suggested that the grain scattering should be a minimum at the eutectoid composition: 0.8% C. At this composition there is neither free ferrite nor free cementite around the pearlite. The apparent conflict with Kamigaki's result on the relative anisotropy of pearlite and iron bears looking into further. Since Kamigaki's steel contained 0.75% C, the composition is probably not at fault. Careful attention should be paid to the grain-size determinations, which were made with different sets of standard overlays.

Papadakis[50,51] found the scattering of longitudinal waves in hardened low-alloy steel (SAE 4150) to be higher by a factor of from 2 to 3 than that in aluminum of the same grain size and lower by a factor of 10 than that in iron. The explanation is that each iron grain is broken up by the martensite upon hardening, thus lowering the anisotropy of the grains as explained in Section 5.2.1.7.

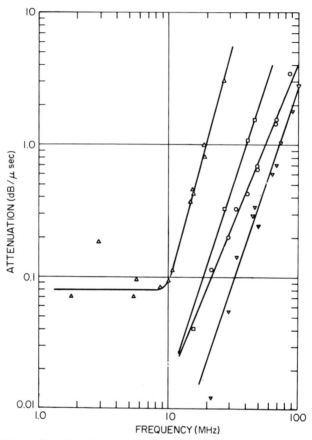

FIG. 28. Attenuation of longitudinal ultrasonic waves versus frequency in SAE 4150 steel with microstructure as a parameter: ($\triangle$) pearlite + ferrite; ($\square$) bainite; ($\bigcirc$) martensite; (∇) tempered martensite. All specimens were austenitized under the same conditions of time and temperature to produce identical austenite grain sizes and grain-size distributions. Various quench rates produced the three upper microstructures. The martensite was subsequently tempered. The steep slopes of these curves indicate considerable Rayleigh scattering (Papadakis,[72] by permission of American Institute of Physics.)

In another paper Papadakis[72] studied the attenuation of longitudinal waves in SAE 4150 steel as a function of microstructure, the prior austenitic grain size being held constant. Three specimens were austenitized together at 845°C. One was quenched in oil to form martensite, one was quenched in molten salt at 400°C for 16 minutes to form bainite, and one

[72] E. P. Papadakis, *J. Appl. Phys.* **35**, 1474–1482 (1964).

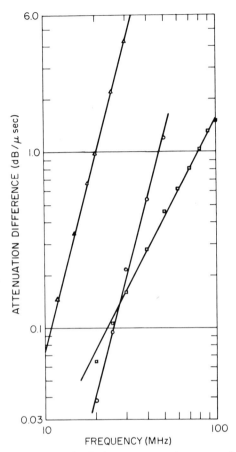

FIG. 29. The difference in longitudinal wave attenuation among the various microstructures of Fig. 28: ($\triangle$) pearlite–bainite; ($\bigcirc$) bainite–martensite; ($\square$) martensite–tempered martensite. From the slopes it is evident that the differences among the primary microstructures is Rayleigh scattering (slope 4 on logarithmic paper), whereas the change on tempering of martensite (martensite–tempered MS) came primarily from an f^2 mechanism (slope 2) (Papadakis,[72] by permission of American Institute of Physics.)

was furnace cooled to form pearlite plus ferrite. The attenuation of longitudinal waves in these differed greatly, as shown in Fig. 28. Tempering changed the attenuation in the martensite considerably. The data in Fig. 28 were analyzed by taking the difference in attenuation between one microstructure and another as shown in Fig. 29. The differences between pearlite and bainite and between bainite and martensite were proportional to the fourth power of the frequency, indicating that the dif-

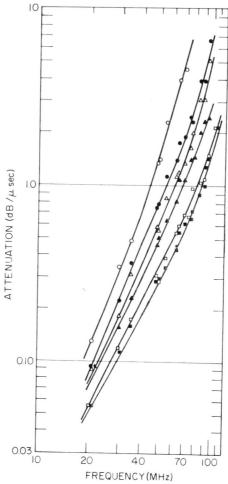

FIG. 30. Ultrasonic attenuation in hardened die steel. Longitudinal waves were propagated in specimens austenitized at various temperatures (degrees centigrade): 760 (○); 790 (■); 815 (□); 845 (▲); 870 (△); 900 (●). The attenuation is minimum for heating slightly above the A_1 temperature (above which austenite forms) and increases for higher temperatures. The attenuation can be described by the equation $\alpha = a_2 f^2 + a_4 f^4$.[36]

ferences in attenuation were caused by differences in the Rayleigh scattering by grains. The difference between martensite and tempered martensite was proportional to the square of the frequency; hence, a different mechanism was involved. Further analysis showed that the attenuation remaining after subtraction of the Rayleigh scattering component was proportional to f^2, not to f, and so was not caused by elastic hysteresis.

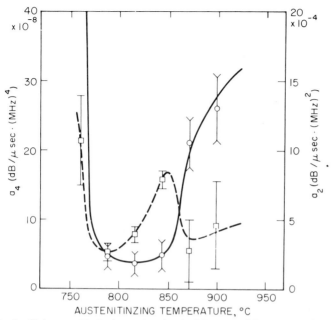

FIG. 31. Coefficients a_2 (dashed curve) and a_4 (solid curve) of the powers of the frequency in the attenuation equation $\alpha = a_2 f^2 + a_4 f^4$. The data in Fig. 30 are analyzed. Just above the A_1 temperature of 770°C the coefficient a_4 is minimal. It rises steeply with grain growth at higher temperatures.[36]

The attenuation was written

$$\alpha = a_2 f^2 + a_4 f^4. \tag{5.2.58}$$

The second coefficient was higher in pearlite than in martensite by a factor of 200, confirming Kamigaki's findings.[34]

A further experiment has demonstrated the propriety of Eq. (5.2.58) for steel. Specimens of SAE type 01 die steel were austenitized at various temperatures between 760 and 870°C and quenched in oil, to form martensite. Data on the attenuation of longitudinal waves in these specimens are presented in Fig. 30. The attenuation increases with austenitizing temperature above 770°C. The specimen heated at 760°C was not austenitized at all, of course, since 760°C is below the minimal austenitizing temperature. As a function of frequency the attenuation varies as f^2 at low frequencies and then bends upward to an f^4 dependence at higher frequencies. The coefficients a_2 and a_4 are plotted in Fig. 31 as functions of the austenitizing temperature. Note that a_4 is very low, until the grains begin to grow at elevated temperatures; then it increases rapidly. This behavior indicates that the grain scattering arises from the austenite grain volumes, as was found by earlier workers.[34,50,51,64] Unfortunately, etching did not reveal the austenite grain size in these type 01 specimens. Very similar results were found in SAE type 52100 steel specimens quenched from various temperatures.[73] The maximum in a_2 at 840°C is unexplained to date.

The problem of the magnitude of the elastic anisotropy of the various transformation products in steel has not yet been solved completely. More work should be done on steel specimens in which the grain-size distribution can be accurately determined. Possibly, reinterpretation of existing data[34,51,64] in terms of the true grain-size distribution (Section 5.2.1.5) in the specimens would be helpful. At present it is known that the scattering is proportional to the prior austenite grain volume and that it is much weaker in martensite than in pearlite and its associated phases, ferrite and cementite. The scattering in pearlitic steel as a function of carbon content should be studied further.

5.2.3.4.5. PRACTICAL APPLICATIONS. Many engineering alloys depend on their microstructure for their mechanical properties. When this fact is known, it becomes evident that changes in ultrasonic attenuation due to grain scattering can be used for nondestructive testing purposes. One test for the heat treatment of hardened steel has already been proposed.[74] In it use is made of the differences in grain scattering among the various microstructures for determining whether a heat treatment process has been carried through properly. Another procedure could be developed for testing the solution treatment and aging of β-titanium, for instance. Certainly there are other alloys that are amenable to testing by ultrasonic attenuation. Even when the grains do not show a complex interior structure, ultrasonic grain scattering is useful as a test of grain size and grain growth. Just as ultrasonic attenuation due to dislocation motion can be used for testing for deformation, recovery, and fatigue,[75] ultrasonic attenuation due to grain scattering can be used for testing for grain size and microstructure.

At times one might like to produce materials of low attenuation as wave propagation media or of high attenuation as absorption media. By adjusting grain size and microstructure (as in steel) one can make either highly attenuating or highly transmitting specimens. Indeed, a steel specimen transmissive at one end and absorptive at the other can easily be fabricated. Some delay-line uses may be found for metals or other poly-

[73] E. P. Papadakis, *Metall. Trans.* **1**, 1053–1057 (1970).

[74] E. P. Papadakis, *Mater. Eval.* **23**, 136–139 (1965).

[75] A. Granato, A. Hikata, and K. Lucke, *Acta Metall.* **6**, 470–480 (1958).

crystals treated so as to be inhomogeneous with respect to ultrasonic attenuation. Additional applications of grain scattering can certainly be found.

5.3. Difficulties to Be Encountered

5.3.1. Anisotropy

5.3.1.1. Relevance and Scope.
During ultrasonic attenuation measurements of polycrystalline media an investigator may encounter anomalous attenuation measurements. Attenuation (in decibels per centimeter, for instance) may seem to vary with the transducer diameter or the specimen length or the pair of echoes used in the measurement; attenuation may also appear to follow no reasonable pattern, instead of being a smooth function of frequency. The echo pattern may deviate from the exponential, sometimes by a large amount; the first echo may even be lower than the second. Why? One cause, diffraction, has been covered in Section 5.2.2.4. Another cause is the simultaneous propagation of more than one mode in a sample. Multiple modes can interfere with one another and cause nonexponential echo patterns. For isotropic samples irradiated by pure-mode transducers there is no problem: only one mode will be excited in the sample. Anisotropic samples are not so simple, for a single-mode transducer will excite three quasipure modes for an arbitrary propagation direction (Musgrave[76]; Mason,[21] pp. 355–373). Only certain directions will sustain pure modes. Along these directions one longitudinal and two transverse modes polarized normal to each other can propagate, so a shear mode transducer can excite two modes simultaneously, if its polarization is not oriented properly with respect to the symmetry of the anisotropic sample. These two shear waves can interfere with each other and cause erroneous attenuation readings. In fact, shear waves of an arbitrary polarization are excellent detectors of small anisotropies in velocity.[77-79] The interference produces an envelope of the form $|\cos \varphi|$ on the echo train. If the phase shift φ between the two modes is $\pi/2$ at the first echo, than the odd-numbered echoes are minimal and the even-numbered echoes are maximal. The minima are nulls if the polarization is at 45° between the pure-mode displacement eigenvectors and the attenuation of both modes is equal.

[76] M. J. P. Musgrave, Proc. R. Soc. London Ser. A 226, 339–355 (1954).

[77] F. A. Firestone and J. R. Frederick, J. Acoust. Soc. Am. 18, 200–211 (1946).

[78] P. F. Sullivan and E. P. Papadakis, J. Acoust. Soc. Am. 33, 1622–1624 (1961).

[79] R. Truell, L. J. Teutonico, and P. M. Levy, Phys. Rev. 105, 1723–1729 (1957).

From the foregoing it is obvious that the anisotropy of a sample for attenuation studies must be investigated and accounted for before valid attenuation measurements can be made. The anisotropy arises from the preferred orientation of the grains, which are themselves elastically anisotropic. The preferred orientation is a result of the symmetry of an operation applied to the polycrystalline aggregate, so the axes of symmetry of the aggregate can be ascertained immediately. Therefore, the pure-mode axis and the polarization directions for pure modes can be obtained by inspection. Hence, ultrasonic attenuation measurements for these modes can be performed without multiple-mode interference problems. Nevertheless, diffraction corrections are available[53,54] only for longitudinal waves along axes of threefold, fourfold, and sixfold symmetry and for isotropic conditions. As a first approximation to diffraction corrections for twofold axes, the isotropic calculation may be used. To compute the corrections one must first find all, or almost all, the effective elastic moduli of the anisotropic polycrystalline aggregate by making a number of velocity measurements with several modes in various directions (Mason,[21] pp. 355–373; Neighbors[80]). Since most operational symmetries produce axes of twofold elastic symmetry, the procedure for making diffraction corrections will be only approximate until the diffraction problem has been solved for twofold axes. Only the centerlines of bars made by drawing, swaging, and extrusion and of billets made by forging either radially or axially are sixfold-symmetry axes; these can be treated analytically for diffraction corrections. No operation yields threefold axes. Only in very special cases, such as that of "cube texture," does one find fourfold axes in polycrystalline aggregates (Barrett,[1] pp. 478–480, 494–502). Neverthelesss, symmetry studies can be of value in attenuation work and can be of interest in themselves.

5.3.1.2. Some Examples of Hexagonal Symmetry. An example of hexagonal symmetry, the zinc bar stock, was treated earlier. Longitudinal wave velocities as a function of the angle of the propagation vector from the bar axis were listed in Table VII of Section 5.2.3.3. The velocity follows the pattern to be expected in a material with hexagonal symmetry on a macroscopic scale. Along the bar axis the longitudinal velocity was only 8% below that in the basal plane of zinc, whereas along the bar diameter it was midway between the values of that in the basal plane of zinc and that normal to it. These velocities differ in a ratio of about 5:3. This information led to the conclusion that the fiber texture of the zinc bar was such that the basal planes of the zinc grains lay essentially parallel to the

[80] J. R. Neighbours, *J. Acoust. Soc. Am.* **26**, 865–869 (1954).

TABLE X.[81]   Elastic Moduli of
Pyrolytic Graphite[a]

5.21	1.88	1.80	0	0	0
1.88	5.21	1.80	0	0	0
1.80	1.80	2.58	0	0	0
0	0	0	0.15	0	0
0	0	0	0	0.15	0
0	0	0	0	0	1.67

[a] All values are to be multiplied by
10^{11} dyn/cm^2.

bar axis. Such a texture is known to occur in the hexagonal metals (Barrett,[1] Chapters 18, 19, 21).

The last-mentioned macroscopic specimen had hexagonal symmetry by virtue of the act that the hexagonal axes of all its grains pointed perpendicular to its own hexagonal axis. The next specimen will be one in which the hexagonal axes of the grains lie parallel to the hexagonal axis of the macroscopic specimen. The material is pyrolytic graphite grown on a flat mandril. The properties of this material are subject somewhat to annealing treatments and other conditions, so the data to be presented are not perfectly valid for all batches of material. The velocities of longitudinal and transverse elastic video pulses were measured[81] for enough modes to determine the effective elastic moduli of the material. The moduli are presented in matrix form in Table X.

The difference of a factor of 2 between c_{11} and c_{33} is large but not astounding. The factor of 11 between $(c_{11} - c_{12})/2 = c_{66}$ and c_{44} is really striking, however. The very low value of c_{44} indicates that the basal planes of the grains of this pyrolytic material are parallel to each other, so that one observes a modulus approaching the true c_{44} of graphite, which gives the small resistance to deformation by the slippage of graphite sheets over each other. The value of c_{33} is low also because it represents the resistance to compression or tension normal to the basal planes of graphite.

As a final example, results of a study on bars of tool steel, made by the author,[40] are presented. The symmetry of bar manufacture yields hexagonal symmetry in the macroscopic sample, although the grains of the material may contain cubic iron, tetragonal martensite, or lamellar pearlite. The bars in question were hardened to produce a high percentage of

[81] E. P. Papadakis and H. Bernstein, *J. Acoust. Soc. Am.* **35**, 521–524 (1963).

TABLE XI.   Elastic Moduli in an
Air-Hardening Hardened
Tool Steel[a]

27.67	11.32	11.24	0	0	0
11.32	27.67	11.24	0	0	0
11.24	11.24	27.24	0	0	0
0	0	0	8.03	0	0
0	0	0	0	8.03	0
0	0	0	0	0	8.18

[a] All values are to be multiplied by
10^{11} dyn/cm^2. From Papadakis[40] © 1964
IEEE.

martensite and a low degree of anisotropy. Specimens of the form shown
previously in Fig. 22 were used, but in this experiment they were sliced at
$10°$ intervals instead of $15°$. From velocity measurements made on these
specimens, the effective elastic moduli of the hardened steel were found.
These are listed in matrix form in Table XI. The shear moduli c_{44} and c_{66}
differ by 2%, whereas the longitudinal moduli c_{11} and c_{33} differ by about
1.6%. This much anisotropy would have virtually no effect on ultrasonic
diffraction corrections,[54] for it produces a value of only -0.0022 for the
anisotropy parameter b. Differences in diffraction are hardly measurable
for values of b between -0.05 and $+0.05$. The expression for b in terms
of the moduli c_{ij} can be found in earlier references.[53,55]

5.3.2. Specimens of Finite Width

Problems arise when one encounters specimens of dimensions not
much larger than the transducer normal to the propagation direction.
This frequently happens in solids where the specimen is a rod, a section of
a plate, or a valuable crystal. Fluids contained in tubes also exhibit the
same problems. One encounters what is called, qualitatively, "sidewall
effects" and technically "multimode guided wave propagation." This
means that although the cylinder or plate are several or even many wavelengths in lateral extent, they are not large enough to support free-field
propagation; rather, they support all the rod modes or plate modes excited by the transducer that are not beyond cutoff due to the geometry.
The modes, being dispersive,[82] interfere with each other as they propagate down the specimen. It is possible under certain circumstances for
destructive interference to be almost complete, canceling the pressure

[82] T. R. Meeker and A. H. Meitzler, in "Physical Acoustics" (W. P. Mason, ed.), Vol. 1,
Part A, pp. 111–167. Academic Press, New York, 1964.

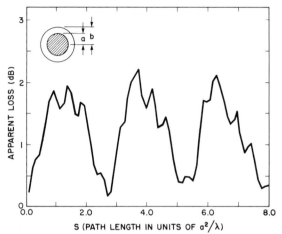

FIG. 32. Apparent loss due to interference among the multiplicity of modes propagating in a rod of finite width. The ratio of transducer to rod width is 2:3. The abscissa is $S = z\lambda/a^2$. Compare the giant excursions to the ripples of fractions of a decibel due to regular diffraction as shown in Fig. 7. (From Carome and Witting,[83] by permission of American Institute of Physics.)

wave over the face of a receiving transducer.[83,84] The interference phenomena are expressible in terms of the normalized distance S-parameter $S = z\lambda/a^2$. Universal curves obtain, with destructive interference at sequential positions along S in the approximate ratio $1:3:5:7$. One example of their calculation is given in Fig. 32. Obviously, attenuation measurements in this regime must be made judiciously because of the apparent loss caused by the interferences. In a pulse-echo experiment it is best to measure loss using echoes at antinodes of the interference pattern.

One spectacular example of a multimode interference pattern was given in a paper on physical properties of delay line materials.[47] (See Fig. 33.) The specimen was a fused quartz plate of dimensions $34.3 \times 20.3 \times 1.3$ cm. The narrow edges were optical flats made perpendicular to the plane of the large faces. The echo pattern is for shear waves polarized perpendicular to the major faces and propagating in the 20.3-cm direction. Transducers were 30-MHz Y-cut quartz plates 0.635×0.940 cm in size with the polarization in the 0.635-cm direction. The echo pattern in Fig.

[83] E. F. Carome and J. M. Witting, *J. Acoust. Soc. Am.* **33**, 187–197 (1961).
[84] E. F. Carome, J. M. Witting, and P. A. Fleury, *J. Acoust. Soc. Am.* **33**, 1417–1425 (1961).

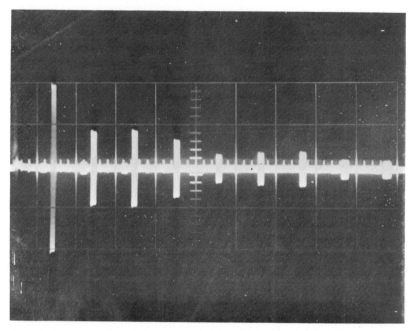

FIG. 33. Example of an echo pattern in a plate of dimensions 34.3 × 20.3 × 1.3 cm with propagation of shear waves polarized in the 1.3-cm direction and propagating in the 20.3-cm direction. The 30-MHz transducer size was 0.635 × 0.940 cm with the polarization in 0.635-cm direction. (Papadakis,[47] by permission of The Institute of Electrical and Electronics Engineers.)

33 shows the difficulties that may be encountered when the specimen is not effectively infinite laterally.

To make the specimen effectively laterally infinite, McSkimin[85] suggested machining threads on a rod-shaped specimen. The resulting helical boundary conditions are incommensurate with coherent reflections at the rod surface to form interferences, so the rod seems laterally infinite from the point of view of multimode guided wave propagation. A simple diffraction loss (Fig. 7) results.

A method of some interest but very limited utility was demonstrated by the author.[86] Steel rods were made inhomogeneous with radial symmetry by hardening only the outside. This was done by austenitizing and quenching a steel of hardenability such that only the outer 3 mm would

[85] H. J. McSkimin, J. Appl. Phys. 24, 988–997 (1953).
[86] E. P. Papadakis, J. Acoust. Soc. Am. 49, 729–745 (1971).

harden with the quench used. Total diameter was 25.4 mm. With this hard case, the longitudinal waves propagated axially from a 12.7-mm transducer met refracting and retarding "boundary conditions" in the hardened outer layer (velocity slightly lower). The multimode interference pattern observed before hardening disappeared upon hardening.

5.4. Summary

The present theories of ultrasonic attenuation due to grain scattering[16-18] yield results in quantitative agreement with experiment when used with the method of grain-size determination proposed by Papadakis.[26] The theories are applicable to single-phase polycrystals with equiaxed single-crystal grains exhibiting no preferred orientation. If preferred orientation exists, then the attenuation will be reduced if the orientation resulted in a lower difference in moduli from grain to grain in the direction of propagation. Some multiple textures could conceivably result in increased scattering if adjacent grains turned unlike directions toward each other. If a complex intragranular structure exists, breaking up the grains, then the scattering loss is reduced because the elastic anisotropy of the grain is reduced through an averaging process over all the microstructural regions. Only in cases in which a microstructural region is a moderate or large fraction of the grain and has a larger anisotropy than the original grain can the scattering increase. In this respect the pearlitic transformation product in steel should be studied further in steel specimens of accurately determined austenitic grain size, so as to determine whether pearlite increases or decreases the scattering loss relative to iron. Further, it can be expected that the scattering loss will change if the grains are elongated or flattened. Preferred orientation produces another effect, namely anisotropic velocity. This condition should be studied in conjunction with attenuation for the purpose of determining the diffraction corrections to be applied to the attenuation in the anisotropic specimen. Experimental methods, equipment, transducers, specimens, and diffraction corrections have been specified and explained.

In reporting experimental results, the investigator should list all relevant parameters, and should provide data on the corrections and on the items of difficulty as well as on the positive results. Examples are as follows:

(1) data on specimen configuration, composition, mechanical and heat treatment, etches, etc.;

(2) modes used and polarizations;

(3)   raw data, diffraction corrections, other corrections, and corrected data;

(4)   parameters for the diffraction corrections:

    (a)   transducer active diameter and operating frequency actually used,

    (b)   specimen length, diameter, and velocity,

    (c)   buffer rod length, diameter, and velocity,

    (d)   anisotropy parameter(s);

(5)   evidence that sidewall effects did not (or did) affect the measured attenuation;

(6)   evidence of specimen isotropy (or anisotropy) through velocity measurements and photomicrographs;

(7)   photomicrographs showing enough grains (~ 200) to permit the calculation of a grain size distribution; show evidence of isotropy and homogeneity (or lack thereof).

Acknowledgments

The author wishes to thank R. E. Caffrey, K. A. Fowler, E. B. Kula, L. C. Lynnworth, and E. L. Reed for consultation and advice on metallurgical and ultrasonic problems encountered in conjunction with this work. J. E. May, Jr., T. R. Meeker, I. E. Fair, and H. J. McSkimin offered valuable suggestions and criticisms. The author wishes to acknowledge the continuing support of C. E. Feltner and M. Humenik, Jr.

6. NONLINEAR PHENOMENA

By James A. Rooney

6.0. Introduction

In discussions of acoustics we deal typically with situations in which the displacement amplitude of the sound wave is relatively small, so that linear or first-order terms that vary sinusoidally in time accurately describe the physical situation. However, there are important acoustic phenomena and experimental methods for exploring such phenomena that can be described only if the nonlinear or higher-order terms of the theory are retained. These nonlinear effects are directly involved in, or can influence, the techniques and accuracy of many experimental methods described in other parts of this volume, including absorption, cavitation, and wave propagation. In addition, nonlinear phenomena are of interest as a subject in themselves and they permit the development and application of some unusual experimental methods for investigation. Such studies will be the subject of this part.

6.1. Nonlinear Propagation of Sound

The propagation of sound waves that have high amplitudes is accompanied by a series of effects that depend on the displacement amplitude of the wave. These effects can be described only if the nonlinear terms of the basic hydrodynamic equations are included. The presence of these nonlinear terms significantly changes the propagation pattern of the sound wave and also its absorption. The subject of this section has been extensively developed by a variety of authors[1-5] and will be only briefly summarized here.

[1] O. V. Rudenko and S. I. Soluyan, "Theoretical Foundations of Nonlinear Acoustics." Plenum, New York, 1977.
[2] K. A. Naugal'nykh, Absorption of finite amplitude waves, in "High Intensity Ultrasonic Fields" (L. D. Rozenberg, ed.). Plenum, New York, 1971.
[3] R. T. Beyer and S. V. Letcher, "Physical Acoustics." Academic Press, New York, 1969.
[4] R. T. Beyer, "Nonlinear Acoustics," Naval Ship Systems Command, US Govt. Printing Office, Washington, D.C., 1974.
[5] L. Bjorno, Ultrason. Int. Proc., p. 110 (1975).

METHODS OF EXPERIMENTAL PHYSICS, VOL. 19

6.1.1. Introduction to the Theory

An extensive literature[1-5] has been developed concerning the theory for describing the nonlinear effects on propagation of sound. We shall only outline certain aspects here in order to place the experimental methods used to study the associated phenomena in perspective. Since the propagation of sound can be described as an isoentropic process, the equation of state can be expanded in a Taylor series in the following manner:

$$P = P_0 + \left(\frac{\partial P}{\partial \rho}\right)_{s,\rho=\rho_0} (\rho - \rho_0) + \frac{1}{2} \left(\frac{\partial^2 P}{\partial \rho^2}\right)_{s,\rho=\rho_0} (\rho - \rho_0)^2 + \cdots .$$

(6.1.1)

It has become customary to keep only the terms shown and express Eq. (6.1.1) as

$$P - P_0 = A \left(\frac{\rho - \rho_0}{\rho_0}\right) + \frac{B}{2} \left(\frac{\rho - \rho_0}{\rho_0}\right)^2 .$$

(6.1.2)

Here A and B are temperature-dependent coefficients with $A = \rho_0 c_0^2$ and the quantity B/A, which is often cited in the literature, has the form

$$B/A = 2\rho_0 c_0 (\partial c/\partial P)_{s,\rho=\rho_0} .$$

(6.1.3)

This expansion leads to a nonlinear equation of motion that has, for the nondissipative case and waves propagating in the $+x$ direction, a solution of the form[6-9]

$$u(x, t) = U_0 \sin \left[\omega t - \frac{\omega x}{c_0} \left(1 + \frac{B}{2A} \frac{u}{c_0}\right)^{-2(A/B)-1}\right] .$$

(6.1.4)

Here u is the particle velocity, c the velocity of sound, and $\omega = 2\pi f$ the angular frequency. This equation shows that large values of u are propagated more rapidly than small values and the wave becomes distorted as it travels through the medium. If viscosity is taken into account in the theory, the predicted distortion of the waveform will include that resulting from the more rapid attenuation of the higher harmonics, since the absorption coefficient is proportional to the square of the frequency. The theoretical analysis of propagation of finite but moderate amplitudes was performed by Airy.[10] An approximate solution for the case of larger am-

[6] G. Stokes, *Trans. Cambridge Philos. Soc.* **8**, 287 (1945).
[7] E. Fubini, *Alta Freq.* **4**, 539 (1935).
[8] S. Earnshaw, *Philos. Trans. R. Soc. London* **150**, 133 (1860).
[9] B. Riemann, "Collected Works." Dover, New York, 1953.
[10] G. B. Airy, Encyclopedia Metropolitana London, Vol. 5 (1845).

plitudes has been given in Fay.[11] Other approximate solutions have been developed by Mendousse[12] and Rudnick.[13] Numerical solutions to the problem have been given by Fox and Wallace[14] as well as by Cook.[15]

Other approaches to the solution of this problem have included attempts to modify the form of the equations for finite-amplitude sound propagation and consider a boundary value problem[16] or to limit regions of solution.[17] In addition, Blackstock[18] has been able to connect the earlier solutions of Fay[11] and Fubini.[7] While we have been discussing only plane waves, the cases of cylindrical and spherical waves of finite amplitude are described in detail by Soluyan and Khokhlov[17] and Blackstock.[18]

6.1.2. Experimental Methods

The experimental methods used to measure the parameters just described have included both electrical and optical techniques. The electrical techniques include both thermal detectors and microphones; the optical techniques depend principally on the diffraction of light.

6.1.2.1. Electrical Methods. Experimenters have long been aware that the use of high intensities leads to falsely high values of the attenuation coefficient. Therefore early observers tended to measure the absorption coefficient as a function of intensity. Unfortunately, they did not also measure the coefficient as a function of distance from the source. In addition, in this earlier work, a ceramic transducer, sensitive principally to the odd harmonics of the incident wave, was used as a detector.[14]

Krasilnikov et al.[19] made improvements in the experimental methods such that the harmonic content of a sound beam with a frequency of 1.5 MHz could be measured. To filter out unwanted harmonics, glass or metal plates were oriented with respect to the beam. The same group has also used thermal detectors to measure the total incident acoustic radiation and determine effective absorption coefficients.

Barnes and Beyer[20] have developed a modification of a radiation pressure microphone to determine the effective absorption coefficient at finite

[11] R. Fay, J. Acoust. Soc. Am. **3,** 222 (1931).

[12] J. S. Mendousse, J. Acoust. Soc. Am. **25,** 51 (1953).

[13] I. Rudnick, J. Acoust. Soc. Am. **30,** 565 (1958).

[14] F. E. Fox and W. A. Wallace, J. Acoust. Soc. Am. **26,** 994 (1954).

[15] B. D. Cook, J. Acoust. Soc. Am. **34,** 941 (1962).

[16] D. T. Blackstock, J. Acoust. Soc. Am. **36,** 534 (1964).

[17] S. I. Soluyan and R. V. Khokhlov, Sov. Phys. Acoust. **8,** 170 (1962).

[18] D. T. Blackstock, J. Acoust. Soc. Am. **36,** 217 (1964).

[19] V. A. Krasilnikov, V. V. Shklovskaya-Kordi, and L. K. Zarembo, J. Acoust. Soc. Am. **29,** 642 (1957).

[20] R. P. Barnes, Jr. and R. T. Beyer, J. Acoust. Soc. Am. **36,** 1371 (1964).

amplitudes. In their experimental arrangement, the carrier wave applied to the transducer is totally modulated by a square wave. As a result, the radiation pressure oscillates at the audio frequency. The signal, whose amplitude is proportional to the intensity, can be easily measured at various distances from the source. Good agreement is obtained between theory and experiment.

6.1.2.2. Optical Methods. Methods using optical diffraction have also been developed for the study of propagation of finite-amplitude waves. The methods use a collimated, monochromatic, light beam directed through a tank containing the liquid in which the ultrasonic beam is propagating. The light is focused on a screen after passing through the tank. Since the sound beam causes a variation in the local index of refraction, which has a periodicity equal to the wavelength of sound used, the system serves as a diffraction grating. A lens then focuses the resulting diffraction pattern on a screen.

The theoretical analysis of light diffraction by a sound beam was initially developed by Raman and Nath.[21] In 1936 Sanders[22,23] performed experiments with optical diffraction by sound and noted an asymmetry between intensities of positive and negative order. Since then, two groups[24,25] have clearly demonstrated finite-amplitude asymmetry in optical diffraction.

In the Hiedemann[25] experiment, which tested a generalization of the Raman and Nath theory, the distorted waves were analyzed in terms of their Fourier components. The experiments were conducted at a frequency at 2 MHz at intensities less than 1 W/cm². The diffraction of each of the components was combined to obtain the total result. Mikhailov and Shutilov[24] used a frequency of 583 kHz and sound intensities up to 20 W/cm². They analyzed the differences between the light intensity in the positive and negative orders of diffraction. In both cases, experimental and theoretical results are in excellent agreement.

6.1.3. Parametric Array

As seen in this section, the passage through a fluid of a sound wave of finite amplitude leads to a distortion of the wave. Because of this distortion, the possibility exists that two acoustic signals of different frequencies may combine to form sum and difference frequency compo-

[21] C. V. Raman and N. S. N. Nath, *Proc. Indian Acad. Sci. Sect. A* **2**, 406 (1935).
[22] F. H. Sanders, *Nature (London)* **138**, 285 (1936).
[23] F. H. Sanders, *Can. J. Res.* **A14**, 158 (1936).
[24] I. G. Mikhailov and V. A. Shutilov, *Sov. Phys Acoust.* **4**, 174 (1958).
[25] E. Hiedemann and K. L. Zankel, *Acustica* **11**, 213 (1961).

nents. The theory and experiments examining this possibility have been summarized by Beyer in Chapters 9 and 10 of Beyer[4] and by Al-Temini.[26]

Probably one of the more useful interactions is for colinear beams of the same intensity and nearly equal frequency. Analysis has been given by Westervelt[27] and Berktay.[28,29] The important characteristic of such a configuration is that it produces a highly directional sound beam whose frequency is predominantly the difference between the two driving frequencies, since the sum frequency is highly attenuated.

Experimental methods for examining these "end fire arrays" have been developed by a variety of investigators. Bellin and Beyer[30] were among the first to study the phenomenon, using two transmitters operating at frequencies of 13 and 14 MHz. The difference frequency was detected using a small barium titanate ceramic microphone and the beam pattern measured. The first sonar system to use such an array was described by Walsh[31]; it used two primary sources with frequencies near 200 kHz and a difference frequency of 12 kHz. The array can also be used as a receiver as described by Barnard et al.[32]

6.2. Radiation Force

6.2.1. Introduction to the Theory

Any object irradiated with an acoustic field will experience a force whose magnitude and direction depend on the intensity of the source and field parameters, as well as the size, shape, and construction materials in the irradiated object. It is this force that has often been misnamed "radiation pressure" for historical reasons. A history of early theory and experiments has been compiled by Post.[33] The observation of the deflection of a target by acoustic radiation force was probably made by Dvorak[34] in 1876, although Eckart[35] believed the phenomena observed by Dvorak to be acoustic streaming. Preliminary discussion of the theory

[26] C. A. Al-Temini, J. Sound Vib. **8**, 44 (1968).
[27] P. J. Westervelt, J. Acoust. Soc. Am. **35**, 535 (1963).
[28] H. O. Berktay, J. Sound Vib. **2**, 435 (1965).
[29] H. O. Berktay and D. J. Leahy, J. Acoust. Soc. Am. **55**, 539 (1974).
[30] J. L. S. Bellin and R. T. Beyer, J. Acoust. Soc. Am. **32**, 339 (1960).
[31] G. M. Walsh, Electron. Prog. **13**, 17 (1971).
[32] G. A. Barnard, J. G. Willette, J. J. Truchard, and J. A. Schooter, J. Acoust. Soc. Am. **52**, 1437 (1972).
[33] E. J. Post, J. Acoust. Soc. Am. **25**, 55 (1953).
[34] V. Dvorak, Poggendorff's Ann. **157**, 42 (1876).
[35] C. Eckart, Phys. Rev. **73**, 68 (1948).

was completed by Rayleigh.[36] It should be noted that indeed two types of radiation force exist, which differ according to the physical situation and boundary conditions applied to the problem. Detailed discussions of the appropriate theory are now available[4,37-39] and will be summarized here.

Historically the first situation considered was that of the Rayleigh radiation force resulting from the nonlinearity of the equation of state. It is applicable to situations that have several restrictions: (1) sound waves are propagating in a closed vessel with a fixed mass of fluid, and (2) regions of the fluid not directly in the sound field can only interact with the field provided that the density and pressure changes in the fluid are such as to maintain equilibrium. For this experimental arrangement and considering plane adiabatic waves, Rayleigh[40] found that the radiation force F could be given by

$$F = \tfrac{1}{2}(1 + \gamma) ES, \qquad (6.2.1)$$

where E is the energy density in the sound field, S the area of the target perpendicular to the sound field, and γ the ratio of specific heats. If second-harmonic terms are also considered, Beyer[4] found that the first coefficient of Eq. (9.2.1) should be $\tfrac{1}{4}$ instead of $\tfrac{1}{2}$.

However, experiments either for investigating the phenomenon of radiation pressure or using it as a technique for dosimetry usually are conducted in open vessels. These free boundary conditions correspond to the Langevin situation. In this case the radiation force is caused by a nonlinearity of the basic Euler equation from which the sound field equations are derived. In the equations for a free sound field, inclusion of the nonlinear terms leads to a lower mean pressure in the fluid as seen by a stationary observer. If the sound beam is surrounded by a medium at rest, then this lowered pressure is compensated by the external medium. For an observer who is moving with the sound beam, this compensation has the same effect as an increased mean pressure in the fluid in the region occupied by the sound field. The difference between the two types of radiation force is discussed in detail in Beyer[4] and Rooney and Nyborg.[37]

Several early investigators[41-43] showed that for the Langevin case the radiation force is simply

$$F = ES. \qquad (6.2.2)$$

[36] Lord Rayleigh, *Philos. Mag.* **3**, 338 (1902).
[37] J. A. Rooney and W. L. Nyborg, *Am. J. Phys.* **40**, 1825 (1972).
[38] J. Zieniuk and R. C. Chivers, *Ultrasonics* **14**, 161 (1976).
[39] R. T. Beyer, *J. Acoust. Soc. Am.* **63**, 1025 (1978).
[40] Lord Rayleigh, *Philos. Mag.* **10**, 364 (1905).
[41] L. Brillouin, *Ann. Phys.* (*Paris*) **4**, 528 (1925).
[42] R. T. Beyer, *Am. J. Phys.* **18**, 25 (1950).
[43] F. E. Borgnis, *Rev. Mod. Phys.* **25**, 633 (1953).

Questions have been raised concerning the effect that nonlinear properties of the propagation medium might have on this expression for the radiation force and the appropriateness of the boundary conditions used in the
early theory.[3,44] Beyer[4,39] and Rooney and Nyborg[37] have reconsidered
the theory and shown again that Eq. (6.2.2) is indeed valid for the
Langevin case. For the Rayleigh case, they[37] obtained

$$F = (1 + B/2A)ES, \qquad (6.2.3)$$

where B/A is a parameter associated with nonlinearities of the medium.[3]
For ideal gases, where $B/A = \gamma - 1$, Eq. (6.1.3) reduces to the Rayleigh
expression given in Eq. (6.1.1). However, Beyer[4,39] showed that if
higher-order terms of the propagating waveform have to be included, then
the magnitude of the force given in Eq. (6.1.3) must be reduced by a factor
of 2. Thus, the correct expressions for evaluating the radiation force may
depend both on the nonlinear properties of the medium and the waveform
of the propagating sound.

6.2.2. Experimental Methods

As expected, different experimental techniques must be used to investigate the two types of radiation force. For the Rayleigh case, a closed
vessel is required while an open vessel typical of dosimetric applications
is used to study the Langevin force.

6.2.2.1. Rayleigh Radiation Force. It is perhaps surprising that detailed quantitative measurements of the Rayleigh radiation force were not
completed until 1962 by Mathoit.[45] The study was conducted by placing
a loudspeaker, driven by a signal generator and an amplifier, and a piezoelectric microphone in a closed vessel. The sound amplitude was measured by the microphone and the pressure within the chamber was determined manometrically. Various gases were introduced into the chamber
and the radiation force was found to agree with Eq. (6.2.1) to within 10%.

A related qualitative experimental arrangement was developed by
Meyer and Neumann.[46] Using manometers they measured the static
pressure in two closed chambers on opposite sides of a loudspeaker
diaphram. When the sound was present, the manometers indicated that
the pressure had been increased on both sides of the diaphram. Thus, we
have an experimental arrangement in which we can easily see that radiation force is a nonlinear phenomenon; while the sound pressure has oppo-

[44] Z. A. Gol'dberg Acoustic radiation pressure, in "High Intensity Ultrasonic Fields" (L.
D. Rozenberg, ed.). Plenum, New York, 1971.
[45] M. Mathoit, C. R. Acad. Sci. Paris 255, 64 (1962).
[46] E. Meyer and E. Neumann, "Physical and Applied Acoustics." Academic Press, New
York, 1972.

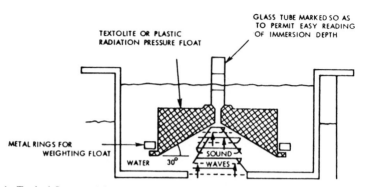

FIG. 1. Typical float used for measurement of radiation force. (Reproduced by permission from Henry.[50])

site phases on the two sides of the diaphragm, the radiation force manifests itself as a pressure increase on both sides of the diaphragm at the same time.

6.2.2.2. Langevin Radiation Force. Either for investigating the phenomenon of radiation force or for using it as a technique for determination of the acoustic power output of ultrasonic transducers, the problem is to design a suitable arrangement for measuring the force that results from the presence of the sound. In the past, researchers have measured this force by noting the deflection of a target in a force field, such as a gravitational field, or by measuring the force required to maintain the target in a given position. Experimental arrangements have included the use of magnetic forces,[47,48] counter weights or mechanical torque,[49] buoyant forces,[50,51] as well as the deflection of fluid surfaces.[52]

6.2.2.2.1. RADIATION FORCE FLOAT. A particularly simple experimental setup has been developed by Henry[50] and modified by Kossoff.[51] It is shown in Fig. 1. The arrangement consists of a float in an equilibrium position with its stem projecting vertically through the surface of the water. When the sound is turned on, the float is deflected upward, its effective weight is increased, and a new equilibrium is reached such that the radiation forces and the decrease in buoyant forces cancel. For calibration, one needs only to place a small known mass on top of the stem and

[47] A. Wemlen, *Med. Biol. Eng.* **6**, 159 (1968).
[48] J. A. Rooney, *Ultrasound Med. Biol.* **1**, 13 (1973).
[49] E. Klein, *J. Acoust. Soc. Am.* **10**, 105 (1938).
[50] G. E. Henry, *IRE Trans. Ultrason. Eng.* **6**, 17 (1957).
[51] G. Kossoff, *Acustica* **12**, 84 (1962).
[52] G. Hertz and H. Mende, *Z. Phys.* **114**, 354 (1939).

note the displacement of the float. A small hole in the center of the float permits escape of air bubbles.

The float should be stabilized against tipping forces so that the stem remains vertical. This can be accomplished by building the float of low-density material and keeping the sinker rings low. It helps if the buoy is also self-centering. For this to occur, the float should have a reflector of sound at its base so that the horizontal component of the radiation force tend to center the float. As indicated on the figure, the preferred angle of the reflector is 30°. For the typical cylindrical geometry, this angle is such that there is a horizontal stabilizing force and yet there is little chance that the sound reflected from one side of the float will impinge upon the other side. If the target surface of the float is an absorber, there is no horizontal component to the radiation force. This fact has been experimentally demonstrated by Herrey,[53] who measured the component of the force parallel to an ultrasonic beam. He found that the force on an absorber in the direction of propagation is independent of the angle of incidence θ, whereas for the case of a reflector, the component varies as $\cos^2 \theta$.

The float system just described provides a simple inexpensive technique for determination of total acoustic power output using radiation pressure. Kossoff[51] improved the sensitivity of the system by using two liquids whose densities were only slightly different, instead of water and air as shown in the figure. At present the limit of the sensitivity appears to be associated with the surface forces on the stem.

6.2.2.2.2. BALANCE ARRANGEMENTS. One typical method for measuring radiation force uses a balance from which an appropriate target whose dimensions are much greater than the wavelength of sound is suspended. Coarse balances[54,55] have adjustments for zeroing and standard weights which permit calibration of the radiation force. For sensitive measurements of radiation force, electromagnetic servomechanisms have been used.[47,48,56]

For investigations of the phenomenon of radiation force and the calibration of ultrasonic medical units, Rooney[48] has developed a sensitive radiometer. The apparatus is shown schematically in Fig. 2 and consists of a Cahn RG electrobalance from one arm of which a cylindrical absorbing target (A) is suspended; appropriate tare weights are suspended from the other arm. The target is hung in the medium of interest within an inner

[53] E. M. J. Herrey, *J. Acoust. Soc. Am.* **27**, 891 (1955).
[54] C. R. Hill, *Phys. Med. Biol.* **15**, 241 (1970).
[55] W. G. Cady and C. E. Gitlings, *J. Acoust. Soc. Am.* **25**, 892 (1953).
[56] G. Kossoff, *J. Acoust. Soc. Am.* **38**, 880 (1965).

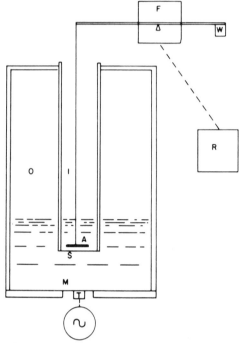

FIG. 2. Schematic diagram of balance system for determination of radiation force, where T refers to the transducer; M, the membrane to which the transducer is acoustically coupled; S, the acoustic streaming shield; A, the absorbing target; O and I, the outer and inner cylinders of test vessel; F, feedback unit of force measurement system; W, counterweight; and R, recorder. (Reproduced by permission from Rooney.[48])

vessel. This small interior vessel is closed at the lower end with a stretched membrane to shield the target from acoustic streaming.

The electrobalance is calibrated by noting the deflections on the recorder which correspond to forces resulting from placing a series of calibrated weights on the balance. The transducer element is brought into contact with a second stretched membrane at the base of the large vessel and a coupling gel is used between the transducer and the membrane. The outer vessel containing water has a large volume so that the thermal fluctuations in the inner vessel are reduced. When the transducer is activated, the target is deflected momentarily from equilibrium. This deflection changes the amount of light reaching a photocell which initiates the response of a magnetic feedback system within the Cahn balance, thus returning the target to its original position. A portion of the feedback voltage to the magnet is recorded on a chart recorder. We note that the use of the feedback system is advantageous because it eliminates errors that result from changes in surface tension forces on the suspending wire and minimizes those that are the result of other "surface" effects. The air gap between the balance and the large vessel is enclosed to prevent disturbing effects of air currents on the suspending wire.

The inner vessel proved to be important for the stability of the system. The vessel should have a diameter only slightly larger than the target in order to minimize drift in the signal which results from convection currents. The response of the system is such that transients decay within 6 sec. The size of the target also proved to be an important consideration for accuracy of the measurements. In considering the appropriate diameter of the target, a compromise had to be made. The diameter of the target must be larger than the diameter of the beam to ensure that the sound is completely absorbed but the volume of the target should be kept to a minimum to reduce excessive noise in the system. Sound-absorbing rubber materials have proven useful; no appreciable standing waves could be detected in the frequency range from 1 to 10 MHz. In addition, the experimenter should realize that building vibrations can set up pressure gradients in the liquid. In the presence of these gradients the target will be subjected to a force proportional to its volume. Therefore, reduction of noise was achieved by using a target of minimum volume to meet the other requirements and isolating the experimental arrangement from the building.

6.2.2.2.2.1. Effect of Nonlinear Parameters. When an absorbing target for the Langevin situation is used, the radiation force equals the total acoustic intensity divided by the velocity of sound even when nonlinearities of the propagation medium and distortion of the waveform are taken into account. Rooney[57] has tested this relationship using the apparatus just described. Determinations were made of the change in effective weight of the target as a function of voltage applied to the transducer for distilled water, *n*-propanol, and ethanol. These liquids were chosen because of their values of the nonlinear parameter B/A are greatly different, while their values of characteristic acoustic impedance are not greatly different. The radiation force was found to be independent of the parameter B/A as predicted.[57]

6.2.3. Calibration and Errors

The functional dependence of radiation force and acoustic power has been measured for self-consistency[57] and comparisons have been made of the absolute magnitude of the radiation force with other absolute calibration techniques. Tarnocy[58] has completed a comparison of the calculated output of a transducer, the radiation force measured on a balance and the

[57] J. A. Rooney, *J. Acoust. Soc. Am.* **54,** 429 (1973).
[58] T. Tarnocy, *Magy. Fiz. Foly.* **2,** 159 (1954).

acoustic power measured with a calorimeter. In the intensity range between 0.5 and 2.5 W/cm^2 he found that the measured radiation force was 20% smaller than that predicted from the theoretical power output of the transducer. However, the radiation force technique produced results closer to the theoretically expected value than calorimetry which gave results 35% smaller than the predicted value. Better agreement has been obtained in other studies. In a comparison between a radiation force measurement and a calorimeter, Wells et al.[59] found agreement between the two techniques to within 6% in the power range from 1 to 2.2 W. The study by Haran et al.[60] of the power output from a transducer showed no statistically significant difference between determinations by radiation force and by an optical system. Based on these comparisons, it appears that indeed there is a good agreement between theory and experiment.

6.2.4. Novel Applications and Techniques

In addition to the more classical type of studies of radiation force, there exist several novel applications of techniques associated with this phenomenon which would be of interest to the experimental physicist.

6.2.4.1. Forces on Liquid Interfaces. The Langevin radiation force can be exerted on the interface between two liquids. This effect can most easily be demonstrated following a classical experimental arrangement developed by Hertz and Mende,[52] which demonstrates an important aspect of radiation force. In their experiment, layers of different liquids, carbon tetrachloride, water, and aniline, were used because of their different densities and velocities of sound. They irradiated the fluids from above using a 5-MHz transducer, reflected the sound from a plate oriented with respect to the transducer, and displayed the image of the transmitted and reflected beams as shown in Fig. 3.

As seen in the figure, the liquid interfaces no longer remain planar but are deformed by the radiation forces of the sound beams. In particular, we note that the boundary between the water and aniline is deflected in a direction antiparallel to that of the sound propagation for the transmitted beam and parallel to the direction of sound propagation for the reflected beam.

This phenomenon can be explained using Langevin radiation force. Because the acoustic characteristics of the liquids are well matched, the acoustic powers and intensities are uniform throughout the vessel. In contrast, the energy density $E = I/c$ is different in the three liquids be-

[59] P. N. T. Wells, M. A. Bullen, and H. F. Freundlich, Ultrasonics **2**, 214 (1969).
[60] M. E. Haran, B. D. Cook, and H. F. Stewart, J. Acoust. Soc. Am. **57**, 1436 (1975).

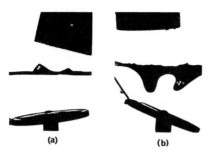

(a) **(b)**

FIG. 3. Radiation forces on liquid interfaces: (a) water over aniline; (b) water over carbon tetrachloride. The transducer is at the top of the figure with the reflecting plate below. (Reproduced with permission from Beyer.[39])

cause of the different velocities of sound. The energy density is less and thus the radiation force is less in water than in aniline because of its larger sound velocity. Therefore the radiation force causes the interface to deflect as shown in the figure. Related results are seen in the figure for the case of carbon tetrachloride and water.

6.2.4.2. Radiation Force on a Vibrating Meniscus. A second novel technique makes use of the radiation force to measure directly the pressure amplitude of a sound field. In the experimental arrangement developed by Nyborg and Rooney,[61] a hemispherical meniscus was formed in water at the end of a small stainless-steel tube filled with air and connected to an air reservoir positioned above a sound source. Upon application of the sound field, the meniscus retreated into the tube and the gas pressure was increased until the meniscus was again hemispherical. The phenomenon can be described using theory for the radiation force on the meniscus and measurement of the excess pressure required to maintain the original shape of the meniscus is a direct measure of the pressure in the sound field.

6.2.4.3. Modulated Radiation Force. In discussions of radiation force, the possibility of a time-varying component of force usually does not arise. However, a method of measuring radiation force that uses modulation has been developed by Greenspan et al.[62] Earlier methods by Belin and Beyer[30] also used modulation. In the Greenspan method the input to the transducer is modulated at a low frequency and the output acoustic power is intercepted by a target that experiences a component of

[61] W. L. Nyborg and J. A. Rooney, *J. Acoust. Soc. Am.* **45**, 384 (1969).

[62] M. Greenspan, F. R. Breckenridge and C. E. Tschiegg, *J. Acoust. Soc. Am.* **63**, 1031 (1978).

radiation force at the modulation frequency. The target is mounted on the armature of an electromagnetic receiver, provided with an independent coil that is fed with a current at the modulation frequency; the frequency is adjusted in amplitude and phase, either manually or automatically by feedback, to arrest the motion of the armature. When the armature is stationary, the force is proportional to the current; the apparatus can be calibrated using weights. This technique avoids the drifts present in dc measurements. Narrowband ac signal processing techniques can be used to improve the sensitivity of the system which is quoted to be 10 μW of acoustic power.[62]

6.3. Acoustic Manipulation of Objects

6.3.1. Introduction

A nonlinear effect of the interaction of a sound field with an object within an acoustic field provides the experimenter with the unique possibility of manipulating an object in a host medium using only the acoustic field. This is a very special application of radiation force, and, in fact, much theoretical and experimental work has been completed concerning the time-averaged force exerted by a sound field on a spherical inclusion.

The first such study was completed by Bjerknes,[63] who investigated the effects on a spherical bubble placed in an oscillating fluid using analogies to electrical and magnetic interactions. A more detailed study of the force on a sphere in a plane wave field was published by King,[64] but the effects of compressibility were neglected. These results were extended by Yosioka and Kawasima[65] to include compressibility. In addition, Gor'kov[66] has used a fluid dynamics approach to obtain similar results for small objects in more general sound fields. Westervelt[67] has derived the more general expression for the radiation force on an object with arbitrary shape, Embleton[68] has extended the work of King to include spherical waves, and Nyborg[69] has calculated the force exerted by a sound field on a small rigid sphere. Also, Dystke,[70] with corrections as noted by Crum,[71] has given a comprehensive treatment of the force exerted on

[63] V. F. K. Bjerknes, "Fields of Force." Columbia Univ. Press, New York, 1906.
[64] L. V. King, *Proc. R. Soc. London Ser. A* **147**, 212 (1934).
[65] K. Yosioka and Y. Kawasima, *Acustica* **5**, 17 (1955).
[66] L. P. Gor'kov, *Soc. Phys. Dokl.* **6**, 773 (1962).
[67] P. J. Westervelt, *J. Acoust. Soc. Am.* **23**, 312 (1951).
[68] T. F. W. Embleton, *J. Acoust. Soc. Am.* **26**, 40 (1954).
[69] W. L. Nyborg, *J. Acoust. Soc. Am.* **42**, 947 (1967).
[70] K. B. Dystke, *J. Sound Vib.* **10**, 331 (1969).
[71] L. A. Crum, *J. Acoust. Soc. Am.* **50**, 157 (1971).

solid, liquid, and gas inclusions in a stationary sound field. Hasegawa has derived theory[72] and conducted experiments[73] on the radiation force on a sphere in a quasi-stationary field.

6.3.2. Theory

In considering the force on an object in a sound field, it is found that two general terms exist in the force equation: the first results from the radiation pressure exerted on a finite-sized inclusion in the second field, while the second arises from the finite compressibility of the object. Thus, for hard spheres, the force is obviously dominated by the first term, while in the case of air bubbles in water, the compressibility term dominates the equation. A liquid droplet lies midway between the two extremes and, for a droplet trapped by a stationary sound field, the terms are of equal magnitude and both must be considered in the theory.

Following the suggestions of Crum,[71] we can obtain the expression for the force on a fluid droplet in a sound field by combining two independent approaches. The result is in agreement with results obtained by Yosioka and Kawasima[65,73a] as well as Gor'kov.[66] Nyborg[69] has developed a theory that can be used to calculate the force F_r exerted on a noncompressible sphere by a stationary sound field, where F_r is given by

$$F_r = V_0[B(\partial\overline{T}/\partial z) - (\partial\overline{U}/\partial z)], \qquad (6.3.1)$$

and $\overline{T}$ and $\overline{U}$ are, respectively, the second-order approximations to the time-averaged kinetic and potential energies and V_0 the equilibrium volume of the sphere. Here $B = 3(\delta - 1)/(2\delta + 1)$ and $\delta = \rho/\rho_0$, where ρ and ρ_0 are, respectively, the densities of the sphere and host medium. Eller[74] has shown that the time-averaged radiation force exerted on a sphere at a position z in the stationary sound field due to its compressibility alone is given by

$$F_c = -\langle V(t)\, \partial P(z, t)/\partial z\rangle, \qquad (6.3.2)$$

where $V(t)$ is the instantaneous volume and $P(z, t)$ the instantaneous pressure along the axis. Adding the two terms and evaluating the quantities for a standing wave acoustic field, one obtains the expression for the net force on the droplet:

$$F = \frac{V_0 P^2 k\, \sin(2kz)}{4\rho c^2} \left[\frac{1}{\delta\sigma^2} - \left(\frac{5\delta - 2}{2\delta + 1}\right)\right], \qquad (6.3.3)$$

[72] T. Hasegawa, J. Acoust. Soc. Am. **65**, 32 (1979).
[73] T. Hasegawa, J. Acoust. Soc. Am. **65**, 41 (1979).
[73a] T. Hasegawa and K. Yosioka, J. Acoust. Soc. Am. **58**, 581 (1975).
[74] A. I. Eller, J. Acoust. Soc. Am. **43**, 170 (1968).

where $k = 2\pi/\lambda$ and σ is the ratio of the velocity of sound c in the droplet to that in the host liquid c_0. This equation describes the cases of a liquid droplet and a rigid sphere but it must be modified for the case of a bubble, as described in Yosioka and Kawasima,[65] Gor'kov,[66] or Eller.[74]

6.3.3. Experimental Arrangements

Experimental testing of the theory for radiation force on solid spheres has been conducted by Embleton[75] and Hasegawa *et al.*,[76] as well as Breazeale and Dunn.[77] For their studies Breazeale and Dunn[77] intercompared three optical techniques for measurement of the intensities of a sound field with the radiation force on a small sphere and a transient thermoelectric method. All experimental curves were within $\pm 10\%$ of the mean values for all techniques. However, this mean was approximately 27% less than the acoustic output calculated from the voltage.

Testing of the theory for the case of a liquid droplet and an air bubble has been conducted by Crum[71] and Crum and Eller,[78] respectively. While qualitative observations of the trapping of bubbles in standing wave acoustic fields had been made by several people, it was only in 1968 that Eller[74] developed techniques for quantitatively testing the theory. The experiment took place in a cylindrical standing wave excited in a vertical tube filled with water and driven by a piston at the bottom. The frequency used was 27 kHz and the wavelength in the vertical direction was 23 cm. The procedure used was to trap the bubble in the sound field above the pressure maximum located about 5.75 cm below the surface. The pressure amplitude necessary to trap the bubble was measured and the bubble size determined from measurement of rise times of the free bubble and Stoke's law. Results from the experiment demonstrated that the average force on a bubble could be described by Eq. (6.3.3) and that these results were consistent with the qualitative results by earlier workers.

For his experiments on levitation of a fluid droplet in a host liquid, Crum[71] used the apparatus shown schematically in Fig. 4. Since pressures on the order of 10 bars were needed to trap the liquid droplets used, it was necessary to use a thin-walled cylindrical vessel coupled to a PZT-4 ceramic transducer. He found that the $(r, \theta, z) = (1, 0, 3)$ resonant cylindrical mode was useful for the experiments. In this case the resonant frequency was 41.7 kHz and the wavelength along the vertical axis was

[75] T. F. W. Embleton, *J. Acoust. Soc. Am.* **26**, 46 (1954).
[76] T. Hasegawa and K. Yosioka, *J. Acoust. Soc. Am.* **46**, 1139 (1969).
[77] M. A. Breazeale and F. Dunn, *J. Acoust. Soc. Am.* **55**, 671 (1974).
[78] L. A. Crum and A. I. Eller, *J. Acoust. Soc. Am.* **48**, 181 (1970).

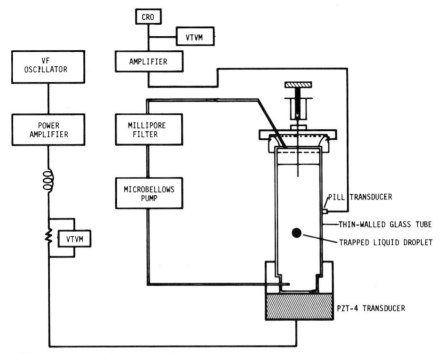

FIG. 4. Apparatus for studies of acoustic levitation. (Reproduced with permission from Crum.[71])

7.8 cm. The acoustic pressure amplitude was measured by a small calibrated ceramic transducer mounted externally to the system. The test
droplet was injected into the system with a syringe and its radius was
measured with a calibrated microscope. Crum then measured the magnitude of the smallest acoustic pressure amplitude at the nearest antinode of
the stationary wave system required to trap a given droplet and compared
the result with the theoretical prediction. Based on his results we can
conclude that it is possible to levitate droplets of liquid in an acoustic stationary wave and that the theory correctly predicts the acoustic pressure
amplitudes required to trap these droplets in the sound field.

The basic problems associated with the levitation apparatus include
choosing and optimizing the appropriate resonance and maintaining the
resonance despite changes that occur in the system. A more sophisticated experimental arrangement for conducting levitation experiments
has been developed by Baxter et al.[79] The design described has the ad-

[79] K. Baxter, R. E. Apfel, and P. L. Marston, *Rev. Sci. Instrum.* **49**, 224 (1978).

vantageous features that any manifestation of resonance can be used as input to the device and that resonances can be scanned manually, viewed, and locked onto by the circuitry.

The experimenter using this technique to trap bubbles for study should be aware of instabilities that may be developed in the bubble under certain conditions. These instabilities were found by Crum and Eller[78] to be of two types. The first is the onset of erratic motion by the bubbles trapped in the standing wave field when the sound–pressure amplitude exceeds a threshold value. The second type of instability is the onset of oscillation of the bubble shape, which also requires that the pressure amplitude exceed a given threshold.

In the same paper, Crum and Eller[78] quantitatively studied another effect that can influence experiments of the types described here. In these experiments they conducted measurements of bubbles moving through standing wave acoustic fields and found that the translational velocity of a bubble smaller than resonance size may be relatively large, perhaps several times greater than the rise velocity of the bubble in the absence of the sound field. In a given sound field, this translational velocity is a function of the bubble position in the field of the bubble radius.

6.3.4. Applications and Special Effects

6.3.4.1. Measurement of Physical Properties of Small Samples of Fluids. Acoustic leviation of one drop of liquid immiscible in a host liquid has been used by Apfel[80-83] to study several properties of fluids. Using an experimental arrangement similar to that shown in Fig. 4, Apfel[80,81] has measured the tensile strength of liquids. This leviation technique has the advantages that only small samples of liquid need to be used and that the possibility of heterogeneous nucleation is greatly reduced. Details of the implications of this type of study to acoustic cavitation are described in Part 7 of this book. Similar techniques have been used by Apfel to measure the adiabatic compressibility, density, and velocity of sound in submicroliter liquid samples.[82,83] The leviation technique for measurement of these parameters has the advantages that since it is a comparison technique the results are independent of droplet size and that the field parameters need not be known; only the voltage applied to the transducers to levitate the droplet to the same position in the field as a reference droplet must be determined.

[80] R. E. Apfel, *J. Acoust. Soc. Am.* **49**, 145 (1971).
[81] R. E. Apfel, *Nature (London) Phys. Sci.* **233**, 119 (1971).
[82] R. E. Apfel and J. P. Harbison, *J. Acoust. Soc. Am.* **57**, 1371 (1975).
[83] R. E. Apfel, *J. Acoust. Soc. Am.* **59**, 339 (1976).

6.3.4.2. Calibration of Small Hydrophone. In a novel application of acoustic trapping of a bubble, Gould[84] was able to demonstrate the capability of using the technique to calibrate small ceramic hydrophones. He used an experimental arrangement similar to that of Eller.[74] After determining the position and size of the trapped bubble, it was removed and the hydrophone to be calibrated was placed in the same position. Then, using the theory tested by Eller, the pressure amplitude at that position in the sound field was calculated, enabling him to calibrate the output of small omnidirectional hydrophones. Gould estimated that the technique is useful in the frequency range from 3 to 90 kHz.

6.3.4.3. Studies of Mass Transfer in Bubbles. The levitation or trapping of droplets or bubbles has proven to be a useful experimental technique for study of mass transfer occurring near gas bubbles. Eller[85] has completed initial quantitative studies. Specifically, he studied the threshold conditions for rectified diffusion and the growth rate of bubbles under nonthreshold conditions. The basic experimental arrangement was that shown in Fig. 4. In these experiments a single air bubble smaller than resonant size was injected and trapped in the standing wave sound field. At selected time intervals the position of the bubble on the axis of the field was measured with a cathetometer. The sound field was then removed and the bubble allowed to rise a given distance so that its radius could be determined from Stoke's law. The bubble was then recaptured by the sound field and returned to its initial position. Thus, this technique with bubbles in air-saturated water allowed Eller to determine the threshold conditions for rectified diffusion and growth rates at acoustic pressures greater than the threshold for these effects. Eller[86] also extended the studies to a frequency of 11 kHz.

In a related study Gould[87] has examined the effect of acoustic streaming on rectified diffusion of bubbles. He used an arrangement similar to that shown in Fig. 4 and directly determined the size of the bubbles as a function of time using a calibrated microscope. The presence of acoustic streaming was observed by using polystyrene spheres in suspension near the trapped bubble. Through these studies Gould[87] was able to demonstrate the important effect that acoustic streaming can have on mass transfer.

6.3.4.4. Acoustic Leviation for Processing in Space. Many of the experiments to be conducted in a laboratory in space require the manipu-

[84] R. K. Gould, *J. Acoust. Soc. Am.* **43**, 1185 (1968).
[85] A. I. Eller, *J. Acoust. Soc. Am.* **46**, 1246 (1969).
[86] A. I. Eller, *J. Acoust. Soc. Am.* **52**, 1447 (1972).
[87] R. K. Gould, *J. Acoust. Soc. Am.* **56**, 1740 (1975).

lation and control of materials. While the magnitude of the forces associated with acoustic levitation (in air for one g of acceleration one W produces a change in weight equal to that associated with 310 mg) may be small in a normal earth environment, it does provide a unique manipulative technique for low-gravity space applications; these include processing such as zone melting, casting, crystal growing, and chemical synthesis. Alteration of the shape of the container alters the acoustic field within the experiment chamber which can cause a melt to assume a particular shape. A movable wall in the resonator can pull crystals by making antinodes of the sound field move apart. Rotating a crystal can degas and control the segregation of substances in it. Laser and ion beams can be used for contactless heating of a melt. Thus, in these processes the materials can be positioned and formed within a container without making contact with its walls. The acoustic method has advantages over electromagnetic methods in that it is not limited to materials that are electrically conducting.

Typical experimental arrangements for acoustic levitation of materials in space have been described by Wang et al.[88-90] as well as others.[91-93] The experimental chamber is nearly cubic with three acoustic drivers fixed rigidly to the center of three mutually perpendicular faces of the chamber. Optical windows are also located in the sides of the chamber for observing the sample. During operation of the chamber each driver excites the lowest-order standing wave along the direction that the driver faces. In this lowest mode, the radiation force is minimum at the center of the chamber causing materials introduced into the system to move toward the central region.

In addition, the chamber has been designed so that two sides have the same length and therefore the fundamental resonant frequencies in these directions are degenerate. If the transducers for these directions are driven 90° out of phase, then a maximum torque will be applied to a sample located at the center of the chamber. This torque, which is in the $x-y$ plane, is proportional to the sine of the phase difference and will cause the sample to rotate about the z axis. For the case of a sphere, lab-

[88] T. G. Wang, M. M. Saffren, and D. D. Elleman, Proc. AIAA Aerosp. Sci. Meeting, 12th, January 30–February 1, Washington, D.C. Paper 74-155 (1974).

[89] T. G. Wang, H. Kanber, and I. Rudnick, Phys. Rev. Lett. 38, 128 (1977).

[90] T. G. Wang, H. Kanber, and E. E. Olli, J. Acoust. Soc. Am. 63, 1332 (1978).

[91] G. Lagomarsini and T. Wang, Proc. AIAA Aerosp. Sci. Meeting, 17th, January 15–17, New Orleans, Louisana, Paper 79-0369 (1979).

[92] N. Jacobi, R. P. Tagg, J. M. Kendall, D. D. Elleman, and T. G. Wang, Proc. Aerosp. Sci. Meeting, 17th, January 15–17, New Orleans, Louisana. Paper 79-0225 (1979).

[93] R. R. Whymark, Ultrasonics 13, 251 (1975).

oratory demonstrations using a 1.25-cm-radius styrofoam ball in a 155-dB (relative to 2×10^{-4} dyne/cm^2) sound field have achieved rotation rates greater than 2000 rpm. Also, oscillations of a sample drop can be produced by modulating the sound pressure of one or more of the transducers.

Since the experiments to be conducted can take place at a variety of temperatures and using different gas environments, it is necessary to provide a way to maintain resonance under many conditions. The required automatic frequency control utilizes a phase-locking loop. The phase-locked loop monitors the driving frequency so that the input signal has at all times at 90° phase lead with respect to the acoustical signal inside the chamber.

6.4. Acoustic Streaming

6.4.1. Introduction

Several phenomena associated with nonlinear acoustics, including enhanced transfer of heat and mass, changes in reaction rates, emulsification, depolymerization, and sonically produced biological effects, are directly related to acoustic streaming. By "acoustic streaming" we mean the time-independent flow of fluid induced by a sound field. In this section we shall briefly discuss the theory for streaming and methods for the production and study of the phenomenon and shall mention some unusual experimental methods for its use.

6.4.2. Theory

To introduce the theory for acoustic streaming we shall outline arguments from the thorough discussion by Nyborg.[94] For a linear, homogeneous, isotropic fluid the dynamical equation is

$$f = \rho[\partial \mathbf{u}/\partial t + (\mathbf{u} \cdot \nabla)\mathbf{u}], \qquad (6.4.1)$$

where

$$f = -\nabla P + [\mu' + \tfrac{4}{3}\mu]\nabla\nabla\mathbf{u} - \mu\nabla \times \nabla \times \mathbf{u}. \qquad (6.4.2)$$

Here the quantities P, ρ, and u are, respectively, the pressure, density, and particle velocity and μ and μ' are the shear and bulk viscosity coefficients, respectively. In addition, we shall make use of the equation of

[94] W. L. Nyborg, Acoustic streaming, in "Physical Acoustics," Vol. 2B. (W. P. Mason, ed.). Academic Press, New York, 1965.

continuity,

$$\partial \rho / \partial t + \nabla \cdot \rho \mathbf{u} = 0. \qquad (6.4.3)$$

Also, we must consider a nonlinear equation of state relating P and ρ.

The usual method of solution of these equations uses the method of successive approximations. Thus, we expand each variable as a series of terms of decreasing magnitude as follows:

$$P = P_0 + P_1 + P_2 + \cdots,$$

$$\rho = \rho_0 + \rho_1 + \rho_2 + \cdots, \qquad (6.4.4)$$

$$\mathbf{u} + \mathbf{u}_1 + \mathbf{u}_2 + \cdots.$$

Here the zero-order quantities P_0 and ρ_0 give the static pressure and density. First-order quantities P_1, ρ_1, and $\mathbf{u}_1$ are the usual solutions of the linear wave equations. Thus, a quantity such as P_1 will vary sinusoidally in time with an amplitude proportional to the source amplitude A. In evaluating the second-order quantities P_2, ρ_2, and $\mathbf{u}_2$ we find that there exist two types of contributions to these terms: the first are second harmonic contributions, which vary sinusoidally in time with a frequency $2f$, while the second are time-independent contribution. The magnitudes of both types of contributions are proportional to A^2.

In this section we are interested only in the time-independent contributions to $\mathbf{u}_2$, which represent the flow generated by the sound. Proceeding with the derivation using approximations appropriate for our applications we obtain

$$\mu \nabla \times \nabla \times \mathbf{u}_2 = \nabla P_2 - \mathbf{F}, \qquad (6.4.5)$$

where $\mathbf{u}_2$, and P_2 are now only the time-independent contributions to the second-order velocity and pressure, while $\mathbf{F}$ is given by

$$\mathbf{F} = -\rho_0 \langle (\mathbf{u}_1 \cdot \nabla) \mathbf{u}_1 + \mathbf{u}_1 (\nabla \cdot \mathbf{u}_1) \rangle, \qquad (6.4.6)$$

where $\langle \ \rangle$ indicates a time-averaged quantity. We note that $\mathbf{F}$ depends only on the first-order velocity field and has the units of force per unit volume. For sound fields where the first-order term $\mathbf{u}_1$ represents a perfectly spherical, cylindrical, or plane wave in an unlimited medium, $\mathbf{F}$ is irrotational and there is no streaming. However, when there are boundaries present on which the wave impinges, boundary layers form, and $\mathbf{F}$ is no longer irrotational; then streaming exists. Numerical techniques have been applied to the problem and further discussion of nonlinear terms that might be included in the theory has been given by Kukarkin and Rudenko[95] and Ostrovskii and Popilova.[96]

[95] A. B. Kukarkin and O. V. Rudenko, *Sov. Phys. Acoust.* **22**, 137 (1976).
[96] L. A. Ostrovskii and I. A. Popilova, *Sov. Phys. Acoust.* **20**, 45 (1974).

6.4.3. Experimental Production of Acoustic Streaming

Experimentally, acoustic streaming can be divided into two general types. The first is associated with plane acoustic waves in the volume of an attenuating medium. This type of streaming is often called "quartz wind," for historical reasons. The second type is associated with either inhomogeneities in a sound field or interactions of a sound field with a boundary or surface.

6.4.3.1. Volume Streaming Configuration. The typical situation for volume streaming is an attenuated plane wave traveling in the positive x direction in an unbounded medium. Experimentally the configuration can be achieved by mounting an ultrasonic disk transducer in one end of a tank whose dimensions are much larger than the transducer element. Liebermann[97] has used this configuration with a diaphragm on the wall opposite the sound source to provide a nonreflecting boundary. He found that streaming existed in the volume of the fluid flowing away from the transducer. The region of outward streaming corresponded closely to the cross-sectional area of the transducer and the return flow was outside that region. Thus, in this case the effective force per unit volume F from Eq. (6.4.6) has a value of $\rho_0 \alpha A^2$ and acts in the direction away from the transducer surface. Here α is the absorption coefficient.

6.4.3.2. Boundary Configurations. Boundary configurations for production of streaming can involve both plane waves and other acoustic field patterns.

6.4.3.2.1. PLANE WAVES. Historically, acoustic streaming was first studied by Rayleigh.[98,99] The experimental conditions were those associated with a standing wave set up in a rectangular channel. He found that acoustic streaming in the form of a regular array of eddy flows, each separated by a quarter-wavelength, occurred along the channel. Typically, changes in the streaming velocities are greatest near the walls and smaller in the central regions of the channel. This type of streaming is not only of historic interest but is of general interest because the motion is largely responsible for the aggregations of dust in the Kundt's tube experiments performed in elementary physics laboratories. This type of streaming can have significant effects on reaction rates and mass transfer occurring in fluid-filled channels,[100,101] although for the case of liquid-filled channels the situation may be complicated by the presence of cavitation.

6.4.3.2.2. OTHER GEOMETRIES. Many experimental arrangements

[97] L. N. Liebermann, *Phys. Rev.* **75**, 1415 (1949).

[98] Lord Rayleigh, *Philos. Trans. R. Soc. London* **175**, 1 (1884).

[99] Lord Rayleigh, "Theory of Sound." Dover, New York, 1945.

[100] H. U. Fairbanks and R. E. Cline, *IEEE Trans. Sonics Ultrason.* **SU-14**, 175 (1967).

[101] H. V. Fairbanks, T. K. Hu, and J. W. Leonard, *Ultrasonics* **8**, 165 (1970).

exist that can be used to produce acoustic streaming in the absence of cavitation. As one can see from Eq. (6.4.6) one important aspect of the "driving force" for acoustic streaming is the spatial variation of the first-order velocity u_1. Such gradients can be created in a variety of ways. Probably the streaming situation which has been most extensively studied theoretically is that of an infinitely long circular cylinder driven transversely with respect to its own axis. Theory for this situation has been developed by Schlichting,[102,103] Holtsmark et al.,[104] Skavlem and Tjotta,[105] Westervelt et al.,[106,107] and Raney et al.[108] All investigators find that streaming occurs in four symmetrical counterrotating circulation cells around the cylinder. Theory for the related case of an oscillating rigid sphere was derived by Lane.[109] The streaming motions are similar to those seen near the transversely vibrating cylinder. Some aspects of the streaming patterns near a sphere were investigated by Rowe and Nyborg,[110] while Sorokodum and Timoshenko[111] have considered the elliptical cylinder.

Another important experimental situation for production of acoustic streaming is that of a vibrating gas bubble. The streaming is generated when a gas bubble is set into oscillation near a surface or if surface wave activity is present on a bubble oscillating in a volume of liquid. Experimental arrangements for studying such streaming include those of Kolb and Nyborg,[112] in which a small bubble is driven by a metal cone, and Elder,[113] who observed a bubble positioned on a surface that was being driven acoustically. In general, the types of patterns set up near the bubble are highly dependent on the viscosity of the fluid, surface condition of the bubble, and driving amplitude of the sound source.

A variety of other experimental arrangements involving solid acoustic-impedance-transforming horns of different configurations can be used to study features of acoustic streaming. Such horns are widely used and have been described by Mason and Wick.[114] These include bringing

[102] H. Schlichting, Phys. Z. 33, 327 (1932).
[103] H. Schlichting, "Boundary Layer Theory." McGraw-Hill, New York, 1955.
[104] J. Holtsmark, I. Johnsen, T. Sikkeland, and S. J. Skavlem, J. Acoust. Soc. Am. 26, 26 (1954).
[105] S. Skavlem and S. Tjotta, J. Acoust. Soc. Am. 27, 26 (1955).
[106] P. J. Westervelt, J. Acoust. Soc. Am. 25, 1123 (1953).
[107] P. J. Westervelt, J. Acoust. Soc. Am. 27, 379 (1955).
[108] W. P. Raney, J. C. Corelli, and P. J. Westervelt, J. Acoust. Soc. Am. 26, 1006 (1954).
[109] C. A. Lane, J. Acoust. Soc. Am. 27, 1082 (1955).
[110] W. E. Rowe and W. L. Nyborg, J. Acoust. Soc. Am. 39, 965 (1966).
[111] E. D. Sorokodum and V. I. Timoshenko, Sov. Phys. Acoust. 19, 597 (1973).
[112] J. Kolb and W. L. Nyborg, J. Acoust. Soc. Am. 28, 1237 (1956).
[113] S. A. Elder, J. Acoust. Soc. Am. 31, 54 (1959).
[114] W. P. Mason and R. F. Wick, J. Acoust. Soc. Am. 23, 209 (1951).

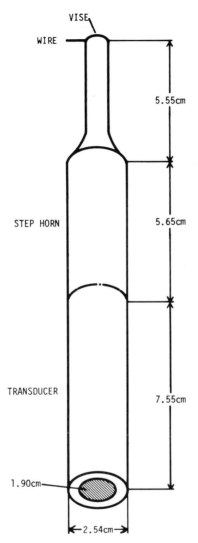

VISE

WIRE

5.55cm

STEP HORN   5.65cm

TRANSDUCER   7.55cm

1.90cm

←2.54cm→

FIG. 5. Method for driving wire transversely.

a vibrating bar near a plane boundary of a rigid solid, as demonstrated by Jackson,[115] and placing a vibrating horn near a membrane, as described by Jackson and Nyborg.[116]   In both cases streaming velocities are particularly high near the edge of vibrating horns.   Acoustic streaming patterns

[115] F. J. Jackson, *J. Acoust. Soc. Am.* **32,** 1387 (1960).
[116] F. J. Jackson and W. L. Nyborg, *J. Acoust. Soc. Am.* **30,** 614 (1958).

near exponential horns have been studied by several investigators, including Hughes and Nyborg[117] and Fill.[118]

A particularly useful arrangement for studies of acoustic streaming and its effects on a variety of systems where only a small volume of sample is available consists of a tungsten wire driven transversely by a transducer with a horn and wire attached as shown in Fig. 5. The vibration pattern for such a driven wire is composed of standing waves, as described by Williams and Nyborg[119] for the case of a transversely driven tube. To achieve maximum displacement amplitude at the tip, the length of the wire is adjusted to an odd multiple of a quarter-wavelength. Results have been obtained using a resonant length of 0.025-cm-diam tungsten wire clamped at one end to the tip of a velocity transformer driven at 20 kHz by a PZT-4 ceramic transducer. In studying the details of streaming patterns that occur near the wire, two important precautions should be taken. As would be expected from Eq. (6.4.6), the velocity of streaming and, in fact, the type of streaming pattern observed is highly dependent on the configuration of the tip of the wire. For well-ordered streaming patterns that permit direct comparison and use of theory, it is necessary to have a hemispherical tip. If the wire tip is merely the cut or jagged end, then the streaming will be controlled by local irregularities of this tip and no comparison or use of theory is possible. Second, orderly, predictable streaming patterns depend on driving the wire only transversely. This is usually accomplished by using only small amplitudes. If other vibrational modes are induced in the wire at high amplitudes, then drastic changes can be seen in the streaming pattern. Evidence for such changes in related systems has been described elsewhere.[120]

6.4.4. Methods for Study of Acoustic Streaming Patterns

Since stable acoustic streaming is a time-independent flow of fluid, methods of studying the patterns produced include all of those typically used for flow visualization. In addition, because of the unique nature of acoustic streaming and the effects produced by it, unusual methods can be used to determine some aspects of the patterns.

The most commonly used techniques for visualizing streaming patterns involve the use of indicators. These include smoke,[121] ink-dyed liquids,[122] aluminum, latex, or similar particles in suspension.[123] An example

[117] D. E. Hughes and W. L. Nyborg, *Science* **138**, 108 (1962).
[118] E. E. Fill, *J. Appl. Phys.* **39**, 5816 (1968).
[119] A. R. Williams and W. L. Nyborg, *Ultrasonics* **8**, 36 (1970).
[120] J. A. Rooney, *Proc. IEEE Ultrason. Symp., Milwaukee, Wisconsin* p. 51 (1974).
[121] U. Ingard and S. Labate, *J. Acoust. Soc. Am.* **22**, 211 (1950).
[122] G. B. Thurston and C. E. Martin, *J. Acoust. Soc. Am.* **25**, 26 (1953).
[123] J. A. Rooney, *Science* **169**, 869 (1970).

FIG. 6. Acoustic streaming pattern near tip of oscillating 250-μm-diameter wire. (Reproduced with permission from Kashkooli.[124])

of the type of patterns that can be visualized is shown in Fig. 6. Here the acoustic streaming was visualized by placing polystyrene spheres 7–10 μm in diameter in a dilute suspension near the vibrating wire. The suspension was lighted from both sides to eliminate shadows and for this particular observation a 4-sec exposure using a Polaroid camera with type 107 film attached to a viewing microscope was used to obtain the photograph. The use of such indicating particles proves to be a convenient technique since time exposures such as that shown in Fig. 6 can demonstrate general features of the streaming, e.g., measurement of the length of the trace allows one to determine the streaming velocity. Frame-by-frame analysis of motion pictures of the movements of the individual particles also permits the experimenter to determine quantitatively the streaming velocities in different parts of the field. Semenova[125] has used different color particles to highlight various aspects of the streaming.

Techniques based on flow birefringence for observing acoustically in-

[124] H. A. Kashkooli, M.S. thesis, Univ. of Maine–Orono (1977).
[125] N. G. Semenova, *Sov. Phys. Acoust.* **20**, 65 (1974).

duced flow have been developed by Hargrove and Thurston[126] and used by Durelli and Clark.[127] In their technique, a solution of milling yellow dye that is optically isotropic when at rest was used. The region of interest can be viewed through crossed polarizing plates, and the fluid becomes anisotropic only when there is motion causing velocity gradients and internal shearing stresses to exist. The flow pattern produced can be photographed or studied visually. A related technique has been described by Nagai et al.[128] In their experiment they measured the amount of birefringence and opacity induced by acoustic streaming in oriented layers of a nematic liquid crystal. The sensitivity of these acousto-optical cells to ultrasound is described in terms of configuration of the liquid crystal and its thickness.

The effects of acoustic streaming can sometimes be used to provide information about the streaming patterns themselves. Thus, Jackson and Nyborg[116] placed vibrating horns of various geometries in contact with a membrane that was coated on the other side with an opaque film of viscous paint. Outlines in the paint obtained with different sizes and shapes of the horns corresponded to regions in the streaming fields where the highest values of shearing stress existed near the surface and removed the paint from the film. In a similar type of experiment Nyborg et al.[129] demonstrated that selective darkening of photographic plates occurred in streaming fields.

6.4.5. Novel Experimental Methods

The phenomenon of acoustic streaming is of interest not only because it is an unusual type of fluid flow but because it has associated with it velocity gradients, shearing stresses, and related characteristics which prove useful in specialized experimental procedures. In this section a few examples will be briefly discussed.

6.4.5.1. Attenuation Coefficient of Liquids Determined by a Streaming Method. Consideration of the theory of acoustic streaming for the case of plane waves indicates that the ultrasonic absorption coefficient can be determined by a measurement of the streaming velocity and the energy density of the sound field. An example of this method is the work of Bhadra and Roy[130] on a variety of liquids. They measured the streaming velocity using tracer particles and the energy density using

[126] L. E. Hargrove, Jr., and G. B. Thurston, *J. Acoust. Soc. Am.* **29**, 966 (1957).

[127] A. J. Durelli and J. A. Clark, *J. Strain Anal.* **7**, 217 (1972).

[128] S. Nagai, A. Peters, and S. Candau, *Rev. Phys. Appl.* **12**, 21 (1977).

[129] W. L. Nyborg, R. K. Gould, F. J. Jackson, and C. E. Adams, *J. Acoust. Soc. Am.* **31**, 706 (1959).

[130] T. C. Bhadra and B. Roy, *Proc. Ultrason. Int.* p. 253 (1975).

radiation force techniques. In comparison to the requirements of other techniques, such as those discussed in Part 2, larger energy densities for the ultrasonic field are needed. The results by streaming yield higher values than those predicted classically but lower than those obtained by pulse and optical methods.

6.4.5.2. A Periodic Cavitation Process. A study completed by Neppiras and Fill[131] demonstrated the complicated interactive nature of nonlinear acoustic phenomena and also described a useful technique for generation of bubbles of defined size for special applications. In their experiment a gas bubble was observed to grow to near resonant size by the accumulation of microbubbles from the field. As the large bubble reached resonant size, it began to oscillate nonlinearly and to produce microbubbles. The process can be made cyclic with the proper choice of experimental conditions because acoustic streaming assists in maintaining the periodicity.

6.4.6. Use of Acoustic Streaming to Study Biological Structure and Function

Experimental methods for producing acoustic streaming have been used to demonstrate that biological effects can be produced by flow and shearing stresses associated with the streaming. Also, they have been used to determine some physical properties of biomaterials and to gain insight into function of biosystems. Although he did not describe it as such, the fact that acoustic streaming could be induced in cellular systems was demonstrated as early as 1929 by Schmitt.[132]

One of the more widely used and successful experimental arrangements for the study of details of interactions of streaming consists of a small exponential or stepped horn connected to a ceramic transducer driven by an oscillator and power amplifier. Typically, these experiments are conducted in the frequency range from 50 to 100 kHz. The tip of the horn is brought into contact with the biological system of interest and motions are observed through a microscope. Induced changes are evaluated using a variety of techniques. Examples of application of this method of investigation include work by Hughes and Nyborg[117] on the hemolysis of erythrocytes, Dyer and Nyborg[133] and Gershoy et al.[134] on the motions induced in plant cells, El'piner et al.[135] and Wilson et al.[136] on deformation

[131] E. A. Neppiras and E. E. Fill, *J. Acoust. Soc. Am.* **46,** 1264 (1969).

[132] F. O. Schmitt, *Protoplasma* **7,** 332 (1929).

[133] H. J. Dyer and W. L. Nyborg, *IRE Trans. Med. Electron.* **ME-7,** 163 (1960).

[134] A. Gershoy and W. L. Nyborg, *J. Acoust. Soc. Am.* **54,** 1356 (1973).

[135] I. E. El'piner, I. M. Fairkin, and O. K. Basurmonova, *Biofizika* **10,** 805 (1965).

[136] W. L. Wilson, F. J. Wiereinski, W. L. Nyborg, R. M. Schnitzler, and F. J. Sichel, *J. Acoust. Soc. Am.* **40,** 1363 (1966).

of cellular structures, and Schnitzler[137] on the changes induced in skeletal muscle.

The vibrating wire described in Section 6.4.3.2.2 has also been used for studies of the interaction of acoustic streaming with biological systems. Specifically, Williams[138-140] has placed such a wire in contact with intact blood vessels and produced a variety of effects including aggregation and rupture of platelets and formation of thrombi. Data on hemolysis obtained by Williams *et al.*[141] using the wire system and related data obtained by Rooney[123,142] using a stable bubbles as a source of streaming provided information concerning the mechanical response of the erythrocyte to shearing stress. Results of hydrodynamic experiments are in reasonable agreement with the ultrasonically obtained results.[143]

6.5. Emulsification and Aggregate Dispersal

6.5.1. Introduction

The problem of formation of stable emulsions has been approached by a variety of experimental methods including some which use sonic or ultrasonic techniques. These techniques are also useful for dispersal of solid or liquid aggregated systems. A related problem, which imposes special constraints on the experimental methods, is that of dispersal of aggregates of biological cells. This section will include discussion of experimental procedures for solving these problems.

6.5.2. Sonic Methods for Emulsification

The two principal ultrasonic techniques used to form emulsions are flowthrough sonifying systems operating in the low-ultrasonic frequency range and liquid whistle arrangements.

6.5.2.1. Sonification Systems. Sonification systems used for emulsification typically consist of a velocity transformer or horn coupled to a transducer capable of oscillating in a longitudinal mode. The horn is immersed in the solutions of interest and activated so that collapse cavitation is produced in the liquids. The intensity of the cavitation, which depends on the power delivered to the horn, must be correctly chosen for

[137] R. M. Schnitzler, Ph.D. thesis, Univ. of Vermont (1969).
[138] A. R. Williams, *J. Physiol.* **257**, 23 (1976).
[139] A. R. Williams, *Ultrasound Med. Biol.* **3**, 191 (1977).
[140] A. R. Williams, *J. Acoust. Soc. Am.* **56**, 1640 (1977).
[141] A. R. Williams, D. E. Hughes, and W. L. Nyborg, *Science* **169**, 871 (1970).
[142] J. A. Rooney, *J. Acoust. Soc. Am.* **52**, 1718 (1972).
[143] J. A. Rooney, *J. Biol. Phys.* **2**, 26 (1973).

efficient treatment of the liquids. Special horns, horn tips, or vessels are available for specific processing applications. These include flowthrough horns, continuous flow attachments, and sealed atmosphere treatment chambers. Typical descriptions of such apparatus, related experimental methods, and research results may be found in articles by Weinstein *et al.*[144] and Genta *et al.*[145]

Hislop[146] and Last[147] have developed sonifying systems for emulsification that use a different technique to concentrate the ultrasonic energy. Their systems use solid focusing horns that concentrate longitudinal ultrasonic waves internally by using a parabolic horn tip. The liquids to be emulsified are passed with their host liquids through a channel that encloses the focus of the parabola.

6.5.2.2. Liquid-Driven Whistles. The other major ultrasonic technique used for emulsification and also dispersal consists of various types of liquid-driven whistles. Whistles are distinguished from other sonification systems by the fact that production of the sound associated with the system is by mechanically induced oscillations instead of the standard electromechanical transduction. One typical type of whistle consists of a resonant chamber of variable depth upon which a jet of liquid formed by a collimation tube impinges. All components of the whistle in contact with the high-pressure jet must be made of hardened steel. An example of this type of whistle is that originally designed by Pohlman.[148,149] In it two jets of the driving liquid cross to impinge on a blade positioned at the cross-over point. This blade is mechanically resonant at a desired frequency and is mounted at its nodal points to minimize damping. Modifications of such whistles include enclosure in chambers designed to confine the emulsification process to the region near the vibrating blade.[148,149] These whistles typically operate at frequencies in the 30-kHz range and can produce acoustic intensities of the order of 2 W/cm^2 in the liquid in the resonant chamber. Such levels are sufficient to produce cavitation in the liquids, thus increasing the efficiency of emulsification.

Other whistle configurations have also been developed; these include the vortex whistles described by Hislop[146] and Last.[147] In these devices the driving liquid enters a large swirl chamber tangentially. As this liquid

---

[144] J. N. Weinstein, S. Yoshikami, P. Henkart, R. Blumenthal, and W. A. Hagins, *Science* **195**, 489 (1977).

[145] V. M. Genta, D. G. Kaufman, and W. K. Kaufman, *Anal. Biochem.* **67**, 279 (1975).

[146] T. W. Hislop, *Ultrasonics* **8**, 88 (1970).

[147] A. J. Last, *Ultrasonics* **7**, 131 (1969).

[148] A. E. Crawford, *Research* **6**, 106 (1953).

[149] B. Brown and J. E. Goodman, "High Intensity Ultrasonics." Van Nostrand-Reinhold, Princeton, New Jersey, 1965.

moves from the large radius of a swirl chamber to the smaller radius of an exit pipe, the liquid is accelerated. The liquid phase to be dispersed or emulsified flows through an annular tube mounted along the axis of the whistle. The tube ends at the exit pipe of the vortex so that the two liquids interact. The geometry and location of the exit orifice and tube are found to be critical for greatest efficiency of emulsification. A somewhat similar device have been developed by Greguss.[150] In this device the driving liquid enters a cavity tangentially by one or more inlets. The liquid is then forced to move from a larger to a smaller radius because the outlet tube is smaller in diameter than the central cavity but larger than the inlet tube. Because of the angular acceleration of the driving liquid, a lower pressure exists in the central cavity and the liquid is forced to the center to be ejected through the outlet into a larger outer region. If the geometric form and dimensions of the inlet, central cavity, and outlet tubes are in proper relation, a sustained vibration develops and acoustic waves whose intensity is sufficient to cause emulsification are produced. The frequency of the waves depends on both the geometric dimensions and liquid driving pressure.

6.5.3. Mechanisms and Efficiency of Emulsification

The mechanisms for the ultrasonic emulsification process are similar to those of other emulsifiers, namely, action of shearing stresses and deformation of the larger droplets in the liquid. El'piner *et al.*[135] have demonstrated that acoustic streaming with its associated shearing stresses can begin the emulsification process. The more subtle details of the emulsification process have been examined both experimentally and theoretically by Hinze[151] and are summarized by Goldsmith and Mason.[152] The important parameters that determine the velocity gradients necessary to cause deformation of larger droplets, leading to their disruption and efficient emulsification, include the liquid viscosities and interfacial tension.

Specifically, Rumscheidt and Mason[153] have shown the relationship developed by Taylor[154] to be valid. Taylor[154] stated that a drop will burst when the shearing stresses become greater than the restoring forces associated with the interfacial tension. Details of the condition for droplet disruption and therefore emulsification are given as

$$\eta_0 G = 8T(P + 1)/b(19P + 16), \tag{6.5.1}$$

[150] P. Greguss, *Ultrasonics* **10**, 276 (1972).

[151] J. O. Hinze, *AIChE J.* **1**, 289 (1955).

[152] H. L. Goldsmith and S. G. Mason, The microrheology of dispersion, *in* "Rheology: Theory and Applications" (F. R. Eirich, ed.). Academic Press, New York, 1967.

[153] F. D. Rumscheidt and S. G. Mason, *J. Colloid Sci.* **16**, 238 (1961).

[154] G. I. Taylor, *Proc. R. Soc. London Ser. A* **138**, 41 (1932).

where P is the ratio of the viscosity η of the disperse medium to that of the continuous medium η_0; G the velocity gradient, T the interfacial tension, and b the radius of the undistorted drop.

Studies of particle size distributions and efficiency of emulsification processes have been conducted by Hislop,[146] Last,[147] and Rajagopol.[155] These studies involve microscopic examination of emulsions after formation by the various techniques described above. Particular care was taken to sample a sufficient number of emulsions to achieve statistically significant results. The droplet sizes in the emulsions were found to be in the order of microns, ranging from 1 to 40 μm in diameter, depending on the procedure and duration of the emulsification process. Neduzhii[156-158] has found familiar results and also demonstrated that the emulsification rate was proportional to the intensity of the sound used above the threshold intensity for the onset of emulsification. It should be noted that prolonged sonification led to an increase in the mean size of the particles and a broadening of the size distribution.

6.5.4. Dispersal of Biological Cell Aggregates

A problem related to emulsification and dispersal of liquid systems is that of dispersal of aggregates of living cells without morphological damage or decrease in their viability. Sonic and ultrasonic techniques have been developed that have many advantages for such an application. Direct adaptation of some experimental methods used for emulsification already discussed was attempted by Mullaney et al.[159] Using leucocytes and Chinese hamster cells they conducted tests of the effectiveness of such sonification procedures. After preparation, cell clumps were dispersed with a Branson sonifier of a type described in Section 6.4.2.1. However, this procedure typically led to an unacceptable increase in the amount of fluorescent debris present in their automated analysis system.

In efforts to develop more controlled ultrasonic methods for cell aggregate dispersal, Williams et al.[160] chose to take advantage of the well-characterized shearing stresses associated with acoustic streaming (see Chapter 6.3). Two methods were developed. In the first, using the vibrating wire method described in Section 6.3.3.2, they dispersed bacterial

[155] E. S. Rajagopol, Proc. Indian Acad. Sci. Sect A **49**, 333 (1959).

[156] S. A. Neduzhii, Sov. Phys. Acoust. **9**, 99 (1963).

[157] S. A. Neduzhii, Sov. Phys. Acoust. **7**, 79 (1961).

[158] S. A. Neduzhii, Sov. Phys. Acoust. **7**, 209 (1961).

[159] P. F. Mullaney, L. S. Cram, and T. T. Trujillo, Acta Cytol. **15**, 217 (1971).

[160] A. R. Williams, D. A. Stafford, A. G. Callely, and D. E. Hughes, J. Appl. Bacteriol. **33**, 656 (1970).

cell aggregates and flocculated sludge. They were able to demonstrate the effectiveness of the technique and showed that the number of single viable cells increased with both sonic displacement amplitude and duration of treatment. Specifically, for an exposure time of 10 min using a displacement amplitude of 15 μm, they found a 260% increase in the number of viable colonies. Of course such dramatic increases do not continue indefinitely with increased sonification but are limited by saturation of the cellular system. Parry et al.[161] have demonstrated that cell aggregates obtained from cervical scrapings could be effectively dispersed using this method. Williams and Slade[162] modified the arrangement by using a long stretched wire driven at 18.5 kHz and demonstrated the successful dispersal of octets of Sarcina lutea. While these vibrating wire methods have proven to be effective, they do have the disadvantages that they treat only small samples to cells at one time and that for the long stretched wire unstable oscillations may occur, which can cause undesirably large shearing stresses near the wire, as demonstrated by Rooney.[120]

Brakey and Rooney[163] have developed experimental methods for cell aggregate dispersal that use a type of vortex whistle similar to those described in Section 6.4.2.2. Liquid containing the cell aggregates to be dispersed is forced by a hydraulically driven syringe to flow through a needle; then it tangentially enters a large swirl chamber, passes through a region of smaller diameter, and finally exits through a small orifice. The acceleration of the liquid in the whistle causes sufficiently high velocities and associated gradients to disperse the cells. The dispersal efficiency was found to be greatly increased by the use of an insert within the swirl chamber, as well as by machining a thread on its surface. These measures limit the trajectories of the cells within the whistle to regions of high shearing stress. The most effective design developed for this whistle is shown in Fig. 7.

In tests of the effectiveness of aggregate dispersal, the vortex whistle was used successfully to disperse 95% of the samples of the aggregates of the algae Scenedesmus and monkey kidney cells grown in tissue culture after a single pass through the whistle. Light microscopy and cell growth studies indicated that the cells were morphologically normal and completely viable. Further experiments on monkey kidney cells demonstrated that cells could be efficiently dispersed without structural changes being detectable when using electron microscopic techniques.

[161] J. S. Parry, B. K. Cleary, A. R. Williams, and D. M. D. Evans, Acta Cytol. 15, 163 (1971).

[162] A. R. Williams and J. S. Slade, Ultrasonics 9, 85 (1971).

[163] M. W. Brakey and J. A. Rooney, Digest Int. Conf. Med. Biol. Eng., 11th, Ottawa p. 80 (1976).

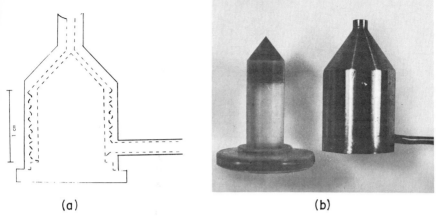

(a)   (b)

FIG. 7. Vortex whistle: (a) Schematic diagram of whistle with insert; (b) photograph of vortex whistle with insert removed. Tangential entrance tube to large diameter swirl chamber shown to the rear of the stainless steel section.

6.6. Atomization and Droplet Formation

6.6.1. Introduction

That it is possible to form a fog or mist ultrasonically was first discovered by Wood and Loomis[164] in 1927. Since then there have been extensive studies and debate about the mechanisms causing the process as well as development of various types of instrumentation for atomization. The principal advantages of the ultrasonic methods for atomization are that the sprays produced have drops that are relatively uniform in size and that the drops have low inertia; therefore they can be easily entrained in a gas flow and transferred from the region of atomization. The methods for ultrasonic atomization, as well as discussion of techniques to investigate the mechanisms, have been summarized by Topp and Eisenklam[165] and by Eknadiosyants.[166]

6.6.2. Methods of Ultrasonic Atomization

Ultrasonic atomizers may be classified into two general types, which include conventional, electrically driven, ultrasonic systems and gas-driven whistles. The former can operate at both high and low frequencies, depending on the experimental arrangement and application. Data on de-

[164] W. R. Wood and A. L. Loomis, *Philos. Mag.* **4**, 417 (1927).
[165] M. N. Topp and P. Eisenklam, *Ultrasonics* **10**, 127 (1972).
[166] O. K. Eknadiosyants, Aerosol production, *in* "Physical Principles of Ultrasonic Technology" (L. D. Rozenberg, ed.) Vol. 1. Plenum, New York, 1973.

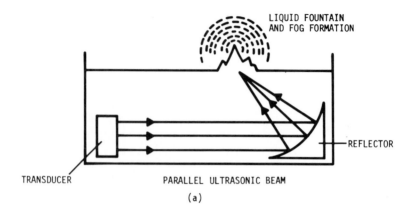

TRANSDUCER   PARALLEL ULTRASONIC BEAM

(a)

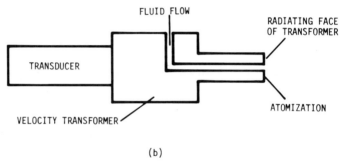

(b)

FIG. 8. Schematic diagrams of two types of ultrasonic atomizers: (a) Focused sound atomizer typical of higher frequency uses; (b) Flowthrough step-horn-type of atomizer used at lower ultrasonic frequencies. [Reproduced from *Ultrasonics* **14**, No. 5 (September 1976), published by IPC Science and Technology Press, Guildford, Surrey, England.]

sign and performance of atomizers that have been used experimentally or are commercially available have been summarized by Topp.[167]

6.6.2.1. Electrically Driven Atomizers. Electrically driven ultrasonic atomizers use both magnetostrictive and piezoelectric transducers. Typically the magnetostrictive laminates are used in the frequency range from 20 to 40 kHz and the piezoelectric ceramics from 30 kHz to 5 MHz. Two types of experimental arrangements are shown in Fig. 8.

For high-frequency applications, the arrangement will be similar to that shown in Fig. 8a. The transducers are usually small at higher frequencies (0.8–5 MHz), but the drop sizes are correspondingly smaller with a range

---

[167] M. N. Topp, *J. Phys. E* **3**, 739 (1970).

typically from 1 to 5 μm in diameter. Because of the relatively small amounts of mist produced by these systems, their use has been for experimental purposes and medical applications, such as nebulizers. For the latter the small-diameter droplets prove to be particularly effective because they can easily penetrate the smaller airways of the lungs.[168]

Although liquids can be atomized directly from the surface of a high-frequency transducer, the amplitude necessary to achieve the same effect at lower frequencies can only be achieved by using high-power magnetostrictive transducers.[169] Therefore, resonant horns must be designed to achieve the necessary displacement amplitudes, typically greater than 3 μm. This requirement of using resonant systems places a restriction on the range of frequencies used for a given arrangement and, therefore, on the droplet sizes that can be produced. A schematic diagram of a low-frequency atomizer is shown in Fig. 8b. A transducer drives a step horn that provides sufficient displacement amplification to cause atomization. The liquid input is made at a nodal position on the horn; the emerging liquid spreads out on the face of the horn by capillary action and is atomized by the longitudinal oscillations of the thin fluid film. These atomization systems have a variety of commercial and experimental uses, including improvement of fuel oil burners, production of metal powders, and processing of liquids.

6.6.2.2. Sonic Whistle Atomizers. A second method for production of sprays uses sonic whistles. These whistles are similar to the liquid-driven ones discussed in Section 6.5.2.2. They are based on the Hartmann whistle or the related stem-jet generator, and an example is shown in Fig. 9. In this type of whistle the vibrations are produced by directing high-pressure gas down the center of the device into a resonant cavity whose dimensions determine the frequency of operation. The liquid to be atomized is injected into the region near the cavity through symmetrically arranged orifices or an annular slit.

Because the atomization efficiency of whistles decreases with increasing frequency, these devices typically operate in the low-ultrasonic range with delibrate effort being made to design nozzles operating above the audible range to reduce the nuisance of noise. Many nozzles have been designed so that atomization can be done on a wide variety of liquids with drop diameters ranging from one to several hundred microns and with flow rates up to 1.5 kg sec^{-1}. Drop size in these systems in a function of

[168] R. F. Goddard, T. T. Mercer, P. O'Neil, R. L. Flores, and R. Sanchez, *J. Asthma Res.* **5**, 355 (1968).

[169] A. E. Crawford, *J. Acoust. Soc. Am.* **27**, 176 (1955).

[170] Heat Systems–Ultrasonics, Inc., Sonomist Ultrasonic Spray Nozzle, Publ. HS-Son-3. Plainville, New York, 1978.

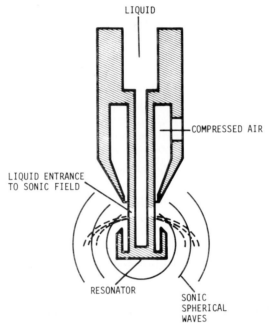

FIG. 9. Sonic whistle atomizer. (Reproduced with permission from Heat Systems–Ultrasonics.[170])

flow rate, driving pressure, orifice size, and resonator position. The atomization process is uniform for a fixed combination of parameters and, therefore, there is closer control of particle size than with other atomization methods.[171]

6.6.3. Investigation of Mechanisms for Ultrasonic Atomization

After the observation of ultrasonic atomization by Wood and Loomis,[164] Sollner[172] was the first to attempt to explain the phenomenon. He believed that cavitation occurred directly below the surface of the liquid that resulted in the formation of the fog or mist. However, work by Bisa et al.[173], Sorokin,[174] and Eisenmenger[175] indicated that instabilities in surface waves were the mechanism involved. Thus, for many years there

[171] E. G. Lierke and G. Greisshamer, *Ultrasonics* **5**, 224 (1967).
[172] K. Sollner, *Trans. Faraday Soc.* **32**, 1532 (1936).
[173] K. Bisa, K. Dirnogl, and R. Esche, *Siemens Z.* **28**, 341 (1954).
[174] V. I. Sorokin, *Sov. Phys. Acoust.* **3**, 281 (1957).
[175] W. Eisenmenger, *Acustica* **9**, 327 (1959).

were two rival theories.   The capillary wave theory is based on a Taylor instability of a liquid–gas interface exposed to a force perpendicular to it, as described by Peskin and Raco.[176]   In this theory the capillary waves produce drops at their crests, whose size is proportional to wavelength. The droplets are essentially monodispersed and have little momentum. In contrast, the cavitation theory developed by Eknadiosyants[177] postulates that the hydraulic shocks associated with collapse cavitation cause the atomization.   These two theories have now been related by Boguslavskii and Eknadiosyants[178] by the suggestion that shocks from cavitation events can interact with the finite-amplitude capillary waves, thereby forming drops in the same way.   In particular, cavitation is important in so-called "fountain atomization" such as used in medical inhalents but not in "liquid-layer atomization" as used in oil-fired burners.   In either case the drop size can be estimated from the equation derived by Lang[179]:

$$D = 0.34(8\pi\gamma/\rho f^2)^{1/3},   (6.6.1)$$

where D is the droplet diameter, γ the surface tension, ρ the fluid density, and f the driving frequency.

Experimental tests of the various theories of atomization have generally used two methods.   The first uses high-speed photomicrography to examine details of the droplet formation.   Experiments using this technique to support the capillary wave theory include those conducted by Lang,[179] as well as by Fogler and Timmerhaus.[180]   Related experiments have been performed by Topp,[181] Bassett and Bright,[182] and Chiba.[183] These latter investigators present photographic evidence that both surface waves and cavitation play roles in atomization.

The second experimental method involved the examination of droplets formed.   For this purpose, Lang[179] atomized molten wax.   The wax droplets solidified quickly in flight and were collected and sized.   Investigations by Lierke and Greisshamer[171] have used a variety of molten metals. The particles obtained were characterized by sieving and both optical and electron microscopy.   Particle sizes measured were in agreement with Eq. (6.5.1).

It should be noted that details of the atomization mechanism occurring

[176] R. L. Peskin and R. J. Raco, *J. Acoust. Soc. Am.* **35**, 1378 (1963).
[177] O. K. Eknadiosyants, *Sov. Phys. Acoust.* **14**, 80 (1968).
[178] Y. Y. Boguslavskii and O. K. Eknadiosyants, *Sov. Phys. Acoust.* **15**, 14 (1969).
[179] R. J. Lang, *J. Acoust. Soc. Am.* **34**, 6 (1962).
[180] H. S. Fogler and K. D. Timmerhaus, *J. Acoust. Soc. Am.* **39**, 515 (1966).
[181] M. N. Topp, *J. Aerosol Sci.* **4**, 17 (1973).
[182] J. D. Bassett and W. W. Bright, *J. Aerosol Sci.* **7**, 47 (1976).
[183] C. Chiba, *Bull. JSME* **18**, 376 (1975).

in whistles are not totally understood. While they are probably similar to those associated with the Taylor instability in capillary waves, Wilcox and Tate[184] have demonstrated that for certain whistle atomizers operating under specified conditions the presence of the sound field played little or no part in the atomization process. In contrast, for fuel oil atomization the presence of the sound field greatly enhances the process.[185]

6.6.4. Droplet Production

For a variety of experiments including studies of the interactions of droplets or sorting various types of materials, it is desirable to have a controlled method for the production of uniform single large droplets rather than a cloud or mist. Ultrasonic methods similar to those discussed in the topic of atomization are capable of such droplet production. The essential ideas are that if periodic oscillations are applied to a nozzle, pressure and velocity fluctuations will be produced at the orifice, which will cause minute perturbations in the surface of the liquid cylinder leaving the orifice. For a velocity of the jet v and period of vibration T, the wavelength of these disturbances λ is described by the usual equation $\lambda = vT$. If λ exceeds the circumference of the jet, then Rayleigh[99,186] has shown that the disturbance grows exponentially until the column divides into detached droplets separated by an interval λ and passing a fixed point with velocity v and frequency $1/T$. Considering the energy relationships, he showed that the instability that grows most rapidly has a wavelength λ_m such that

$$\lambda_m = 9.016b, \tag{6.6.2}$$

where b is the radius of the jet. The mass of fluid M associated with this disturbance can be expressed as

$$M = \pi b^2 \rho \lambda, \tag{6.6.3}$$

where ρ is the density of the fluid in the jet. Therefore, the radius r of the droplet formed is given by

$$r = (3b^2\lambda/4)^{1/3}. \tag{6.6.4}$$

Upon substitution of λ_m, we find that the radius of the droplets formed is

[184] R. L. Wilcox and R. W. Tate, *AIChE J.* **11,** 69 (1965).
[185] A. Clamen and W. H. Gauvin, *Can. J. Chem. Eng.* **46,** 223 (1968).
[186] J. W. S. Rayleigh, *Proc. London Math Soc.* **10,** 4 (1979).

just

$$r = 0.946A, \qquad (6.6.5)$$

where A is the diameter of the fluid jet. Equation (6.6.5) has been found to be accurate to within 1/3% by Lindblad and Schneider.[187]

A variety of experimental methods for droplet production have been developed. These include the method of Lindblad and Schneider,[187] who used a capillary tube orifice driven by a PZT bimorph transducer at audio frequencies to produce drops of various sizes and corresponding rates. In the higher-frequency range Fulwyler[188] ultrasonically drove a nozzle at 40 kHz to produce 40,000 uniform droplets per second. An additional option available to the experimenter is to charge the droplets formed and selectively deflect them at some later time using electric fields. Such systems have provided the basis for biological cell sorting methods,[189] as well as for the jet ink printer.[190]

When considering the use of such methods, the experimenter should be aware of the influence of other parameters on the systems involved. For example, work by Stovel[191] has demonstrated the influence that particles in the liquid have on jet breakoff. When using a 50-μm-diam nozzle driven at 40 kHz, he found that particles in the liquid whose diameters were about 2 μm had no effect on droplet formation, those 20 μm in diameter totally disrupted normal breakoff, and those in the range from 7 to 10 μm in diameter caused erratic droplet formation.

Klopovsky and Fridman[192] have shown that the variation of the driving frequency from that predicted by the Rayleigh relation can also affect the efficiency of drop formation. At the optimal frequency jet disintegration is periodic and monodispersed drops are formed at a distance of from 2 to 3 cm away from the nozzle. An increase in frequency of forced vibrations by 15–25% will result in the formation of drops of uniform size. However, they begin to move in pairs and may coalesce, becoming nonuniform. A decrease in the driving frequency by the same amount will result in the formation of small "satellite" droplets between the larger ones. The formation of these satellites causes the major drops to differ from spherical form and to change their form in flight. The amount of deviation from optimal driving frequency permissible will depend on the particular liquid used.

[187] N. R. Lindblad and J. M. Schneider, *J. Sci. Instrum.* **42**, 635 (1965).
[188] M. J. Fulwyler, *Science* **150**, 910 (1965).
[189] L. A. Herzenberg, R. G. Sweet, L. A. Herzenberg, *Sci. Am.* **234**, 108 (1976).
[190] J. D. Beasley, *Photogr. Sci. Eng.* **21**, No. 2 (1977).
[191] R. T. Stovel, *J. Histochem. Cytochem.* **25**, 813 (1977).
[192] B. A. Klopovsky and V. M. Fridman, *Ultrasonics* **14**, 107 (1976).

6.7. Acoustic Agglomeration

6.7.1. Introduction

In 1874 Kundt and Lehmann[193] observed that sound can cause aggregation of particles suspended in a fluid medium. Observations of agglomeration of particles in air were made by Carwood and Patterson[194] in
1931. Detailed studies of the mechanisms involved in the phenomenon
were conducted by Brandt and Hiedemann,[195] as well as by Sollner and
Bondy.[196] Perhaps the best summary of work in the area was prepared by
Mednikov.[197] The development of the field has also been reviewed by
Shirokova.[198] In this section we briefly consider experimental methods
for studying the phenomenon and possible mechanisms for sonic agglomeration.

6.7.2. Experimental Approaches to the Study of Agglomeration

The experimental methods used to investigate both the physical mechanisms involved in agglomeration and empirical improvement of its efficiency are similar to those used in related nonacoustic studies. The degree of agglomeration and characterization of physical properties of the
particles formed have been studied by measurement in changes in transmitted light intensities, high-speed photography, and both optical and
electron microscopy. Systems studied range from fogs (nearly monodispersed, 0.2 μm) to sulfuric acid mist (particle sizes from 5 to 100 μm).
Also, a variety of frequency and sonic exposure conditions have been
studied; results are reviewed in Mednikov[197] and Shirokova.[198] In this
section we shall merely indicate some of the breadth of experimental
methods applied to this nonlinear phenomenon.

A typical example is that of Volk and Moroz.[199] The investigators were
interested in determining the behavior of carbon black particles in a sound
field. Separate experiments were conducted to determine the effect of
variation of parameters including sound pressure level, frequency, particle mass-loading of the air, and exposure time. Using several of the
techniques already mentioned, they found that increasing mass-loading,

[193] A. Kundt and O. Lehmann, *Ann. Phys.* (Leipzig) **153**, 1 (1874).

[194] W. Carwood and H. S. Patterson, *Nature (London)* **127**, 150 (1931).

[195] O. Brandt and E. Hiedemann, *Trans. Faraday Soc.* **32**, 1101 (1936).

[196] K. Sollner and C. Bondy, *Trans. Faraday Soc.* **32**, 616 (1936).

[197] E. P. Mednikov, "Acoustic Coagulation and Precipitation of Aerosols." English trans-
lation by Consultants Bureau, New York, 1966.

[198] N. L. Shirokova, Aerosol coagulation, in "Physical Principles of Ultrasonic Technology" (L. D. Rosenberg, ed.). Plenum, New York, 1973.

[199] M. Volk, Jr. and W. J. Moroz, *Water Air Soil Pollut.* **5**, 319 (1976).

sound pressure level, and exposure time resulted in increased agglo-
merate size. Optimum growth of agglomerates occurred at a frequency
of 3 kHz.

A study to characterize the structure and physical parameters of particle aggregates of industrial aerosols was conducted by Belen'kii et al.[200]
Using optical and electron microscopy, as well as x-ray analysis techniques, they demonstrated that there is a linear decrease in the actual density of the aggregates as their size increases. In fact, the density can
differ by two orders of magnitude from the true material density of the initial particles, thus significantly altering the mobility of the aggregates and
their displacement amplitudes in the sound field.

A different approach to the problem has been taken by Scott,[201] who
has investigated the efficiencies of agglomeration using various types of
acoustic waveforms. In particular, he explored the advantages of
pulse-jet sound generation. His studies using large-scale facilities indicate that the acoustic technique may be economically feasible.

Acoustic agglomeration not only occurs at low frequencies but can indeed cause important effects at higher frequencies. For example, several
investigators[202-204] have reported the clumping red cells within ultrasonic
fields in the megahertz frequency range. Gould and Coakley[205] investigated the mechanisms involved in these and related interactions in
sound fields of small particles in aqueous systems using optical microscopy and photography techniques. Reasonable agreement was obtained
between the predicted and observed thresholds for particle striation and
for rate of migration of polystyrene spheres. The role that radiation force
plays in the banding of erythrocytes has been studied by ter Haar and
Wyard.[206] They derived the equation of motion for a particle in a stationary volume of fluid in a standing wave field to be

$$X = \frac{1}{k} \arctan \left[\tan kX_0 \exp \left(- \frac{k\Omega t}{6\pi\eta a} \right) \right], \qquad (6.7.1)$$

$$\Omega = \frac{VP^2 k}{4\rho_0 C_0^2} f \left(\frac{\rho}{\rho_0} \right), \qquad (6.7.2)$$

$$f \left(\frac{\rho}{\rho_0} \right) = \left[\frac{\rho_0 C_0^2}{\rho C^2} - \left(\frac{5\rho - 2\rho_0}{3\rho + \rho_0} \right) \right], \qquad (6.7.3)$$

[200] V. A. Belen'kii', V. A. Sopronov, and V. I. Timoshenko, Sov. Phys. Acoust. 22, 275
(1976).

[201] D. S. Scott, J. Sound Vib. 43, 607 (1975).

[202] M. Dyson, B. Woodward, and J. B. Pond, Nature (London) 232, 572 (1972).

[203] U. Abdulla, D. Talbot, M. Lucas, and M. Mullarkey, Br. Med. J. 3, 797 (1972).

[204] N. V. Baker, Nature (London) 239, 398 (1972).

[205] R. K. Gould and W. T. Coakley, Proc. Symp. Finite-Amplitude Wave Effects in Fluids,
Copenhagen p. 252 (1973).

[206] G. ter Haar and S. J. Wyard, Ultrasound Med. Biol. 4, 111 (1978).

where X_0 is the particle position at time $t = 0$, P the pressure amplitude, V the volume of sphere of radius a, k the wave number, η the fluid viscosity; ρ, ρ_0, C, C_0 are, respectively, the densities and velocities of sound for the spherical particle and suspending medium.

Studies of the aggregation of polystyrene spheres in the megahertz frequency range have also been performed by Miller.[207]

6.7.3. Mechanisms Relevant to Agglomeration

A variety of mechanisms have been suggested for important roles in the process of acoustic agglomeration. One is orthokinetic differential motion. An orthokinetic interaction occurs between two suspended particles of different aerodynamic sizes when they are located with an initial separation approximately equal to the displacement amplitude of the sound field in the suspending medium and their relative motion is parallel to the direction of vibration. Brandt *et al.*[208] suggest that because of the differential fluid and inertial forces they experience, the particles vibrate with different amplitudes and phases. Such differential motion greatly increases the probability of collision and, therefore, of agglomeration. Radiation forces that cause particles to move toward velocity nodes or antinodes have been suggested by St. Clair[209] as possibly playing a role in agglomeration because increased concentration in such regions would enhance orthokinetic interactions. Acoustic streaming and its associated mass flow have been suggested by Statnikov and Shirokova[210] as playing a role in agglomeration. The theory for hydrodynamic particle interaction has been developed by Timoshenko,[211] while other investigators including Richardson[212] have emphasized the role of turbulence in promoting interactions between the particles. Thus there are many postulated mechanisms and probably more than one of these mechanisms play important roles in the agglomeration process.

6.8. Acoustic Drying

6.8.1. Introduction

With the reporting of the use of ultrasound to dry quartz by Burger and Sollner[213] in 1936, a new nonlinear phenomenon and experimental tech-

[207] D. L. Miller, *J. Acoust. Soc. Am.* **62**, 12 (1977).
[208] O. Brandt, H. Freund, and E. Hiedemann, *Kolloid Z.* **77**, 103 (1936).
[209] H. W. St. Clair, *Ind. Eng. Chem.* **41**, 2434 (1949).
[210] Y. G. Statnikov and N. L. Shirokova, *Sov. Phys. Acoust.* **14**, 118 (1968).
[211] V. I. Timoshenko, *Appl. Acoust.* **1**, 200 (1968).
[212] E. G. Richardson, *Acustica* **2**, 141 (1952).
[213] F. J. Burger and K. Sollner, *Trans. Faraday Soc.* **32**, 1598 (1936).

nique became known. Since then a series of investigators have found that the use of ultrasound during the drying process can increase the rate of drying, aid the drying of heat-sensitive materials, and decrease the final moisture content of the sample. However, because of high power demands, ultrasound is currently useful only for the drying of high-cost materials for experimental purposes. A summary of the scientific work as well as practical applications of acoustic drying has been prepared by Borisov and Gynkiva.[214]

6.8.2. Experimental Methods

Experimental objectives associated with ultrasonic drying are of three basic types: empirical determination of efficient drying parameters, careful studies of drying processes to determine the mechanisms involved, and determination of the optimal combination of sound with a conventional drying techniques. We shall briefly describe examples of experimental methods, but the reader is referred to the work of Borisov and Gynkiva[214] for a review of experiments.

Fairbanks[100,101,215] has conducted several studies of the acoustic drying of coal dusts. His experiments are done at 12 kHz and at sound levels up to 150 dB (relative to 2×10^{-4} dyne/cm²). He has studied the efficiency of drying as a function of intensity, comparing the rates of drying to those of the nonsonated control sample. His work has included studies of the use of sound with various types of drying apparatus, including tunnel and rotary drying. He concludes that ultrasound increases the rate of drying of coal dust and that its efficiency increases as the size of the coal particles decreases.

An example of the experimental methods used to investigate the mechanisms involved in sonic drying is that of Borisov and Gynkiva.[214] They conducted their investigations in standing wave fields at a frequency of 2.5 kHz in a rectangular tube of 4×4 cm cross section. The samples to be dried were disks or cylinders of kaolin ceramic. To study the role of streaming they observed the rate of drying of their samples when placed in different positions within the sound field and also as a function of orientation of the samples. These experiments clearly demonstrated the important role that acoustic streaming plays in the drying process.

Experiments by Shatalov[216] are typical of those that study sound combined with drying processes. In his experiments he used 20-kHz, 150-dB (relative to 2×10^{-4} dyne/cm²) ultrasound to increase the drying rate of

[214] Y. Y. Borisov and N. M. Gynkiva, Acoustic drying, in "Physical Principles of Ultrasonic Technology" (L. D. Rosenberg, ed.) Vol. 2, Plenum, New York, 1973.
[215] H. V. Fairbanks, Proc. Ultrason. Int. p. 43 (1975).
[216] A. L. Shatalov, Electrochem. Ind. Process. Biol. (GB) 5, 37 (1976).

silica gel samples by microwaves. He found that the rate of ultrasonically assisted drying was 35% greater than that of the nonsonated control.

6.8.3. Possible Mechanisms of Acoustically Assisted Drying

Although the mechanisms for acoustic drying are not fully understood,
several theories have been advanced to explain what occurs in the acoustically assisted drying process. Drying of a material in which the moisture is held only by physical absorption can be described by using two distinct drying periods. The first period is characterized by a single drying
rate and continues until a critical moisture content of the material is
reached. The second is one in which the rate decreases continuously in
time, and it begins when the surface from which the moisture is evap-
orating begins to become dry. Thereafter, the area of moist surface continues to decrease until moisture for further evaporation comes solely
from within the capillaries of the material being dried.

As expected, during the constant rate period, the drying rate is independent of the material being dried. However, the rate does depend on
the diffusion rate of vapor through the liquid interface into the air, relative
humidity and velocity of the air above, and temperature at the evap-
orating surface. Boucher[217] and Soloff[218] have shown that during this
first part of the drying sequence the rate depends on the area exposed to
drying, a mass-transfer coefficient, and the ratio $\Delta P/P$, where ΔP is the
difference between the saturation vapor pressure of water at the liquid
temperature and the partial pressure of water vapor in air, with P the gas
pressure in the surrounding atmosphere.

When considering ultrasonically assisted drying, different mechanisms
probably dominate the effect in the two different drying periods. Historically, Boucher[217] and Soloff[218] have described the accelerated drying by
considering changes in the constant rate period of drying. Soloff[218] believed that the increased velocity of the air above the wet surface caused
changes in the mass-transfer coefficient, which resulted in increased
drying rates. Boucher[217] favored the explanation that the sound influenced the term $\Delta P/P$, but Borisov and Gynkiva[214] showed that this effect is small and that acoustic streaming plays an important role in the first
phase of drying.

Ultrasound can also increase the rates of the second phase of the drying
process and the mechanisms for this effect are not well understood.
Greguss[219] believes that the acceleration of the second phase is due to

[217] R. M. G. Boucher, *Ultrason. News* **3**, 14 (1959).
[218] R. S. Soloff, *J. Acoust. Soc. Am.* **36**, 961 (1964).
[219] P. Greguss, *Ultrasonics* **1**, 83 (1963).

sonically caused reduction in viscosity and increase in diffusivity of the liquid phase, heating of small bubbles within capillaries in the solid phase, which creates a pressure to force out the moisture, or creation of radiation pressure gradients across water filaments causing them to move toward the surface. On the other hand, Borisov and Gynkiva[214] tend to emphasize sonically produced temperature elevations as the important mechanism for drying.

6.9. Ultrasonic Fatigue Testing

6.9.1. Introduction

Although the phenomenon of fatigue plays a large role in the majority of failures of materials in use, studies of the details of fatigue failure have been hampered by the time required to conduct fatigue tests. Most fatigue tests are made at cyclic frequencies in the range from 25 to 200 Hz. Conventionally, fatigue is expressed by relating the number of cycles before failure to the stress range used. For example, if a sample has a life of 10^6 cycles, it would take nearly 3 hr to fail at 100 Hz. Thus, in order to study a material that has a life of 10^8 cycles, it would be desirable to have an experimental method for gathering data quickly, particularly since variations in crack initiation and propagation require repetition of the tests to achieve statistically significant results. Therefore, ultrasonic fatigue testing methods have been developed. Problems exist in that the life of the sample may depend on the testing frequency and fatigue mechanisms may be frequency dependent. However, materials are tested at both the low- and high-frequency ranges in order to understand fatigue mechanisms and to provide efficient testing of materials.

6.9.2. Experimental Methods

6.9.2.1. Experimental Arrangements.
One problem in making measurements of fatigue using ultrasound is in obtaining sufficient strains in the materials being tested. An arrangement developed by Mason[220] to overcome this difficulty is shown in Fig. 10. It consists of a PZT-4 ceramic transducer resonant at 22 kHz driven in a longitudinal mode. A pickup electrode measures the strain in the ceramic. Cemented to the transducer is a mechanical transformer to increase the displacement amplitudes and strains in the material to be tested. The test specimen is attached to the far end of the transformer as shown in the figure. The com-

---

[220] W. P. Mason, *J. Acoust. Soc. Am.* **28**, 1207 (1965).

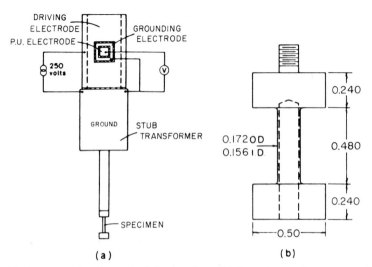

FIG. 10. Apparatus for ultrasonic fatigue testing. (a) Ceramic transducer, step-velocity transformer, and sample (b) Ceramic transducer, exponential horn, and sample. (Reproduced with permission from Mason.[221])

bined configurations of transformer and test specimen produce a gain of over 60 in strains that can be produced in the sample. For the configuration shown, the ratio of the voltage applied to the ceramic to the pickup voltage can be used to measure the internal friction of the entire system. The strain in the sample is directly related to the pickup voltage and can be calibrated by attaching a strain gage to the sample. The basic measurements to be made on the sample using this system include determination of the resonant frequency and Q as a function of strain. The modulus defect, defined to be the change in resonant frequency as a function of the magnitude of the strain, is related to the average increase per cycle of the anelastic strain due to dislocation displacements. The internal friction is just the reciprocal of the Q of the sample. It is related to the energy losses in the sample and measured by measuring the resonant frequency and frequencies at which the pickup voltage is 3 dB less than that at resonance. The dislocation distribution can also be evaluated from the internal friction data if it assumes the existence of a discrete number of dislocation loop sizes arranged in logarithmic intervals. Details of such measurements have been described by Mason.[220–222] Similar techniques

[221] W. P. Mason, Low- and high-amplitude internal-friction measurements in solids and their relationship to imperfection motions, in "Microplasticity" (C. J. McMahon, Jr., ed.), pp. 187–363. Wiley (Interscience), New York, 1968.
[222] W. P. Mason, J. Acoust. Soc. Am. 28, 1197 (1956).

have been described by Hockenhull *et al.*,[223] as well as Neppiras,[224] and reviewed in .[225] Related work on nonlinear elastic properties and damping at high strains has been done by Kuzmenko.[226]

6.9.2.2. Typical Experimental Results. A considerable amount of research has been completed on ultrasonic fatigue testing of materials since a large number of cycles can be obtained in a short time. The question arises whether ultrasonic measurements give the same results as those obtained by lower-frequency methods.

The general opinion is that ultrasonic measurements give higher fatigue stresses than those obtained at lower-frequency testing. For iron, however, the result is the opposite.[227] For brass, measurements have been made at both low and high frequencies, and it has been demonstrated that details of the fatiguing mechanisms are different for the two frequency regimes.[228] Results obtained on other metals have shown that ultrasonic vibrations are more damaging than those of lower frequencies. This result was confirmed by Werner[229] for the case of steel used in railway wheels. A fatigue mechanism involving the movement and pinning of dislocations has been found for titanium and other high-strength materials.[230,231]

In summary, ultrasonic fatigue testing has been demonstrated to be a useful experimental technique. The method permits the study of fundamental fatigue processes as well as routine evaluation of fatigue properties of materials. Further studies, such as those of Bajons *et al.*,[232] are required to develop simplified equipment and to demonstrate the practical limits of this method.

6.9.3. Mechanisms of Fatigue

The models used to describe fatigue in metals have been discussed by Mason.[220-222] The fatigue process consists of three phases. The first phase is within the elastic limits of the material and all elastic strains as well as those associated with dislocations are in phase with the applied stress.

[223] B. S. Hockenhull, C. N. Owston, and R. G. Hacking, *Ultrasonics* **9**, 26 (1971).

[224] E. A. Neppiras, *Proc. ASTM* **59**, 691 (1959).

[225] B. S. Hockenhull and R. G. Hacking, *Proc. Int. Conf. High Power Ultrason.*, 1st p. 8 (1970).

[226] V. A. Kuzmenko, *Ultrasonics* **13**, 21 (1975).

[227] W. A. Wood and W. P. Mason, *J. Appl. Phys.* **40**, 4514 (1969).

[228] W. P. Mason, *Eng. Fract. Mech.* **8**, 89 (1976).

[229] K. E. T. R. Werner, *Eisenbahntech. Rundsch.* **4**, 1 (1973).

[230] W. Kromp, K. Kromp, H. Bitt, H. Langer, and B. Weiss, *Proc. Ultrason. Int.* p. 238 (1973).

[231] S. Purushothaman, J. P. Wallace, and T. K. Tien, *Proc. Ultrason.* p. 224 (1973).

[232] P. Bajons, K. Kromp, W. Kromp, H. Langer, B. Weiss, and R. Stickler, *Proc. Ultrason. Int.* p. 95 (1975).

The second phase is associated with anelastic behavior of the material. Here, coupled dislocations modeled by Frank–Read loops[233] are formed but are restrained by dislocations in other slip planes, impurity atoms, or similar obstructions. During this phase all loops formed are relieved during the low-stress parts of the sound cycle and there is no permanent change in the material. The distribution of dislocation loop lengths as a function of loop length may be calculated from the shape of the internal friction versus strain amplitude plot. Results from the ultrasonic data agree well with other methods of evaluating this distribution.

In the third phase, the ultrasonically produced strains are of sufficient magnitude that the coupled dislocations are no longer restrained and defects, such as vacancies or interstitial siting of atoms, are produced. When a sufficient number of such vacancies are formed, they coalesce to produce a fatigue crack.[234] It appears that the high mobility necessary for the diffusion of defects during the time of a cycle of the sound field is a result of the lowering of the activation energy for the process by the applied stress.

6.10. Ultrasonic Processing of Materials

6.10.1. Introduction

The use of high-amplitude ultrasound with displacement amplitudes up to 150 μm has proven to be practical for a variety of both commercial and experimental methods to improve the processing of materials. Ultrasound has been demonstrated to be useful in the welding of plastics and other materials, the machining of special materials, as well as other specialized applications. Some of these methods have been explored in detail while others are at the experimental stage. In this section we shall provide a brief introduction to these specialized ultrasonic methods.

6.10.2. Ultrasonic Welding

Ultrasonic welding has been used to join a variety of materials, as summarized by Kholopov.[235] The application of ultrasound produces welds of high quality and for many applications the method is commercially feasible. Ultrasonic welding can replace other mechanical methods of jointing and adhesion bonding of certain polymers. The weldability of these polymers has been determined to be directly proportional to the

[233] W. T. Read, "Dislocations in Crystals." McGraw-Hill, New York, 1953.
[234] N. F. Mott, *J. Phys. Soc. Jpn.* **10**, 650 (1955).
[235] Y. V. Kholopov, *Weld. Prod. (USSR)* **22**, 56 (1975).

modulus of elasticity, heat conductivity, and frictional coefficient and inversely proportional to the density, heat capacity, and melting temperature. Welding efficiency also depends on the geometric shape and dimensions of the components, on the gap between the components, their distance from the welding tips, and the shape of the contact zone.

6.10.2.1. Plastics Welding. The basic mechanism involved in plastics welding is ultrasonically produced heat, which causes the plastics to melt to form the weld. Ideally, the heat is produced only at the interface. The process is therefore efficient and causes minimal distortion and material degradation.

Most thermoplastics have characteristics that are suitable for ultrasonic welding and these have been summarized by Shoh.[236] Such properties as the ability to transmit and absorb vibration, low melting temperature, and low thermal conductivity are important for ultrasonic welding. Welding of parts at a distance from the ultrasonic horn requires that the material not have excessive attenuation. In order to increase the efficiency of the welding process, parts to be welded are designed so that some of the parts are melted away and the melted plastic flows along the desired interface.[236,237]

The equipment used for plastics welding consists of standard transducers, velocity transformers to attain large displacement amplitudes, and various types of horn configurations to increase the efficiency of the welding process.[236] The power densities used are typically on the order of 1000 W cm^{-2}. The equipment also uses automatic frequency control, methods to maintain a constant displacement amplitude,[236] and techniques for regulating the temperature in the region of the weld.[238] Welding equipment can use normal longitudinal motion of the horn tip or, as seen in Fig. 11, shear motion. Details of the welding process have been elucidated by experimenting with different materials, as well as high-speed film studies, such as those conducted by Mozgovi et al.[239] Discussion of welding problems associated with their plastic films has been given by Bogdashevaskii et al.[240] Ultrasonic welding has proven to be an efficient method of joining thermoplastics. In addition, the technique can be used for staking, insertion of metal parts in plastic, spot welding, scan welding, and the joining of woven and nonwoven synthetic fibers.[236]

[236] A. Shoh, *Ultrasonics* **14**, 209 (1976).
[237] A. N. Sovetov, S. S. Volkov, Y. N. Orlov, and B. Y. Chernyak, *Weld. Prod. (USSR)* **22**, 54 (1975).
[238] I. V. Mozgovi and L. V. Gonashevskii, *Autom. Weld. (USSR)* **28**, 62 (1975).
[239] I. V. Mozgovi et al., *Autom. Weld. (USSR)* **28**, 16 (1975).
[240] A. V. Bogdashevaskii, M. A. Karamishev, and V. S. Nemshilova, *Weld. Prod. (USSR)* **19**, 63 (1972).

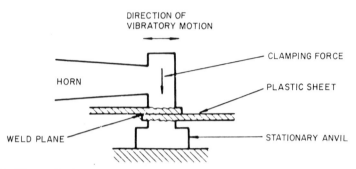

FIG. 11. Ultrasonic shear mode welding. (Reproduced with permission from Shoh.[236])

6.10.2.2. Welding Other Materials. Ultrasound has not only been used to assist conventional welding techniques but also has been shown to be capable of welding a variety of materials purely ultrasonically. Ultrasonic welding has the advantage that it does not produce many unwanted effects such as crack formation or warping. Absence of appreciable heating effects makes it a useful method for working with thin films or electronic components. Another advantage of the ultrasonic technique is the low residual stress level in the weld. In general, the bond formed has higher performance and better mechanical characteristics than conventionally welding bonds. These characteristics of ultrasonic welding of metals are described by Mitskevich.[241]

The basic equipment used for welding of materials other than plastics is similar to that discussed in the previous section. Variations on the methods discussed there include work by Pohlman and Leven[242] in which the anvil as well as the horn oscillate. This method has the advantage that better mechanical properties can be produced for the same energy input. The advantages of the ultrasonic techniques have been demonstrated by metallurgical methods, including tensile shear tests, scanning electron microscopy, and metallographic sectioning. Typical detailed discussion of these methods can be seen in Harthoorn's[243] report of ultrasonic joining of aluminum.

6.10.3. Ultrasonic Machining

Ultrasonic machining provides a unique experimental method to fabricate structures from hard brittle materials, heat resisting alloys, cemented carbides, etc., that are not machinable using conventional techniques.

[241] A. M. Mitskevich, Ultrasonic welding of metals, in "Physical Principles of Ultrasonic Technology" (L. D. Rozenberg, ed.) Vol. 1. Plenum, New York, 1973.

[242] R. Pohlman and D. Leven, Schweissen Schneiden 25, 81 (1973).

[243] J. L. Harthoorn, Ultrason. Int. Conf. Proc. p. 43 (1973).

The method has also proven useful for materials where deformation due to cutting forces is not desirable, such as glass or semiconductors. The equipment used is similar to that described in Section 6.10.2. Longitudinal vibrations produced by a transducer are communicated to the tool by using a velocity transformer, thus causing the tool to vibrate against the work piece. An abrasive slurry is used between the work piece and tool and a static force applied. As a result of the reciprocating action of the tool, the abrasive particles remove the material in precisely the shape of the working tool. The machining speeds depend on the amplitude of vibrations, material properties, type of abrasive used, and the rate of supply of the abrasive slurry to the work area.

It is not possible to summarize all the results of studies conducted on mechanisms involved in machining or optimization of ultrasound parameters. Historically, both kinetic bombardment of abrasive particles against the workpiece and cavitation erosion were considered as important mechanisms in the machining process. Now studies using high-speed photography of the machining process have elucidated the mechanism. It has been shown that the direct mechanical impact of the tool on abrasive particles in contact with the workpiece is the dominant cutting mechanism. Thus there is direct transmission of forces from the tool to the workpiece. Only slight plastic deformation occurs at the tool–particle interface since tools are made of tough ductile materials. However, if the workpiece is a brittle material, fracturing and chipping occurs caused by the forces applied by the abrasive particles. The machining is the result of the many abrasive particles and high frequency of impact. Neither the kinetic bombardment of particles as such nor cavitation erosion plays a significant role in the machining process.

Studies of the optimization of machining parameters including vibrational amplitude and frequency, the type and rate of flow of slurry, as well as magnitude of the static force used have been summarized by Graff.[244] The major advantages of ultrasonic cutting are the ability to cut hard and/or brittle materials and the ability to cut patterns of irregular shape. The disadvantages are the very slow cutting rate and the reduction of penetration rate with depth.

Developments on the topic of ultrasonic machining have been summarized by Graff,[244] Kazantsev,[245] and Markov.[246] Research efforts are typically aimed at examination of the importance of such parameters as static load, machining time, and type of abrasives. Detailed studies of

[244] K. F. Graff, *Ultrasonics* **13**, 103 (1975).

[245] V. F. Kazantsev, Ultrasonic cutting, *in* "Physical Principles of Ultrasonic Technology" (L. D. Rozenberg, ed.), Vol. 1. Plenum, New York, 1973.

[246] A. I. Markov, "Ultrasonic Machining of Intractable Materials." Iliffe Books, London (1966).

this type have been conducted by Adithan and Venkatesh,[247] Paustovskii et al.,[248] and Kubota et al.[249] Ultrasonic methods have also been developed to assist other machining techniques. Examples are the work of Komasa,[250] who employed ultrasonic cavitation to flush the dielectric solution in electric-discharge machining, and ultrasonically assisted twist drilling as summarized by Graff.[244]

6.10.4. Other Applications

The diversity of application of ultrasonic methods to ultrasonic processing of materials can best be demonstrated by citing some examples from several areas. Techniques for ultrasonic soldering have been described by Hunicke.[251] Kholopov[252] has used ultrasonic methods to relieve residual stresses in welded joints in metals. Walker and Walker[253] have demonstrated that ultrasonic methods can change the hardness of metals immersed in and electrodeposits formed in ultrasonically agitated solutions. Ultrasonic methods have proven useful in production of metal plasticity with applications in wire drawing and tube bending as discussed by Langenecker and Vodep[254] and summarized by Eaves et al.[255] Finally, Johnston and Reid[256] have developed ultrasonic methods for improved preparation of carbon films for electron microscopy.

6.11. Concluding Remarks

The field of nonlinear acoustics and methods used both to study the phenomena themselves or to use the phenomena to study other systems is too large to be completely covered here. The investigator should, of course, examine the current literature concerning the experimental methods of interest. He or she should be aware, however, that nonlinear acoustic methods are unique in that by using them one not only gains information about the system on which they are used, but quite frequently

[247] M. Adithan and V. C. Venkatesh, Wear **40**, 309 (1976).
[248] A. V. Paustovskii, V. T. Onyshko, and S. A. Shvoh, Sov. Powder Metall. Met. Ceram. **15**, 149 (1976).
[249] M. Kubota, Y. Tomura, and N. Shomamura, J. Jpn. Soc. Precis. Eng. **42**, 197 (1976).
[250] E. Komasa, IBM Tech. Discl. Bull. **16**, 3429 (1974).
[251] R. L. Hunicke, Ultrason. Int. Conf. Proc., London p. 32 (1975).
[252] Y. V. Kholopov, Weld. Proc. (USSR) **20**, 34 (1973).
[253] R. Walker and C. T. Walker, Ultrason. Int. Conf. Proc., London p. 28 (1975).
[254] B. Langenecker and O. Vodep, Ultrason. Int. Conf. Proc., London p. 202 (1975).
[255] A. E. Eaves, A. W. Smith, W. J. Waterhouse, and D. H. Sansome, Ultrasonics **13**, 162 (1975).
[256] H. S. Johnston and O. Reid, J. Microsc. (Oxford) **94**, 283 (1971).

the investigator has the opportunity to gain insight into the physics associated with the nonlinear phenomenon. With such diversity of nonlinear phenomena it should not be a surprise that such methods will find applications in many fields and that these are only limited by our own ingenuity.

7. ACOUSTIC CAVITATION

By Robert E. Apfel

List of Symbols

A	nondimensional acoustic pressure $= P_A/(P_0 - P_V)$; Eq. (7.3.3)
c	sound velocity in liquid
C	nondimensional viscosity $= 4\pi/R_0[(P_0 - P_V)/\rho]^{1/2}$; Eq. (7.3.3)
C	relative gas saturation; Eq. (7.3.7)
C_p; $C_{p,v}$; $C_{p,L}$	heat capacity per unit mass at constant pressure; subscripts v and L refer to vapor and liquid, respectively
d	length parameter $= 2\sqrt{2}\sigma/3P_V$, Eq. (7.3.11)
D	nondimensional surface tension $= 2\sigma/R_0(P_0 - P_V)$; Eq. (7.3.3)
D_L, D_v	thermal diffusity in liquid or vapor $= k/\rho C_p$
E	nondimensional gas pressure $= P_G/(P_0 - P_V)$; Eq. (7.3.3)
f	frequency
f_D, f_r	resonance frequency of bubble of radius R_D or R_r, given by Eq. (7.3.5)
f_e	evaporation–condensation resonance frequency, given by Eqs. (7.3.9) and (7.3.10) divided by 2π.
k	thermal conductivity (subscripts v and L refer to vapor and liquid, respectively)
l_L, l_v	thermal diffusion length in liquid (L) or vapor (v); $= (2D_{L,v}/\omega)^{1/2}$
M_{eff}	effective liquid mass $(= 4\pi R^3 \rho_L)$ felt by bubble of radius R
p	$= P_A/P_0$
P_A	acoustic pressure amplitude
P_B	"Blake" pressure threshold for static nucleation of a spherical bubble.
P_D	acoustic pressure threshold for rectified diffusion
P_G	gas pressure
P_i	internal pressure of bubble
P_T	minimum acoustic pressure amplitude at which transient bubble will reach supersonic collapse velocity; "transient" threshold.
P_V	vapor pressure of liquid
P_0	liquid ambient pressure
ΔP	time-averaged difference between internal bubble pressure and external pressure (ambient plus acoustic)
R	bubble radius
R_B	minimum bubble radius that will grow according to "Blake" threshold; Eq. (7.2.2)
R_D	minimum bubble radius that will grow by rectified diffusion; Eq. (7.3.7)
R_I	approximate initial bubble size above which inertial effects predominate; Eq. (7.3.14)
R_M	approximate maximum size to which bubble will grow if inertial effects not controlling; Eq. (7.3.23)
R_r	radius of bubble resonant at frequency f_r; Eq. (7.3.5)

355

METHODS OF EXPERIMENTAL PHYSICS, VOL. 19

R_T radius of bubble which will reach supersonic collapse velocity when subjected
          to acoustic pressure amplitude P_T
R_0 initial size of bubble
R_1 bubble size reached while pressure difference across bubble is positive
R^* size of bubble that goes unstable when acoustic pressure P_A is applied; Eq.
          (7.2.1) and preceding
S stiffness term for vapor bubble resonance
t time
t_1 inertial start-up time
t_m time interval during which pressure difference across bubble is positive,
          $= t_1 - t_2$
t_V viscous start-up time; see text preceding Eq. (7.3.14)
T temperature
T acoustic period
V bubble wall velocity
X_i $= 2\sigma/P_0 R_1$, where i is B, D, r
β nondimensional radius $= R/R_0$; Eq. (7.3.3)
γ ratio of specific heats
Γ nondimensional parameter $= P_V \phi/T$; Eq. (7.3.12)
η non-dimensional parameter $= \rho_L C_{p,L} \phi$; Eq. (7.3.13)
μ liquid viscosity
ϕ reciprocal slope of liquid's vapor pressure curve
ρ, ρ_L density (L for liquid)
σ surface tension
τ nondimensional time $= t[(P_0 - P_v)/\rho]^{1/2}/R_0$; Eq. (7.3.3)
τ_0 nondimensional acoustic period $= T[(P_0 - P_V)/\rho]^{1/2}/R_0 = [(P_0 - P_V)/\rho]^{1/2}/fR_0$
ω $= 2\pi f$
ω_e $= 2\pi f_e$; Eqs. (7.3.9) and (7.3.10)
Ω_L $= l_L/2R_0$; see Eq. (7.3.9) and preceding text

7.1. Introduction

7.1.1. Nomenclature

7.1.1.1. Cavitation—General Definition. *Cavitation* is defined as the
formation of one or more pockets of gas (or "cavities") in a liquid. Here
the word "formation" can refer, in a general sense, both to the creation of
a new cavity or the expansion of a preexisting one to a size where macroscopic effects can be observed. The cavity's gas content refers to the liquid's vapor, some other gas, or combinations thereof. Sometimes these
cavities are referred to as "bubbles" or "voids," depending on the relative amount and type of gas.

Cavitation usually occurs in response to a reduction of the pressure
sufficiently below the vapor pressure of the liquid or to the elevation of
the temperature above the boiling point, although chemical, electrical, and
radiation-induced phenomena can be important.

7.1.1.2. Acoustic Cavitation. Cavity formation in response to an alternating pressure field (i.e., an acoustic field) is called *acoustic cavitation*. One can generalize this definition to encompass any observable activity involving a bubble or population of bubbles stimulated into motion by an acoustic field. Henceforth in this chapter, "cavitation" will mean acoustic cavitation.

During the negative part of the acoustic cycle, when the local pressure falls below the ambient pressure, a preexisting bubble may begin to grow. When the acoustic pressure turns positive, the bubble growth slows and is finally reversed. The degree of growth and collapse during a given acoustic cycle and the lifetime of the bubble in the acoustic field depend on the acoustic pressure amplitude, the ambient pressure, the acoustic frequency, the acoustic duty cycle (if not a continuous wave), and characteristics of the liquid and dissolved gases.

7.1.1.3. Motivation for Cavitation Studies. Cavitation tests are one way of characterizing the liquid medium. In very carefully cleaned liquids the cavitation threshold may approach or even equal the tensile strength of the liquid. For less clean liquids, the rate and type of cavitation can tell us much about impurities in the liquid, about the liquid's gas content, and about the history of the liquid. Cavitation events in a liquid can also be a measure of ionizing radiation incident on the sample, thereby providing a way of learning about the radiation–matter interaction for known sources of radiation and providing the possibility of an instrument for radiation monitoring.

There are also a number of basic problems relating to the fluid mechanics and heat transfer associated with acoustically induced bubble dynamics. The results of basic studies have direct bearing on the practical applications of acoustic cavitation (see Section 7.4.2) as well as the solution of problems resulting from unwanted cavitation (see Section 7.4.3).

The importance of understanding the mechanisms of acoustic cavitation stems in large part from the observation that relatively low energy density, low particle velocity sound waves can result in cavity collapse that is sometimes so rapid as to compress adiabatically the cavity's noncondensible gas contents to a pressure of thousands of atmospheres and to a temperature of thousands of degrees kelvin. Cavitation can thus be thought of as a means of energy concentration in liquids.

The study of acoustic cavitation shares much physics with that of hydrodynamic cavitation phenomena, superheat and boiling phenomena, and supersaturation phenomena. Whether one is talking about cavitation limits to long-range sonar communication, ship propeller damage, a sudden explosion when water and molten metal come into contact, bubble chambers, decompression sickness in animals, or effervescence in beer, one must know much the same physics and may very well apply the same mathematics.

7.1.1.4. Other References on Cavitation. Several review articles and books, many stressing theoretical aspects of the subject, are available in the literature. Most notable of these are by Flynn,[1] Plesset,[2] Rozenberg (reviewing much of the Russian work),[3] Plesset and Prosperetti (recent theoretical advances),[4] and Coakley and Nyborg (emphasis on gas bubble dynamics and biological implications).[5]

7.1.2. Types, Stages, and Effects of Acoustic Cavitation

7.1.2.1. Types of Cavitation. Flynn has attempted to bring some order to the variety of cavitation phenomena by adopting nomenclature which most investigators in the field now apply. The two basic types of acousti-cally induced cavities are: (1) a stable cavity, which oscillates many times about its equilibrium radius, and (2) a transient cavity, which undergoes much greater variations from its equilibrium size in relatively few acoustic cycles and which may terminate in a fairly violent collapse (the interface velocity can approach or exceed the speed of sound in the liquid).

Stable cavities are usually gas bubbles excited by relatively low acoustic pressures (a fraction of an atmosphere to a few atmospheres). Transient cavities are often generated by higher acoustic pressures (and/or lower frequencies); their growth to maximum size is rapid, and their gas content plays only a small role until the violent collapse is rap-idly decelerated by the entrapped gas.

Stable and transient cavities are, of course, not exclusive categories; some cavities behave like stable cavities in their early lifetime and then, when size excursions become sufficiently great, change to transient cavi-ties.

7.1.2.2. Stages of Cavitation. There are two basic stages of acoustic cavitation: the inception or initial formation of the cavity and the subse-quent cavity dynamics involving growth and collapse.

[1] H. Flynn, Physics of acoustic cavitation in liquids, in "Physical Acoustics" (W. P. Mason, ed.), Vol. IB, pp. 57–172. Academic Press, New York, 1964.
[2] M. Plesset, Bubble dynamics, in "Cavitation in Real Liquids" (R. Davies, ed.), pp. 1–18. Amer. Elsevier, New York, 1964.
[3] L. D. Rozenberg (ed.), "High Intensity Ultrasonic Fields." Plenum, New York, 1971.
[4] M. S. Plesset and A. Prosperetti, Bubble dynamics and cavitation, Ann. Rev. Fluid Mech. 9, 145–185 (1977).
[5] W. T. Coakley and W. L. Nyborg, Cavitation; dynamics of gas bubbles; applications, in "Ultrasound: Its Applications in Medicine and Biology" (F. J. Fry, ed.), Part I, pp. 77–159. Amer. Elsevier, New York, 1978.

7.1.2.2.1. INCEPTION. The term "cavitation threshold" has been used in numerous ways to describe the minimum conditions—usually the minimum acoustic pressure—necessary to initiate cavitation of one of the types discussed in Section 7.1.2.1.

As the varied results of the preponderance of experiments illustrate, the threshold of any given type of cavitation is usually not a measure of a property of a pure liquid but more likely a measure of the impurities (especially undissolved ones) in the liquid, and the liquid's history (i.e., method of liquid preparation).

Furthermore, whether or not the cavitation threshold is reached may depend on the means by which the experimenter observes the liquid. For instance, at 10 MHz a cavitation bubble may form, oscillate for a few cycles, and disappear. It will not be visible and may not produce much noise. The observer may conclude that a threshold has not been reached, because an event of sufficiently macroscopic character has not been produced. Clearly this observer is basing his view of cavitation thresholds on the dynamics of bubbles and on the effects produced rather than the initiation phase. As this example suggests, the conditions and data of each experiment must be detailed carefully and the particular type of cavitation must be specified if the results are to have meaning.

7.1.2.2.2. CAVITY DYNAMICS. As mentioned, the inception of a cavity is usually the least observed part of cavitation phenomena. In response to a pressure reduction, the cavity grows. This growth is slowed and eventually reversed during the positive pressure phase of the acoustic cycle. Whether the cavity is short lived, i.e., lasts for a few acoustic cycles or less, or whether it survives a large number of cycles depends on a number of factors, such as the acoustic pressure amplitude, whether the gas in the cavity is a condensible vapor or an inert gas, and the relations of the temperature to the normal boiling temperature and the static pressure to the vapor pressure.

The dynamics of cavities are rarely as simple as even the more complicated theoretical models would allow. The collapse becomes asymmetric, surface waves grow into shape instabilities, single bubbles shatter into pulsating microbubbles that interact hydrodynamically, and bubbles translate as well as pulsate in pressure gradients. Considering this complexity, it is indeed remarkable and a tribute to experimenter and theoretician that there is as much agreement as there is.

7.1.2.3. Effects Produced by Cavitation. The high pressure and temperature of a violently collapsing cavity may result not only in mechanical effects (such as erosion of solid surfaces) but also in chemical reactions and luminescence of the trapped gas (see Section 7.3.4.3).

The gentler, stable cavitation also produces interesting mechanical effects, especially when the bubble is near its resonance frequency. Here the pressure disturbance introduced by the sound wave is transformed into a velocity source, with the action of the flowing liquid at the microscopic scale being responsible for beneficial effects (e.g., ultrasonic cleaning) and damaging effects (e.g., when bubble-induced shear stresses cause hemolysis of red blood cells).

7.2. Cavitation Inception

7.2.1. Cavitation Threshold Measurements

7.2.1.1. Some Basic Considerations. Three golden rules of any cavitation experiment, which are especially important for threshold measurements, are: "Know thy liquid" (including its solid impurities); "Know thy sound field"; and "Know when something happens." The first of these, of course, determines the threshold; the second determines whether measurements are accurate; the third relates to the observation that cavitation inception is usually observed indirectly through the effects of cavitation. For example, at low acoustic frequencies, cavitation events are dramatic and obvious; at higher frequencies, they may go unnoticed although effects at the microscopic scale have occurred.

If a cavitation threshold measurement is to be indicative of the tensile strength of the pure liquid, there must be no foreign site from which a cavity can be initiated more readily than in the pure liquid.

The degree of cleanliness required for a liquid to be considered "pure" for tensile strength experiments can be illustrated by the following simple example. Consider tap water: The concentration of solid particles or "motes" might be on the order of $50,000-100,000$ cm^3. (Yes, you drink this!) How can all of these particles be removed without introducing others? Filtering? Yes and no. One would have to fill by filtering into a perfectly (?) clean and outgassed glass vessel. Experience with cavitation has shown us that cleaning volumes more than a few cubic centimeters usually does not eliminate all motes.

7.2.1.2. The "Clean" Experiments of Greenspan and Tschiegg. The best job of cleaning for acoustic cavitation studies was done by Greenspan and Tschiegg of the National Bureau of Standards.[6] (Interestingly, Briggs, also of NBS, has recorded the highest tensile strengths for triply distilled water using a centrifugal technique.[7]) They had a closed-loop system consisting of a sample vessel of water, a pump, a rectan-

[6] M. Greenspan and C. E. Tschiegg, *J. Res. Natl. Bur. Stand. Sect. C* **71**, 299 (1967).
[7] L. J. Briggs, *J. Appl. Phys.* **21**, 721 (1950).

gular viewing tube, and a filter; the system had no liquid–gas interfaces. The viewing tube was observed with a dark-field microscope. In the early stages the scattered light (Tyndall effect) appears as a snow storm in our experience, but eventually the cleaning yields no scattered light and a very tired observer (know thy liquid). Measured cavitation thresholds in water at 42.9 kHz near 30°C, as observed directly or via very audible "pops" (know when something happens), ranged from 160 bars (16 × 10⁷ Pa) for periods of a minute or so to as high as around 210 bars for periods on the order of a few seconds. But the predictions of the theoretical tensile strength using homogeneous nucleation theory exceed 1000 bars. The authors presume that the tensile strength of water was not reached and that a stray mote or an "adventitious" neutron was responsible for the cavitation event.

Greenspan and Tschiegg achieved their high acoustic pressures using a cylindrical resonator excited in the $(r, \theta, z) = (3, 0, 1)$ mode (know thy sound field). Figure 1 repeats the figure from their paper. The high acoustic pressures are generated at the center, away from solid boundaries where nucleation sites abound. The sensitive volume—that is, the region in the liquid where significant negative pressures occur—is relatively small at low acoustic pressures (maybe a few cubic centimeters) and grows with increases in the acoustic pressure amplitude; but the volume in which a nucleation site might exist or be produced (perhaps by radiation) is close to the volume of the container (approximately 600 cm³).

7.2.1.3. Threshold Measurements Using the Drop Levitation Technique. Is there a technique by which the probability of nucleation from

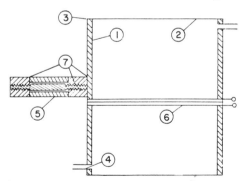

Fig. 1. Cylindrical resonator used in experiments of Greenspan and Tschiegg. Prestressed driving transducer (5) epoxied to stainless steel cylinder (1) and closed by thin stainless steel sheets (2). Silicone-rubber adhesive joints (3), filling tubes (4), wire strain gauge (6), and epoxy joints (7) are also shown. [from M. Greenspan and C. E. Tschiegg, J. Res. Natl. Bur. Stand. Sect. C **71**, 299 (1967)].

foreign sources can be minimized? Returning to hypothetical tap water containing 50,000 motes/cm³, we inquire into how many parts we have to subdivide our 1-cm³ sample so that about half have no motes. The answer is about 100,000, and if these all have the same size, then the drops would each have a 0.27-mm diameter. In order to transmit acoustic stress to these drops, they must be suspended in an immiscible host liquid, which simultaneously serves as an ideal, "smooth" container, thereby avoiding cavitation problems at the interface. One problem remains: several hundred atmospheres of tension must be sustained by the host if the tensile strength of pure liquids at room temperature is to be measured. One can side step this problem if the experiment is performed at elevated temperatures, where predicted tensile strengths are lower.

At the "limit of superheat" of a liquid, the tensile strength is zero; i.e., our drop will spontaneously vaporize within a reasonable waiting time. (Predictions of the limit of superheat are very insensitive to the waiting time.) The degree to which a liquid can be superheated at one atmosphere of pressure is rather remarkable: For water at atmospheric pressure, it is at least 180°C above the boiling point (i.e., to 280°C); for ether (bp 35°C), the limit of superheat is 144°C. For temperatures below this "limit of superheat," the liquid sustains tensile stress, with predicted tensile strength increasing with decreasing temperature in typical organic liquids by about 1 bar/°C as predicted by homogeneous nucleation theory.[8,9]

The only experiments in which results are consistent with this theory involve the acoustical levitation of drops.[10] The apparatus for measuring the tensile strength of superheated drops is shown in Fig. 2a. (A recent version of this apparatus, shown in Fig. 2b, includes a piston with a thin, air-backed diaphragm, which assures strong reflections and, therefore, an intense standing wave field.) A drop of the sample is injected into the host in the lower tube (Fig. 2a), where the temperature is below the sample's boiling point. As the sample rises, due to its buoyancy, it is heated by the surrounding host liquid and a heating coil wrapped around the glass column. The drop then enters the acoustic resonator section in which the (3, 0, n) mode (at about 50 kHz) is excited (n = 3, 4, or 5). The sound field, established by the cylindrical piezoelectric transducer epoxied to the Pyrex tube, serves not only to subject the drop to high acoustic stress, but it also levitates the drop at a position slightly above

[8] M. Volmer, Kinetik der phasenbildung, in "Die Chemische Reaktion," Band IV. Steinkopff, Dresden and Leipzig, 1939.

[9] V. P. Skripov, Russ. J. Phys. Chem. **38**, 208 (1964).

[10] R. E. Apfel, J. Acoust. Soc. Am. **49**, 145 (1971); Nature (London) **233**, 119 (1971); Sci. Am. **227**, 58 (1972).

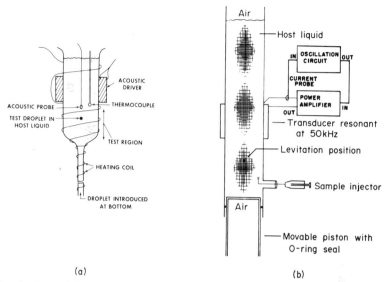

FIG. 2. Acoustic levitation apparatus: (a) early apparatus for superheated drop experiments with drop introduced at low temperature in smaller tube and heated as it rises. (b) improved apparatus having piston bottom with thin, sound-reflecting diaphragm.

one of the positions of the acoustic pressure maxima. The levitation occurs due to acoustic radiation forces which balance the buoyancy force (see Part 6). As the voltage to the transducer is increased, the drop moves to a new equilibrium position closer to the acoustic pressure maximum. Eventually the acoustic stress is sufficient to cause the rapid explosive vaporization of the drop. The peak acoustic pressure at the position of the drop minus the hydrostatic pressure is the tensile strength of the sample at the temperature of the test. The test is repeated several times at the same temperature before changing the temperature. The fact that the result is repeatable within 5–10% or better, with few premature nucleation events, strongly suggests that the nucleation was not heterogeneously initiated. Also, the tensile strength measured in this way is independent of acoustic frequency because the period of the sound (2×10^{-5} sec) is large compared to the time for the inception of a cavity ($\sim 10^{-10}$ sec). Since the liquid is superheated, a cavitation event leads to the complete vaporization of the sample. Cavitation dynamics do not enter into the determination of thresholds.

The temperature dependence of the tensile strengths of ether and hexane heated in a host of glycerin are shown in Fig. 3 along with the theoretical predictions. Note that the maximum acoustic pressure produced

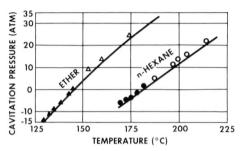

FIG. 3. Tensile strength of ether (▲) and hexane (●) heated in a host of glycerin. ○ and △ represent corresponding results for the limit of superheat at positive pressures. Solid line represents theoretical predictions based on homogeneous nucleation theory.

was about 20 bars. This was the highest pressure generated without worrying about cavitation in the glycerin. The host was prepared by outgassing the vessel, back-filling heated glycerin through a 0.45-μm pore filter, and cavitating under reduced pressure and elevated temperature before returning to atmospheric pressure. This procedure reduces the gas content of the glycerin while encouraging the wetting of solid surfaces, thereby deactivating potential nucleation sites in the host. The moral of this tensile strength story might be summed up by saying that cleanliness is next to godliness—or in less religious and more mathematical terms: One mote is infinitely worse than none.

7.2.2. Cavitation and Dirt

For the reasons to which we just alluded, few cavitation experiments yield results for the pure liquid. A general requirement for understanding cavitation in real liquids is to understand how bubbles can be established in liquids. This stabilization requirement stems from the instability of a free bubble; i.e., a free bubble in a liquid will rise due to buoyancy, will collapse, or will grow.

The collapse of a gas-filled bubble, however, is limited by diffusion. Assuming that slowly dissolving gas bubbles do exist in the liquid, one can ask what minimum value of acoustic pressure P_A would be responsible for the growth of a gas-filled bubble of radius R_B. Following Blake[11] and Neppiras and Noltingk,[12] we note that if the bubble is in mechanical equilibrium, then the pressure inside P_i must equal $P_0 + 2\sigma/R_B$. If the bubble then changes size due to a decrease in the absolute external pres-

[11] F. G. Blake, Jr., Technical Memo 12, Acoustics Research Laboratory, Harvard Univ., Cambridge, Massachusetts (1949).

[12] E. A. Neppiras and B. E. Noltingk, *Proc. Phys. Soc. London Sect. B* **64**, 1032 (1951).

sure (by the application of a sound field), then the mechanical equilibrium condition becomes

$$(P_0 + 2\sigma/R_B)(R_B/R)^3 = P_0 - P_A + 2\sigma/R. \tag{7.2.1}$$

The term on the left represents the gas pressure in the bubble associated with isothermal expansion. On the right, P_A is the peak negative value of the acoustic pressure. If we desire to find the minimum value of external pressure $(P_A - P_0)$ for growth, we take $\partial(P_0 - P_A)/\partial R = 0$. This gives $P_A - P_0 = 4\sigma/3R^*$, where $R^* = [3(P_0 + 2\sigma/R_B)R_B^3/2\sigma]^{1/2}$; or, solving for P_A in terms of the initial bubble size, we find what has been termed the "Blake" threshold pressure (hence the use of the subscript B on the initial size):

$$
\begin{aligned}
P_B &= P_0 + \frac{8\sigma}{9} \left[\frac{3\sigma}{2(P_0 + 2\sigma/R_B)R_B^3} \right]^{1/2} \\
&= \begin{cases}
P_0 + 0.77 \dfrac{\sigma}{R_B}; & \dfrac{2\sigma}{R_B} \gg P_0 \\[2ex]
P_0 + \dfrac{8\sigma}{9} \left[\dfrac{3}{2} \dfrac{\sigma}{P_0 R_B^3} \right]^{1/2}; & \dfrac{2\sigma}{R_B} \ll P_0.
\end{cases}
\end{aligned} \tag{7.2.2}
$$

The most obvious feature of this derivation is the neglect of dynamic factors (i.e., inertial and viscous effects). The limitations imposed by those considerations will be discussed in Sections 7.3.4.2 and 7.3.5. We also assumed that gas diffusion occurs slowly compared to the mechanical growth of the bubble. (It is estimated that a 10-μm air bubble in air-saturated water will take approximately 7 sec to dissolve.[13])

As will be seen in Section 7.3.2.1.1, acoustic thresholds for bubble growth lower than that defined in Eq. (7.2.2) can exist via a process called *rectified diffusion*. Furthermore, under some conditions (see Sections 7.3.4.2 and 7.3.5), inertial forces prevent significant growth in a time period comparable to half an acoustic cycle.

If free bubbles do not exist for appreciable periods of time in a liquid, then, in order to understand why cavitation thresholds are low, one must find a mechanism by which gas "nuclei" are stabilized. A microscopic crack or crevice in a container surface or an imperfectly wetted dirt particle is a suitable site for bubble entrapment. That certain sites (or "nuclei") nucleate bubbles is a common occurrence in observing water boiling in a pot or gas bubbles forming on the wall of a glass of beer.

[13] P. S. Epstein and M. S. Plesset, *J. Chem. Phys.* **18**, 1505 (1950).

Harvey et al.,[14] Strasberg,[15] Apfel,[16] and others have considered simple models for gas stabilization in cracks and crevices, trying to determine how cavitation thresholds might be influenced by gas saturation in the liquid, hydrostatic pressure, advancing and receding contact angles, pretreatment of the sample (e.g., temporary pressurization), temperature, and size of the crevice. Such "crevice" models seem to explain why in many circumstances the measured threshold pressure is highly dependent on the equilibrium gas saturation pressure P_G, whereas Eq. (7.2.2) is independent of P_G, except that the size of the gas pockets present in the liquid will be influenced by the gas saturation pressure (see also Section 7.2.5).

By filtering, one can change the nucleus distribution, thereby decreasing the rate of acoustic cavitation, but only extensive filtering of the type described earlier[6] will alter substantially the acoustic cavitation threshold, which depends on the size of the largest nuclei. (See Section 7.4.3.2 for other techniques for raising cavitation thresholds.)

Another model for the stabilization of bubbles involves the presence of contaminants or a "skin" on the bubble surface. Although a model of this type originally proposed by Fox and Herzfeld[17] was later abandoned by Herzfeld,[18] the debate has not ended.[19] Perhaps this model has viability for certain liquids or systems (e.g., biological) and certain types of contaminants. There has been insufficient experimental work, however, to resolve this question.

7.2.3. Radiation-Induced Cavitation

If a liquid has been strengthened by filtering, outgassing, and/or hydraulic pressurization, it may still fail at tensions less than its ultimate tensile strength. The origin of this failure can sometimes be traced to the excitation of atoms of the liquid by incident-ionizing radiation. The precise mechanism of such interaction is in doubt, although it is now generally agreed that Seitz's "thermal spike" model is most appropriate.[20] In this model it is proposed that the motion of a positive ion, resulting from the radiation–matter interaction, excites neighboring atoms thereby creating,

[14] E. N. Harvey, A. H. Whitely, W. D. McElroy, D. C. Pease, and K. W. Cooper, J. Cell. Comp. Physiol. 24, 1, 23 (1944).

[15] M. Strasberg, J. Acoust. Soc. Am. 31, 163 (1959).

[16] R. E. Apfel, J. Acoust. Soc. Am. 48, 1179 (1970).

[17] F. E. Fox and K. F. Herzfeld, J. Acoust. Soc. Am. 26, 984 (1954).

[18] K. F. Herzfeld, Proc. Symp. Naval Hydrodynam., 1st (F. S. Sherman, ed.), p. 319. National Academy of Sciences, Publ. 515, Washington, D.C., 1957.

[19] D. E. Yount, J. Acoust. Soc. Am. 65, 1429 (1979).

[20] F. Seitz, Phys. Fluids 1, 2 (1958).

in effect, a thermal spike and subsequently a vapor bubble that grows sufficiently to be expanded by the tension present in the liquid. The same considerations apply to superheated liquids, as the bubble chamber demonstrates.

Several authors have investigated radiation-induced cavitation. The results of these studies once again vary greatly. As an example, let us return to the work of Greenspan and Tschiegg on cavitation thresholds.[6] They irradiated water samples with 10-MeV neutrons and found a reduction of cavitation thresholds to about 50 bars—about 110 bars below the theshold of clean water samples not irradiated. Their results point strongly to single-particle events; that is, a single neutron interaction causes a cavitation event, a result similar to that of West and Howlett,[21] who used a pulsed neutron source, produced by an accelerator, which interacted with trichloroethylene. These investigators saw no memory effect; that is, an irradiated liquid did not remain "weak" after the source of radiation was removed. Sette et al.,[22] Lieberman,[23] Finch,[24] and Barger[25] have seen a memory effect. Finch, for example, observed a time lapse of about $\frac{1}{2}$ hr for the full threshold lowering effect of 14-MeV neutrons incident on degassed water; and once the radiation was turned off, it appeared that an approximately equal time lapse was required before the threshold returned to the unirradiated value.[24] Sette et al. have observed the reduction of cavitation thresholds in water from 1.3 bars to about 1 bar in gassy water. Note that these thresholds are far below those of Greenspan and Tschiegg, prompting the latter authors to comment, "Perhaps dirt is more important than gas." [6]

In order to check thermal spike theories for their appropriateness, a great deal of data is needed on the temperature dependence of cavitation thresholds of liquids exposed to known and, preferably, monoenergetic radiation.

The acoustic levitation technique discussed in Section 7.2.1.3 (or another technique in which a gel is used for the immobilization of superheated drops[26]) appears to be a prime candidate for amassing such data. In this case, drops of known volume and degree of superheat are exposed to the radiation source. If an interaction occurs, the drop vaporizes

[21] C. West and R. Howlett, J. Phys. D 1, 247 (1968).

[22] D. Sette and F. Wanderlingh, Phys. Rev. 125, 409 (1962).

[23] D. Lieberman, Phys. Fluids 2, 466 (1959).

[24] R. D. Finch, J. Acoust. Soc. Am. 36, 2287 (1964).

[25] J. E. Barger, Technical Memo No. 57, Acoustics Research Laboratory, Harvard Univ., Cambridge, Massachusetts (1964).

[26] R. E. Apfel, Detector and Dosimeter for Neutrons and Other Radiation. Patent 4,143,274.

explosively because it was above its normal boiling point. There is no ambiguity, as might be the case with short-lived acoustic cavitation events in a liquid below its boiling point. Furthermore, the small sample size allows for far greater superheating. The pressure can be varied hydrostatically or acoustically. In the latter case, negative pressure variations can be produced.

Results from these experiments should not only provide fundamental information about the interaction of radiation and matter but might well be applied to the detection of radiation. Furthermore, since energy thresholds for radiation-induced cavitation depend on temperature and pressure, practical energy spectrometers based on these fundamental experiments may be feasible.[27]

7.2.4. Unsolved Problems

7.2.4.1. Cavitation Thresholds for Liquid Helium.
It is fairly safe to say that the ultimate tensile strength of any of the forms of liquid helium has not been measured. It is not because there is an agreed upon theoretical estimate, since for all helium liquids the quantum-mechanical, zero-point energy per atom is significant in comparison to $k_B T$ (the atom's thermal energy), and, therefore, classical theory may be inappropriate; rather, it is the lack of repeatability among different experiments that casts doubts on the achievement of absolute limits.

Liquid ^{4}He at temperatures above the λ transition (2.172°K at SVP) is known as helium I, and it is hydrodynamically classical. Above 4°K, classical homogeneous nucleation theory predicts that helium I will not support tension; few measurements, however, of the tensile strength of helium I at any temperature have been reported to confirm or disprove this prediction.

Liquid helium II, the form of helium that exists below the λ point, will support tension. Homogeneous nucleation theory predicts a tensile strength of between 4 and 5 atm at 2.1°K.[28] Beams[29] used both centrifugal and linear deceleration methods for producing negative pressures and found tensile strengths of about 0.15 bar, more than an order of magnitude below predictions. (Results for argon, nitrogen, and oxygen are similarly low, suggesting that helium is not unique in this respect.) This discrepancy has been attributed to impurity nuclei in the liquid or at the container walls or to the presence of ions due to radioactivity or cosmic rays. But liquid helium freezes out all foreign gases, thus assuring that no gas pockets exist for nucleation sites; also surface tension and the superfluid

[27] R. E. Apfel, *Nucl. Instrum. Methods* **162**, 603 (1979).
[28] J. W. Beams, *Phys. Fluids* **2**, 1 (1959).
[29] J. W. Beams, *Phys. Rev.* **104**, 880 (1956).

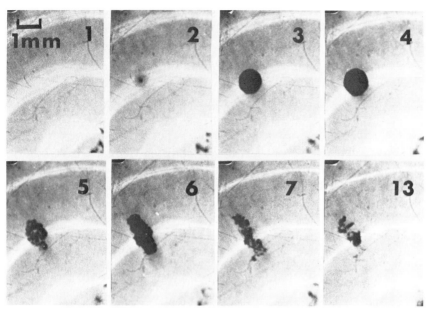

FIG. 4. Growth of a vapor cavity in helium II. The frame rate is 6000 frames/sec and the number of each frame in the sequence is indicated. The bubble first becomes visible in frame 2 when it is located at a pressure maxima in the 50.6-kHz standing wave. In frame 3, which is 0.17 msec later, it has expanded to a diameter of 0.94 mm. The growth rate may have been aided by rectified heat transport in the superfluid. The surface of the bubble appears unstable in frames 4 and 5 and fracturing is visible in frame 13. The most negative pressure achieved at the location of the initial visible bubble was estimated as −0.6 bar and is a measure of the tensile strength [from Tensile strength and visible ultrasonic cavitation of superfluid helium-four, Part II of Ph.D. Thesis of Philip L. Marston, Stanford Univ. (1976)].

properties of helium II encourage the wetting of all solid surfaces. In addition, vanishing viscosity means that motes cannot stay in suspension in helium at rest. Furthermore, Fairbank et al. have noticed that the radiation-induced bubble tracks found upon decompression of helium at temperatures above the λ point are absent at temperatures below the λ point.[30] Still another possible reason for the discrepancy between theory and experiment is the suggestion that quantum rather than thermal effects predominate in heterophase fluctuations responsible for bubble nucleation.[31] Theory predicts, however, that these quantum effects do not become important until the temperature is reduced below about 0.5°K.[32]

[30] W. M. Fairbank, J. Leitner, M. M. Block, and E. M. Harth, "Problems of Low Temperature Physics and Thermodynamics," Vol. 1, pp. 45–54. Pergamon, Oxford, 1958.
[31] I. M. Lifshitz and Yu. Kagan, Sov. Phys. JETP 35, 206 (1972).
[32] V. A. Akulichev and V. A. Bulanov, Sov. Phys. Acoust. 20, 501 (1975).

Studies of acoustic cavitation in liquid helium have also resulted in little success in measuring a real tensile strength.[33,34]   In fact, cavitationlike noise has been detected even though the absolute pressure remained positive, the acoustic pressure amplitude being less than the hydrostatic pressure.   However, Marston has measured peak tensile strengths of between 0.3 and 1.2 bars in helium II at 2.09°K by using a clever optical technique to infer the acoustic pressure amplitude.[35]   The cavitation occurred in the focal region of a cylindrical transducer identical to that described in Section 7.2.1.3; photographs of vapor cavity expansion are shown in Fig. 4. Some of Marston's other observations shed some light on the low noise thresholds in helium II.   In particular, he noticed that when the acoustic transducer was tuned to its resonance rather than to the frequency associated with the standing wave resonance in the liquid, the noise signal was most evident.   Since the transducer dissipates more energy at its resonance than at the liquid standing wave resonance, it is thought that the heat generated at the transducer surface was responsible for the noise rather than cavitation in the bulk liquid.   Whether or not small, short-lived bubbles formed on the transducer surface remains to be determined. The role of rectified heat transport (RHT) also remains to be clarified.[33]

One nonacoustic experiment with pulsed superheating in liquid helium I has lead to heating 0.45°K above the boiling temperature of 4.0°K.[36]   This result is comparable to the predictions of classical homogeneous nucleation theory.   The authors acknowledge their uncertainty about the appropriateness of this theory, but we were left with the impression that an ultimate limit had, indeed, been reached.

It may turn out that the failure of acoustical techniques to achieve analogous pressure limits is traceable to problems similar to those found in acoustical cavitation studies of relatively large volumes of common liquids.   For the time being, however, uncertainties as to the origins of bubbles remain.   Suggestions that bubbles are velocity rather than pressure induced[37] and that bubbles may originate from vortex lines[38] are intriguing possibilities indicated by some data but are yet to be confirmed by definitive measurements.

7.2.4.2. Effect of Electric Fields on the Tensile Strength of Liquids.
There has been some controversy about whether or not the presence of an

[33] R. D. Finch and E. A. Neppiras, Cavitation nucleation in liquid helium, *Ultrason. Int. Conf. Proc.* pp. 73–80 (1973).

[34] P. D. Jarman and K. J. W. Taylor, *J. Low Temp. Phys.* **2**, 389 (1970).

[35] P. L. Marston, *J. Low Temp. Phys.* **25**, 383 (1976).

[36] L. C. Brodie, D. N. Sinha, J. S. Semura, and C. E. Sanford, *J. Appl. Phys.* **48**, 2882 (1977).

[37] T. Vroulis, E. A. Neppiras, and R. D. Finch, *J. Acoust. Soc. Am.* **59**, 255 (1976).

[38] P. M. McConnell, M. L. Chu, Jr., and R. D. Finch, *Phys. Rev. A* **1**, 411 (1970).

electric field in a pure dielectric liquid will decrease its tensile strength or its ability to be superheated. Results of some experiments suggest that the electric field may reduce the strength,[39] but the authors have reversed a sign on an additional energy term in homogeneous nucleation theory and, therefore, have incorrectly produced a result consistent with their experiments.[40] That the electric field *inhibits* cavitation can be shown by the simple observation of bubble motion in a dielectric liquid toward regions of lower electric field. Thus additional positive work must be done in bubble production—an effect that can be thought of in terms of a positive, field-induced pressure increment to the liquid pressure. Presumably, the reported electric-field-induced cavitation can be attributed to a dielectric breakdown or heterogeneous effects, such as those which might be found at the electrode–liquid interface.

7.2.5. Concluding Remarks on Cavitation Thresholds

Cavitation thresholds have been measured as a function of many variables: for example, gas content of liquid, acoustic frequency, and temperature. Some of the results appear to be contradictory. For example, several investigators have found that as the gas content is reduced, cavitation thresholds increase. Galloway's highest attainable acoustic pressure ranged from 1 to 200 bars as the air content of water was reduced from 100% saturation to about 0.05% saturation.[41] However, for carefully cleaned liquids, Greenspan and Tschiegg have found no dependence on gas content so long as the liquid is at least a little undersaturated.[6] Apfel,[16] employing a model for the stabilization of gas pockets in conical crevices as originally developed by Harvey *et al.*,[14] has shown that this lack of dependence on gas content is the case if the size of imperfectly wetted impurities is sufficiently small, whereas a large dependence on gas saturation results if the liquid contains large impurities.

Several investigators have observed that acoustic cavitation thresholds increase with frequency.[25,42] Here the meaning of cavitation may play an important role. Crum and Nordling have shown that at low frequencies a transient acoustic cavitation event is not a single bubble event but rather a multiple bubble event, possibly initiated by the shattering of the initial bubble into microcavities during its collapse.[43] The multiple-bubble event can be observed as a darting, white, cometlike cavity that rapidly

[39] D. S. Pamar and A. K. Jalaluddin, *Phys. Lett. A* **42**, 497 (1973); *J. Phys. D.* **8**, 971 (1975).

[40] P. L. Marston and R. E. Apfel, *Phys. Lett. A* **60**, 225 (1977).

[41] W. J. Galloway, *J. Acoust. Soc. Am.* **26**, 849 (1954).

[42] R. Esche, *Acustica* **2**, 208 (1952).

[43] L. A. Crum and D. A. Nordling, *J. Acoust. Soc. Am.* **52**, 294 (1972).

moves away from the point of high acoustic pressure. As the frequency increases, the maximum size of a bubble decreases and multiple bubble effects are diminished. At 10 MHz, cavitation is not visible to the eye under normal conditions and the acoustical signal is much reduced. Furthermore, the time for a bubble to free itself of the mote from which it was initiated reduces the time available during the negative part of the acoustic cycle for bubble growth.

This restriction on the time allotted for inception and growth is also relevant when discussing pulsed acoustic signals. Here the frequency of the tone burst, the number of cycles per burst, and the duty cycle influence not only inception and growth but also multiple bubble effects. Such considerations have great importance to ultrasonic systems used in medicine (see also Section 7.4.3.5).

From the preceding discussion it should be obvious why no list of cavitation thresholds for liquids is included here. Methods of liquid preparation, techniques for cavitation detection, the conditions of the liquid, and the measurement procedure vary so greatly that such a list, except for true tensile strengths, would be misleading.

7.3. Cavitation Dynamics

7.3.1. General Considerations

The dynamics of acoustically initiated cavities depend on many factors, such as acoustic pressure amplitude, acoustic frequency, gas content of the cavity, temperature (which influences the vapor pressure), hydrostatic pressure, and the material properties of the liquid (e.g., density, viscosity, surface tension, heat conductivity, compressibility, and specific heat).

The chapters by Flynn,[1] Plesset,[2] Plesset and Prosperetti,[4] and Coakley and Nyborg[5] and the book by Rozenberg[3] (with chapters on cavitation by Akulichev, Sirotyuk, and Rozenberg) review the equations most often used to describe the spherically symmetric motion of a single cavity. Probably the most used and reasonably successful equation is similar to one that Rayleigh originally developed to discuss the collapse of a void in a fluid[44] and to which Noltingk and Neppiras,[45] Plesset,[46] and Poritsky[47]

[44] Lord Rayleigh, *Philos. Mag.* **34**, 94 (1917).
[45] B. E. Noltingk and E. A. Neppiras, *Proc. Phys. Soc. London Sect. B* **63**, 674 (1950).
[46] M. S. Plesset, *J. Appl. Mech.* **16**, 277 (1949).
[47] H. Poritsky, The collapse or growth of a spherical bubble or cavity in a viscous liquid, *in Proc. U.S. Nat. Congr. Appl. Mech., 1st* (E. Sternberg, ed.), p. 813. American Society of Mechanical Engineers, New York, 1952.

contributed in its development. We present a relatively simple derivation for bubble motion.

The kinetic energy of the mass of fluid surrounding a pulsating sphere of radius R is given by $\frac{1}{2}M_{\text{eff}}(R')^2$, where $R' \equiv dR/dt$ and M_{eff}, the effective mass "felt" by the sphere, given by three times the mass of liquid that would fill the sphere (assuming the acoustic wavelength is large compared to R); that is, $M_{\text{eff}} = 3\rho(4\pi/3)R^3$, where ρ is the liquid density. This kinetic energy minus the energy dissipation at the surface due to viscous effects is equal to the work done by surface tension σ and internal and liquid pressures, P_i and P_0, respectively.

$$\frac{1}{2} M_{\text{eff}} \left(\frac{dR}{dt}\right)^2 - \int_{R_0}^{R} \left(-\frac{4\mu}{R}\frac{dR}{dt}\right) 4\pi R^2 \, dR$$

$$= \int_{R_0}^{R} \left(P_i - P_0 - \frac{2\sigma}{R}\right) 4\pi R^2 \, dR, \qquad (7.3.1)$$

where μ is the viscosity of the liquid and R_0 the initial size. Differentiation with respect to R and division by $4\pi R^2 \rho$ yield

$$RR'' + \frac{3}{2}(R')^2 + \frac{4\mu}{\rho}\frac{R'}{R} + \frac{2\sigma}{\rho R} + \frac{P_0 - P_i}{\rho} = 0. \qquad (7.3.2)$$

(A more rigorous derivation of this equation by Poritsky[47] elucidates the way in which the viscous term enters.) This equation facilitates the study of single bubbles, even though multiple-bubble effects are of practical importance in many of the applications and problems associated with acoustic cavitation (see Section 7.3.2.3). The first two terms of Eq. (7.3.2) are just the inertial terms; the third term indicates the effects of viscous stresses at the surface; and the remaining terms relate to surface tension and pressure effects. The internal pressure can usually be broken down into a condensible part P_V (V for vapor) and a noncondensible P_G (G for gas). The external pressure is composed of a static part P_0 and a time-varying part $P(t)$. For continuous, single-frequency, sinusoidal excitation, we can write $P(t) = -P_A \sin \omega t$, assuming the negative half-cycle is first.

The basic bubble motion Eq. (7.3.2), can be put in nondimensional form by writing

$$\beta \equiv \frac{R(t)}{R_0}; \qquad \tau \equiv \frac{t[(P_0 - P_V)/\rho]^{1/2}}{R_0};$$

$$C \equiv \frac{4\mu}{R_0[(P_0 - P_V)/\rho]^{1/2}}; \qquad D \equiv \frac{2\sigma}{R_0(P_0 - P_V)}; \qquad (7.3.3)$$

$$E(\beta) \equiv -\frac{P_G(\beta)}{P_0 - P_V}; \qquad A(\tau) \equiv \frac{P(\tau)}{P_0 - P_V}.$$

These substitutions give

$$\beta\ddot{\beta} + \frac{3}{2}\dot{\beta}^2 + C\frac{\dot{\beta}}{\beta} + \frac{D}{\beta} + E(\beta) + 1 = -A(\tau); \qquad \dot{\beta} \equiv \frac{d\beta}{d\tau}. \quad (7.3.4)$$

Equation (7.3.4) applies to an incompressible liquid—an assumption that breaks down in the final stages of collapse of a cavity if the interfacial velocity approaches the speed of sound. Other assumptions are that there are no large thermal effects (large viscous or latent heat effects), no mass transport (the gas content remains sensibly constant), and constant internal pressure throughout the cavity at any instant in time, which is valid if the wavelength of sound in the gas is sufficiently larger than the bubble diameter. When any of these assumptions breaks down, a more comprehensive theory must be substituted. (See Flynn's review for several "higher-order" equations.[1])

Assuming constant internal pressure, Rayleigh derived Eq. (7.3.2) without terms two and three and solved it analytically for the case of the collapse of a cavity in a liquid. His solution demonstrated the high interface velocities and liquid pressures that were necessary to explain cavitational damage.

Noltingk and Neppiras solved Eq. (7.3.2) numerically for cases in which viscous effects were negligible; the noncondensible gas pressure was assumed to obey Boyle's law ($P_G R^3$ = const) and the external pressures included a simple harmonic term.[45] They found that for small, steady-state oscillations, Eq. (7.3.2) reduced to the form of a simple, sinusoidally excited, spring-mass oscillator.

The resonance frequency of a bubble of radius R_r for such a system is given by

$$f = \frac{1}{2\pi R_r}\left\{\frac{3\kappa P_0}{\rho}\left[1 + X_r\left(1 - \frac{1}{3\kappa}\right)\right]\right\}^{1/2}, \quad \text{where} \quad X_r = \frac{2\sigma}{P_0 R_r}, \quad (7.3.5)$$

$$f \simeq \frac{1}{2\pi R_r}\left(\frac{3\kappa P_0}{\rho}\right)^{1/2} \qquad \text{for} \quad X_r \ll 1. \quad (7.3.6)$$

This result is based on the assumption that the gas pressure follows PV^κ = const, with κ being the polytropic constant which varies between γ (the ratio of specific heats) and 1, the limits for adiabatic and isothermal conditions, respectively. Equation (7.3.6) was originally derived by Minnaert,[48] while the derivation of the more complete expression by Richardson[49] was illustrated clearly by Devin.[50] Equation (7.3.6) gives

[48] M. Minnaert, *Philos. Mag.* **16**, 235 (1933).
[49] J. M. Richardson, as quoted in H. B. Briggs, J. B. Johnson, and W. P. Mason, *J. Acoust. Soc. Am.* **19**, 664 (1947).
[50] C. Devin, *J. Acoust. Soc. Am.* **31**, 1654 (1959).

frequencies within 5% of Eq. (7.3.5) for R_r greater than 10 μm ($f < 300$ kHz) for $P_0 = 1$ bar.

Most investigations of the dynamics of acoustic cavitation can be separated into two major categories: (1) cases in which the fractional change in cavity size during one acoustic cycle is reasonably small (say not exceeding 50%) and (2) cases in which the cavity size changes dramatically during one acoustic cycle (e.g., when the acoustic pressure amplitude is greater than several atmospheres). Equation (7.3.4) does suprisingly well in predicting the behavior of bubbles in both cases, even when the velocity of the collapsing cavity is approaching the speed of sound (see Section 7.3.4.2). We shall now discuss these two classes of bubble dynamics.

7.3.2. Gas Bubbles; Noncatastrophic Dynamics

7.3.2.1. Small Fluctuations.
Small oscillations of a bubble occur for acoustic pressure amplitudes ranging from a small fraction of an atmosphere to a few atmospheres (1 atm = 10^5 Pa). The noncondensible gas, usually air, acts as a spring to limit large changes from the equilibrium radius of the bubble. Since bubbles are mechanically unstable in a liquid, they will slowly change equilibrium size at a rate determined by the diffusion of gas across the interface.

7.3.2.1.1. RECTIFIED DIFFUSION. This diffusion is accelerated, however, by the presence of the sound field. In a process, first discussed by Harvey et al.[14] and later studied and named "rectified diffusion" by Blake,[11] more gas diffuses into the bubble during the negative part of the acoustic cycle than leaves during the positive part, owing to the slightly larger surface area during the half-cycle of expanded size. That is, gas is acoustically "pumped" into the bubble. Safar[51] has given a general result for the acoustic pressure threshold for rectified diffusion P_D for a bubble of radius R_D and relative gas saturation C:

$$P_D/P_0 = [3\kappa(1 + X_D) - X_D][1 - (f/f_D)^2][1 - C + X_D]^{1/2}/[6(1 + X_D)]^{1/2},$$

(7.3.7)

where $X_D \equiv 2\sigma/P_0R_D$ and f_D is the frequency for which a bubble of radius R_D is in resonance [see Eq. (7.3.6)]. This expression reduces to the simpler expressions of Blake,[11] Strasberg,[16] and Eller[52] for $X_D \ll 1$.

Strasberg,[15] Eller,[52] and Gould[53] have observed experimentally both threshold and growth effects. Each has levitated a bubble smaller than

[51] M. H. Safar, *J. Acoust. Soc. Am.* **43**, 1188 (1968).
[52] A. I. Eller, *J. Acoust. Soc. Am.* **52**, 1447 (1972).
[53] R. K. Gould, *J. Acoust. Soc. Am.* **56**, 1740 (1974).

resonant size near an acoustic pressure maximum. With the assumption of an appropriate drag law, size changes were deduced from the rise speed of the bubble when the sound field was turned off. An acoustic pressure threshold for rectified diffusion was determined for many bubbles by observing the conditions that produced growth and the conditions that were not able to arrest the bubble dissolution (for which surface tension is responsible). Measured threshold pressures are on the order of tenths of atmospheres (the total pressure never goes negative), in reasonable agreement with theory, but the rate of growth often exceeds predictions,[52] suggesting that the predicted gas concentration gradient near the bubble, in the assumption of spherical symmetry, is not appropriate. Gould has noticed that acoustic streaming could be avoided if bubble asymmetries of the bubble shape were not excited and that, in this situation, predicted growth rates were in reasonable agreement with theory.[53]

One uncertainty in predicting the threshold for rectified diffusion relates to whether the bubble behaves adiabatically or isothermally. For acoustic wavelengths in the gas that are large compared to the bubble size, one would expect that an important parameter would be the ratio of the bubble size to the thermal diffusion length of the gas. If this ratio is much larger than one, the bubble behaves adiabatically; if much smaller than one, isothermally. These tendencies are consistent with Eller's experimental results (see also Section 7.3.2.1.3).[52]

When a bubble that is growing by rectified diffusion reaches a size that will make the bubble resonant at close to the Minnaert frequency, the bubble undergoes large oscillations. In most experimental situations the bubble has grown near a region of local acoustic pressure maximum, because, as many have shown, forces due to acoustic radiation pressure (see Part 6) compel the bubble to move to these positions. However, for a bubble that continues to grow slowly by rectified diffusion and becomes slightly larger than the size of resonance, the driving frequency becomes nearly 180° out of phase with the motion of the bubble (as would be the case for an undamped, spring-mass oscillator driven above resonance), and then the bubble is forced away from the local pressure maximum and toward an acoustic pressure minima; bubble growth ceases.

7.3.2.1.2. BUBBLE DAMPING. The undamped resonance frequency, as predicted by Minnaert, must be modified because of damping. Devin,[50] Hsieh and Plesset,[54] and Prosperetti[55] are among those who have considered the various mechanisms for damping, including heat conduction, viscous losses, and acoustic radiation of energy away from the

[54] D.-Y. Hsieh and M. S. Plesset, *J. Acoust. Soc. Am.* **33**, 206 (1961).
[55] A. Prosperetti, *J. Acoust. Soc. Am.* **61**, 17 (1977).

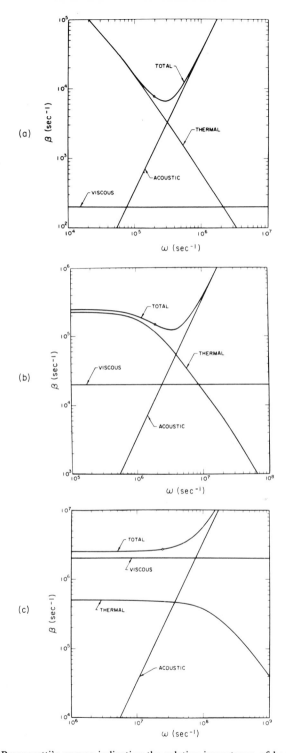

FIG. 5. Prosperetti's curves indicating the relative importance of heat conduction, viscosity, and acoustic radiation in bubble damping: (a) $R_0 = 10^{-2}$ cm; (b) $R_0 = 10^{-3}$ cm, $\beta_{vis} = 0.02$; (c) $R_0 = 10^{-4}$ cm, $\beta_{vis} = 0.02$ [from A. Prosperetti, *J. Acoust. Soc. Am.* **61,** 17 (1977)].

cavity. Prosperetti's predictions for the relative importance of these mechanisms are based on a model that assumes small amplitude oscillations of stable, spherical bubbles; his principal results for the dimensional damping coefficient (in reciprocal seconds) are repeated in Figs. 5a–c, which show how the relative importance of damping mechanisms depends on bubble size.

7.3.2.1.3. ISOTHERMAL AND ADIABATIC BUBBLES. Hsieh and Plesset[54] and Prosperetti[55] also considered the effective "spring" constant of the bubble. They found, as expected, that for bubbles large compared to the thermal diffusion length $[l = (2k/\rho C_p \omega)^{1/2}]$ in the gas, adiabatic behavior is predicted, and for small bubbles, isothermal behavior results. (Here k is the thermal conductivity, C_p the specific heat per unit mass, and $\omega = 2\pi f$.) For cases in which the wavelength of sound in the liquid is large compared to the bubble size but the wavelength of sound in the gas is comparable to the bubble size, the effective elastic constant of the bubble varies in a manner depending on the relative phase of pressure and particle velocity at the bubble surface and may even take a value outside the limits set by the thermodynamically defined adiabatic and isothermal limits. Experimental work on the thermal behavior of large bubbles has been carried out by Jensen, and adiabatic behavior of such bubbles has been observed, as predicted.[56]

7.3.2.2. Nonlinear Effects: The Case of Periodic Motion of Gas Bubbles. It has been observed that acoustically excited bubbles, when excited at one frequency, will radiate at that frequency and others. This is not surprising in view of the nonlinear equations that model bubble motion. For instance, if the driving frequency is an integral multiple of the bubble resonance frequency, energy may be produced at this resonance, which is a subharmonic of the driving frequency. If the resonance frequency is an integral multiple of the driving frequency, then a harmonic of the driving frequency may be observed. As several theoretical papers have demonstrated, the possibilities are restricted not only to subharmonics and harmonics but also to sum and difference components and ultraharmonics. (See references 4–15 in Lauterborn.[57]) Lauterborn[57] probably has claim to the most prodigous output of numerical (computer) solutions of Eq. (7.3.4). Not only were radius versus time curves forthcoming but also the spectral content of periodic bubble oscillations were retrieved, as demonstrated in Fig. 6.

The problem with the variety of results is that there are so many possi-

---

[56] F. B. Jensen, Report 30, Danish Center for Applied Mathematics and Mechanics, Tech. Univ. of Denmark (July 1972).

[57] W. Lauterborn, *Acustica* **23**, 74 (1970); *J. Acoust. Soc. Am.* **59**, 283 (1976).

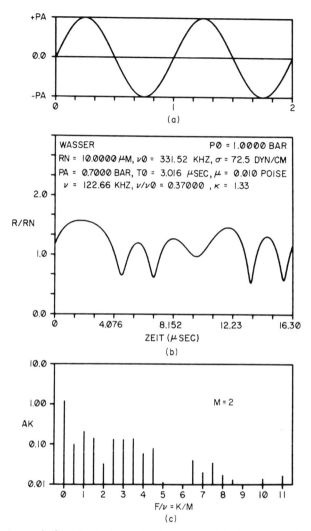

FIG. 6. An example from Lauterborn of information obtainable from periodic solutions of Eq. (7.3.4). Shown are: (a) acoustic pressure; (b) bubble radius versus time curve; and (c) frequency spectrum analysis of the periodic bubble motion [from W. Lauterborn, *J. Acoust. Soc. Am.* **59**, 283 (1976)].

bilities but no definitive experimental studies that confirm or disprove a particular model. It has been suggested, for example, that the observation of the first subharmonic component could be used as an indicator of cavitation. This would be particularly convenient because then one would merely set an analyzer at half the driving frequency and increase the amplitude of the driving signal until a signal at one-half of the driving frequency is observed. There are several caveats to this approach: For instance, subharmonic generation itself has a threshold. This threshold would have to be below the cavitation threshold. Another criticism is that bubbles that would participate in generating the subharmonic would grow, presumably by rectified diffusion, and reach their resonance size for the given driving frequency before growing to the larger size associated with the subharmonic resonance. These resonant bubbles would presumably become unstable and break up. (Lauterborn has shown that a subharmonic signal can also arise from ultraharmonic oscillations— especially the $\frac{3}{2}$ one—a fact that negates this criticism.[57]) Still another, often neglected, fact is that subharmonics can be generated in the absence of bubbles, solely due to the nonlinear behavior of the liquid itself.[58] In Section 7.3.4.3.1 subharmonic generation resulting from transient cavitation will be discussed.

The uncertainties associated with the interpretation of theories of single-bubble dynamics and their relation to cavitation of the noncatastrophic variety (i.e., stable cavitation) suggest that for this type of cavitation, thresholds are less meaningful than measures based on the *effects* of cavity dynamics. These effects often result from nonsymmetric bubble motion and multiple bubble phenomena.

7.3.2.3. Bubble Fields. 7.3.2.3.1. MUTUAL "BJERKNES" FORCES. One often-observed cavitation phenomenon is the coalescence of bubbles in a sound field. A bubble in an intense sound field will move because of acoustic radiation pressure, as discussed in Section 7.2.1.3 for acoustic levitation of drops. Each pulsating bubble also generates its own "secondary" radiation field. Two bubbles of smaller than resonance size caught in each others' secondary radiation fields will attract each other, a result first noted by C. A. Bjerknes and V. F. K. Bjerknes (father and son),[59,60] later revived in the context of cavitation phenomena by Kornfield and Survorov,[61] Blake,[62] and Rozenberg,[63] and recently checked quantitatively in experiments by Crum.[64] Crum and Nordling[44] photographed stroboscopically violent transient cavitation events and observed that the cometlike cavity moving rapidly from the pressure maximum might be composed of several little cavities in close proximity (see Fig. 7). In another

[58] N. Yen, *J. Acoust. Soc. Am.* **57**, 1357 (1975).
[59] V. F. K. Bjerknes, "Die Kraftfelder." Vieweg, Braunschweig, 1909.
[60] V. F. K. Bjerknes, "Fields of Force." Columbia Univ. Press, New York, 1906.
[61] M. Kornfeld and L. Savorov, *J. Appl. Phys.* **15**, 495, (1944).
[62] F. G. Blake, *J. Acoust. Soc. Am.* **21**, 551(L) (1949).
[63] M. D. Rozenberg, Technical Memo No. 26, Acoustical Research Laboratory, Harvard Univ., Cambridge, Massachusetts (1953).
[64] L. A. Crum, *J. Acoust. Soc. Am.* **57**, 1363 (1975).

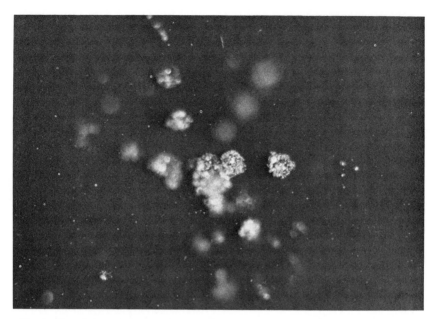

FIG. 7. Transient cavity appears, in this stroboscopically illuminated scene, as a cluster of oscillating bubbles moving together away from the region of high acoustic pressure where cavitation was initiated [L. A. Crum and D. A. Nordling, *J. Acoust. Soc. of Am.* **52**, 294 (1972)].

manifestation of these mutual Bjerknes forces, Apfel and Harbison[65] have noticed that the explosive vaporization of a superheated drop suspended in another liquid will result in many bubbles in the absence of a sound field but in only one major bubble in the presence of the sound field. Presumably, the collapsing vapor bubble, which begins to come apart because of growing surface asymmetries, is "held together" because of attractive Bjerknes forces.

7.3.2.3.2. STREAMERS. A line of bubbles along a path in a sound field is sometimes referred to as a *cavitation streamer*. This tendency of collective motion is most likely a result of the combined effects of radiation pressure, which moves bubbles smaller than resonant size toward acoustic pressure maxima, and mutual Bjerknes forces as just discussed. Bubbles reaching resonance size will then be pushed away from pressure maxima and will collapse, often shattering into microcavities, which then redissolve into the liquid.[66]

[65] R. E. Apfel and J. P. Harbison, *J. Acoust. Soc. Am.* **57**, 1371 (1975).
[66] W. L. Nyborg and D. E. Hughes, *J. Acoust. Soc. Am.* **42**, 891 (1967).

7.3.2.3.3. DYNAMICS OF BUBBLE FIELDS—HOLOGRAPHIC METHODS. Crum was able to observe the motion of bubbles easily by using low-frequency, 60-Hz, shaker-table excitation and reduced pressure[64]; this has the added advantage that bubble sizes are relatively large. Ebeling and Lauterborn[67-69] have sought to study the dynamics of bubble fields with much greater temporal and spatial resolution. This was achieved using high-speed holocinematography in which up to eight holograms of the bubble fields could be recorded at maximum repetition rates of 20 kHz. In their particular experiments, the bubble fields have been produced by focusing a giant pulse from a ruby laser into water (see also Section 7.3.4.2). Another ruby laser, Q-switched, was used to produce coherent light pulses for exposing the hologram. In reconstructing the image from the hologram not only was the incoherent light which was produced during the laser-induced breakdown of the liquid not recorded, but, also, by focusing in different planes of the hologram, different planes of the bubble field could be distinguished.

7.3.2.4. Asymmetric Bubble Oscillation. Strasberg and Benjamin,[70] Eller and Crum,[71] and Gould,[53] among others, have observed the onset of asymmetric shape oscillations with acoustically levitated air bubbles. If the pressure amplitude on such a bubble, growing by rectified diffusion, is increased beyond a certain threshold value, an erratic dancing motion of the bubble is observed. This motion is attributed to shape oscillations. Benjamin and Strasberg,[72] Eller and Crum,[71] Hsieh and Plesset,[54] and Hsieh[73] have used perturbation techniques to provide theoretical estimates for the threshold pressure; Hsieh has also used variational principles to study the more general problem of the dynamics of nonspherical bubbles and liquid drops.[74] He points out that a subharmonic signal can be generated, but notes, as has Storm who observed oscillating bubbles in a gel,[75] that the radiated pressure for subharmonics is very weak, owing to its origin in surface rather than volume oscillations.[76] Much stronger sub-

[67] K. J. Ebeling, *Optik* **48**, 383, 481 (1977).
[68] K. J. Ebeling and W. Lauterborn, *Opt. Commun.* **21**, 67 (1977).
[69] W. Lauterborn and K. J. Ebeling, *Appl. Phys. Lett.* **31**, 663 (1977).
[70] M. Strasberg and T. B. Benjamin, *J. Acoust. Soc. Am.* **30**, 697(A) (1958).
[71] A. I. Eller and L. A. Crum, *J. Acoust. Soc. Am.* **47**, 762 (1970).
[72] T. B. Benjamin and M. Strasberg, *J. Acoust. Soc. Am.* **30**, 697(A) (1958).
[73] D.-Y. Hsieh, *J. Acoust. Soc. Am.* **56**, 392 (1974).
[74] D.-Y. Hsieh, *Proc. Symp. Finite-Amplitude Wave Effects Fluids, Copenhagen, 1973* (L. Bjorno, ed.), pp. 220–226. IPC Science and Technology Press, Guildford, England, 1974.
[75] D. L. Storm, *Proc. Symp. Finite-Amplitude Wave Effects Fluids, Copenhagen, 1973* (L. Bjorno, ed.) pp. 220–226. IPC Science and Technology Press, Guildford, England, 1974.
[76] M. Strasberg, *J. Acoust. Soc. Am.* **28**, 20 (1956).

harmonics do arise directly from nonlinear, spherical oscillations, as discussed in Section 7.3.2.2.

As we shall see in Section 7.3.4.3.3, the asymmetric collapse of "transient" cavities is the culprit in much of the damage produced by acoustic and hydrodynamic cavitation.

7.3.3. Vapor Bubble Dynamics

Evaporation and condensation at a bubble interface introduce new elements into the consideration of the dynamics of bubbles in a sound field. Here were can include condensible vapors of composition different from the liquid or the vapor of the liquid itself.

Several articles have been written on this subject in conjunction with interest in interpreting cavitation in liquid helium and other cryogenic liquids. Marston has recently obtained approximations for describing the resonance of vapor bubbles and has prepared the following summary[77]:

One of the most interesting features of the solution of the equations for vapor bubble dynamics is the prediction that there are *two* distinct frequencies for resonance of a vapor bubble. One of these is similar to the familiar spring-mass resonance frequency associated with inert gas bubbles [see Eq. (7.3.6)], which is derived under the adiabatic assumption that the bubble is large compared to the thermal diffusion length l_v in the vapor: $l_v = (2D_v/\omega)^{1/2}$; $D_v = k_v/\rho_v C_{p,v}$, where k_v, ρ_v, and $C_{p,v}$ correspond to the vapor's thermal conductivity, density, and heat capacity per unit mass as constant pressure, respectively, and $\omega = 2\pi f$ and $f =$ the acoustic frequency. The discussion of the second resonance, which follows, also uses the diffusion length l_L and diffusivity D_L of the *liquid,* which are computed from the corresponding liquid properties.

Finch and Neppiras[78] and Wang[79] found there should be a second resonance which is apparently associated with evaporation and condensation of vapor at the bubble's surface. Their mathematical treatments of the steady-state response of a vapor bubble for small acoustic pressure amplitudes assume that the temperature oscillations of the *vapor* are uniform (T is independent of r for $r < R_0$) which is a good approximation provided that $R_0 < l_v$. The second resonance appears in Finch and Neppiras' calculation as a second root to the characteristic equation:

$$S - \omega^2 M_{eff} = 0 \qquad (7.3.8)$$

[77] See also, P. L. Marston, *J. Acoust. Soc. Am.* **66,** 1516 (1979).
[78] R. D. Finch and E. A. Neppiras, *J. Acoust. Soc. Am.* **53,** 1402 (1973).
[79] T. Wang, *Phys. Fluids* **17,** 1121 (1974).

where the effective mass M_{eff} is $4\pi R_0^3 \rho_L$, and the stiffness S is a function of R_0 and ω; the ω dependence is due to the condensibility of the vapor. They give plots of ω versus R for the two roots, which were obtained by numerical methods, and predict an enhancement in the radial oscillation amplitude when the condition for either resonance is satisfied.

Approximations for the frequency of the new root ω_e reveal the physically significant parameters which relate ω_e and R_0. The self-consistency of the most readily obtainable approximations depends on the resultant magnitude of $\Omega_L = l_L/2R_0$, where l_L is to be computed from the approximated ω_e. When $\Omega_L \gg 1$, the approximation for ω_e is[79]:

$$\omega_e = D_L(d\Gamma\eta)^{2/3}R^{-8/3}[1 + O(\Omega_L^{-1})], \qquad (7.3.9)$$

and when $\Omega_L \ll 1$, it is:

$$\omega_e = D_L(d\Gamma\eta)^2 R^{-4}[1 + O(\Omega_L)], \qquad (7.3.10)$$

where d, which has the dimensions of length, is:

$$d = \frac{2\sqrt{2}\sigma}{3P_V}. \qquad (7.3.11)$$

σ is the surface tension and, following the notation of Finch and Neppiras,[78] the dimensionless parameters Γ and η are:

$$\Gamma = P_V\phi/T \qquad (7.3.12)$$

$$\eta = \rho_L C_{p,L}\phi \qquad (7.3.13)$$

where ϕ is the reciprocal slope of the liquid's vapor pressure curve. The mean vapor pressure in the bubble P_V, which appears in Eqs. (7.3.11) and (7.3.13), should be specified at the quasi-stable equilibrium conditions which can be maintained because of rectified heat transport; however, the vapor pressure of the liquid at the bath temperature may be used provided the estimated $d \ll R_0$.

Equations (7.3.9) and (7.3.10) were derived by noting that the dependence of ω_e on R is dominated by S (see Fig. 1 of Finch and Neppiras[78]); for the purpose of estimating ω_e, one can find the roots of $S = 0$. The approximations used in solving $S = 0$ are: $\eta\Omega_L \gg 1$; $\Gamma\eta\Omega_L \gg 1$; $\Gamma\eta\Omega_L \gg \gamma\Gamma^2/(\gamma - 1)$ (where γ is the polytropic constant); and (as assumed by Finch and Neppiras) $\Omega_v = l_v/2R_0 > 1$. In spite of these requirements, Eqs. (7.3.9) and (7.3.10) are applicable to many liquids. They show a strong dependence for ω_e on σ; this is to be expected since the plots of roots of Eq. (7.3.8) given by Finch and Neppiras show that the R_0 of a bubble which resonates at a prescribed ω is typically much *smaller* than the R_0 associated with the usual (Minnaert) resonance. Hsieh,[80] in a first attempt at a physical explanation of the condensation–evaporation resonance, has suggested an approximation for ω_e which is apparently inconsistent with the above results since it neglects all σ dependence. (See Hsieh[80] for an update.)

Equation (7.3.9) is appropriate for estimating $f_e = \omega_e/2\pi$ of vapor bubbles in Helium I. For example, using liquid and vapor properties at 3.0°K, for $R_0 = 3$ μm, it gives: $f_e \approx 66$ Hz, $\Omega_L \approx 3$, and $\Omega_v \approx 6$. This is to be compared with roots plotted by Finch and Neppiras which for the new resonance is ≈ 110 Hz and for the Minnaert resonance is ≈ 1.2 MHz. Equation (7.3.10) is the appropriate one for bubbles of water vapor in water. For example at 350°K with $R_0 = 250$ μm, it gives: $f_e \approx 5$ Hz, $\Omega_L \approx 0.2$, and $\Omega_v \approx 3.6$. The plotted new and Minnaert roots are 12 Hz and 1.8 kHz, respectively. The power law dependence between ω_e and R, given in Eqs. (7.3.9) and (7.3.10), also shows close correspondence with the numerical results of Finch and Neppiras.

There is need for a definitive experimental verification of the existence of the new resonance. Marston and Greene[81] observed stable vapor bubbles in insonified He I with R_0 several decades smaller than the radius corresponding to the Minnaert resonance. Furthermore, these small bubbles appeared to be acted on by acoustic radiation pressure. They interpreted their observations as evidence for the new resonance and suggested that quantitative measurements of the radiation pressure could test for the predicted enhancement of the radial oscillation amplitude.

7.3.4. Transient Cavitation

7.3.4.1. Introduction. Transient cavitation is characterized by bubble motion in which the characteristic dimension of the cavity changes greatly during one acoustic period. This may result from an originally stable bubble growing by rectified diffusion to resonance size and then collapsing strongly or from a cavity that is initiated at a fairly high acoustic pressure (greater than the ambient static pressure) and at a frequency that is sufficiently low to allow adequate time for growth. The cavity grows to at least two to three times its original size during the negative part of the acoustic cycle, it may oscillate a few times, and it then collapses irreversibly.

[80] D.-Y. Hsieh, *Phys. Fluids* **19**, 599 (1976) [Hsieh has revised his formulation and now obtains a physical formulation that leads to the results of Marston.[77]]

[81] P. L. Marston and D. B. Greene, *J. Acoust. Soc. Am.* **62**, 319 (1978).

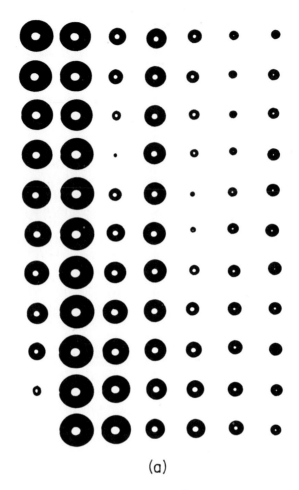

(a)

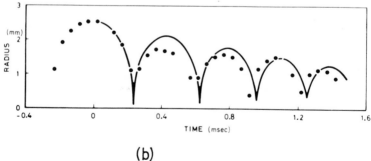

(b)

FIG. 8. Dynamics of laser-induced bubble motion in silicone oil: (a) experimental observations at a frame rate of 75,000 frames/sec. (b) theoretical predictions (solid curve) compared to experimentally observed radius versus time curves, showing good agreement as to the amplitude and period of the oscillations [W. Lauterborn, *Proc. Symp. Finite Amplitude Wave Effects Fluids, Copenhagen, 1973* (L. Bjorno, ed.), pp. 272–276. IPC Science and Technology Press, Guildford, England, 1974].

7.3.4.2. Applicability of Eq. (7.3.4). Most of the type of bubble motion just discussed can be described by Eq. (7.3.4), which assumes an incompressible liquid and no gas diffusion or heat exchange. It is more than adequate for describing the transient motion of a cavity up to the point when the collapsing interface approaches the speed of sound. Upon expansion, the velocity will never exceed $(2 \Delta P/3\rho)^{1/2}$, where ΔP does not exceed the peak difference between internal and external pressure (ΔP will be taken as positive when the internal minus external pressure is positive.) This velocity does not approach the speed of sound of most liquids until ΔP approaches 10,000 atm. Lauterborn has filmed the dynamics of single, laser-initiated bubbles in silicone oil in the absence of a sound field.[82] Results of one of these experiments are shown in Fig. 8. The solid lines produced by the numerical solution of Eq. (7.3.2) attest to the adequacy of the theory through most of the bubble motion, predicting with reasonable accuracy the maximum bubble size on each oscillation and the oscillation period. One interesting feature of this work with silicone oil is the symmetrical rebound of the bubble owing to the stabilizing influence of the liquid's viscosity (4.85 P). Bubbles in water (viscosity $\simeq$ 0.01 P) do not rebound symmetrically owing to the underdamped instabilities established on the bubble surface.

Apfel[83] and Akulichev[84] have attempted to give a little more generality to numerical solutions of Eq. (7.3.4) by considering it in nondimensional forms and solving it for a series of nondimensional parameter values. Curves of this type by Apfel are given in Fig. 9 for the case of an air bubble behaving adiabatically and for sinusoidal excitation with peak pressure amplitude P_A; the parameters β, τ, C, D are defined in Eq. (7.3.3). Akulichev's curves, which are for zero viscosity, are very similar in character to these curves. Figures 9a–c display another feature of bubble dynamics—their similarity as τ_0, the nondimensional period, is varied over a factor of 10, a result predicted from the analysis of Akulichev for cases in which the resonance frequency associated with the initial bubbles size [see Eq. (7.3.5)] is greater than the driving frequency and the ratio P_A/P_0 is large.

For high driving pressure and low acoustic frequencies, viscous effects

[82] W. Lauterborn, *Proc. Symp. Finite-Amplitude Wave Effects Fluids, Copenhagen, 1973* (L. Bjorno, ed.), pp. 272–276. IPC Science and Technology Press, Guildford, England, 1974.

[83] R. E. Apfel, Technical Memo. 62, Chapter 4. Acoustics Research Lab., Harvard Univ., Cambridge, Massachusetts (1970); Some new results on cavitation thresholds and bubble dynamics, *in* "Cavitation and Inhomogeneities" (W. Lauterborn, ed.), pp. 79–83. Springer-Verlag, Berlin and New York, 1980.

[84] V. A. Akulichev, Pulsations of cavitation voids, *in* "High Intensity Ultrasonic Fields" (L. D. Rozenberg, ed.), pp. 203ff. Plenum, New York, 1971.

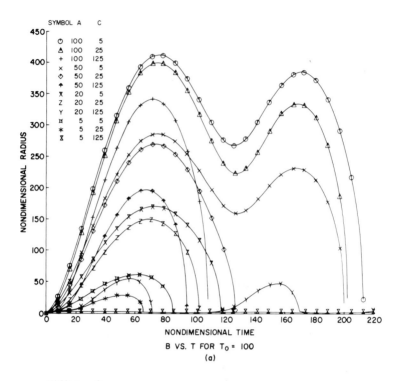

B VS. T FOR T_0 = 100

(a)

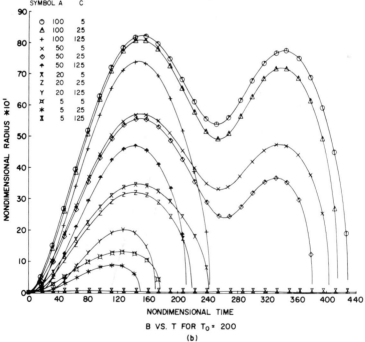

B VS. T FOR T_0 = 200

(b)

FIG. 9.   Numerical solutions of Eq. (7.3.4) giving nondimensional radius versus nondimensional time for sinusoidal excitation starting with the negative half-cycle.   Parameters A, C, and τ_0 are defined in Eq. (7.3.3).   Shown are nondimensional surface tension $D = 1$,

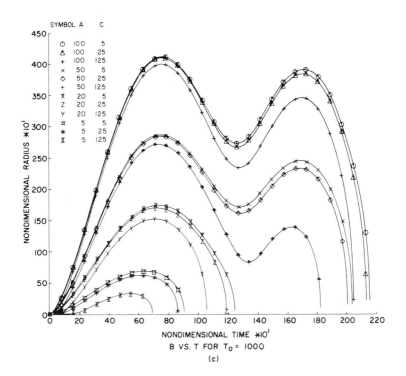

B VS. T FOR T_0 = 1000

(c)

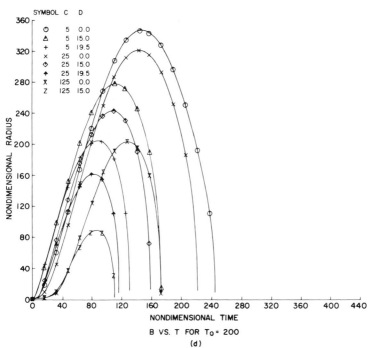

B VS. T FOR T_0 = 200

(d)

nondimensional gas pressure $E = 1$, and $\tau_0 = 100$ (a), 200 (b), and 1000 (c); note the $\times 10$ on some of the scales. Part (d) shows solutions for nondimensional acoustic pressure $A = 20$, gas pressure $E = 1$, with C and D varying, and $\tau_0 = 200$.

play a relatively small role in the overall dynamics of cavity motion, except that a cavity that might oscillate two or three times in a low viscosity liquid might oscillate once or twice in a liquid of high viscosity. Moreover, for high collapse velocities and small size, the $\dot{\beta}/\beta$ term becomes very significant.

For low driving pressures and higher frequencies, the viscous time constant may represent a significant fraction of a half acoustic period, leading to a strong retarding effect on the initial growth. This viscous time constant t_V can be estimated by equating the viscous and driving terms in Eq. (7.3.2). One finds that $t_V \simeq 4\mu/\Delta P$, where ΔP is approximately the average driving pressure difference across the bubble interface (see the following discussion) and is independent of initial size and liquid density. (See also Persson.[85]) As an example, consider a bubble in water ($\mu = 0.01$ P) subjected to a sound field of peak acoustic pressure amplitude 1.3 bar. Then ΔP, as will be discussed, is $\simeq 0.2$ bar, giving $t_V \simeq 0.2$ μsec; viscous effects are not important until the frequency is above 1 MHz, corresponding to $t_V > T/5$, where T is the acoustic period. For an oil, such as a light machine oil, viscous effects could be important at or above the 10–50-kHz range. For highly viscous materials (e.g., tissue, as in Chapter 7.4), viscous effects prevent significant growth of bubbles without mass transport unless driving pressures are very high.

Nor can one neglect the inertial "start-up" time t_I, which might be defined as the time required for a cavity to reach about 75% of its ultimate velocity ($d\beta/d\tau = \sqrt{\frac{2}{3}}$) under a constant driving pressure difference ΔP. This can be estimated from Eq. (7.3.4), neglecting viscous and surface terms; it is $t_I \simeq 2R(\rho/\Delta P)^{1/2}/3$. This is an important parameter because it represents the minimum time before significant growth by dynamic means can get started. If a bubble is to grow significantly during less than half an acoustic cycle during which the absolute pressure is negative, then the inertial time must be less than about one-fifth of a period: $t_I < T/5 = 1/5f$, or $R < 0.3(\Delta P/\rho)^{1/2}/f$. One can, therefore, define an "inertial radius" as

$$R_I = \frac{0.3}{f} \left(\frac{\Delta P}{\rho} \right)^{1/2} \qquad (7.3.14)$$

As an example, consider a 20-μm bubble in water subjected to a sound field of peak acoustic pressure amplitude 1.3 bars ($\Delta P = 0.2$ bars, as before). Then $t_I \simeq 3$ μsec. If the acoustic frequency were 20 kHz, the

[85] B. Persson, Proc. Symp. Finite-Amplitude Wave Effects Fluids, Copenhagen, 1973 (L. Bjorno, ed.), pp. 268–271. IPC Science and Technology Press, Guildford, England, 1974.

maximum size would be about 170 μm, independent of the initial size [see Eq. (7.3.23)]. However, for a 170-kHz sound field, the start-up time is comparable to half an acoustic period. Clearly, a 100-μm bubble at 20 kHz will require greater pressure amplitude for significant growth by dynamic means, provided that this amplitude is above the "Blake" threshold given by Eq. (7.2.2).

The general picture of the growth and initial collapse of a transient cavity is particularly simple when "start-up" times associated with inertial and viscous effects are small compared to the acoustic period. In this case we can, in fact, make a crude analytic estimate of the maximum size R_M a bubble will achieve, thereby providing ways of estimating how R_M is influenced by P_0, P_A, ρ, R_0, and f.

Consider the growth of a bubble in two stages: (1) while the driving pressure difference across the bubble is positive and (2) while the pressure difference is negative, until the bubble momentum goes to zero.

While the pressure difference is positive, an average bubble velocity can be estimated by $V = (2 \, \Delta P/3\rho)^{1/2}$, where ΔP is the time-averaged pressure difference. During a time t_m (which is less than a half acoustic period), the bubble will grow to a size Vt_m, where t_m is the time interval during which the pressure difference remains positive. In finding ΔP, we neglect the internal gas pressure (which drops rapidly as $(R_0/R)^{3\kappa}$). Therefore, $\Delta P = \langle P_A \sin \omega t - P_0 \rangle$, while this pressure difference is positive ($\langle \, \rangle$ indicates time average). The Taylor expansion to second order about $\omega t = \pi/2$ gives for the pressure difference

$$P_A[1 - (\omega t - \pi/2)^2/2] - P_0. \tag{7.3.15}$$

Setting this equal to zero and solving for time, with $p \equiv P_A/P_0$, gives

$$t_{1,2} - (1/\omega)\{(\pi/2) \pm [2(p - 1)/p]^{1/2}\}, \tag{7.3.16}$$

The time interval during which the pressure difference is positive is $t_1 - t_2$ or

$$t_m = (2/\omega)[2(p - 1)/p]^{1/2}. \tag{7.3.17}$$

The time-averaged pressure difference during this time is

$$\Delta P = \frac{1}{t_m} \int_{t_2}^{t_1} (P_A \sin \omega t - P_0) \, dt. \tag{7.3.18}$$

To second order this has the magnitude

$$\Delta P = \tfrac{2}{3} P_0(p - 1), \tag{7.3.19}$$

which is less than 5% from the exact value even when $p \gg 1$.

From Eqs. (7.3.17) and (7.3.19) we obtain an estimate of the bubble radius R_1 at the end of the first stage,

$$R_1 = Vt_m = (4/3\omega)(p - 1)[2P_0/\rho p]^{1/2}. \qquad (7.3.20)$$

The bubble of this size still has outward momentum, even though the driving pressure has gone negative. This negative pressure difference starts at zero and will go through $-P_0$ to a greater negative value as the acoustic pressure goes positive. The calculation of the additional growth is particularly complicated if one includes the time-varying acoustic pressure. Therefore, we neglect it, leaving the justification for later, and just assume that the negative pressure difference remains constant at a value equal to the ambient pressure P_0. Then all we have to do is equate the initial kinetic energy of the fluid surrounding the bubble with the change in the potential energy as the bubbles reaches its maximum (zero kinetic energy) size R_M; that is,

$$\tfrac{1}{2}M_{eff}V^2 = P_0 \tfrac{4}{3}\pi(R_M{}^3 - R_1{}^3), \qquad (7.3.21)$$

where M_{eff}, as for Eq. (7.3.1), is $4\pi R_1{}^3\rho$. With $V^2 = 2\,\Delta P/3\rho$ and Eq. (7.3.18) for ΔP, we have $V^2 = 4P_0(p - 1)/9\rho$; substituting into Eq. (7.3.21) gives

$$R_M/R_1 = [1 + \tfrac{2}{3}(p - 1)]^{1/3}. \qquad (7.3.22)$$

Note that this ratio varies over a factor of less than two for a tenfold variation in p, suggesting that our assumption of constant pressure during this stage will not lead to a large error in R_M/R_1. Substituting for R_1 into Eq. (7.3.22) now gives, for the maximum radius,

$$R_M = (4/3\omega)(p - 1)(2P_0/\rho p)^{1/2}[1 + \tfrac{2}{3}(p - 1)]^{1/3} \qquad (7.3.23)$$

$$= (4/3\omega)(\tfrac{2}{3})^{1/3}(2P_0/\rho)^{1/2}p^{5/6} \qquad \text{for } p \gg 1. \qquad (7.3.24)$$

Equation (7.3.23) will tend to overestimate the maximum radius because its derivation has neglected viscous and inertial start-up effects. It is noteworthy that this equation is independent of initial bubble size, although it is implicitly assumed that the bubble has at least doubled its original size (otherwise t_m will approach t_v and T_l).

In fact, a transient cavitation threshold $P_A = P_T$ might well be defined by $R_M = 2.3R_0$, the condition that assures that a bubble will just reach a supersonic collapse velocity in water[86] (in the incompressible assumption). (See the appendix of Noltingk and Neppiras.[45])

---

[86] W. Lauterborn, *Acustica* **22**, 48 (1969/70).

Then, substituting $2.3R_0$ for R_M in Eq. (7.3.23) yields, with $p_T \equiv P_T/P_0$ and $R_0 = R_T$,

$$R_T = (0.82/\omega)(p_T - 1)(P_0/\rho p_T)^{1/2}[1 + \tfrac{2}{3}(p_T - 1)]^{1/3}. \quad (7.3.25)$$

For $p_T \gg 1$, this reduces to

$$R_0 = (0.72/\omega)(P_0/\rho)^{1/2}p_T^{5/6}. \quad (7.3.26)$$

The derivation of Eq. (7.3.25) did not take inertial effects into account. We can determine approximately at what acoustic pressure inertial effects begin to predominate by setting R_T from Eq. (7.3.25) equal to R_I from Eq. (7.3.14), with $\Delta P = \tfrac{2}{3}P_0(p - 1)$ from Eq. (7.3.19). One finds that this occurs at about $p \approx 11$. Thus above $p = 11$, Eq. (7.3.14) with $\Delta P = \tfrac{2}{3}P_0(p - 1)$ defines the transient threshold condition. Note, therefore, that much above $p = 11$, the threshold goes as f^2 rather than the $f^{6/5}$ suggested by Eq. (7.3.26). This is consistent with the observations of Barger,[25] Esche,[42] and others.

The physical significance of the transient cavitation threshold is a much discussed point. It may well be that it closely corresponds to the conditions for surface instabilities and thus bubble break up in low viscosity liquids.

One can reasonably ask, however, about the relationship between the transient cavitation threshold and the violence (or ability to do damage) of the collapse. For example, will a 10-μm bubble that grows to 23 μm collapse with about the same "violence" as a 100-μm bubble that grows by the same factor to 230 μm? Equation (7.3.23) suggests that even though each bubble will just be transient, by the preceding definition, the larger one will store far more energy in the expanded bubble and will give back far more energy in the kinetic energy of collapse.

Thus the "violence" of a transient event is more appropriately tied to the maximum size achieved by the bubble R_M, which corresponds to the energy stored in the liquid. This maximum size of Eq. (7.3.23) is independent of initial size and only depends on acoustic pressure and frequency. Therefore, for a given frequency, a given degree of violence will occur for a given acoustic pressure.

This acoustic pressure does not correspond to a "threshold" for transient–violent cavitation, because we have not identified a critical maximum bubble size above which violent events occur and below which events are not violent. However, Eq. (7.3.23) does allow us to estimate the acoustic pressure required for comparably violent events at different frequencies.

7.3.4.3. Violent Collapse of Cavities.
Acoustical, mechanical,

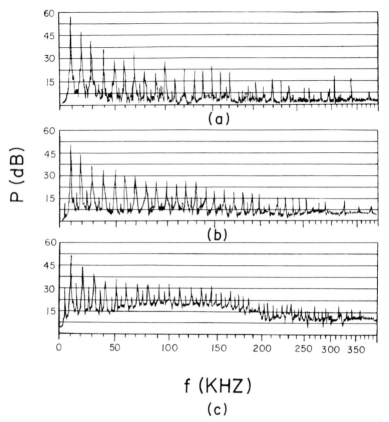

FIG. 10. Frequency spectrum from cavitation-produced acoustic emission in water. The driving frequency was 10 kHz and the peak acoustic pressures were (a) 0.4, (b) 0.6, and (c) 0.8 atm. With increasing amplitude, half harmonic multiples appear and nonperiodic noise is produced [L. Rosenberg (ed.), "High Intensity Ultrasonic Fields," p. 248. Plenum, New York, 1971].

thermal, optical, and chemical effects result from the violent collapse of cavities, as we shall now discuss.

7.3.4.3.1. ACOUSTICAL EMISSION. General observations of the acoustic spectrum of cavitation noise point to both pure tone and broadband signals. The line spectra come from steady-state oscillations of bubbles (harmonics, subharmonics, ultraharmonics as discussed in Section 7.3.2.2). The broadband noise is attributed to the violent collapse of transient cavities.

Illustrating this, Akulichev presented the graphs shown in Fig. 10 for the spectral components of acoustic pressure produced when cavitation was excited at 10 kHz in fresh tap water at peak acoustic pressures of (a) 0.4, (b) 0.6, and (c) 0.8 atm.[84] In Fig. 10b, the onset of half-harmonic components at 15, 25, 35 kHz, etc., is seen. These are even larger in Fig. 10c; superimposed upon the line spectrum is broadband noise. The origin of these half-harmonic components is clear from Fig. 9, where it is seen that transient bubbles often survive two or three acoustic cycles corresponding to half- and third-subharmonic signals. The sound created is rarely from a single bubble but rather from the initial bubble and its progeny of cavities created by instabilities during its collapse.

The use of certain characteristics of the acoustic signal produced by cavitation to characterize the threshold and intensity of cavitation will be discussed further in Chapter 7.4.

7.3.4.3.2. SONOLUMINESCENCE AND SONOCHEMISTRY. The high temperature and gas pressure generated within a collapsing, transient cavity are believed responsible for producing light[45] in a process called *sonoluminescence*,[87] although electric discharge models have not been ruled out. Furthermore, it is known that water is decomposed during intense cavitation; the role of free radicals and their recombination in cavitation-induced luminescence and chemical reactions is, however, uncertain.[88] It has been demonstrated that sonoluminescence can be observed in a 95% solution of glycerin in water from arrays of bubbles undergoing relatively large oscillations but not shattering, presumably because of the stabilizing influence of viscosity[89] (see, for example, Fig. 8).

The thermal mechanism for sonoluminescence is supported by the observation that dissolved gases of low thermal conductivity (e.g., argon) produce greater sonoluminescence and sonochemical yields than dissolved gases of high conductivity (e.g., helium and hydrogen).[90] These yields, however, also increase monotonically with the inverse of the ionization potentials of the atoms.

The reader is referred to articles by El'Piner,[91] Finch,[92] and Coakley and Nyborg[5] in which this subject is extensively reviewed. One may conclude, as in our earlier discussion of cavitation inception, that the wide variety of results and the residual uncertainty of mechanisms can be attributed, in part, to the incomplete specification of the type of cavitation and the state of the liquid at the time of the test.

[87] H. Frendel and H. Schultes, Z. Phys. Chem. **27**, 421 (1934).

[88] M. Del Duca, E. Yeager, M. O. Davies, and F. H. Hovorka, J. Acoust. Soc. Am. **30**, 301 (1958).

[89] T. K. Saksena and W. L. Nyborg, J. Chem. Phys. **53**, 1722 (1970).

[90] F. R. Young, J. Acoust. Soc. Am. **60**, 100 (1976).

[91] I. E. El'Piner, Sov. Phys. Acoust. **6**, 1 (1960).

[92] R. Finch, Ultrasonics **1**, 87 (1963).

7.3.4.3.3. ASYMMETRIC COLLAPSE NEAR BOUNDARIES: CAVITATION EROSION. The asymmetric collapse of a bubble undergoing relatively high-amplitude oscillations is probably the rule rather than the exception in low-viscosity liquids, undermining the ultility of theories that assume spherical symmetry.

The problem of symmetric versus asymmetric collapse of spheres has important implications in other areas. For example, the symmetric and adiabatic collapse of laser-bombarded deuterium and tritium held in spherical glass shells is relied on for the production of the temperatures and pressures capable of sustaining nuclear fusion. Asymmetric collapse would, of course, lessen the chance of success. The collapse of a cavitation bubble shares some physics and mathematics with the fusion problem.

Not only are surface instabilities generated by small shape perturbations on periodically oscillating bubbles, as discussed in Section 7.3.2.4, but also strong shape deformations occur for transient cavities near boundaries producing concentrated mechanical and/or thermal effects on a nearby surface. Over the years there has been a controversy as to whether the main effect was caused by the strong shock wave produced when the collapsing cavity suddenly decelerates or by jet formation due to asymmetric cavity collapse.

Because of some recent experimental and theoretical work, the jet theory now has wide favor. This is not to say that shock waves cannot play an important role in cavitation erosion but just that the shock waves associated with the deceleration of spherically symmetric cavities do not occur when the cavities are near boundaries. Asymmetries develop in these cases which can lead to jets; here the jet and associated shock waves can act as partners in the erosion process.

Benjamin and Ellis[93] elucidated the asymmetric character of cavity collapse in their elegant experiments employing single, electrolysis-generated bubbles. In Fig. 11, we see a bubble formed far away from container surfaces; a jet, attributed to asymmetrics generated by the hydrostatic gradient, grows as the bubble rebounds. In Fig. 12 we see a vapor cavity grown near a surface by negative hydrostatic pressure generated when the box containing the experiment was struck downward and fell, free fall (i.e., no gravity effects to worry about). During the collapse the side of the bubble farthest from the wall becomes involuted (frame C), and the jet has passed through the cavity in frame D at a velocity estimated at 10 m/sec. (If pressures were scaled up to atmospheric levels, i.e., by a factor of 25, the jet velocity would scale up by a factor of 5.)

---

[93] T. B. Benjamin and A. T. Ellis, *Philos. Trans. R. Soc. London Ser. A* **260,** 221 (1966).

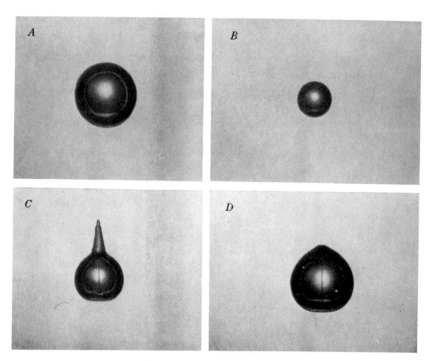

FIG. 11. Example of asymmetric motion of a transient cavity. Jet formation can result from hydrostatic pressure gradients or proximity to boundaries [from T. B. Benjamin and A. T. Ellis, *Philos. Trans. R. Soc. London Sect. A* **260**, 221 (1966)].

These experiments implicate jet formation in the erosion of solid surfaces by cavitation.

Ellis, working with Felix,[94] produced cavities by laser-induced liquid breakdown. At about the same time Lauterborn was performing similar experiments. Single bubble motion was filmed at framing rates up to 300,000 frames/sec, and bubble motion was evaluated with the aid of a digital computer. Figure 13 shows one sequence at 75,000 frames/sec.[95] In another sequence Lauterborn measured tip jet velocities in excess of 100 m/sec.

Theories of the asymmetric motion of collapsing cavities are quite complex. A remarkable comparison between Plesset and Chapman's jet theory[96] and Lauterborn's observations[95] is shown in Fig. 14 for the initial phase of a typical collapse sequence.

[94] M. P. Felix and A. T. Ellis, *Appl. Phys. Lett.* **19**, 484 (1971).
[95] W. Lauterborn and H. Bolle, *J. Fluid Mech.* **72**, 391 (1975).
[96] M. S. Plesset and R. B. Chapman, *J. Fluid Mech.* **47**, 283 (1971).

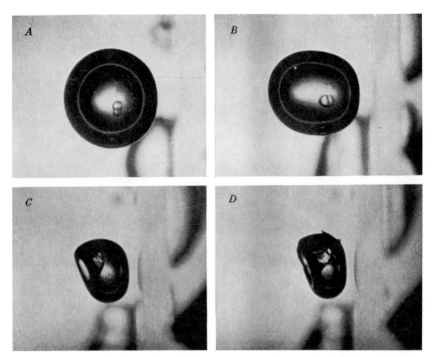

FIG. 12. Jet formation near a boundary. During the collapse, the side of the bubble far-thest from the wall becomes involuted (frame C) and in frame D the jet has passed through the cavity [from T. B. Benjamin and A. T. Ellis, *Philos. Trans. R. Soc. London Sect. A* **260**, 211 (1966)].

7.3.5. Synthesis of Some of the Theoretical Results

In Chapters 7.2 and 7.3 we have dealt with cavitation inception and dynamics. One of the goals of those who experience cavitation, whether as a desired outcome or a by-product, is to know what type of cavitation occurs in a given acoustic pressure range for a given frequency and gas saturation and what effects can be expected. Attempts at this type of synthesis have seen some success in the work of Esche,[42] Neppiras and Noltingk,[12] Strasberg,[16] Flynn,[1] and Akulichev.[84]

In the remainder of Chapter 7.3 we shall combine some of the ingredients found in these studies with that which has preceded this section to give cavitation prediction charts.

   7.3.5.1. A Cavitation Predictor.   Different cavitation thresholds— Blake, rectified diffusion, and transient thresholds defined in Eqs. (7.2.2), (7.3.7) and (7.3.25), respectively—contain different ingredients. Since

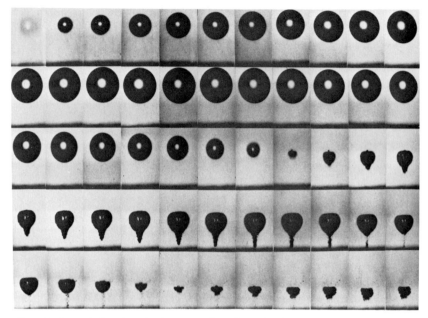

FIG. 13. Laser-induced cavitation near a boundary showing a jet formation sequence at 75,000 frames/sec. [Photograph supplied by W. Lauterborn].

the Blake threshold, Eq. (7.2.2), and the derivation of the maximum radius, Eq. (7.3.23), do not take inertial effects fully into account, we must also include in our overall predictions the bubble radius defined by Eq. (7.3.14) above which inertial effects become predominant. These four expressions are summarized in Table I.

In Figs. 15 and 16 each radius [normalized to the resonance radius of Eq. (7.3.5)] is plotted against the acoustic pressure amplitude (normalized to the ambient pressure) for the case of water at 20°C at an ambient pressure of 1 bar. Figure 15 shows the case for 20 kHz, a frequency near which many practical applications of acoustic cavitation occur. Figure 16 displays the plots for a frequency of 1 MHz, near which therapeutic and diagnostic ultrasonic units are operated. For both cases the water is assumed to be saturated with air.

The intersections of these curves define regions of different types of cavitational activity.

Region A: Only rectified diffusion will lead to bubble growth. Bubbles reaching resonance size will undergo more violent oscillations (see Sections 7.3.2.2 and 7.3.2.4) and may break up into microcavities. Because of the cushioning effect of entrapped gas, it is unlikely that cavita-

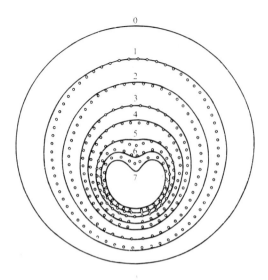

///////////////////// Solid boundary ///////////////////////////

FIG. 14. Plesset's and Chapman's theoretical predictions for the initial stages of jet formation near a boundary (solid lines) compared with experimental results of Lauterborn and
Bolle [from W. Lauterborn and H. Bolle, *J. Fluid Mech.* **72**, 391 (1975)].

TABLE I.   Summary of Cavitation Thresholds

Thresholds	Equation in text	Expression
"Blake" nucleation threshold P_B for bubble of radius R_B	(7.2.2)	$\dfrac{P_B}{P_0} = 1 + \dfrac{4}{9} X_B \left[\dfrac{3 X_B}{4(1 + X_B)}\right]^{1/2}$ where $X_B = 2\sigma/P_0 R_B$
Threshold for rectified diffusion P_D for bubble of radius R_D[a]	(7.3.7)	$\dfrac{P_D}{P_0} = \dfrac{[3\kappa(1 + X_D) - X_D][1 - (f/f_D)^2]}{[6(1 + X_D)]^{1/2}[1 - C + X_D]^{-1/2}}$ where $X_D = 2\sigma/P_0 R_D$ and f_D is given by Eq. (7.3.5)
Transient threshold P_T for bubble of radius R_T	(7.3.25)	$R_T = \begin{cases} \dfrac{0.13}{f}\left(\dfrac{P_0}{\rho}\right)^{1/2}\left\{\dfrac{p-1}{\sqrt{p}}\left[1 + \dfrac{2}{3}(p-1)\right]^{1/3}\right\} \\ \quad \text{for } p \lesssim 11 \\[2ex] \dfrac{0.3}{f}\left(\dfrac{P_0}{\rho}\right)^{1/2}\left[\dfrac{2}{3}(p-1)\right]^{1/2} \\ \quad \text{for } p \gtrsim 11 \\ \quad \text{where } p \equiv P_T/P_0 \end{cases}$

[a] After M. H. Safar, *J. Acoust. Soc. Am.* **43**, 1188 (1968).

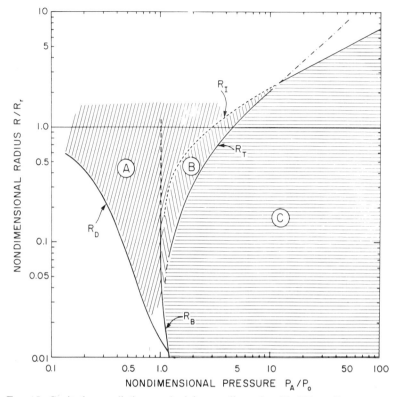

FIG. 15. Cavitation prediction graph giving nondimensional bubble radius versus nondimensional acoustic pressure for a frequency of 20 kHz, with $R_r = 170 \, \mu$m and $C = 1$ (saturation); see text for discussion.

tion of this type will lead to the erosion of solid surfaces. It is interesting to note that at reduced gas saturation the R_D curve is much closer to the resonance radius size; that is, only the very small number of bubble nuclei near resonance size are likely to participate in growth by rectified diffusion.

Region B: Growth by rectified diffusion and/or by direct mechanical means (i.e., with little gas transport) may occur but the initial bubble will not be transient. If a bubble reaches resonance size, it may eventually shatter into microbubbles; if some of these are smaller than the initial size of the original bubble, they may be in region C and, therefore, may go transient. Note that R_T and R_I curves are not really valid for $p < 2$.

Region C: This region, defined by the transient threshold, the "Blake" threshold, and the condition for inertially controlled bubbles, indicates which bubbles will be transient.

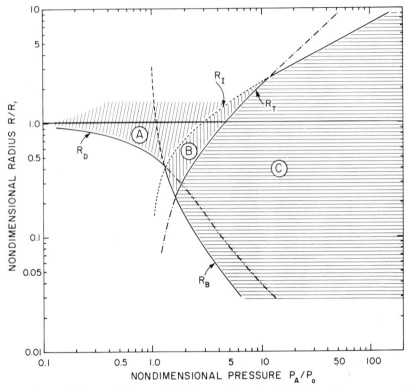

FIG. 16. Cavitation prediction graph giving nondimensional bubble radius versus nondimensional acoustic pressure for a frequency of 1 MHz with $R_r = 3.25 \mu$m and $C = 1$ (saturation); see text for discussion.

We could have also drawn in two vertical lines in the transient regions in Figs. 15 and 16 to indicate approximate equality of the energy of the collapse. For instance, the maximum energy of collapse corresponding to an energy of 0.1 erg occurs for a bubble with a maximum radius of about 29 μm. At 20 kHz a pressure of about 1.2 bars is required, using Eq. (7.3.23). (This number is probably an overestimate because we are using this equation outside its range of validity.) At 1 MHz, a pressure of about 19 bars is required to produce the same growth and approximately equivalent collapse.

7.3.5.2. Important Caveats. Those who have studied cavitation are well aware of the difficulty of generalizing about the many phenomena that are described under its heading. The plots of Figs. 15 and 16 are meant to be taken semiquantitatively. Regions are not as well defined as they appear.

Viscous effects have been neglected because the viscous time constant

$t_v \sim 4\mu/\Delta P$ is not important below 2 MHz except when $P_A \simeq P_0$, where rectified diffusion effects are most important. For liquids of higher viscosity, the effect can be much more important (e.g., tissue as discussed in Section 7.4.3.5).

Also obvious from the preceding discussion is that time-averaged approximations of the dynamic situation tend to miss certain features of bubble dynamics. For example, rather than a single expansion and collapse, transient bubbles may oscillate a few times owing to the phase difference between pressure and bubble motion. In a good many situations, however, the severity of collapse will be close to predicted even after two or three oscillations.

For instance, in Section 7.4.3.5.2 an example that might be appropriate to diagnostic ultrasound units is discussed. A bubble of an initial radius of 2.5 μm in water is subjected to a few cycles of a sound wave of peak pressure 10 bars and frequency $f = 2.25$ MHz. According to Eq. (7.3.23) this bubble will grow to a maximum size of about 7.25 μm. Since this is 2.9 times the original size, the bubble is transient (according to the discussion in Section 7.3.4.2).

If one were to solve Eq. (7.3.4) numerically for the preceding conditions, then one would have for the nondimensional period, viscosity, surface tension, and gas pressure: 1.78, 0.16, 0.58, and 1.0, respectively. One would find that the nondimensional acoustic pressure must be about 11.9 for a transient event and that the violent collapse takes place during the second, not first, cycle. Moreover, the maximum radius during growth is about 2.8 times the initial radius. (If, in the numerical solutions, the nondimensional viscosity and surface tension are set equal to zero, as assumed in our crude analytic theory, then the minimum nondimensional acoustic pressure for a transient event is about 9.5 and the maximum radius is about 2.7 times the initial radius.)

Thus, although one has used several approximations in the analytic formulations of growth (Section 7.3.4.2) and collapse (see Noltingk and Neppiras[45]), useful information about growth and collapse has been gathered which is in reasonable agreement with numerical solutions, while baring the explicit dependences on acoustic pressure and frequency.

7.4. Acoustic Cavitation: Applications and Problems

7.4.1 Introduction

In Chapters 7.2 and 7.3 we have described cavitation inception and dynamics, the latter being ordered in terms of the intensity of the effects produced. These effects are useful in producing or enhancing several processes—from cleaning to emulsification to biological extraction—but are an undesirable side effect of others—such as in long-range sonar transmission or in therapeutic or diagnostic uses of ultrasonics. In this section we shall look at some of the practical aspects of the contrary goals of cavitation promotion and inhibition. The literature in this area is plentiful. The reader is referred to the reviews of Neppiras,[97,98] Webster,[99] Shoh,[100] Flynn,[1] and Coakley and Nyborg[5] for more detail and many references to the primary sources on this subject.

7.4.2. Promoting Cavitation: Activity Measures for and Applications of Cavitation

When acoustic cavitation is used as a means to an end, it is desirable to evaluate the extent to which cavitation is promoting this end. For this purpose several activity measures have been devised, and these have been reviewed and compared in the literature.[98] In the design stage such a measure (or measures) can be used to optimize the role cavitation plays; and in the operation stage such a measure (or measures) allows for quality control; that is, it assures that the process continues to work as designed. Ideally, one would like a single, simple, reliable, and inexpensive measure that could be applied to many applications; but with the great variety of phenomena that come under the umbrella category "cavitation," it is not surprising that this ideal is not met. As expected, the best measures are those chosen for the particular application.

Activity measures generally fall into two categories: probe methods, such as those using small, nonperturbing hydrophones that give very detailed information at a point but which require tedious scanning for more complete information, and simple, average measures that integrate effects over space and time.[98]

One measure that has been shown to be strongly correlated to the degree of cavitation is its white noise output (see Section 7.3.4.3.1). As the input to the driving transducer increases, this white noise, presumably generated by the collapse of transient cavities, increases in power and the measured intensity of the fundamental decreases, owing to the shielding effect of bubbles. This suggests that the ratio of white noise to fundamental outputs would correlate well with cavitation activity.[98] As a measure of the acoustic cavitation field, this ratio has been shown to be more

[97] E. A. Neppiras, *Ultrasonics*, p. 10 (January–March 1965); *Ultrasonics*, p. 10 (January 1972).

[98] E. A. Neppiras, *IEEE Trans. Sonics Ultrason.* **SU-15**, 81 (1968).

[99] E. Webster, *Ultrasonics*, p. 39 (January–March 1963).

[100] A. Shoh, *IEEE Trans. Sonics Ultrason.* **SU-22**, 60 (1975).

effective than measures based on the subharmonic output[98] (see discussion in Section 7.3.2.2).

Other general measures, such as those relating to the energy lost in the cavitation process and the total volume of bubbles in the cavitation field, have been suggested but not generally adopted, probably because they do not offer a simple correlate to the cavitation-induced product of the particular application. Some of these applications will be briefly reviewed next and, where appropriate, cavitation measures will be described.

7.4.2.1. Processes Promoted by Stable Cavitation.[97-99]   As discussed in Chapter 7.3, rectified diffusion, bubble coalescence, and acoustic streaming can result from relatively moderate acoustic pressures (on the order of tenths of atmospheres or more) in liquids saturated or nearly saturated by gas.

Cavitation greatly speeds the process of *degassing* because of the combined effects of rectified diffusion and bubble coalescence.

Gas diffusion and *heat transfer due to conduction* may also be greatly enhanced by cavitation, because the bubbles created move in nonuniform pressure fields (see Part 6), thereby producing fluid flow that breaks down uniform diffusional and thermal boundary layers. Similarly, *electroplating* uniformity and rates of deposition are increased when the stirring action of cavitation bubbles breaks down the ion barrier that shields the electrode.

Bubbles that reach resonance size can be exceptionally strong sound sources; nonlinear effects, such as radiation pressure and streaming, may thereby cause the motion of material in an homogeneous medium (see Part 6). The implications of this observation to biological media will be discussed in Section 7.4.3.5.

Since bubbles tend to form preferentially at imperfectly wetted cracks and crevices on solid particles or container surfaces, it stands to reason that their pulsations conveniently occur at the best place for *mechanical "scrubbing."* The *degreasing* ability of a solvent or of water containing an appropriate surfactant is also greatly enhanced by even the mild, stable form of cavitation.

7.4.2.2. Processes Promoted by Transient Cavitation.[98,101,102]   More difficult *cleaning* tasks require the more vigorous fluid motion generated by transient cavitation. Since ultrasonic cleaning is the most popular use of cavitation, a few comments about activity measures to evaluate cleaning are appropriate. The most straightforward measure is the cleaning of some standard probe coated with some standard composition.

[101] T. J. Bulat, *Ultrasonics*, p. 59 (March 1974).
[102] R. Pohlman, B. Werden, and R. Marziniak, *Ultrasonics*, p. 158 (July 1972).

Probe size, materials, surface condition, time of treatment, temperature, ambient pressure, degree of aeration, composition of liquid cleaner, and liquid height and volume all must be recorded if the tests are to be reliable and comparable. Then a method of assessing the degree of cleaning must be chosen. Several indicators have been suggested, including (1) a radioactive tracer in the dispersed material, (2) a chemical tracer or dye, (3) a fluorescent material, or (4) a ferromagnetic powder. Other methods include weighing the probe or, for dispersed material on transparent probes, optical density tests. The erosion permanently recorded on thin, stretched foils provides a convenient measure as long as the erosion is not so extensive as to cause relatively large pieces of foil to fall out of the sample.

As suggested by these measures, one application of transient cavitation is the *dispersion of solid particles in liquid.* Examples of such dispersed systems include cosmetics and paints. The dispersion of one liquid phase as drops in another liquid, *emulsification,* occurs with the high shear rates accompanying transient cavitation. The details of this process are not fully understood. It is clear, however, that like many other processes employing acoustic cavitation, emulsification can occur in the absence of an externally imposed acoustic field (e.g., the flow from a high-pressure to low-pressure region past a sharp edge). Acoustic cavitation superimposed on this process, however, may often yield smaller and more uniform drops. Note that there are limits to the acoustic intensities that can be employed in emulsification and homogenization not only because of heating but also because the forces due to radiation pressure tend to promote the coalescence of drops (see Section 7.3.2.3.1 and Part 6).

The shearing action required for emulsification is also employed in *depolymerization* and *cell disruption* (the membrane being rent apart). The latter of these can aid in the *extraction* of the cell contents or the *killing of bacteria;* cavitation is also employed in tinning and soldering, chemical extraction processing, filtration, and atomization.

7.4.3. Inhibiting and Avoiding the Effects of Cavitation

7.4.3.1. Avoiding Stable Cavitation and Its Effects. As is clear from Section 7.4.2.1, degassing is promoted by cavitation employed when the liquid is under reduced pressure; conversely, once the degassing has taken place and the static pressure is returned to ambient, the liquid is essentially strengthened against the gaseous, stable form of cavitation. Container and solid impurities are more completely wetted thereby raising thresholds from tenths of atmospheres (for rectified diffusion) to a few atmospheres or more (for transient cavitation). The results of Galloway[41] dramatically illustrate the effects of gas content on thresholds (see Section 7.2.5), whereas the work of Greenspan and Tschiegg[6] shows that thresholds for mote-free water are independent of gas content and composition (section 7.2.1.2).

7.4.3.2. Preventing the Initiation of Transient Cavitation. There are several methods of preventing the initiation of cavitation:

(1) Raising the ambient pressure so that the instantaneous pressure is never less than the vapor pressure assures cavitation suppression in degassed liquids.

(2) Harvey *et al.*[14] have shown that if the liquid is hydraulically pressurized temporarily (say to 500 atm for 20 min), container and solid impurity surfaces are better wetted as gas pockets are forced into solution. When returned to ambient pressure, the liquid is greatly strengthened against cavitation. The same procedure can be used for increasing the limit of superheat or limit of supersaturation of liquids.

(3) The addition of surfactant to liquids of high surface tension prior to cavitation-assisted degassing will lead to better wetting of solid surfaces and, therefore, high transient cavitation thresholds. The addition of surfactant after degassing may produce the opposite effect (see Section 7.4.3.4).

(4) Filtering itself will cut down on cavitation by simply altering the distribution of nuclei but is unlikely to reduce thresholds significantly unless almost all motes are removed (see Section 7.2.2).

7.4.3.3. Lessening the Effects of Strong Cavitation. Since the effects of transient cavitation increase with increasing collapse velocity, steps taken to decrease that velocity will promote the desired end. This can be achieved by decreasing the external pressure P_0 or by increasing P_V through heat or the addition of a more volatile component to the liquid (ether in water has been shown to be effective[103]). An increase in the inert gas concentration will shorten the time during which the collapse proceeds nearly unimpeded and will, therefore, lessen the inertial forces. Increasing the acoustic frequency lessens the time allowed during each cycle for growth and collapse and, therefore, lessens the effects of cavitation. Pulsed ultrasound can also lessen the effects of cavitation as long as the peak acoustic pressure in the pulse is not correspondingly increased to compensate for the lower duty cycle (see Section 7.4.3.5). Also, the effects of cavitation can be lessened by exposing hard, smooth, hydrophilic surfaces to the effects of collapsing bubbles. It is not surprising that some of these suggestions have been shared among those working in hydrodynamic and acoustic cavitation.

[103] T. F. Hueter and R. H. Bolt, "Sonics," p. 239. Wiley, New York, 1955.

7.4.3.4. Apparent Paradoxes.   In the preceding discussion we have suggested both increasing P_0 to prevent the *inception* of cavitation and decreasing P_0 to lessen its *effects*. What is one to do? If one follows the former suggestion, then one must be assured that P_0 is *always* greater than the peak acoustic pressure; otherwise the effects will be more severe. In some emulsification procedures using cavitation, this higher P_0 is desirable because high shear rates are required. (In fact, for a given acoustic pressure, there is usually an optimum value of P_0 for a peak in cavitation activity.) If one does not mind some cavitation, as long as its effects are not too strong, then the reduction of the ambient pressure is an appropriate step.

We have also suggested using a surfactant to increase thresholds, whereas a concomitant feature is the reduction of surface tension and, therefore, the lowering of the tensile strength of the pure liquid. Since the tensile strengths of even low surface tension liquids tend to be over 100 atm (e.g., ether, $\sigma = 12–13$ dyn/cm, tensile strength $\doteq 200$ atm), the addition of a surfactant should be effective as long as the resulting threshold does not exceed on the order of tens of atmospheres.

7.4.3.5. Bioeffects of Ultrasound and Cavitation. Some of the most important applications of ultrasound to medicine have occurred in the last two decades: these applications include acoustic doppler devices for the measurement of blood flow and the detection of bubbles, ultrasonic imaging in the body for diagnosis, which encompasses pulse-echo and holographic techniques, and the therapeutic use of ultrasound in physical therapy and in the destruction of certain tissues.

As with any form of radiation in the body, ionizing or not, there is concern that its use should not produce undesirable effects. This means achieving an understanding of the mechanisms of interaction of the radiation with the biological system, determination of thresholds, if any, for any observed effects, and the adoption of guidelines or standards that assure appropriately low risk. Both for ionizing and acoustic radiation, the mechanisms are not clearly understood, and guidelines tend to change, almost always in a more conservative direction.

7.4.3.5.1. PHYSICAL MECHANISMS FOR DAMAGE PRODUCTION BY ULTRASOUND. Damage to tissue may occur by a number of different mechanisms. A little input energy can go a long way if it encounters a gas bubble and if the frequency is at the bubble's resonance. The bubble becomes a secondary source of sound and can exert forces on objects in its immediate environment by nonlinear acoustic effects such as radiation pressure (including aggregation), streaming, and torque production (see Part 6).

These effects have been elegantly demonstrated by a number of studies reported from Nyborg's laboratory at the University of Vermont.[104–109] Mason horn excitation techniques have been convenient to use because they have allowed direct microscopic examination (as employed first by Harvey et al.[110]) and cinephotography of cells (even single cells) while they are being sonated. Because bubbles with diameters on the order of micrometers, which resonate in the convenient low-megahertz range, tend to dissolve, Nyborg has employed "cylindrical" bubbles trapped in the 10-μm pores of hydrophobically treated Nucleopore, polycarbonate filter membranes. Nyborg et al. have observed the whole spectrum of bubble-induced motions; furthermore, they have observed blood platelet damage at an intensity of only 125 mW/cm^2 for 15 min at 1 MHz, a time-averaged level that some diagnostic units may approach.[109]

Since these trapped bubbles are responsible for the effects produced, it is important to know where such bubbles may reside in living systems. Nyborg, Gershoy, and Miller have found them in plant tissues and have illustrated how ultrasound interacts with them.[106] Miller has found gas trapped in intercellular channels of Elodea leaves. The excitation of these regions with relatively low-intensity ultrasound can lead to cell death. Miller found that such disruption requires higher delivered energy for low-intensity, continuous ultrasound than for higher-intensity, pulsed ultrasound, confirming that both intensity and the time history of the application of sound influence the bubble-cell dynamics in an important way.[108]

Natural bubbles in animal tissue have been far less easy to find, leading some to question whether they exist at all. Their existence or nonexistence has important implications not only for diagnostic ultrasound but also for studies of the origins of decompression sickness. The work of Harvey et al. is a primary source for early advances in this area.[14] Walder and Evans have suggested that the seeds for decompression bubbles may originate via the natural radiation background of nuclear fis-

[104] W. L. Nyborg, J. Acoust. Soc. Am. **44**, 1302 (1968).

[105] W. L. Nyborg, Proc. Symp. Finite-Amplitude Wave Effects Fluids, Copenhagen, 1973 (L. Bjorno, ed.), pp. 245–251. IPC Interscience and Technology Press, Guildford, England, 1974.

[106] A. Gershoy, D. L. Miller, and W. L. Nyborg, in "Ultrasound in Medicine" (D. White and R. Barnes, eds.), pp. 501–511. Plenum Press, New York, 1976.

[107] W. L. Nyborg, D. L. Miller, and A. Gershoy, in "Fundamental and Applied Aspects of Nonionizing Radiation" (S. M. Michaelson and M. M. Miller, eds.), pp. 277–299. Plenum, New York, 1975.

[108] D. L. Miller, Ultrasound Med. Biol. **3**, 221 (1977).

[109] W. L. Nyborg, A. Gershoy, and D. L. Miller, in "Ultrasound International," pp. 19–27. IPC Science and Technology Press, Guildford, England, 1977.

[110] E. N. Harvey, E. B. Harvey, and A. L. Loomis, Biol. Bull. **55**, 459 (1928).

sion in the body.[111]   Gramiak *et al.* have found evidence suggesting the possibility of gas microbubbles in human cardiac chambers.[112]

In the absence of seed bubbles, sufficiently intense acoustic waves, such as those used in therapeutic applications, can produce destructive biological effects by mechanical or thermal means, or both, although there are some unresolved issues on this subject. (See Gavrilov[113] and Reid and Sikov[114] and their references for a more complete discussion.)

The possible physical effects of ultrasound can be discussed more concretely in the context of a specific example: diagnostic units for visualization in the body.

7.4.3.5.2. ACOUSTICAL PARAMETERS FOR DIAGNOSTIC UNITS. Diagnostic units for visualization within the body operate in a pulse-echo mode. Typically a highly damped transducer will produce about two cycles at its resonance frequency (generally 2.25 MHz) at a pulse repetition rate of about 1 kHz. These units may put out a time-averaged power in the range 10–100 mW. For a 2-MHz transducer, this corresponds to an instantaneous peak power of 10–100 W, which, if spread over an area of 1 cm^2, gives an instantaneous intensity in the range 10–100 W/cm^2. This corresponds to a peak acoustic pressure in a plane traveling wave of about 5–15 atm. This pressure can be higher if the transducer focuses to smaller than 1 cm^2 (as is common).

For the case of a few cycles at 10 atm, one can show, from Section 7.3.4.2, that the viscous start-up time for a bubble in water is small compared to an acoustic period and that bubbles of initial radius less than about 3.3 μm are not inertially limited. This implies that significant growth could occur in water to a size maximum, according to Eq. (7.3.23), of about 7–7.5 μm if the initial size were less than about 3 μm. For a bubble of this initial size, the collapse will just be transient, as defined in Section 7.3.4.2. Using the equations in the appendix of Noltingk and Neppiras,[45] we can estimate that the pressure of a bubble that starts at 3 μm (1 atm) and grows to 7 μm will peak at greater than 200 atm at a minimum size of less than 1 μm, which is comparable to intracellular dimensions. The internal temperature, continuing our assumptions of adiabatic collapse of a spherical bubble in an incompressible liquid, will be in excess of 1000°C. The energy of about 1.4 milliergs stored in the fluid when the bubble is at a maximum size of 7 μm is concentrated into a volume of about 2×10^{-12} cm^3, giving a local energy density of about 10^9

[111] D. N. Walder and A. Evans, *Nature (London)* **252**, 696 (1974).
[112] R. Gramiak and P. M. Shah, *Radiology* **100**, 415 (1971).
[113] L. R. Gavrilov, *Sov. Phys. Acoust.* **17**, 287 (1972).
[114] J. M. Reid and M. R. Sikov (eds.), Interaction of ultrasound and biological tissues, *Proc. Workshop, Battelle Seattle Res. Center, Seattle, Washington, November 8–11, 1971.* U.S. Dept. of HEW, Publ. (FDA) 73-8008 1972.

ergs/cm^3 = 100 J/cm^3. As crude as these estimates are, they show, nevertheless, the order of magnitude of the energy concentration due to cavitation.

For tissue-type materials the effective viscosity is over 100 times that of water and bubble motion is limited by viscous effects. Rectified diffusion could occur for continuous wave excitation but is unlikely for the duty cycle associated with diagnostic units since the diffusion process is so slow and since gas diffusion rates are likely to be lower in tissues than in aqueous solutions.

The potential, but as yet unsubstantiated, dangers due to diagnostic ultrasound, therefore, appear in theory to be restricted to cavitation in lower-viscosity tissues (e.g., blood) or second-order, flow-induced effects in the vicinity of shock-excited, preexisting bubbles (should they be present).

Additional controlled experiments in the 1–200-mW range with typical diagnostic parameters are called for.

7.5. Final Remarks

This part has been meant as a practical guide to those employing or trying to avoid the effects of cavitation in any of its varied forms. Many subjects have been treated briefly, and some have been neglected either because of space limitations or the author's unfamiliarity with the material. For instance, Chincholle's work on the influence of electrical phenomena on the rapid movement of cavitation bubbles and its implications and applications in the biomedical domain have not been discussed.[115] Also, the experimental work of Kuttzuff and Radek[116] and Radek[117] on pressure pulses generated by cavitation bubbles gives further insight to cavitation dynamics and erosion. A recent review on acoustic cavitation by Neppiras complements this article.[118] These are left for the reader to pursue.

Acknowledgments

The author, in addition to acknowledging his reliance on the excellent work quoted in the references, extends his appreciation to P. L. Marston and E. Neppiras for their helpful discussions and suggestions on many of the topics in this manuscript, especially with regard to Sections 7.3.3, 7.3.4.2 and 7.3.5, where new material has been introduced.

[115] L. Chincholle, *Onde Electrique* **56**, 28 (1976).
[116] H. Kuttruff and U. Radek, *Acustica* **21**, 253 (1969).
[117] U. Radek, *Acustica* **26**, 270 (1972).
[118] E. A. Neppiras, *Phys. Rep.* **61**, 159 (1980).

8. ACOUSTIC MEASUREMENTS IN SUPERFLUID HELIUM

By Joseph Heiserman

List of Symbols

A	resonator cross-sectional area
B	resonator perimeter
C	acoustic compliance
C_v	specific heat at constant volume
C_1	velocity of first sound
C_2	velocity of second sound
C_4	velocity of fourth sound
d	depth of superleak
E	ionization energy
E_1	energy density of first sound
E_2	energy density of second sound
f	frequency
Δf	bandwidth
$G(T)$	thermodynamic correction term for fourth sound doppler shift [see Eq. (8.3.13)]
$\hbar$	Planck's constant divided by 2π
J_1	first-order Bessel function
K	thermal conductivity
k	wave number or the ratio of the circular frequency to the acoustic velocity
k_B	Boltzmann's constant
k_T	electromechanical coupling constant
L	resonator depth
l	resonator length
M	acoustic inertance
m	mass per unit area
n	empirical acoustic index of refraction
P	superleak porosity
p	pressure
Q	quality factor
q	heat energy
R	resistance
r	radius
s	entropy per gram
s'	area
T	temperature
T_c	superconducting transition temperature
T_λ	Lambda temperature $T_\lambda = 2.172°K$

METHODS OF EXPERIMENTAL PHYSICS, VOL. 19

t	time
V	voltage
$\mathbf{v}$	velocity
$\mathbf{v}_n$	normal fluid velocity
$\mathbf{v}_s$	superfluid velocity
X_c	electrical reactance
Z	acoustic impedance
α	temperature coefficient of resistance
β	isobaric expansion coefficient
γ	attenuation coefficient
δ_{ij}	Kronecker delta function
ϵ	dielectric coefficient
η	viscosity
λ	viscous penetration depth
λ_T	thermal penetration depth
μ	chemical potential
ρ	density
ρ_E	electrical resistivity
ρ_n	normal fluid density
ρ_s	superfluid density
ϕ	azimuthal angle
Ω	radial dependence factor [see Eq. (8.3.20)]
ω	circular frequency

8.1. Introduction

8.1.1. Liquid and Superfluid Helium

Helium, the second lightest element, exists in two stable forms, ^{3}He and ^{4}He. The common isotope ^{4}He condenses at the extraordinarily low temperature of 4.2°K. The liquid formed at this temperature behaves as a classical fluid and continues to do so until it is cooled to the lambda temperature, which at its vapor pressure is $T_\lambda = 2.17$°K. At this temperature the character of the liquid abruptly changes completely; above T_λ the liquid is designated He I, and below, He II or superfluid helium.

He II behaves as though it were composed of two fluids, a superfluid component and a normal fluid component. The mass density of the liquid can be written $\rho = \rho_s + \rho_n$, where ρ_n and ρ_s are the densities of the normal and superfluid components, respectively. The momentum density of the liquid is $\rho\mathbf{v} = \rho_s\mathbf{v}_s + \rho_n\mathbf{v}_n$, where $\mathbf{v}$, $\mathbf{v}_s$, and $\mathbf{v}_n$ are the velocities of He II, the superfluid component and normal fluid component, respectively. The superfluid and normal fluid densities are found to vary with temperature as shown in Fig. 1. The superfluid component has no viscosity

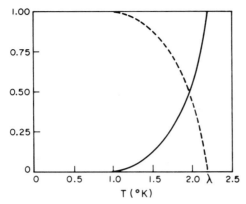

FIG. 1. Densities of the normal (solid curve) and superfluid (dashed curve) components from the two-fluid model.

and, remarkably, carries no entropy. In the two-fluid theory, this leads to assignment of the entire superfluid to a single quantum (ground) state.[1]

The thermohydrodynamic description of He II was first given by Landau.[2] The Landau theory can be obtained by assuming that eight independent variables, for example v_s, v_n, ρ, and the entropy per gram s constitute a complete set for He II. Application of the principles of classical physics (the laws of thermodynamics, conservation of mass and entropy, Newton's laws, and the principle of Galilean relativity) then yields the following description of superfluid helium, valid when dissipation is ignored[3,4]:

$$\frac{\partial \rho}{\partial t} + \nabla \cdot \rho v = 0 \qquad \text{(mass conservation)}, \qquad (8.1.1)$$

$$\frac{\partial \rho s}{\partial t} + \nabla \cdot \rho s v_n = 0 \qquad \text{(entropy conservation)}, \qquad (8.1.2)$$

where s is the entropy per gram,

$$D_s v_s / Dt = -\nabla \mu, \qquad (8.1.3)$$

[1] L. Tisza, *Nature (London)* **141**, 913 (1938).

[2] L. D. Landau, *J. Phys. (Moscow)* **5**, 71 (1941).

[3] L. D. Landau, "Fluid Mechanics," Chapter 16. Pergamon, Oxford, 1959.

[4] S. J. Putterman, "Superfluid Hydrodynamics," p. 19ff. North-Holland Publ., Amsterdam, and Amer. Elsevier, New York, 1974.

which is the equation of motion for the superfluid component, with

$$\frac{D_s}{Dt} = \frac{\partial}{\partial t} + (\mathbf{v}_s \cdot \nabla), \qquad (8.1.4)$$

the hydrodynamic derivative, and

$$d\mu = -s\ dT + \frac{1}{\rho}\ dp - \frac{\rho_n}{2\rho}\ d(\mathbf{v}_n - \mathbf{v}_s)^2, \qquad (8.1.5)$$

the differential of the chemical potential. The symbols dT and dp are the differentials of the temperature and pressure. Next,

$$\frac{\partial}{\partial t}(\rho_n v_{ni} + \rho_s v_{si}) + \frac{\partial \Pi_{ij}}{\partial x_j} = 0 \quad \text{(momentum conservation)}, \quad (8.1.6)$$

where

$$\Pi_{ij} = \rho_n v_{ni} v_{nj} + \rho_s v_{si} v_{sj} + \delta_{ij} p. \qquad (8.1.7)$$

The indices run over all coordinates, and δ_{ij} is the Kronecker delta. Finally, Landau restricted the motion of the superfluid with the condition

$$\nabla \times \mathbf{v}_s = 0. \qquad (8.1.8)$$

These equations, together with an equation of state and appropriate boundary conditions, have been found in many experiments to give a description of the dynamical behavior of He II. However, it has been found that, although for low enough velocities the Landau theory describes superfluid helium flow, there exists in every geometry a critical velocity v_{sc} above which frictionless superfluid flow does not occur. Although several descriptions for this breakdown have been advanced, no single proposal has been found to be completely satisfactory. In 1941 Bijl et al.[5] noted that from experimental data v_{sc} was approximately proportional to the inverse of the cross-sectional channel size d^{-1}. The numerical value of the constant of proportionality was found to be close to $\hbar = h/2\pi$, Planck's constant divided by 2π.

Consider now a helium atom of mass m in a channel of width d. If the particle is traveling with a velocity v_s, the angular momentum with respect to the wall is of the order of mv_sd. From quantum collision theory, for interaction with the wall to occur, the angular momentum with respect to the wall must be at least $\hbar$. Thus the particle cannot interact with the wall if $v_s < \hbar/md$. If we now take the two-fluid view and picture the superfluid as occupying a single quantum state, then the result for a

[5] A. Bijl, J. De Boer, and A. Michels, *Physica* (Utrecht) **8**, 655 (1941).

single particle becomes applicable to the entire superfluid collectively and the maximum velocity for frictionless flow is given by

$$v_{sc} \approx \hbar/md. \qquad (8.1.9)$$

This then is Bijl's understanding of the experimental observation that the critical velocity increases as the channel size decreases. Although several other empirical and theoretical relationships have been advanced to explain the dependence of the critical velocity on channel size, at this time no approach has had significantly more success than the proposal of Bijl in 1941.

A particularly dramatic demonstration of frictionless flow is given by the persistent current. For example, if an annular or toroidal cavity filled with helium is cooled through T_λ while rotating about its symmetry axis and then stopped at some low temperature, the flow of the superfluid is found to persist for long periods after the flow is stopped. Currents formed in this way often exhibit little or no decay for long periods of time.

Further discussion of the thermohydrodynamics of superfluid helium and its limitations and extension of these results to include dissipation can be found in the work of Putterman.[4]

^{3}He has also been found to exhibit superfluid behavior below about 3×10^{-3}°K. Although ^{3}He is a more complicated superfluid and exhibits many effects not found in superfluid ^{4}He, much of the discussion of this section applies equally well to this new superfluid, and an acoustic probe has been used to investigate the nature of ^{3}He superfluidity.[6,7]

8.1.2. The Sounds of Helium

When the Navier–Stokes or Euler equations for a classical fluid are linearized and propagating wave solutions are sought, the result is the acoustic solution that propagates at a characteristic velocity, the velocity of sound. The Landau equations also possess acoustic solutions; however, because of the two-fluid nature of He II, more than one type of wave solution exists. There are three propagating bulk hydrodynamic modes in He II; they are called *first, second, and fourth sound*. To linearize the Landau equations we start by following the development given by Rudnick[8] and writing

^{6} J. C. Wheatley, *Rev. Mod. Phys.* **47**, 415 (1975).

^{7} H. Kojima, D. N. Paulson, and J. C. Wheatley, *Phys. Rev. Lett.* **32**, 141 (1974).

^{8} I. Rudnick, *Proc. Int. School Phys.*, "*Enrico Fermi*" —*Course LXIII—New Directions Phys. Acoust.* (D. Sette, ed.), p. 139ff. North-Holland Publ., New York, 1976.

$$\rho = \rho_0 + \delta\rho, \qquad s = s_0 + \delta s, \qquad p = p_0 + \delta p,$$

$$T = T_0 + \delta T, \qquad \mathbf{v}_n = \delta\mathbf{v}_n, \qquad \mathbf{v}_s = \delta\mathbf{v}_s, \qquad (8.1.10)$$

where the zero subscript refers to equilibrium values and the second terms represent the infinitesimal variations introduced by the propagating wave. We assume no steady flow; hence $\mathbf{v}_{n0}$ and $\mathbf{v}_{s0}$ are zero. The Landau equations then become, to first order in infinitesimal quantities,

$$\frac{\partial\delta\rho}{\partial t} + \rho_s\nabla \cdot \mathbf{v}_s + \rho_n\nabla \cdot \mathbf{v}_n = 0, \qquad (8.1.11)$$

$$\rho\frac{\partial\delta s}{\partial t} + s\frac{\partial\delta\rho}{\partial t} + s\rho\nabla \cdot \mathbf{v}_n = 0, \qquad (8.1.12)$$

$$\frac{\partial\mathbf{v}_s}{\partial t} + \frac{1}{\rho}\nabla\delta p - s\nabla\delta T = 0, \qquad (8.1.13)$$

$$\frac{\partial\mathbf{v}_n}{\partial t} + \frac{1}{\rho}\nabla\delta p + \frac{\rho_s}{\rho_n}s\nabla\delta T = 0, \qquad (8.1.14)$$

where Eq. (8.1.14) results from subtracting the linearized versions of Eqs. (8.1.3) and (8.1.6). Now, from Eqs. (8.1.11), (8.1.13), and (8.1.14), we obtain

$$\partial^2\delta\rho/\partial t^2 = \nabla^2\delta p \qquad (8.1.15)$$

and from Eqs. (8.1.11)–(8.1.14),

$$\partial^2\delta s/\partial t^2 = (\rho_s/\rho_n)s^2\nabla^2\delta T. \qquad (8.1.16)$$

We can expand $\delta\rho$ and δs thermodynamically as

$$\delta\rho = (\partial\rho/\partial p)_T\,\delta p + (\partial\rho/\partial T)_p\,\delta T, \qquad (8.1.17)$$

$$\delta s = (\partial s/\partial T)_p\,\delta T + (\partial s/\partial p)_T\,\delta p. \qquad (8.1.18)$$

The second term in Eq. (8.1.17) is related to the isobaric expansion coefficient $\beta \equiv -(1/\rho)(\partial\rho/\partial T)_p$. At low temperatures and also at about 1.15°K, β vanishes and (8.1.15) becomes

$$\partial^2\delta\rho/\partial t^2 = C_1{}^2\nabla^2\delta\rho, \qquad (8.1.19)$$

where $C_1{}^2 = (\partial p/\partial\rho)_s$ since when $\beta = 0$, $(\partial p/\partial\rho)_s = (\partial p/\partial\rho)_T$. Further, by a Maxwell relation when $\beta = 0$, $(\partial s/\partial p)_T = 0$ and Eq. (8.1.16) becomes

$$\partial^2\delta s/\partial t^2 = C_2{}^2\nabla^2\delta s \qquad (8.1.20)$$

with $C_2{}^2 = (\rho_s/\rho_n)Ts^2/C_v$, where C_v is the specific heat.

Equation (8.1.19) is the wave equation for a propagating pressure–

density wave of velocity C_1, called *first sound*. In our present approximation there is no variation in temperature or entropy. From Eqs. (8.1.13) and (8.1.14) with $\delta T = 0$, we find that for first sound $v_n = v_s$. First sound was observed by Burton and by Findlay *et al.* in 1938.[9] Equation (8.1.20) describes a propagating temperature–entropy wave of velocity C_2 called *second sound*. There are no variations in pressure or density at this level of approximation. From Eqs. (8.1.13) and (8.1.14) with $p = 0$, $\rho_s v_s = -\rho_n v_n$ for second sound. Second sound was observed by Peshkov in 1946.[10] This purely thermal propagating mode is unique to superfluids and contrasts sharply with the diffusion of heat in non-superfluids.

For temperatures at which $\beta \neq 0$, Eqs. (8.1.15) and (8.1.16) cannot be simplified to Eqs. (8.1.19) and (8.1.20), and first sound involves small changes in temperature and entropy while second sound has small density and pressure variations. First and second sound still exist as propagating modes but modifications of the calculated velocities C_1 and C_2, which can be as large as 10%, arise. The greatest corrections occur near T_λ and at elevated pressures. For a discussion that includes these effects, the reader is referred to the works of London[11] and Putterman.[4]

Imagine a tube or cavity filled with He II and packed with a fine powder. Will any propagating modes exist in the fluid under these conditions? The normal fluid will be locked by viscous forces, but the zero viscosity superfluid will be free to flow. If v_n is set equal to zero in the Landau equations, the system is overdetermined and three equations must be discarded. Since the tube and packed powder are assumed to be held rigidly, the momentum equations (8.1.6) are no longer valid and can be discarded. The five remaining equations are linearized to obtain

$$\frac{\partial \delta \rho}{\partial t} + \rho_s \nabla \cdot \mathbf{v}_s = 0, \tag{8.1.21}$$

$$\frac{\partial \rho s}{\partial t} = 0, \qquad \rho s = \text{const}, \tag{8.1.22}$$

$$\frac{\partial \mathbf{v}_s}{\partial t} + \nabla \delta \mu = 0. \tag{8.1.23}$$

It is convenient to choose ρ and ρs (the total entropy per unit volume) to be independent thermodynamic variables. Then (8.1.23) becomes

[9] E. F. Burton, *Nature* (*London*) **141**, 970 (1938); J. C. Findlay, A. Pitt, H. Grayson-Smith, and J. O. Wilhelm, *Phys. Rev.* **54**, 506 (1938).

[10] V. P. Peshkov, *J. Phys.* (*Moscow*) **10**, 389 (1946).

[11] F. London, "Superfluids," Vol. 2. Dover, New York, 1961.

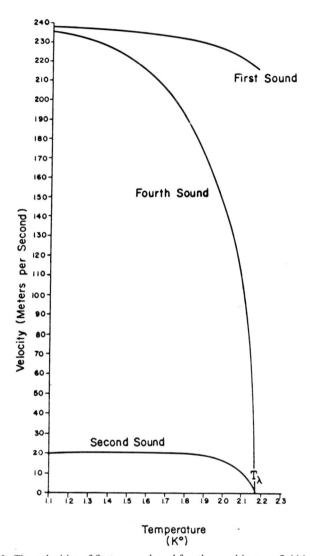

FIG. 2. The velocities of first, second, and fourth sound in superfluid helium.

$$\frac{\partial \mathbf{v}_s}{\partial t} + \left(\frac{\partial \mu}{\partial \rho}\right)_{\rho s} \nabla \delta \rho = 0, \tag{8.1.24}$$

which can be combined with Eq. (8.1.21) to yield a wave equation

$$\frac{\partial^2 \delta \rho}{\partial t^2} - C_4^2 \nabla^2 \delta \rho = 0, \tag{8.1.25}$$

TABLE I.   Summary of Properties of the Sounds of He II

Sound	Normal component velocity	Superfluid component velocity	Type of wave	Wave velocity
First	$\rightarrow$	$\rightarrow$	Pressure density	$C_1 = \left[\left(\dfrac{dp}{d\rho} \right)_s \right]^{1/2}$
Second	$\rightarrow$	$\leftarrow$	Temperature entropy	$C_2 = \left[\dfrac{\rho_s}{\rho_n} \dfrac{Ts^2}{c_V} \right]^{1/2}$
Fourth	Zero	$\rightarrow$	Pressure wave in superleak	$C_4 = \left[\dfrac{\rho_s}{\rho} C_1^2 \right]^{1/2}$

where $C_4^2 = \rho_s (\partial\mu/\partial\rho)_{\rho s}$. The new mode is called *fourth sound* and is a propagating wave in the superfluid only. The velocity can be written

$$C_4^2 \approx (\rho_s/\rho)C_1^2. \tag{8.1.26}$$

The error made in this approximation is maximum at about 2°K and is at most 2%. Fourth sound was first observed by Rudnick and Shapiro in 1962.[12] The temperature dependence of the velocities of first, second, and fourth sound from 1.1°K to T_λ is shown in Fig. 2.   Between 1 and 1.9°K, C_2 is about one-tenth as large as C_1. Table I summarizes some properties of the three modes.

Third sound is a surface wave in thin superfluid films and will not concern us here.[13] Additional discussion of all the sounds of helium can be found in the very useful review by Rudnick.[8]

8.2. Transducers

The special nature of the sounds of helium and the extremely low temperatures involved in superfluid helium research impose certain constraints on the transducers used to generate and detect first, second, and fourth sound.   Nonetheless, a rich variety of devices has been developed for use over a wide range of frequencies.   In the following we shall limit ourselves to devices for the generation and detection of coherent acoustic waves.

8.2.1. First Sound

Since first sound is a pressure wave, pressure transducers are used to generate and detect it.   First sound has been studied over a remarkably

[12] I. Rudnick and K. A. Shapiro, *Phys. Rev. Lett.* **9**, 191 (1962).
[13] K. R. Atkins and I. Rudnick, *Prog. Low Temp. Phys.* **6**, Chapter 2 (1970).

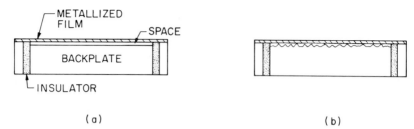

(a)   (b)

FIG. 3. (a) An idealized capacitive transducer with a metallized film stretched over a smooth backplate leaving a space d cm thick; (b) Transducer with roughened backplate (see text).

wide frequency range, from the lowest audio frequencies up to about 1 GHz. Choice of transducer is dictated by the frequency range of interest.

8.2.1.1. Capacitive Transducers. At frequencies below about 100 kHz capacitive transducers are widely employed. These are constructed by stretching a dielectric film, for example, a 12-μm-thick mylar sheet, coated on its outer side with a thin metallic layer over a metal backplate as shown in Fig. 3. The backplate may be relatively smooth or deliberately roughened by sandblasting or grooving, depending on the maximum frequency of operation.

Suppose an ac voltage $V \cos \omega t$ of frequency ω is applied between the thin metallic outer electrode and the backplate of the transducer. A force will be developed proportional to the square of the applied voltage and will lead to a pressure variation in the adjacent fluid also proportional to the square of the voltage

$$p \propto V^2 \cos^2 \omega t \qquad (8.2.1)$$

or by a trigonometric identity

$$p \propto \tfrac{1}{2}V^2(1 + \cos 2\omega t). \qquad (8.2.2)$$

Thus the applied voltage at frequency ω leads to a pressure variation at frequency 2ω. If, however, we apply a large polarizing voltage V_0 to the transducer, then to first order the pressure will vary at frequency ω as

$$p \propto (V_0 + V \cos \omega t)^2 \approx V_0^2 + V_0 V \cos \omega t \qquad (V_0 \gg V). \quad (8.2.3)$$

In practice V_0 is usually several hundred volts. The polarized transducer is a reciprocal device; a pressure wave incident on the diaphragm will cause a change in thickness of the gap generating an ac voltage proportional to the variations in the spacing.

If the dielectric film is lightly stretched and the gap is kept small, then the transducer behaves like a mass spring system or Helmholtz resonator

with the stiffness of the helium layer and the film elasticity acting as restoring forces for the mass of the dielectric film. Consider the design of
Fig. 3a. Assume the film is loosely stretched so that all the restoring
force comes from the stiffness of the liquid layer between the film and
backplate. The resonance of the system will occur at the frequency[14]

$$f = (1/2\pi)(1/MC)^{1/2},$$ (8.2.4)

where $1/C$ represents the stiffness (reciprocal of acoustic compliance), M
is the acoustic inertance, $M \equiv m/s'^2$,[14] and s' is the area of the transducer. The compliance can be written

$$C = s'd/\rho_0 C_1^2,$$ (8.2.5)

where d is the thickness of the liquid layer, ρ_0 the density of helium, and
C_1 the velocity of first sound. With these definitions Eq. (8.2.4) becomes

$$f = (1/2\pi)(\rho_0 s' C_1^2/md)^{1/2}.$$ (8.2.6)

For the polymer dielectrics of interest to us here, with a thickness of 12
μm, $m \approx 2 \times 10^{-3}$ gm/cm^2. Taking $\rho_0 = 1.45 \times 10^{-1}$ gm/cm^3, $C_1 =$
0.23×10^5 cm/sec, $d = 1 \times 10^{-3}$ cm, and $s' = 1$ cm^2, we obtain $f \approx 1$
MHz. The Helmholtz resonator model of the condenser microphone is
valid for frequencies well below resonance and in the absence of damping
where the response is expected to be independent of frequency.[15]†

Dielectric films for condenser transducers are chosen on the basis of
tensile strength, flexibility, and thinness. Mylar, a polyester, has been a
popular choice. Matsugawa has studied condenser microphones made
with Mylar films over a smooth backplate in air at frequencies up to 500
kHz.[16] In many applications electret films have replaced these polarized
transducers.[17] Electrets are ferroelectric and can be poled by applying a
large voltage across them; they then acquire a quasipermanent polarization.[18] Since then the dielectric already possesses a displacement field,
there is no need to use high-voltage polarizing supplies, which are inconvenient and can be a source of excess noise. Sessler has considered, in
detail, the operation of an electret microphone and derived an expression

---

[14] L. Kinsler and A. Frey, "Fundamentals of Acoustics," 2nd ed. Wiley, New York,
1962.
[15] E. P. Cornet, Ph.D. Dissertation, Appendix, Univ. of Texas at Austin, 1972 (unpublished).
[16] K. Matsuzawa, *Jpn. J. Appl. Phys.* **17**, 451 (1978).
[17] G. M. Sessler and J. E. West, *J. Acoust. Soc. Am.* **40**, 1433 (1966); **53**, 1589 (1973).
[18] G. M. Sessler and J. E. West, *J. Appl. Phys.* **43**, 922 (1972).

---

† Cornet[15] also gives some comments on the geometry of Fig. 3b.

giving the sensitivity at low frequencies.[19]   Teflon FEP has been used to construct electret transducers for use in superfluid helium.[20]   Garrett has measured the sensitivity of an electret microphone in a waveguide filled with superfluid helium and found a sensitivity of about -70 dB re 1 V/μbar after losses in coaxial cable have been subtracted out.[21]

8.2.1.2. Impedance Matching of Capacitive Microphones. Capacitive transducers for use in liquid helium are frequently small because of the limited space available in the cryogenic system.   When used at low frequencies (for instance, to excite and detect sound in a resonator), such transducers can have a large electrical impedance, sometimes as great as 100 MΩ, due to the small capacity.   When used with small, high-capacitance, coaxial cables between the transducer and room-temperature electronics, these high impedance levels can be a source of trouble.   The shunt capacitance of the cable acts to attenuate the signal level from the detector and is a source of microphonic noise.   To avoid these effects, a transimpedance amplifier can be fitted directly to the transducer in the liquid helium.   After the source impedance has been transformed down to a relatively low level, the cable problems are no longer serious.   Since bipolar transistors do not function at liquid helium temperatures, field effect transistors have been chosen and successfully used in this application.[22]   MOSFET transistors are most common, but certain J FETs may offer some advantages.[23]

8.2.1.3. Piezoelectric Transducers. Piezoelectric materials are useful at frequencies above about 100 kHz.   Much below this frequency their size becomes prohibitive when used at resonance.   The piezoelectric effect is only weakly temperature dependent in many materials and thus a wide variety of suitable materials exists.   The piezoelectric properties of several materials at liquid helium temperatures are listed for comparison in Table II.   A detailed discussion of transducer design and matching techniques can be found in Part 1.

At frequencies below about 100 MHz, disk transducers can be used in their fundamental mode coupled directly to liquid helium.   For frequencies above 100 MHz, thin disk transducers must be used at higher harmonics with the attendant loss in bandwidth and efficiency.   Above

[19] G. M. Sessler, *J. Acoust. Soc. Am.* **35,** 1354 (1963).

[20] J. Heiserman and I. Rudnick, *J. Low Temp. Phys.* **22,** 481 (1976).

[21] S. Garrett, Ph.D. Dissertation, Univ. of California, Los Angeles, 1977 (unpublished).

[22] B. Lengler, *Cryogenics* **14,** 439 (1974); J. Heiserman, Ph.D. Dissertation, Appendix 1, Univ. of California, Los Angeles, 1975 (unpublished); J. Goebel, *Rev. Sci. Instrum.* **48,** 389 (1977).

[23] K. W. Gray and W. N. Hardy, Technical Report SCTR-69-28, Science Center, North American Rockwell Corp., Thousand Oaks, California (1969).

TABLE II. Comparison of Low-Temperature Properties of Transducer Materials

Material	Orien-tation	$k_T(2°K)$	$k_T(2°K)/k_T(300°K)$	$\dfrac{\varepsilon^S(2°K)}{\varepsilon_0}$	$\dfrac{\varepsilon^S(2°K)}{\varepsilon^S(300°K)}$	Z^a	Reference
Quartz	X	0.1	1.0	4.5	1.0	17	24
CdS	Z	0.15	1.0	9.5	1.0	21	25
ZnO	Z	0.28	1.0	8.8	1.0	36	25,26[b]
PZT-4[c]	—	0.45	0.8[d]	100.0	0.14	34	27
PZT-5H[c]	—	0.38	0.6[d]	270.0	0.16	34	27
PVF$_2$[c]	TE[e]	0.18	1.0	3.2	0.5	4	28
	LE[f]	0.02	0.13	3.6	0.2	2.5	28

[a] $\times 10^{-5}$ gm/cm^2 sec.
[b] Note that the value of k listed by Tokarev et al.[26] is lower than in Reeder and Winslow.[25] We are interested in the temperature dependence.
[c] Poled ferroelectric.
[d] Poled ferroelectrics are temperature dependent and this ratio varies considerably from material to material.
[e] Thickness extensional.
[f] Length extensional.

several hundred megahertz even this mode is unacceptable and thin disks give way to vacuum-deposited, thin-film transducers. These devices are too fragile to be used free standing and thus must be deposited on a buffer rod the other end of which contacts the liquid helium. A cadmium sulfide (CdS) thin film deposited on a quartz buffer rod has been used to generate plane waves in liquid helium at 1 GHz.[29] The author has used an oriented zinc oxide (ZnO) film of fused quartz, which has a larger piezoelectric coupling coefficient k_T than CdS. Although the coupling coefficient in this thin film is not as large as in bulk material, it is still independent of temperature.†

The major difficulty encountered with thin-film transducers on buffer rods is the large acoustic mismatch at the buffer rod–helium interface. Because of the very low acoustic impedance of liquid helium ($Z_{He} \approx 0.03 \times 10^5$ gm/cm^2 sec) the transmission coefficient from any solid of impedance Z_s to liquid helium is extremely small. The standard expres-

[24] W. G. Cady, "Piezoelectricity." Dover, New York, 1964.
[25] T. M. Reeder and D. K. Winslow, IEEE Trans. Microwave Theory Tech. MTT-17, 927 (1969).
[26] E. F. Tokarev et al., Sov. Phys. Solid State 17, 629 (1975).
[27] H. Jaffe and D. Berlincourt, Proc. IEEE 53, 1372 (1965).
[28] H. Ohigashi, J. Appl. Phys. 47, 949 (1976).
[29] J. S. Imai and I. Rudnick, J. Acoust. Soc. Am. 46, 1144 (1969).

† See Reeder and Winslow,[25] Table 3.

TABLE III.   Properties of Various Solids for Use as Buffer Rods in Liquid Helium

Material	$Z \times 10^{-5}$ gm cm^2/sec	Thermal expansion coefficient $(L_2 - L_{273})/L_{273} \times 10^4$	T (dB)	Reference
Aluminum	17	-43	-22	30,31
X Quartz	14	-5	-21	30,32
Fused silica	13	$+1$	-20	30,31
Magnesium	10	-50	-19	30,31
Polyethylene	1.7	-150	-11	30,31

sion for the power transmission coefficient in decibel notation[14]

$$T = 10 \log[4Z_s Z_{He}/(Z_s + Z_{He})^2] \qquad (8.2.7)$$

yields the power transfer from various solids to liquid helium shown in Table III.   Although plastics offer the best transmission, they are usually unsuitable for use as buffer rods.   Because of their large coefficients of thermal expansion, separation of the piezoelectric films is likely to occur on cooling.   In addition, they tend to be quite lossy, typically thousands of decibels per centimeter at high frequencies.   Both fused and crystalline quartz have been used as buffer rods at frequencies up to 1 GHz in superfluid helium.

If narrowband operation is acceptable, it should be possible to fabricate quarter-wave matching layers to reduce transmission losses.   These are commonly used in room-temperature acoustics, but have not been used in liquid helium.[33]   Optimum power transmission is obtained when a quarter-wave matching layer of impedance $Z_l = (Z_s Z_{He})^{1/2}$ is used.   However, because of the small value of Z_{He}, it is difficult to find a material with sufficiently low impedance to serve as a matching layer.   Alternatively, two layers can be used to form a double matching layer, which allows more flexibility of choice in the impedances of the two layers.   At high frequencies this can be accomplished by using deposited films on the end of the buffer rod.   At center frequency the transmission coefficient with two isotropic quarter-wave matching layers is given by

$$T_l = 10 \log\left[\frac{4Z_{He}(Z_2/Z_1)^2 Z_s}{(Z_{He} + (Z_2/Z_1)^2 Z_s)^2}\right], \qquad (8.2.8)$$

[30] "American Institute of Physics Handbook." McGraw-Hill, New York, 1972.
[31] R. B. Scott, "Cryogenic Engineering." Van Nostrand-Reinhold, Princeton, New Jersey, 1959.
[32] A. C. Rose-Innes, "Low Temperature Laboratory Techniques." Pergamon, Oxford, 1973.
[33] C. DeSilets, J. C. Fraser, and G. S. Kino, IEEE Trans. Sonics Ultrason. SU-25, 115 (1978).

TABLE IV. Quarter-Wave Matching Sections
for Fused Silica to Helium Match

Layer 1	Layer 2	Transmission coefficient (dB)
None	None	-20
Au	Al	-9
Au	SiO_2	-6
Au	Mg	-5
W	Mg	-2

where Z_1 and Z_2 are the impedances of the two layers. Calculated results for several possible double matching sections on fused quartz are listed in Table IV. Of course the increase in transmission coefficient occurs at the expense of bandwidth. The calculated bandshape for a gold–glass double matching section is shown in Fig. 4. The bandwidth is approximately 4%.

At frequencies below about 200 MHz, free-standing transducers can be fabricated and operated in direct contact with liquid helium. The very low impedance of helium implies light loading of the transducer, which leads to a number of interesting consequences. For example, it is of

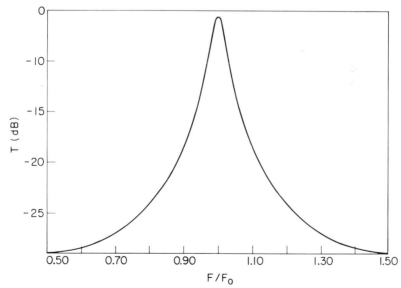

FIG. 4. Bandshape of gold–glass, double quarter-wave, matching layers between fused quartz and liquid helium.

interest to evaluate the electrical input impedance presented at the leads of the piezoelectric transducer immersed in liquid helium. Let the transducer be freestanding and symmetrically loaded by liquid helium and operating at half-wave resonance. The electrical input impedance at resonance for these boundary conditions is

$$Z = \frac{4k_T{}^2}{\pi\omega_0 C_0} \frac{Z_T}{Z_{He}} + \frac{1}{j\omega_0 C_0}, \tag{8.2.9}$$

where ω_0 is the resonant frequency, C_0 the capacitance of the transducer disk, and Z_T and Z_{He} the mechanical impedances of the transducer material and liquid helium, respectively. The first term is the radiation resistance and the second is the static capacitance of the transducer material. For a transducer loaded on one side only, the expression is the same except for a factor of 2 in place of the factor of 4. This expression can be rewritten in a particularly simple way for a circular transducer if the transducer radius is normalized to the wavelength of sound in the transducer material as $r = n\lambda$. Then

$$Z = A/n^2 + (1/j)B/n^2, \tag{8.2.10}$$

where

$$A = (k_T{}^2/\pi^3 v\epsilon)Z_T/Z_{He} \quad \text{and} \quad B = 1/4\pi^2\epsilon v,$$

ϵ and v are the dielectric constant and longitudinal acoustic velocity of the transducer material, respectively, and A and B depend only on material constants and not on frequency. Thus for any given transducer material, a disk transducer can be made that has an electrical resistance of any desired value. Values of A and B for several materials are given in Table V. Also given is n, the transducer size in wavelengths for a 50-Ω resistive match and the resulting electrical reactance X_c. (Note that the transducer size listed here is in wavelengths in the transducer material, not in the helium.) Because of the very light loading the helium makes on the transducer, the reactance for a 50-Ω resistance is always small and is smallest for high k_T materials.

Another consequence of the low impedance of helium is a very high acoustic quality factor Q_a. For a symmetrically loaded transducer Q_a is

TABLE V. Values of Constants in Eq. (8.2.10) for Several Transducer Materials

Material	A ($\times 10^{-6}$) (Ω)	B ($\times 10^{-6}$) (Ω)	n ($R = 50\ \Omega$)	X_c ($R = 50\ \Omega$) (Ω)
Quartz	0.66	10	110	7.7
CdS	1.3	6.7	160	2.1
ZnO	5.6	5.1	340	0.45

given by[33]

$$Q_a = \tfrac{1}{4}\pi Z_T / Z_{He}.$$ (8.2.11)

In helium, Q_a ranges from 10^2 to 10^3, corresponding to bandwidths of less than 1%.

Most low-frequency measurements in superfluid helium have been made with quartz as the piezoelectric material. As indicated, materials with a larger value of k_T offer some advantages.

8.2.1.4. Poled Ferroelectrics. Certain poled ferroelectric materials are suitable for use in liquid helium. Finch *et al.* have used PZT-4 at frequencies near 100 kHz for caviation studies in superfluid helium.[34] This is a poled ceramic with a large k_T available in thicknesses and shapes that make it useful in the 50-kHz to 10-MHz range. The various material constants for PZT-4 (and most other ferroelectrics) are temperature dependent between room temperature and liquid helium temperatures. Some poled ferroelectrics with large values of k_T at room temperature are much less attractive in helium (for example, k_T for PZT-5H is 40% larger than PZT-4 at 300°K but is 16% smaller at 2°K).[27]

Considerable attention has been paid in the literature to the low-temperature properties of the polymer film polyvinylidene fluoride PVF_2 in an effort to understand its properties.[28,35,36] It is a polymer ferroelectric and has been poled to make ultrasonic transducers, which have been used to propagate sound in a quartz rod at low temperatures.[37] Its low-temperature properties are listed in Table II. To the author's knowledge it has not been used for sound propagation in liquid helium to date, but its mechanical flexibility, low acoustic impedance, ease of handling, and fairly large coupling constant may make it useful in some special applications. It should be noted that only k_{TE} is large at low temperature.

8.2.2. Second Sound

As discussed previously, second sound is principally a temperature–entropy wave and thus heaters and thermometers are appropriate generators and detectors of second sound. This was not at first appreciated, and early efforts to observe second sound failed because pressure transducers were employed.[38] Second sound has been observed over the frequency range from a few hertz to about 15 MHz using the transducers that will be described.

[34] R. D. Finch, R. Kagiwada, M. Barmatz, and I. Rudnick, *Phys. Rev.* **134**, A1425 (1964).

[35] E. Fukuda and S. Takashita, *Jpn. J. Appl. Phys.* **8**, 960 (1969).

[36] H. Sussner, *Phys. Lett A* **58**, 426 (1976).

[37] H. Sussner *et al., Phys. Lett. A* **45**, 475 (1973).

[38] E. M. Lifshitz, *J. Phys. (Moscow)* **8**, 110 (1944).

8.2.2.1. Resistive Films as Generators and Detectors of Second Sound. Probably the most popular transducer for second sound is a thin, resistive film applied to a smooth, thermally insulating backplate. Application of a sinusoidal current $I_0 \cos \omega t$ to a film of resistance R leads to a heating

$$q = I^2 R = \tfrac{1}{2} I_0^2 R(1 + \cos 2\omega t) \qquad (8.2.12)$$

with a sinusoidal component. A third of the power appears at frequency 2ω in the oscillating component and the other two-thirds appears as steady heat. If the film is biased with a constant current I_B, then

$$q = (I_B + I_0 \cos \omega t)^2 R = I_B^2[1 + 2(I_0/I_B) \cos \omega t]R \qquad (I_B \gg I_0) \quad (8.2.13)$$

and most of the oscillatory component appears at the frequency ω. Thus the resistive film transducer can generate second sound at ω or 2ω. The biased film can also be used as a detector of second sound if it possesses a nonzero temperature coefficient of resistance $\alpha \equiv (1/R) \, \partial R/\partial T$. If a second sound wave of amplitude δT strikes the (low thermal mass) biased film, then a voltage

$$\delta V = I_B \delta R = I_B(\partial R/\partial T) \, \delta T \qquad (8.2.14)$$

will be produced. In terms of the dc power $P = I_B^2 R$,

$$\delta V = (P/R)^{1/2} \, (\partial R/\partial T) \, \delta T. \qquad (8.2.15)$$

We are actually interested in the signal-to-noise ratio, so, assuming that the Johnson noise $\langle V_n^2 \rangle^{1/2} = (4k_B T \, \Delta f \, R)^{1/2}$ of the film is the dominant noise source,

$$\frac{V}{\langle V_n^2 \rangle^{1/2}} = \left(\frac{P}{4k_B T \, \Delta F} \right)^{1/2} \frac{1}{R} \frac{\partial R}{\partial T} \delta T \qquad (8.2.16)$$

where k_B is the Boltzmann constant and Δf the bandwidth of the measurement system. To maximize the signal-to-noise ratio we choose as our transducer the material with the greatest α.

At very low temperatures the thermal boundary (or Kapitza) resistance must also be taken into acount in determining the response of a second sound transducer.[39] Although not well understood, this effect is basically due to the large acoustic mismatch between liquid helium and solids, which impedes the progress of thermal phonons across the boundary.

The frequency response of the second-sound transducer is determined in principle by the thermal response time of the resistive film and substrate. In practice, due to the very large heat capacity of superfluid he-

[39] H. A. Fairbank and C. T. Lane, *Rev. Sci. Instrum.* **18**, 525 (1947).

lium, the details of the substrate are always unimportant. The response time of the film is determined by the thermal penetration depth

$$\lambda_T = (2K/\omega\rho C_v)^{1/2}, \qquad (8.2.17)$$

where K is the thermal conductivoty and ρC_v the specific heat per unit volume of the transducer. When λ_T is a few times the transducer thickness or less, the response begins to fall off. The response time of thin-film transducers is further limited by the lead capacitance of coaxial cables used to connect them to room-temperature electronics.

Resistive film transducers can be made of carbon films,[39,40] which have been used at frequencies up to 300 kHz.[40] Epitaxial silicon on sapphire bolometers have been constructed and should make good detectors and generators of second sound at megahertz frequencies.[41] Doped germanium has also been used as a second sound transducer.[42] For semiconducting materials, the resistivity ρ_E can be roughly expressed in terms of an ionization energy E as

$$\rho_E = \rho_0 \exp(E/k_B T) \qquad (8.2.18)$$

and thus α can be written

$$\alpha = -E/k_B T^2$$

and is expected to be temperature dependent.

Resistive metal films have been used as generators of second sound up to 25 MHz.[43] Since α is small or zero for most metals, they do not make good detectors. Garrett and Rudnick have used thin wires to construct a tuned, dipole driver for low-frequency second sound.[44]

8.2.2.2. Superconducting Edge Bolometers. A particularly sensitive detector of second sound is a superconducting metal film biased at its superconducting transition temperature T_c. These devices have been used extensively as infrared detectors.[45,46] The temperature dependence of resistance in a pure superconductor such as tin is shown in Fig. 5. For zero applied magnetic field the resistance passes abruptly from some value typical of the normal metal to zero upon reaching the transition temperature T_c. By setting the temperature at T_c and biasing the device with a small current, a large temperature sensitivity results. Application of a mag-

[40] W. B. Hanson and J. R. Pellam, *Phys. Rev.* **95**, 321 (1954).
[41] S. Early, *Rev. Sci. Instrum.* (to be published).
[42] H. A. Snyder, *Rev. Sci. Instrum.* **33**, 467 (1962).
[43] H. Notarys, Ph.D. Dissertation, California Institute of Technology, 1964.
[44] S. Garrett, S. Adams, S. Putterman, and I. Rudnick, *Phys. Rev. Lett.* **41**, 413 (1978).
[45] J. Clarke *et al., J. Appl. Phys.* **48**, 4865 (1977).
[46] M. Maul and M. Strandburg, *J. Appl. Phys.* **40**, 2822 (1969).

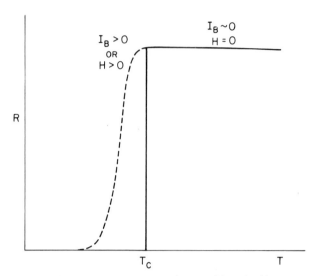

FIG. 5. Resistance of a pure type I superconductor with and without a current bias or magnetic field.

netic field (or large biasing current) shifts the transition down in temperature as shown. Thus the bolometer can be "tuned" to the desired temperature. The sensitivity of pure superconducting material tends to be too large for practical use, so it is desirable to broaden the transition. In addition, no elemental superconducting materials with a transition temperature in the range 1.2–2.2°K exist, so special techniques are used to prepare bolometers with the properties desirable for second-sound detectors. Two techniques are especially useful. It is known that thin aluminum films prepared in the presence of a small amount of oxygen gas can exhibit transition temperatures in the range 1.1–2.2°K with relatively broad transitions.[47,48] Such a bolometer with an appropriately high T_c can be tuned to most temperatures in the second-sound range. A composite of a layer of tin evaporated on a layer of nonsuperconducting gold makes use of the proximity effect in superconductors to lower the transition temperature of the tin below its bulk value of 3.5°K.[49] With an appropriate choice of film thicknesses, a bolometer can be made with a transition temperature of about 2.2°K, which can then be tuned to lower temperatures as desired. Sensitivities of superconducting bolometers are generally high; however, they suffer from a limited temperature range and some

[47] R. W. Cohen and B. Abeles, *Phys. Rev.* **168**, 444 (1968).
[48] R. B. Pettit, *Phys. Rev. B* **13**, 2865 (1976).
[49] G. Laguna, *Cryogenics* **16**, 241 (1976).

complication in the tuning procedure. Because of their thinness and high sensitivity, superconducting edge bolometers are suited for high-frequency second-sound studies. Notarys has used a tin–gold bolometer as a second-sound receiver at frequencies up to 25 MHz.[43]

8.2.2.3. Capacitive Transducers for Second Sound. As pointed out in Section 8.1.1, there are two ways of describing a second-sound wave: either as a temperature–entropy wave or by specifying that the momentum density ρv of the wave is zero, i.e.,

$$\rho_s \mathbf{v}_s = -\rho_n \mathbf{v}_n. \tag{8.2.19}$$

These can be shown to be equivalent descriptions since the entropy of helium II is carried totally by the normal fluid. This second description of second sound forms the basis of a useful and imaginative second-sound transducer.[50,51] Imagine an oscillating membrane with holes so small that the normal fluid cannot flow through them at the frequency of interest (i.e., that the viscous penetration depth is larger than the hole diameter). Such a membrane is an example of a superleak, because only the inviscid superfluid can flow through the holes. Let v_1 and v_2 be the particle velocities of first and second sound, respectively. In the limit of a very large number of holes the boundary conditions at the surface of the superleak are[51]

$$v_n = v_1, \qquad v_s = 0. \tag{8.2.20}$$

Now the energy densities E_1 and E_2 of first and second sound can be written

$$E_1 = \tfrac{1}{2}\rho_n v_{n1}^2 + \tfrac{1}{2}\rho_s v_{s1}^2 = \tfrac{1}{2}\rho v_{n1}^2, \qquad E_2 = \tfrac{1}{2}\rho_n v_{n2}^2 + \tfrac{1}{2}\rho_s v_{s2}^2 = \tfrac{1}{2}(\rho_n\rho/\rho_s)v_{n2}^2. \tag{8.2.21}$$

The energy–density ratio is then

$$E_2/E_1 = (\rho_n/\rho_s)v_{n2}^2/v_{n1}^2 \tag{8.2.22}$$

and finally, by the boundary conditions for the superleak,

$$E_2/E_1 = \rho_s/\rho_n. \tag{8.2.23}$$

Thus the oscillating superleak transducer generates both first and second sound with the relative intensities determined by the ratio ρ_s/ρ_n, which in turn is a function of temperature. An oscillating superleak transducer can be made as a condenser microphone using a piece of Millipore or Nu-

---

[50] R. Williams et al., Phys. Lett. A **29**, 279 (1969).
[51] R. A. Sherlock and D. O. Edwards, Rev. Sci. Instrum. **41**, 1603 (1970).

cleopore plastic film.[52]   These materials can be obtained with various hole sizes and densities and then aluminized on one side. Of course, in a practical superleak the porosity is never unity; with Millipore films porosities as high as 80% are available with Nucleopore up to 20%. With porosities less than 100%, these estimates of sound generation efficiencies are not valid because the superfluid velocity will not be zero. It can be shown that nonzero superfluid velocity decreases the energy–density ratio (8.2.23).

A second type of superleak transducer is possible where the superleak remains stationary and an oscillating piston pushes superfluid through the superleak from behind. Analysis of these boundary conditions yields[51]

$$E_2/E_1 = \rho_n/\rho_s, \tag{8.2.24}$$

so that such a geometry is a more efficient generator of second sound at higher temperatures.

Many other configurations that produce both first and second sound are possible. For example, a capacitive microphone with an electret film as the active element with only the ends (not the sides) clamped has been used to generate first and second sound simultaneously.[53]

8.2.3. Fourth Sound

Fourth sound has been studied from subaudio frequencies up to about 100 kHz. Capacitive pressure transducers like those used to generate and detect first sound are also commonly used for fourth sound. The transducer is mounted in contact with a superleak material as will be described in Section 8.3.1.1. Fourth sound has also been generated thermally by using a chromium heater,[54] and Rudnick has pointed out that a porous membrane should also be useful as a transducer of fourth sound.[8] The Doppler shift of fourth sound in an annular channel has been used to determine the velocity of persistent currents. This method is nondestructive and can be employed to measure, among other things, the velocity of persistent currents in both rotating and stationary annuli. This topic will be discussed more fully in Section 8.3.1.5. It should also be noted that if the velocity of first sound is known, then, using Eq. (8.1.26), a measurement of C_4 at temperature T yields a value of ρ_s/ρ at that temperature. Thus fourth sound probes the superfluid density directly.

In a practical superleak such as a packed powder, the porosity P, de-

---

[52] Millipore Corp., Bedford, Massachusetts 01730, and Nucleopore Corp., Pleasanton, California 94566.
[53] Suggested by R. Williams; see also Heiserman and Rudnick.[20]
[54] H. Wiechert and R. Schmidt, *Phys. Lett.* A **40**, 421 (1972).

fined as the ratio of open volume to total volume, is always less than one. In such a superleak fourth sound is scattered by the powder grains and suffers velocity dispersion. Because of the high density of particles, the scattering problem is difficult to treat theoretically; however, it has been determined that an empirical index of refraction $n = (2 - P)^{1/2}$ gives agreement with experiment to within 3%. The relation was found to be independent of temperature and frequency in the audio-frequency range and applies for grain diameters from 1 μm to 500 Å and for porosities from 0.42 to 0.94.[55]

8.3. Measurement Techniques

The transducers described in the preceding section can be used to construct acoustic systems for use as probes of the properties of superfluid helium. Both resonant and pulse methods have been employed and each offers advantages.

8.3.1. Resonant Methods

Resonant techniques have been employed extensively in the lower range of frequencies for all three bulk modes of helium. Because a single mode can be selectively excited, resonators are, by their very nature, acoustically monochromatic devices and are thus suited to dispersion and attenuation studies. By using phase-locking techniques, data on the temperature and pressure dependences of acoustic velocity and attenuation may be obtained. The precision of resonator measurements depends on the quality factor Q and the parallelism of the endplates, which determines the precision in length. The Q may be written, in terms of the appropriate sound velocity C and angular frequency ω,

$$Q = \omega/2\gamma C. \tag{8.3.1}$$

The constant γ is the attenuation coefficient. Attenuation is usually present as absorption in the fluid itself. Additional contributions to the attenuation can arise from boundary losses and other mechanisms associated with the resonator geometry.

8.3.1.1. Simple Resonators. A hollow cylindrical tube closed at both ends by flat plates forms a resonator for plane acoustic waves. If the length of the cavity l is much greater than the diameter d, then the low-

[55] M. Kriss and I. Rudnick, *J. Low Temp. Phys.* **3**, 339 (1970).

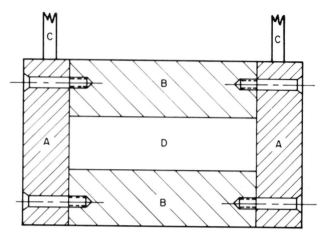

FIG. 6. Schematic of a simple plane wave resonator (from Shipiro and Rudnick[57]): A—end plates; B—resonator body; C—support tubes; D—space for helium (and superleak in fourth-sound experiment).

frequency modes occur at the half-wave $\lambda/2$ resonant frequencies

$$f_n = nC/2l, \qquad n = 1, 2, 3, \ldots . \qquad (8.3.2)$$

Above frequencies for which $d \simeq \lambda$, radial and azimuthal modes arise.[56]

In such a resonator the endplates are usually designed to be transducers for the desired type of sound with one transducer used as a source and the other as a detector. In the case of fourth sound, the cavity can be tightly packed with a fine powder superleak to lock the normal fluid. Figure 6 shows the main features of such a resonator. The endplates contain insulated contacts which serve as one electrode for the transducers. A diaphragm stretched over the endplates is chosen to form an appropriate capacitive transducer for the sound desired as described in the previous section. For the case of second sound, the diaphragm may be replaced with a thin-film thermometer or heater deposited directly on the endplates. Generally, even a tightly assembled resonator will fill with helium through the crack between the endplate and resonator body at temperatures below T_λ; however, it is sometimes desirable to fill the cavity through a small capillary. For measurements under pressure, the endplates can be sealed to the body using indium O-rings.

Such a resonator has been selected for several studies using first, second, and fourth sound at saturated vapor pressure,[8] as well as elevated

[56] P. Morse, "Vibration and Sound," 2nd ed. McGraw-Hill, New York, 1948.
[57] K. Shapiro and I. Rudnick, *Phys. Rev.* **137**, 1383 (1965).

pressure. Especially interesting results have been obtained regarding the nature of the lambda transition itself. Studies using first and second sound near T_λ have yielded critical exponents and provided tests of universality and scaling theories.[58]

Near a second-order phase transition a set of relations known as the *Ehrenfest equations* relate values of various thermodynamic parameters near the critical temperature. A corresponding set, known as the *Pippard–Buckingham–Fairbank (PBF) relation*, applies to lambda transitions as found in liquid helium. Measurements of the low-frequency velocity of sound near T_λ serve as input for PBF from which critical values of other thermodynamic properties such as specific heat and thermal expansion coefficient can be determined.[59] The velocities of all three sounds have been measured simultaneously as a function of temperature and pressure using an apparatus in which four simple resonators were fabricated as an integrated unit.[60] From the high-precision data thus obtained, it was possible to calculate precise and consistent values for the thermodynamic properties of helium in the $P-T$ plane.[61]

Kojima *et al.* employed a simple fourth-sound resonator to demonstrate the superfluid nature of ^{3}He at temperatures below 3 mK.[6] From their data the temperature dependence of the superfluid density of ^{3}He (ρ_s/ρ) was first determined using Eq. (8.1.26), a scattering correction (see Section 8.2.3), and measurements of the first-sound velocity at these temperatures. It would be especially interesting to examine the various phase transitions observed in superfluid ^{3}He using techniques analogous to those described for ^{4}He.

8.3.1.2. Loss Mechanisms in Simple Resonators. The Q of a simple resonator is lowered by losses that arise from many sources. The absorption in the fluid imposes a fundamental limit, but in practice the Q observed in a resonator is usually further lowered by other loss mechanisms. Resonator leakage and mechanical losses in the transducer diaphragm are minor mechanisms usually observable only in very high Q systems. A more common loss mechanism found in first- and second-sound resonators arises from relative motion of the viscous normal fluid and the walls of the resonator. In the case of fourth sound in a resonator filled with a coarse-grained or loosely packed superleak, incomplete locking of the normal fluid by the grains of the superleak is a similar source of loss. Imagine a smooth-walled, first-sound resonator of cross-sectional area A and perimeter B. If viscous surface losses are the dominant source of

[58] D. Greywall and G. Ahlers, *Phys. Rev. A* **7**, 2145 (1973).
[59] M. Barmatz and I. Rudnick, *Phys. Rev.* **170**, 224 (1968).
[60] J. Heiserman, J. P. Hulin, J. Maynard, and I. Rudnick, *Phys. Rev. B* **14**, 3682 (1976).
[61] J. Maynard, *Phys. Rev. B* **14**, 3868 (1976).

loss, then $Q = Q_1$ can be written[20]

$$Q_1 = (\rho/\rho_n)(2A/B)1/\lambda, \tag{8.3.3}$$

where λ is the viscous penetration depth, which can be written, in terms of the viscosity η,

$$\lambda = (2\eta/\rho_n\omega)^{1/2}.$$

For the case of a circular cross section, $A/B = r/2$ and

$$Q_1 = (\rho/\rho_n)r/\lambda. \tag{8.3.4}$$

Similarly, for a second-sound resonator of the same geometry,[20]

$$Q_2 = (\rho/\rho_s)(2A/B)1/\lambda. \tag{8.3.5}$$

From Eqs. (8.3.3) and (8.3.5) it is clear that viscous surface losses are a minimum at low temperatures for first sound while for second sound the loss is smallest near T_λ.

In cases where it is necessary to minimize viscous surface losses, the radial modes of a spherical resonator may be useful, since all normal fluid motion is perpendicular to the wall.

8.3.1.3. Helmholtz Resonators. Consider a flask of volume V with a neck of length l and cross-sectional area A as shown in Fig. 7. At low frequencies a sound mode known as a *Helmholtz resonance* is observed at a circular frequency ω given by

$$\omega^2 = C_n^2 A/Vl, \tag{8.3.6}$$

where C_n is the velocity of sound appropriate to the medium in the flask.

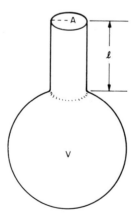

FIG. 7. Schematic of a Helmholtz resonator, where V is the volume of the chamber and A the cross-sectional area of the neck (from Rudnick[8]).

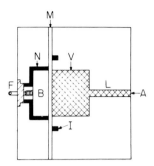

Fɪɢ. 8. Schematic of an experimental Helmholtz resonator: M—metallized Mylar diaphragm; B—insulated backplate; I—indium seal. For fourth sound measurements, V and L are packed with superleak (from Kriss and Rudnick[62]).

Superfluid Helmholtz resonances have been observed for first, second, and fourth sound. Kriss and Rudnick have shown that the resonance occurs as predicted by (8.3.6), where C_n is the velocity of first, second, or fourth sound.[62] Figure 8 shows a schematic of an experimental Helmholtz resonator used by Kriss and Rudnick to observe first- and fourth-sound Helmholtz resonances. A receiving transducer outside the neck was used to identify the frequency of resonance. Viscous surface losses in the neck and radiation losses due to sound radiated out of the neck are sources of loss in a Helmholtz resonator. Kriss and Rudnick reported a Q of 500 for their apparatus (Fig. 8), operated as a fourth-sound resonator.

The linear dimensions of a Helmholtz resonator are usually much smaller than its resonant wavelength. This feature allows experiments to be performed at very low frequencies, which is useful when measurements in the thermodynamic (zero-frequency) limit are needed. One such application has already been cited in Section 8.3.1.1. Rudnick has pointed out that a second-sound Helmholtz resonator may be useful in observing second sound in ^{3}He (which has not yet been observed).[8] At these temperatures the normal component of ^{3}He is very viscous and second sound will suffer attenuation by the mechanism of Section 8.3.1.2, which increases with increasing frequency.

8.3.1.4. Mechanically Driven Fourth-Sound Resonator. Hall, Kiewiet, and Reppy have developed a novel scheme for observing fourth sound that departs from standard acoustical practice.[63] The apparatus as used by Yanoff and Reppy is shown schematically in Fig. 9a.[64] A cavity

[62] M. Kriss and I. Rudnick, Phys. Rev. **174**, 326 (1968).
[63] H. Hall, C. Kiewiet, and J. Reppy, Phys. Rev. Lett. **35**, 631 (1974).
[64] A. Yanof and J. Reppy, Phys. Rev. Lett. **33**, 631 (1974).

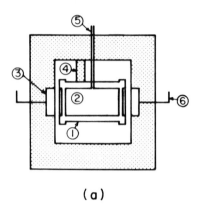

(a)

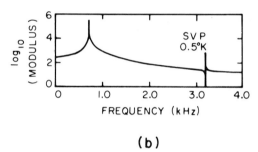

(b)

FIG. 9. (a) Fourth sound cell of Yanoff and Reppy[64]: 1—sample chamber, 2—superleak, 3—electrostatic driving electrode, 4—suspension rod, 5—fill capillary, 6—electrostatic sensing electrode; (b) Frequency spectrum of resonator in (a). The low-frequency peak is due to the simple harmonic resonance of the suspension and the smaller peak is the fourth sound resonance.

containing superleak material is suspended from an elastic support. The cell is driven electrostatically and the displacement is observed with a capacitive detector. Two main resonances are observed as shown in the spectrum of Fig. 9b. At a low frequency a simple harmonic resonance of the mechanical system is observed. At somewhat higher frequency in this case, the fourth-sound resonance is observed through its effect on the amplitude of oscillation of the entire cell. Yanoff and Reppy employed their apparatus to make an independent determination of the superfluid density of ^{3}He which agrees with the data of Kojima *et al.*[7]

8.3.1.5. Annular Resonators. To study superfluid flow using the Doppler shift of sound, Rudnick and co-workers developed an annular

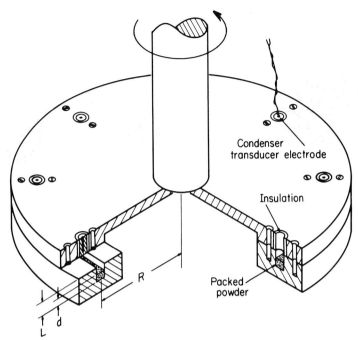

FIG. 10. A schematic diagram of an annular resonator. For fourth sound, $d = L$. Only four of the eight electrodes are shown for purposes of illustration (from Heiserman and Rudnick[20]).

resonator[65] as shown in Fig. 10. (For the moment, assume $d = L$ so the channel is completely filled with superleak.) Small-capacitive transducers located at intervals around the annulus serve as generators and detectors of sound. The general case of an annulus of unspecified inner and outer radius (r_i and r_o) is treated by Morse,[66] but in the limit of ($r_o - r_i$) $\ll r_i$ and low frequencies the curvature of the annular cavity may be neglected and an approximate solution can be obtained by treating a cavity of length

$$l \approx 2\pi[(r_i + r_0)/2]$$

and applying cyclic boundary conditions. Low-frequency resonances

[65] H. Kojima, W. Veith, E. Guyon, and I. Rudnick, *J. Low Temp. Phys.* **8,** 187 (1972).
[66] P. Morse and U. Ingard, "Theoretical Acoustics," p. 603. McGraw-Hill, New York, 1968.

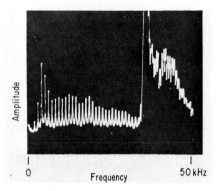

FIG. 11. Resonances in air of an annular cavity of radius $R = 5.25$ cm and height and width 0.5 cm.

then occur at frequencies given approximately by

$$f_n = nC/\pi(r_i + r_o). \qquad (8.3.7)$$

When the depth or width of the annulus is a half-wavelength, $L, w \approx \lambda/2$, additional "height" and radial modes occur. Actually, traveling waves generated by the source propagate in both clockwise and counterclockwise directions, leading to two standing wave fields; however, in the absence of fluid flow the two fields are degenerate and the annulus exhibits only one set of resonances, given by Eq. (8.3.7). Figure 11 shows the spectrum in air at room temperature of the annular resonator shown in Fig. 10 without a superleak, in which the height and width of the channel are equal and much smaller than the radius of the annulus. As expected, equally spaced azimuthal modes occur at low frequencies with height and radial modes entering at a higher frequency. Of course any of the three sounds of helium can be observed in an annular resonator with the proper transducers and use of a superleak when appropriate. In fact, by partially packing the annulus of Fig. 10 to a depth d with a superleak material, hybrid modes can be observed that are combinations of first, second, and fourth sound. In the situation just described, two modes are observed.[67] One, called C_{II}, is basically a hybrid of second and first sound. The magnitude of C_{II} as a function of temperature and packing fraction (i.e., d/L) is shown in Fig. 12. In the limit of no superleak, $C_{II} \rightarrow C_2$, the velocity of second sound, as it must. In the opposite limit of 100% packing, C_{II} is nonpropagating. A second hybrid mode C_{14} is also observed in the partially packed annulus. It is a hybrid of first and fourth sound and its

[67] J. Rudnick, I. Rudnick, and R. Rosenbaum, *J. Low Temp. Phys.* **16**, 417 (1974).

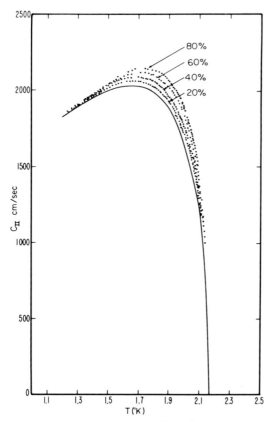

FIG. 12. C_{II} versus T with various percentages of the annulus packed with superleak material. The solid line is C_2 (from Heiserman and Rudnick[20]).

velocity as a function of temperature and packing fraction is shown in Fig. 13. In the limit of no packing and 100% packing, C_{14} becomes C_1 and C_4, respectively. The annular resonators previously discussed have proved useful as probes of persistent currents and critical velocities in superfluid helium by observing the Doppler shift of sound in the resonator.

When sound propagates in a moving classical fluid it is Doppler shifted. If the velocity of sound in the static medium C_0 is known, the Doppler-shifted velocity C can be used to determine the velocity of the medium v. For flow in one dimension,

$$C = C_0 \pm v, \qquad (8.3.8)$$

where the plus and minus signs correspond to sound propagating with and against the motion of the fluid, respectively.

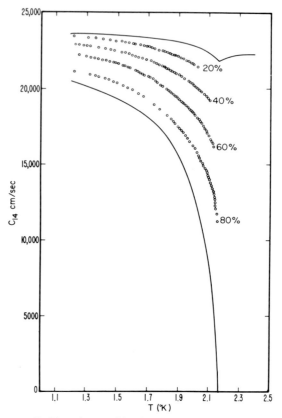

FIG. 13. C_{14} versus T with various packing fractions.   Solid lines are C_1 and C_4 (from Heiserman and Rudnick[20]).

Similarly, the sounds of helium are Doppler shifted by moving He II. In general the values of the Doppler-shifted velocities in helium depend on the velocities of both the normal fluid and superflud, and thus a measurement of the Doppler shift does not determine v_n or v_s but only some function of the two velocities.   Expressions for the Doppler shift of first, second, and fourth sound follow from the linearized Landau equations when the normal and superfluid are allowed to have steady nonzero velocity fields.

Suppose the flow is in the direction of sound propagation.   The expression for the Doppler shift of first sound in analogous to the classical result[68]

---

[68] I. M. Khalatnikov, *Zh. Eksp. Teor. Fiz.* **30,** 617 (1956) [*English transl.: Sov. Phys. JETP* **2,** 73 (1956)].

$$C_1 = C_{10} + \left(\frac{\rho_s}{\rho} v_s + \frac{\rho_n}{\rho} v_n \right), \tag{8.3.9}$$

where the zero subscript denotes a quantity measured in the stationary medium. The Doppler shift of second sound involves the entropy per gram s,[68]

$$C_2 = C_{20} + \left(\frac{\rho_s}{\rho} v_s + \frac{\rho_n}{\rho} v_n \right) + (v_n - v_s) \left[\frac{2\rho_s}{\rho} - \frac{s}{\rho_n} \left(\frac{\partial \rho_n}{\partial T} \right)_p \left(\frac{\partial T}{\partial S} \right)_p \right]. \tag{8.3.10}$$

The second term is the shift in velocity due to motion of He II as in first sound. However, the third term is new and is sensitive to differences in superfluid and normal fluid velocities. Equation (8.3.10) can be simplified using the fact that the normal fluid carries all the entropy of He II,

$$\rho s = \rho_n s_n, \tag{8.3.11}$$

and the Tisza approximation that s_n is approximately a constant independent of T. This is valid in the range of the temperature $1.1°K$ to T_λ where s_n changes only 20%. Then (8.3.10) becomes

$$C_2 = C_{20} + \frac{\rho_s}{\rho} v_s + \frac{\rho_n}{\rho} v_n + (v_n - v_s) \left(\frac{\rho_s}{\rho} - \frac{\rho_n}{\rho} \right). \tag{8.3.12}$$

The Doppler shift of second sound is not zero when the He II velocity is zero as is the case for first sound.

The Doppler shift for fourth sound involves only v_s because the normal fluid is always locked in a superleak[69]:

$$C_4 = C_{40} + [\rho_s/\rho + G(T)] v_s, \tag{8.3.13}$$

where

$$G(T) = \frac{s [\partial(\rho_n/\rho)/\partial T]_p}{(\partial s/\partial T)_p} - \frac{\rho [\partial(\rho_n/\rho)/\partial T]_T}{(\partial \rho/\partial T)_T}.$$

The function $G(T)$ is imperfectly known at high temperatures. At temperatures below $1.4°K$, $G(T)$ is small, and neglecting it results in an error of less than 2%.[69]

In the case where the Doppler shift is caused by a persistent current, the normal fluid velocity is zero and the equations for the Doppler shifts are simplified to

[69] I. Rudnick, H. Kojima, W. Veith, and R. S. Kagiwada, *Phys. Rev. Lett.* **23**, 1220 (1969).

$$C_1 = C_{10} + (\rho_s/\rho)v_s, \tag{8.3.14}$$

$$C_2 = C_{20} + (\rho_n/\rho)v_s, \tag{8.3.15}$$

$$C_4 = C_{40} + [\rho_s/\rho + G(T)]v_s. \tag{8.3.16}$$

The Doppler shifts of first and fourth sound are greatest at low temperatures while that of second sound is greatest nearer T_λ. These Doppler shifts can all be expressed as a constant E_l times v_s, where

$$E_1 = \rho_s/\rho, \tag{8.3.17}$$

$$E_2 = \rho_n/\rho, \tag{8.3.18}$$

$$E_4 = \rho_s/\rho + G(T). \tag{8.3.19}$$

Consider now an annular resonator such as the one in Fig. 10. If the annular cavity is filled with superfluid helium and a persistent current is created by cooling the cell through T_λ while rotating and then stopping at a low temperature, the low-frequency resonances of first and second sound can be observed and, in principle, will be Doppler shifted [although the effect may be too small to measure in large geometries because of Eq. (8.1.9)]. Similarly, fourth sound will be Doppler shifted in an annulus filled with fine powder.

The pressure field of the lth sound in an appropriate annulus can be written[69]

$$P_l = \cos \phi\Omega(k_l r) \cos \omega_l t, \tag{8.3.20}$$

where $l = 1,2,4$ for first, second, or fourth sound, ϕ is the azimuthal angle, k the wave number, r the radial coordinate, ω the frequency, t the time, and $\Omega(k_l r)$, which describes the radial dependence, is the difference between a first-order Bessel function and a normalized Neumann function. If there is a persistent current in the annulus, (8.3.20) is the pressure field in a frame rotating with the velocity of the superfluid or from (8.3.14)–(8.3.16) at an angular velocity $E_l\omega_s$. Thus in the stationary frame

$$P_l = \cos(\phi + E_l\omega_s t)\ \Omega(k_l r) \cos \omega_l t$$

$$= \tfrac{1}{2}\{\cos[(\omega_l + E_l\omega_s)t + \phi] + \cos[(\omega_l - E_l\omega_s)t - \phi]\}\Omega(k_l r). \tag{8.3.21}$$

Thus, due to the presence of the persistent current, the two paths from the emitter to the receiver are not equivalent and the degeneracy in the resonant modes is lifted. The result is a frequency doublet where the separation between the two resonant peaks is related to the persistent current angular velocity ω_s by

$$\Delta\omega_l = 2n_l E_l\omega_s, \tag{8.3.22}$$

where n_l is the harmonic number. This can be expressed in terms of the observed splitting of the resonance and the resulting linear velocity of the persistent currents. Using (8.3.14)–(8.3.20) in the open channel, we obtain, for first and second sound, respectively,

$$v_s = (\pi R/n_1)(\rho/\rho_s)\, \Delta f_1, \qquad (8.3.23)$$

$$v_s = (\pi R/n_2)(\rho/\rho_n)\, \Delta f_2, \qquad (8.3.24)$$

where R is the resonator radius and Δf_l the frequency splitting of the nth harmonic of the appropriate sound. In a powder-packed channel the persistent current is given by the splitting of fourth sound as

$$v_s = \frac{\pi R}{n_4} \frac{1}{[(\rho_s/\rho) + G(T)]}\, \Delta f_4. \qquad (8.3.25)$$

Similar considerations lead to expressions for the Doppler shifts of C_{14} and C_{II} in a partially filled annulus.[70]

With the application of different boundary conditions, superfluid helium can be made to exhibit a variety of different propagating modes. Williams et al. and Gelatis et al. described a propagating thermal mode in a clamped superfluid with a pressure release boundary, dubbed *fifth sound*.[71] Putterman et al. have discussed a hybrid mode of superfluid helium in equilibrium with its saturated vapor.[72] Fifth sound may have application in the study of critical velocities and persistent currents in ^{4}He and as a probe of thermal properties of superfluid ^{3}He.[73]

8.3.2. Pulse Methods

Standard pulse methods (as described in Part 2) have been used in several investigations of the sounds of helium. A useful feature of pulse techniques is that by limiting the repetition rate the amount of heat dissipated at low temperatures can be limited as required. Pulse methods can be classified as either pulse echo or variable path (interferometer) techniques. Additionally a single transducer reflection or two transducer transmission apparatus may be chosen. In an idealized transmission pulse echo experiment, the velocity of sound is determined from the known transducer spacing d and the time delay between successive echoes t by

$$C = 2d/t. \qquad (8.3.26)$$

[70] J. Heiserman and I. Rudnick, *Phys. Rev. B* **12**, 1739 (1975).

[71] G. A. Williams, R. Rosenbaum, and I. Rudnick, *Phys. Rev. Lett.* **42**, 1282 (1979); G. J. Gelatis, R. A. Roth, and J. D. Maynard, *ibid.* **42**, 1285 (1979).

[72] S. Putterman, D. Heckerman, R. Rosenbaum, and G. Williams, *Phys. Rev. Lett.* **42**, 580 (1979).

[73] I. Rudnick, J. Maynard, G. Williams, and S. Putterman, *Phys. Rev. B* **20**, 1934 (1979).

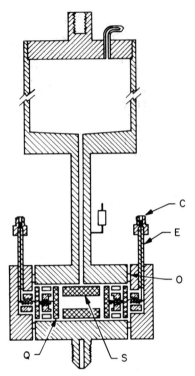

FIG. 14. Acoustic cell. See discussion in text. (From B. Abraham, Y. Eckstein, J. B. Ketterson, and J. Vignos, *Cryogenics* **9**, 274 (1969), published by IPC Science and Technology Press Ltd., Guildford, Surrey, UK.

The attenuation coefficient γ is given by

$$\gamma = (1/2d)\, \ln(A_n/A_{n+1}), \qquad (8.3.27)$$

where A_n and A_{n+1} are the observed amplitudes of successive echoes. For the interferometer method the velocity can be obtained by changing the spacing of the transducers by a distance $\Delta d = d_2 - d_1$ and observing the change in time delay Δt for a selected echo. The attenuation is given by

$$\gamma = [1/2(d_2 - d_1)]\, \ln[A_n(d_1)/A_n(d_2)], \qquad (8.3.28)$$

where $A_n(d_1)$ and $A_n(d_2)$ are the amplitudes of the same echo at the positions d_1 and d_2.

Each of these pulse techniques suffers from a variety of spurious loss mechanisms. At low frequencies, diffraction effects occur, which lead to significant loss due to beam spreading.[74]   At high frequencies, nonparallelism of the transducers is more important.   Abraham, Eckstein, Ketterson, and Vignos have shown that for a transmission, pulse-echo experiment the amplitude of the nth pulse echo is reduced due to nonparallelism by a factor[75]

$$[\lambda/\pi(2n - 1)r_0\theta]J_1[2\pi(2n - 1)r \, \theta/\lambda], \qquad (8.3.29)$$

where λ is the wavelength, r_0 the radius of the transducer, θ the angle of tilt, and J_1 the first-order Bessel function.   If the transducers are well matched to the helium, an additional loss can occur in the pulse-echo technique due to power extraction at the transducer.   This loss does not affect the variable-path method, since the same echo is always used.   The experiment by Abraham, Eckstein, Ketterson, and Vignos serves as a good example of the pulse-echo method.[75]   Using their apparatus the velocity and attenuation of first sound was measured from 0.1 to 3°K and at several frequencies from 12 to 208 MHz.   Their cell is shown in Fig. 14. A fused quartz spacer S determines the spacing between two quartz transducers Q.   The cell is sealed by a lead O-ring O for use at high pressures. Coaxial cables and connectors E and C provide electrical contact to the transducers.   The large volume at the top serves as a ballast chamber. Practical experiments do not make use of the simple data collection methods described previously but rather use a comparator circuit for higher precision.   A schematic is shown in Fig. 15.   A master oscillator generates a sine wave at the desired frequency, which is split into two signals.   Each signal is switched simultaneously to produce an rf pulse or tone burst.   One signal is routed to the acoustic cell while the other is passed through a variable attenuator and variable delay line.   The two signals are then summed and amplified.   By adjusting the attenuator and delay line to achieve a null on the oscilloscope, the delay and attenuation suffered in the helium channel can be determined from which the velocity and attenuation of sound in the helium can be calculated.

At 1 GHz the first-sound wavelength is about 2000 Å, less than the wavelength of visible light.   To make attenuation measurements at this frequency, Imai and Leonard conceived of the interferometer apparatus shown in Fig. 16.[29]   Provision is made for correcting the parallelism of the transducers with three spring-loaded adjustment screws.   The transducers are cadmium sulfide thin films deposited on two crystalline quartz delay rods.   Electronics similar to Fig. 15 are used to generate and detect acoustic signals.   Data are taken by displacing one crystal and recording

[74] H. Seki, A. Granato, and R. Truell, *J. Acoust. Soc. Am.* **28**, 230 (1956).
[75] B. Abraham, Y. Eckstein, J. B. Ketterson, and J. Vignos, *Cryogenics* **9**, 274 (1969).

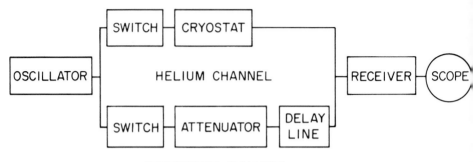

REFERENCE CHANNEL

FIG. 15. Schematic of an ultrasonic pulse comparator. (From B. Abraham, Y. Eckstein, J. B. Ketterson, and J. Vignos, *Cryogenics* **9**, 274 (1969), published by IPC Science and Technology Press Ltd., Guildford, Surrey, UK.)

the change in receiver output as a function of position. Figure 17 shows raw data taken by demodulating the receiver output and sampling with a boxcar integrator. As the crystals are separated, the output from the acoustic cell V_A changes phase and amplitude and beats with the signal on the reference channel V_E. The resulting voltage varies from a maximum of $|V_A + V_E|$ to a minimum of $|V_A - V_E|$ each time the spacing increases by a half-wavelength. The exponential decrease in the envelope of Fig. 17 reflects the exponential decrease in V_A as the crystals are separated. If the change in separation of the delay rods is accurately known, the acoustic velocity can be determined by counting nulls. Conversely, if the change in spacing is not well known, knowledge of the velocity allows it to be determined.

Measurements of the attenuation of sound at high frequencies test theories of relaxation in the critical region near T_λ. These measurements have revealed that the peak in attenuation commonly associated with T_λ actually occurs 3.3×10^{-3} °K *below* T_λ. The process responsible for this shift is still a matter of speculation.[76]

Pulse methods have also been used in investigations of second and fourth sound up to about 100 kHz.[54,77,78]

8.3.3. Calibration Techniques

Frequently it is necessary to obtain data regarding absolute sensitivities of transducers used in measurements on liquid helium. Such calibrations

[76] J. Imai and I. Rudnick, *Phys. Rev. Lett.* **22**, 694 (1969).
[77] D. Osborne, *Nature (London)* **162**, 213 (1948).
[78] J. Pellam, *Phys. Rev.* **75**, 1183 (1949).

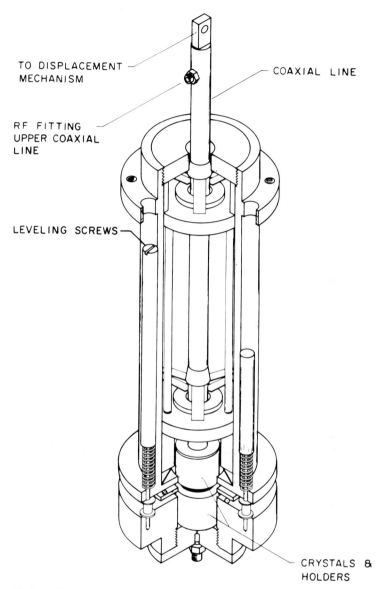

TO DISPLACEMENT
MECHANISM

COAXIAL LINE

RF FITTING
UPPER COAXIAL
LINE

LEVELING SCREWS

CRYSTALS &
HOLDERS

FIG. 16. Experimental apparatus of Imai and Rudnick for pulse measurements of attenuation in superfluid helium at 1 GHz. Spring-loaded adjustment screws are for correcting parallelism of transducers (from Imai and Rudnick[29]).

FIG. 17. Raw data from the apparatus of Fig. 16 using a phase comparator and boxcar integrator (from Imai and Rudnick[29]).

are possible using the reciprocity theorem.[79] Free-field techniques requiring only two reversible transducers have been described.[80] Rudnick has developed a method for calibration of two transducers in a resonator cavity.[81] Barmatz and Rudnick[82] and Garrett[21] (pp. 126, 145) have used these techniques to make absolute sensitivity determinations in superfluid helium. It is also possible to self-calibrate a transducer at high frequencies using short pulses.[14] Due to the great generality of the reciprocity theorem, calibrations can also be established for second-sound transducers.

Acknowledgments

Part of this work was supported by the Office of Naval Research under Contract N00014-77-C-0412. The author would also like to thank Steven Garrett, who made many useful comments and suggestions during preparation of this manuscript.

[79] J. Rayleigh, *Proc. London Math. Soc.* **4**, 357 (1873).
[80] W. Maclean, *J. Acoust. Soc. Am.* **12**, 140 (1940).
[81] I. Rudnick, *J. Acoust. Soc. Am.* **63**, 1923 (1978).
[82] M. Barmatz and I. Rudnick, *Phys. Rev.* **170**, 224 (1968).

9. ACOUSTO-OPTIC PHENOMENA

By G. I. A. Stegeman

List of Symbols

Acoustic Parameters

A	acoustic polarization unit vector	DE_0	amplitude of reference field
I_a	acoustic power per unit area	E	optical field scattered on reflection
k	acoustic wave vector		into air
$k_\parallel$	component of the acoustic wave vector parallel to the surface	E'	optical field scattered in the material bulk
L	width of acoustic beam (measured along the optical wave vector)	E_I	optical field incident from air side
		E_I'	optical field transmitted through
P_a	surface wave power per unit length along the wave front		the surface into the material bulk
P_R	surface power normalization parameter	E_0	amplitude of the incident optical field in the air
S	acoustic strain tensor	E_0'	amplitude of the incident optical
u	acoustic displacement field		field in the material bulk
u_0, u_1, u_2	acoustic amplitudes	$E_{\pm s}$	amplitude of the $\pm s$ diffraction order on reflection
u_{oz}	maximum surface ripple	$E_{\pm s}''$	amplitude of the $\pm s$ diffraction order in the medium
v	longitudinal sound wave velocity		
X	position vector	f_1, f_2	focal lengths of lenses
x_0	distance from optical probe to acoustic source	$\hat{f}$	effective Fabry–Perot finess
		$\hat{f}_F$	flatness finesse
η	angle of incidence of acoustic wave onto a surface	$\hat{f}_R$	reflectivity finesse
		h'	distance from the 0 to the ± 1
Λ	acoustic wavelength		diffraction orders on the screen
ρ	material density	H, H'	scattering efficiency of the ± 1
ϕ	electrostatic potential associated with sound waves in piezo-electric media		order
		I_0	incident light intensity
ψ, ψ_1, ψ_2	acoustic phase	$I_{\pm 1}, I_1$	scattered field intensity of the ± 1 order
Ω	acoustic frequency	L_1, L_2	lenses
		M_2	acousto-optic parameter characterizing scattering by bulk sound waves
Optical Parameters			
		n	refractive index of air
c	Fabry–Perot interferometer contrast	n'	refractive index of the bulk medium
d	Fabry–Perot plate separation	p	elasto-optic tensor

455

$\bar{p}$	effective elasto-optic coefficient	$\Delta\theta_{\pm1}$	change in direction for the ±1
q	integer, Fabry–Perot order		diffraction order; i.e.,
	number		$\Delta\theta_{\pm1} = \theta_{\pm1} - \theta$
Q	Raman–Nath parameter	θ'	angle of refraction in the bulk
r	electro-optic tensor		medium
R	Fabry–Perot plate reflectivity	$\theta'_{\pm s}$	scattering angle for the $\pm s$ dif-
s	diffraction order number		fraction order in the material
s'	lens to screen distance		bulk
V	scanning velocity of optical probe	κ	optical wave vector of light
W	width (diameter) of the incident		incident from the air
	optical probe	λ	optical wavelength in air
$\delta\omega$	Fabry–Perot interferometer	ζ	angle of optical incidence on
	linewidth		detector relative to the surface
$\delta\epsilon$	modulated dielectric tensor		normal
ε_0	dielectric constant of free space	ω_0	angular frequency of the incident
θ	optical angle of incidence (in air)		light
	measured from the surface	Ω_m	modulation frequency of the
	normal		reference beam
$\theta_{\pm s}$	scattering angle for the $\pm s$ diffrac-	$\Delta\Omega$	Doppler shift due to laser probe
	tion order on reflection		scanning

9.1. Introduction

The acousto-optic interaction has been under investigation by both physicists and engineers for approximately half a century. This phenomenon was first suggested by Brillouin[1] in 1922, and extensive experimental and theoretical studies of light scattering by sound waves (primarily of low frequency and transducer generated) were reported[2] in the 1930s. Interest then waned until the 1960s, when the invention of the laser provided, for the first time, a coherent, collimated, intense, and monochromatic light source. Since then progress has been rapid and the acousto-optic interaction has been used to study the acoustic properties of various materials,[3-5] to evaluate the performance of acoustical devices,[6] and to perform some signal processing functions.[6-9] Of particular

[1] L. Brillouin, *Ann. Phys.* (*Paris*) **17**, 88 (1922).
[2] L. Bergmann, "Ultrasonics." Bell and Sons, London, 1938.
[3] B. Chu, "Laser Light Scattering." Academic Press, New York, 1974.
[4] J. R. Sandercock, *Festkörperprobleme* **15**, 183 (1975).
[5] P. A. Fleury, in "Physical Acoustics," (W. P. Mason and R. N. Thurston, ed.), Vol. 6, p. 2. Academic Press, New York, 1970.
[6] G. I. Stegeman, *IEEE Trans. Sonics Ultrason.* **SU-23**, 33 (1976).
[7] I. C. Chang, *IEEE Trans. Sonics Ultrason.* **SU-23**, 2 (1976).
[8] R. V. Schmidt, *IEEE Trans. Sonics Ultrason.* **SU-23**, 22 (1976).
[9] R. W. Damon, W. T. Maloney, and D. H. McMahon, in "Physical Acoustics," (W. P. Mason and R. N. Thurston, ed.), Vol. 7, p. 273. Academic Press, New York, 1970.

interest here is that it is now possible to measure virtually every acoustic parameter of interest.

The investigation and use of light scattering by sound waves can be separated into two distinct fields, which require different instrumentation. In what is traditionally called "Brillouin scattering," the origin of the acoustic modes is the thermal acoustic "noise" in the medium with the result that the scattered light signals are weak. This technique[3-6] has been used to study phonons in the frequency range 0.5–100 GHz in gases, liquids, and solids. "Acousto-optics" has come to mean the diffraction of light by generated sound waves in liquids and solids and will be the subject of this part, specifically with respect to experimental techniques. Since light scattering from sound waves in liquids is primarily a simplification of the same phenomenon in solids, only diffraction of light by bulk and surface waves in solids will be discussed in detail.

One of the distinct advantages[9] of light scattering as a probe for sound waves is that it does not perturb the acoustic field. This is due to the inherent quantum amplification process that occurs, i.e., one photon interacts with one phonon in the acousto-optic interaction. Since the velocity of sound is much less than that of light and the wave vectors can be comparable, the phonon energy is at least five orders of magnitude smaller than that of the photon. As a result, usually an insignificant fraction of phonons are removed from the sound wave in the scattering process.

The sophistication of the instrumentation used in acousto-optics depends on the amount and quality of acoustic information desired. The amplitude, phase, frequency, and direction of the scattered light contain the equivalent information about the sound wave responsible for the scattering. Simply by measuring the direction and intensity of the scattered light it is possible to detect the presence of sound waves and to evaluate their propagation direction, wave vector (and therefore velocity), and amplitude. Interferometric techniques, based on Fabry–Perot interferometry, can be used to obtain the same information in standing wave geometries and in cases where multiple acoustic frequencies are present. If information about the acoustic phase is also required, heterodyne methods, which are the most powerful and sophisticated tools currently available in acousto-optics, are needed. All these techniques will be discussed in this part, with most of the examples drawn from the area of light scattering by surface waves, where most of the recent activity has occurred. There will be no attempt to give an exhaustive table of references; rather we shall list a few representative and particularly instructive papers.

9.2. Review of Light Scattering Theory

The theory of the acousto-optic interaction involving bulk and surface acoustic waves is well known, and a thorough discussion (and further references) can be found in review articles by Stegeman,[6] Chang,[7] Schmidt,[8] and Damon *et al.*[9]   A brief review will be presented here in order to facilitate the discussion of the various experimental techniques treated in subsequent sections.

9.2.1. Scattering Mechanisms

There are two light scattering mechanisms in the bulk of a material, namely, the elasto-optic and electro-optic effects.   At a material surface, an acoustically produced surface corrugation also scatters light.

9.2.1.1. Elasto-optic and Electro-optic Effects.   The elasto-optic effect is the primary scattering mechanism in most acousto-optic interactions.   This phenomenon has a simple origin in a liquid and arises from the spatial and temporal modulation of the dielectric constant by the density fluctuations associated with a longitudinal sound wave.   The relation between the modulated dielectric constant and the acoustic fields is of a tensorial nature in a solid, and scattering can take place from both shear and longitudinal waves.   In piezoelectric media the electrostatic potential fields ϕ associated with a sound wave also induce changes in the dielectric constant via the electro-optic effect.   For the simple case of an optically isotropic medium of refractive index n', the acoustically induced fluctuation in the dielectric constant $\delta\epsilon_{ij}$ is given by

$$\delta\epsilon_{ij} = -\epsilon_0 n'^4[p_{ijkl}S_{kl} - r_{ij}\,\partial\phi/\partial X_k], \qquad (9.2.1)$$

where $\mathbf{p}$ is the elasto-optic tensor,[10] $\mathbf{r}$ the electro-optic tensor,[11] and the subscripts i, j, k, l each correspond to one of the coordinate axes x, y, and z.   The details of the acoustic wave (longitudinal, shear, or surface) are contained in ϕ and the strain term $S_{kl} \equiv \frac{1}{2}(\partial u_k/\partial X_l + \partial u_l/\partial X_k)$, where $\mathbf{u}$ is the mechanical displacement field associated with the sound wave and X defines the vector $\hat{\imath}x + \hat{\jmath}y + \hat{k}z$.   In the standard method of analysis this modulated dielectric constant appears as a source term in Maxwell's equations and the scattered fields are calculated.   (Details of the analysis and the generalization to optically anisotropic materials can be found in the work of Stegeman,[6] Chang,[7] Schmidt,[8] Damon *et al.*,[9] and Pinnow[10]).

---

[10] D. A. Pinnow, *in* "CRC Handbook of Lasers" (R. J. Pressley, ed.).   Chem. Rubber Publ. Co., Cleveland, Ohio, 1971.

[11] I. P. Kaminov and E. H. Turner, *in* "CRC Handbook of Lasers" (R. J. Pressley, ed.). The Chemical Rubber Co., Cleveland, Ohio, 1971.

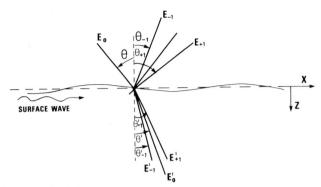

FIG. 1. Geometry for light scattering by surface waves on reflection from and transmission through the acoustic propagation surface.

9.2.1.2. Corrugation Effect.

The corrugation effect is important when the incident light is reflected from or transmitted through an interface corrugated by the presence of surface or bulk waves (reflecting from the surface). For example, a surface ripple of the form $u_z = u_{0z} \cos(\Omega t - k_{\parallel} x)$ is created by surface acoustic waves traveling along the x axis of a surface defined by $z = 0$ in the absence of sound waves; see Fig. 1. Here Ω and $k_{\parallel}$ refer to the acoustic frequency and the wave vector component parallel to the surface, respectively. Bulk waves with wave vectors in the $x-z$ plane and reflecting off the $z = 0$ plane (Fig. 2a) can also produce a corrugated surface. The scattering phenomenon is similar to that of a traveling diffraction grating, i.e., Doppler-shifted light is scattered into various "diffraction orders." The diffracted fields are calculated via the diffraction[12] or Helmholtz[13] integrals or by matching electromagnetic boundary conditions[14] across the rippled interface.

9.2.2. Light Scattering from Bulk Waves

The two common interaction geometries for scattering from bulk waves are shown in Fig. 2. The incident light is usually either s- or p-polarized, i.e., the optical electric field is either orthogonal to or parallel to the plane defined by the optical and acoustical wave vectors, respectively. For example, considering s-polarized light of frequency ω_0 and wave vector κ (in vacuum or air) incident from the air side ($n = 1$ in Fig. 2) at an angle θ rel-

[12] R. J. Hallermeier and W. G. Mayer, *J. Appl. Phys.* **41**, 3664 (1970); *J. Acoust. Soc. Am.* **51**, 161 (1972).

[13] E. G. H. Lean, *in* "Progress in Optics" (E. Wolfe, ed.). North-Holland Publ., Amsterdam, 1973.

[14] G. I. Stegeman, *J. Appl. Phys.* **49**, 5624 (1978).

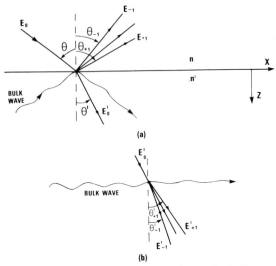

FIG. 2. Light–sound interaction geometries for scattering by bulk waves: (a) scattering from acoustic waves reflecting off a surface; (b) scattering in the material bulk.

ative to the surface normal, the optical electric field is written

$$\mathbf{E}_1 = \tfrac{1}{2}E_0\hat{\jmath} \exp[i(\omega_0 t - \kappa \sin \theta x - \kappa \cos \theta z)] + \text{cc} \qquad (9.2.2)$$

and

$$\mathbf{E}_1' = \tfrac{1}{2}E_0'\hat{\jmath} \exp[i(\omega_0 t - n'\kappa \sin \theta' x - n'\kappa \cos \theta' z)] + \text{cc} \qquad (9.2.3)$$

in the air $\mathbf{E}_1$ and material $\mathbf{E}_1'$, respectively. Henceforth all parameters in the material are identified by a prime. Here $\sin \theta = n' \sin \theta'$, the amplitude ratio E_0'/E_0 is given by the appropriate Fresnel coefficient, $\hat{\jmath}$ is the unit vector along the y axis, and cc denotes the complex conjugate term. In order to simplify the formulas for the fields scattered via the elasto-optic and electro-optic mechanisms, the bulk accoustic waves are assumed to propagate along the x axis and the ith component of the acoustic displacement field is written

$$u_i = \tfrac{1}{2}A_i u_0 \exp[i(\Omega t - \mathbf{k}x + \psi)] + \text{cc}, \qquad (9.2.4)$$

where $\mathbf{A}$ is a unit vector that describes the polarization of the sound wave, u_0 the acoustic amplitude, and ψ the acoustic phase. (Note that at a surface the reflected and possibly mode-converted acoustic waves must also be included when computing, for example, the surface corrugation.)

9.2.2.1. In Bulk Media. For s-polarized incident light (as just defined), the scattered fields can be either s-polarized (polarized scattering) or p-

polarized (depolarized scattering). For example, the total s-polarized optical field leaving the acousto-optic interaction region for scattering by longitudinal waves is of the form

$$\mathbf{E}' = \frac{1}{2} \hat{j} \sum_s E'_{\pm s} \exp\{i[(\omega_0 \pm s\Omega)t$$

$$- (n'\kappa \sin \theta' \pm sk)x - n'\kappa \cos \theta'_{\pm s} z \pm s\psi]\} + \text{cc} \quad (9.2.5)$$

and contains contributions at the frequencies $\omega_0 \pm s\Omega$, where s is the order of scattering. For example, $s = 2$ indicates that the light is scattered at least twice by the sound wave and that the scattered field occurs at the frequencies $\omega_0 \pm 2\Omega$. The direction of the scattered light is different from that of the incident light and appears at the scattering angles

$$\sin \theta'_{\pm s} = \sin \theta' \pm sk/n'\kappa \quad (9.2.6)$$

measured in the medium of refractive index n'. For small values of the ratio $sk/n'\kappa$, the $\pm s$ orders are symmetrically displaced about the transmitted, unscattered beam designated by $s = 0$. An example of multiple-order scattering on reflection from surface waves is shown in Fig. 14 of Part 10. The amplitude ratio $E'_{\pm s}/E'_0$ depends on details of the scattering geometry[7,9] such as the direction, polarization, and frequency of the sound wave and the incident and scattered light, the material elasto-optic coefficients, and the dimensions of the scattering volume. If $|E'_{\pm 1}| \ll |E'_0|$, which is usually the case of interest for the acousto-optic probing of sound waves,

$$\frac{E'_{\pm 1}}{E'_0} = H \frac{\sin(\beta_{\pm 1} L/2)}{(\beta_{\pm 1} L/2)}, \quad (9.2.7)$$

where

$$\beta_{\pm 1} = \mp \frac{2k}{\cos \theta' + \cos \theta_{\pm 1}} \left(\sin \theta' \pm \frac{k}{n'\kappa} \right) \quad (9.2.8)$$

and L is the width of the acousto-optic interaction region measured along $\kappa'_{\pm 1}$, the scattered optical wave vector. The details of the scattering efficiency are combined into the one parameter H. For depolarized scattering (from y-polarized shear waves), the scattered fields are of the same form as previously discussed with the exception that the polarization and H for the scattered field are different. Details of these and more general cases are found in the work of Chang,[7] Schmidt,[8] and Damon et al.[9]

The number of scattered fields, their relative amplitudes, and their variation with the angle of incidence depend on the value of the Raman–Nath parameter[15] $Q = k^2 L/n'\kappa \cos \theta'$. For $Q \gg 1$ (Bragg limit), $E'_{+1} = E'_0 H$

[15] W. R. Klein and B. D. Cook, *IEEE Trans. Sonics Ultrason.* **SU-14**, 123 (1967).

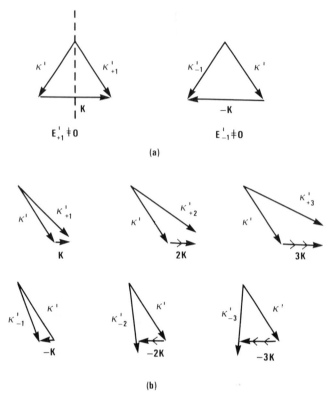

(a)

(b)

FIG. 3. The wave vectors of the incident light κ', the acousto-optically scattered light $\kappa'_{\pm s}$, and the acoustic wave $\mathbf{k}$ in: (a) the Bragg limit, (b) the Raman–Nath limit.

and $E'_{-1} \approx 0$ if the Bragg condition $\sin \theta' = -k/2n'\kappa$ is satisfied. If $\sin \theta' = k/2n' \kappa$, then $E'_{\pm 1} \approx 0$ and $E'_{-1} = E'_0 H$. These two cases are shown in Fig. 3a and correspond to total wave vector conservation, i.e., $\kappa'_{\pm 1} = \kappa' \pm \mathbf{k}$. On the other hand, if $1 \gg Q$ (Raman–Nath limit),

$$\frac{E'_{+1}}{E'_0} = \frac{E'_{-1}}{E'_0} = H \frac{\sin(kL \tan \theta'/2)}{(kL \tan \theta'/2)} \tag{9.2.9}$$

and scattering is obtained for essentially all angles of incidence. Furthermore, as indicated in Fig. 3b, multiple scattering, i.e., $s > 1$, can occur in this limit as long as the effect of wave vector mismatch is not too large over the scattering volume. For $5 \geqslant Q \geqslant 0.2$, multiple scattering with $E'_{+s} \neq E'_{-s}$ takes place and Eq. (9.2.7) is still approximately valid for the first-order fields. This last case occurs often experimentally (see, for example, Fig. 5a) and care must be taken in interpreting the results.

In many experiments it is the scattered light intensity that is measured.

The maximum diffraction efficiency can be written, in terms of the incident acoustic power per unit area I_a,

$$I_1/I_0 = \tfrac{1}{8} M_2 (\kappa L)^2 I_a \qquad (9.2.10)$$

for approximately normal incidence and small scattering angles where $I_{\pm 1}$ has been abbreviated as I_1. The material dependence has been condensed into the parameter $M_2 = (\bar{p}n'^3)^2/\rho v^3$, where $\bar{p}$ is the effective elasto-optic constant characteristic of the acoustic modes and the polarization of the incident and scattered fields. Furthermore, ρ is the material density and v the sound velocity. Typical values[7,9] are given in Table I, and it is evident that this parameter varies from material to material over several orders of magnitude. For example, $I_1/I_0 \sim 2 \times 10^{-4}$ for a 0.63-μm HeNe laser beam illuminating a 1 W/cm^2 acoustic beam, 1 mm wide in fused silica.

9.2.2.2. At Surfaces. Consider the interaction geometry of Fig. 2a again. The surface is rippled according to $u_z = u_{oz} \cos(\Omega t - k_{\parallel} x + \psi)$ by incident and reflected (possibly with mode conversion) bulk waves. (Note that acoustic modes polarized strictly in the plane of the surface, i.e., y-polarized shear waves, do not ripple the surface and therefore do not scatter light via this mechanism.) For $ku_{oz} \ll 1$, $\kappa u_{oz} \ll 1$, and hence small scattering angles, the scattered fields obtained on reflection from the surface have the same polarization as the incident fields. For example, with s-polarized incident light [Eq. (9.2.2)]

$$\mathbf{E} = \frac{1}{2} \hat{\jmath} \sum_s E_{\pm s} \exp\{i[(\omega_0 \pm s\Omega)t - (\kappa \sin\theta \pm sk)x$$

$$+ \kappa \cos\theta_{\pm s} z \pm s\psi]\} + \text{cc} \qquad (9.2.11)$$

and

$$\frac{E_{\pm s}}{E_0} = \frac{-i^s}{s!} (\kappa \cos\theta u_{oz})^s. \qquad (9.2.12)$$

The scattering angles are given by $\sin\theta_{\pm s} = \sin\theta \pm sk/\kappa$ and the diffraction orders appear on both sides of the specularly reflected light. If $k/\kappa \sim 1$ (large scattering angles), there is an additional contribution from the elastooptic effect just under the surface.[14]

Scattering also occurs on transmission through a corrugated surface. However, it is usually negligible in comparison to the elasto-optic mechanism for a bulk wave if the acoustic beam is more than a few acoustic wavelengths wide.

It is difficult to define appropriate general parameters (such as M_2 for scattering from bulk waves) for this scattering phenomenon since the cor-

TABLE I.   Typical values of the Acousto-Optic Parameter M_2
for Longitudinal Waves[a]

Material	n'	v (10^5 cm/sec)	$M_2 \times 10^{18}$
Fused quartz	1.46	5.95	1.51
As_2S_3	2.61	2.6	433
YAG	1.83	8.6	0.073
GaP	3.31	6.3	44.6
α-Al_2O_3	1.76	11.2	0.34

[a] Values were computed at $\lambda = 0.633$ μm with M_2 expressed in cgs units (from Damon et al.[9]).

rugation depends on the details of the acoustic reflection at the interface. In order to obtain an order of magnitude estimate, consider longitudinal waves directed at an angle η onto a water–air interface upon which light is incident in a direction perpendicular to the surface, i.e., $\theta = 0$ in Eq. (9.2.2). The diffraction efficiency is

$$\frac{I_1}{I_0} = 8 \cos^2 \eta \frac{\kappa^2}{k^2} \frac{I_a}{\rho v^3}, \qquad (9.2.13)$$

which gives $I_1/I_0 \simeq 1.4 \times 10^{-4}$ for $\eta = 45°$, $\kappa/k \sim 100$, and $I_a = 1$ W/cm². Note that the diffraction efficiency is proportional to Ω^{-2} ($\equiv kv)^{-2}$ and therefore decreases rapidly with increase in frequency.

9.2.3. Light Scattering from Surface Waves

The two common interaction geometries for the scattering of light by surface acoustic waves are shown in Fig. 1. The analysis of the fields scattered on reflection is the same as that discussed previously for bulk waves. The scattered fields are given by Eqs. (9.2.11) and (9.2.12). Typical scattering efficiencies can be estimated from[14]

$$I_1/I_0 = (\kappa^2 \cos^2 \theta / k P_R) P_a, \qquad (9.2.14)$$

where P_a is the surface wave power per unit length along the acoustic wavefront and P_R a constant characteristic of the material, propagation direction, and surface cut. For most materials P_R varies from 10^{+14} to 10^{+16} W/m². For example, for surface waves propagating along the z direction on the z–x plane of lithium niobate, $P_R = 0.5 \times 10^{+15}$ W/m² and $I_1/I_0 \sim 10^{-4}$ for $P_a \sim 1$ W/cm HeNe red light (0.633 μm) at normal incidence and an acoustic frequency of 100 MHz.

Scattering on transmission takes place via both the corrugation and elasto-optic effects and the total scattered field is a coherent sum of the two contributions. The dependence of the scattered field on the angle of incidence is complicated and the reader is directed to Stegeman[6,14] for further details. Typically the scattering efficiency is up to one order of magnitude larger than for the reflection case.[14]

9.2.4. Characteristics of the Scattered Light

The simplest experiments to interpret in acousto-optics are those involving the presence of only one acoustic wave in a sample. However, for many practical situations a partial standing acoustic wave or harmonics at integer multiples of the incident sound frequency are present and the light scattered into a given direction in space contains contributions from more than one sound wave.

9.2.4.1. Unidirectional Sound Waves. As indicated by Eqs. (9.2.5) and (9.2.11), the first-order ($s = \pm 1$) scattered fields contain a great deal of information about the sound waves. For small scattering angles (i.e., $k/n'\kappa \ll 1$), the diffracted fields are symmetrically displaced to either side of the unscattered beam and the resulting changes in direction are a direct measure of the acoustic wave vector. Light scattered toward the direction of the wave vector $\mathbf{k}$ results in an upshifted frequency of $\omega_0 + \Omega$, and conversely deflection toward $-\mathbf{k}$ corresponds to a frequency of $\omega_0 - \Omega$ for the scattered light. Thus the acoustic frequency can be evaluated by directly measuring the optical frequencies of the scattered and unscattered light or by beating the two light beams on a square-law detector. The latter measurement can also be used to determine the local acoustic phase ψ. Furthermore, the ratio of the amplitudes (or intensities) of the scattered to unscattered light yields the local acoustic amplitude (or intensity). Point-by-point mapping of the acoustic fields in the propagation region of the sound waves can give information about the sound wave attenuation as well as the diffraction spreading of the acoustic beam. Finally, the scattering efficiency depends on a number of factors, principally on the geometry and properties of the medium, and the direction and polarization of the optical and acoustical fields.

The simplest parameter to evaluate acousto-optically is the sound wave velocity, i.e., $v = \Omega/k$. This yields the elastic constants (solids) or compressibility (liquids) that can be measured as a function of direction, pressure, temperature, etc. In turn, other phenomena such as phase transitions, etc., can be investigated.

Other acoustic parameters commonly measured are the sound wave amplitude and phase. These are useful in mapping out acoustic fields for the purpose of evaluating transducers, the reflection of sound at boundaries and irregularities, etc.

It must be emphasized that an optical probe samples the acoustic field along the optical path. Therefore, when studying bulk waves we obtain only a line average that includes the acoustic fringe fields. Furthermore, certain incident field polarizations and acousto-optic geometries sample different acoustic field components with different weighting factors. Nevertheless, multiple line probes along different directions in space and subsequent computer analysis can be used to map out an acoustic field completely. However, because of this complexity, the acoustic field is usually modeled as being uniform over some region in space and the calculations are thus simplified.

The analysis of the light diffracted by surface acoustic waves depends on the scattering geometry. Scattering on reflection from the propagation surface samples the surface corrugation only, and the fringe fields at the edges of the wave front can thus be measured separately. On the other hand, scattering on transmission samples the average of a complicated strain field, and detailed knowledge of the structure of the acoustic fields and of the details of the acousto-optic interaction is needed to interpret the results.

9.2.4.2. Standing Waves. Optical fields originating from sound waves of the same frequency and polarization but traveling in opposite directions (a partial standing wave) contain information about both sound waves. Consider the case in which the total standing wave displacement field consists of a wave of amplitude u_1 traveling along $+x$ and a wave of amplitude u_2 propagating along $-x$, i.e.,

$$\mathbf{u} = \tfrac{1}{2}\mathbf{A}\{u_1 \exp[i(\Omega t - kx + \psi_1)] + u_2 \exp[i(\Omega t + kx - \psi_2)]\} + \text{cc.}$$

$$(9.2.15)$$

The ± 1 diffraction orders generated by scattering from these bulk waves are

$$E'_{\pm 1} \propto H'E'_0(\exp\{- i[(n'\kappa \sin \theta' + k)x + n'\kappa \cos \theta'_{+1}z]\}$$

$$\times \{u_1 \exp[i(\omega_0 + \Omega)t + i\psi_1] + u_2 \exp[i(\omega_0 - \Omega)t + i\psi_2]\})$$

$$+ H'E'_0(\exp\{- i[(n'\kappa \sin \theta' - k)x + n'\kappa \cos \theta'_{-1}z]\}$$

$$\times \{u_1 \exp[i(\omega_0 - \Omega)t - i\psi_1] + u_2 \exp[i(\omega_0 + \Omega)t - i\omega_2]\}),$$

$$(9.2.16)$$

respectively, where the scattering angles $\theta'_{\pm 1}$ are defined relative to the direction of the forward traveling wave. The $+1$ order due to the forward propagating wave coincides in space with the -1 order from the backward traveling wave. Similarly the -1 and $+1$ orders, respectively, are codirectional. Therefore, some form of frequency analysis or mixing of the scattered fields is required in order to identify the individual contributions from the two acoustic fields.

9.2.4.3. Harmonic Sound Waves. The case of scattered fields containing information[16] about multiple sound waves also occurs if harmonic acoustic waves are investigated in an acousto-optic interaction geometry characterized by multiple scattering processes (Raman–Nath limit). This situation arises frequently[16] with surface waves since the power densities are so large that multiple harmonics are easily generated and observed. For example, the s-polarized field scattered on reflection from a second harmonic surface wave is given by

$$\mathbf{E} = \frac{1}{2}\hat{\jmath}\sum_s E_{\pm s}\exp\{i[(\omega_0 \pm 2s\Omega)t - (\kappa\sin\theta \pm 2sk)x$$

$$+ \cos\theta_{\pm s}\kappa z \pm s\psi_2)]\}. \tag{9.2.17}$$

Comparison with Eq. (9.2.11) shows that the ± 2 orders scattered from the fundamental are coincident in direction and frequency with the $s = \pm 1$ orders due to the harmonic. Since there is a fixed phase relationship between the harmonic and the fundamental from which it evolved, the complex amplitude (not intensity) of the second-order fields scattered by the fundamental must be subtracted from the total field in order to evaluate the harmonic acoustic field amplitude.

9.3. Classification of Experimental Techniques

It proves convenient to classify the common experimental techniques in terms of how the scattered light is processed. The most sensitive devices for detecting light are photoelectric in nature (photodiodes, photomultiplier tubes, etc.) and the output signal (usually a current) is proportional to the square of the total incident radiation field.

Consider the general case in which one of the acousto-optically scattered fields is incident along with a reference field onto a photoelectric device. This reference field has an amplitude DE_0, is modulated at the frequency Ω_m, and is written

$$DE_0\exp[i(\omega_0 + \Omega_m)t - i(n'\kappa\sin\theta' + k)x - in'\kappa\cos\theta'_{+1}z], \tag{9.3.1}$$

where it has been assumed that this field is codirectional (parallel wave fronts) with the $+1$ order scattered from a partial standing wave, i.e., the scattered field given by Eq. (9.2.16). This modulation can be achieved

[16] T. H. Neighbors III and W. G. Mayer, *J. Appl. Phys.* **42**, 3670 (1971).

with either a commercial acousto-optic or electro-optic modulator. For $\Omega_m = 0$, this reference field can correspond to a fraction of the incident light obtained via, for example, a beamsplitter, the zeroth diffraction order, or stray light. The detector signal (due to the overlap of the $+1$ order and the reference field) is proportional to

$$D^2 E_0{}^2 + E_0{}^2 u_1{}^2 H'^2 + E_0{}^2 u_2{}^2 H'^2 + 2E_0 D u_1 H' \sin[\Omega_m - \Omega)t + \psi_1]$$
$$+ 2E_0 D u_2 H' \sin[(\Omega_m + \Omega)t - \psi_2]$$
$$+ 2E_0{}^2 u_1 u_2 H'^2 \cos(2\Omega + \psi_1 - \psi_2). \tag{9.3.2}$$

The various experimental techniques differ in the processing of the light prior to detection and in the treatment of the detector output.

Three methods of analysis will be discussed here. In the first the acoustic information is evaluated by measuring one or both of the dc terms, i.e., $E_0{}^2 u_1{}^2 H'^2$ and $E_0{}^2 u_2{}^2 H'^2$. The second approach relies on using high-resolution optical interferometry to filter out one of the two dc terms prior to detection since these scattered fields occur at the frequencies $\omega_0 + \Omega$ and $\omega_0 - \Omega$, respectively. In the third method the time-varying terms in Equation (9.3.2) are analyzed.

The absolute sensitivities of the various experimental techniques depend on many parameters in a nontrivial way. For example, in scattering from bulk waves, the scattered fields depend on the acoustical and optical polarization, scattering geometry and volume, effective elasto-optic coefficient, etc. However, scattering on reflection from surface acoustic waves has a relatively simple form and hence is used to compare the sensitivities of the techniques discussed in subsequent sections. Furthermore, it is assumed that the light source is a 1-mW HeNe laser operating at $\lambda = 0.6328$ μm and that the photomultiplier tube has an appropriate bandwidth and a typical noise figure.

9.3.1. Direct Current Detection

In the direct current detection type of experiments only the first three terms of Eq. (9.3.2) are analyzed. This is usually accomplished by making the transient response of the detector electronics too slow to respond to the beat frequencies. The individual amplitude components of a standing wave can be measured with this technique: (a) if the standing wave ratio u_2/u_1 is known or (b) if one of the waves originates from an acoustic reflection of the other wave at a sample boundary and the two acousto-optically scattered signals can subsequently be time resolved by using short acoustic pulses. The reference beam for this case usually corresponds to stray light and should either be eliminated (if possible) or subtracted by using phase-sensitive detection (which requires an additional slow modulation of the acoustic beam). These techniques are usually used for studying unidirectional sound waves.

9.3.2. Interferometric Techniques

In the case of interferometric techniques the different frequency components of the scattered light are isolated prior to detection with a Fabry–Perot interferometer. This optical device is basically a very high-resolution tunable optical filter which transmits only one frequency component at a time. Thus u_1 and u_2 can be measured independently although the detection is essentially dc in character.

9.3.3. Heterodyne Techniques

Heterodyne systems use the last three terms of Eq. (9.3.2) and rely on the mixing of multifrequency optical waves at the detector. (This also includes the case where $\Omega_m = \pm \Omega$ and the beat signals are made to occur at zero frequency.) The time constant of the detector and electronics must be small enough to pass the beat frequencies of interest and the mixing fields must overlap at the detector surface. Since usually $D \gg u_1 H'$ and/or $D \gg u_2 H'$, this technique has inherently a greater sensitivity than the dc methods. Note also that the individual acoustic amplitudes and phases can be recovered electronically.

9.4. Direct Current Detection

The applicability of the techniques described in this chapter depends on separating out in space the scattered and unscattered fields so that each may be individually detected. Usually the incident light beam is made as parallel as possible with a beam diameter comparable to the acoustic beam cross section. The required spatial separation can be achieved with lenses or by moving the "detector" a sufficiently large distance from the sample. (For a parallel beam of width W, the minimum usable distance between detector and sample is of the order of $W/\Delta\theta_{\pm 1}$ for a diffraction angle $\Delta\theta_{\pm 1}$.) The $I_{\pm 1}$ and I_0 fields (of Figs. 1 and 2) are: (1) projected onto a screen with a superimposed grid for making accurate position measurements; (2) photographed directly; or (3) detected by a photoelectric device.

The surfaces that the light encounters must all be of optical quality, preferably with flatnesses of $\lambda/4$ or better. The worse the surface finish is, the more stray light is obtained, and the harder it becomes to observe the diffraction orders of interest for low-frequency sound waves. Opposing sample faces should be wedged in order to avoid coincident multiple optical reflections.

It is assumed throughout that the transducers that generate the sound waves are excited at only one frequency.

9.4.1. Observation of Diffraction Orders

The simplest measurements that can be made are the observation of the presence of the scattered light (and therefore the verification of the existence of the sound) and the evaluation of the scattering angle (and hence the acoustic wave vector and velocity). Note that nonobservation of diffraction orders does not necessarily imply the absence of sound waves since the scattering efficiency depends on the direction of the incident light (especially in the Bragg limit) and the polarization of both the sound and light waves.

The scattering geometry can be measured for the acousto-optic interaction in both the Raman–Nath and Bragg regimes. In the Bragg limit the experimental geometry tends to be very well specified since appreciable scattering is obtained only if the Bragg condition is satisfied. Scattering from bulk waves in the Raman–Nath limit or from surface waves is more difficult to analyze since scattering can take place with comparable intensities for a range of acoustic propagation directions.

In the early work (as summarized, for example, by Bergmann[2]), the diffraction patterns were photographed directly. Assuming small diffraction angles and near normal incidence of light onto the acoustic waves, the scattering angles are given by

$$\Delta\theta_{\pm 1} = (\Omega/\kappa)1/v, \tag{9.4.1}$$

where v is the sound wave velocity. More recent studies[17] have used lens systems such as the one shown in Fig. 4 in order to increase the accuracy and visibility of the diffracted light. The displacement h' on the screen is

$$h' = \left(\frac{s'}{f_2} - 1\right) f_1 \frac{\Omega}{\kappa} \frac{1}{v}, \tag{9.4.2}$$

where the parameters f_1 and f_2 refer to the focal lengths of the two lenses and s' is the separation of the screen from lens 2.

The most spectacular results are obtained in the Raman–Nath limit with sample geometries for which acoustic reflections, transducers, etc., create sound waves traveling in essentially all directions in a given plane of a sample.[17] In the typical result shown in Fig. 5a the undiffracted ($s =$

---

[17] W. S. Goruk, R. Normandin, P. J. Vella, and G. I. Stegeman, *J. Phys. D* **9**, 999 (1976).

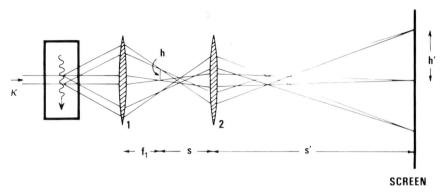

SCREEN

FIG. 4. Lens system used for imaging the light diffracted by sound waves and for measuring the diffraction angles. (From Goruk et al.[17] Copyright, the Institute of Physics.)

0) light has been blocked and diffraction spots up to the fourth order (i.e., $s = 4$) due to the scattering of light by yz-polarized shear waves propagating in the $y-z$ plane of LiNbO$_3$ are recorded. (The acoustic modes were generated by interdigital transducers and by the mode conversion of surface waves at the sample boundaries.)

The application of the lens system shown in Fig. 4 to this technique is illustrated in Fig. 5b. The curves are indicative of scattering from all the acoustic modes associated with a given plane in a material, i.e., one surface wave and three bulk waves. (This particular example corresponds to the $x-z$ plane of lithium niobate.[17]) The outermost curve is due to diffraction by the slowest mode (surface waves) and the innermost one is produced by the fastest waves (quasi-longitudinal sound waves). These contours are commonly called *cuts* of the inverse phase velocity surfaces and their analysis yields the velocity of sound for all four modes in all directions in a given plane. The variation in scattering intensity with direction is an indication of the corresponding acousto-optic scattering efficiency. Similar curves are obtained on reflection from a sample surface and have been used to acousto-optically measure elastic constants, etc., in opaque materials.[2]

The accuracy of the elastic constants and sound velocities evaluated in this way depends on the precision with which the scattering angle can be measured. For frequencies of ~ 100 MHz, the velocities can easily[2,17] be measured to an absolute accuracy of $\sim 1\%$. (Better accuracies can be obtained by scanning across the contours with a photodiode in order to evaluate more precisely the positions of maximum scattering.) In general, the higher the acoustic frequency, the larger the diffraction angle and the better the absolute accuracy possible.

FIG. 5. (a) Multiorder light scattering in the Raman–Nath limit from bulk waves propagating in the y–z plane of lithium niobate; (b) light scattering in the Raman–Nath limit from surface and bulk waves in the x–z plane of LiNbO$_3$ using the apparatus shown in Fig. 4. (From Goruk et al.[17] Copyright, The Institute of Physics.)

9.4.2. Schlieren Studies

Optical diffraction experiments in which a substantial fraction of a sample is illuminated with a uniform optical beam can provide useful information[18] about acoustic power flow. Visualization of the sound waves can be obtained by using the simple optical apparatus illustrated in Fig. 6. A well-collimated incident light beam is required, and the lens is placed so that in the absence of the spatial filter the crystal is imaged on the screen. Light scattered by the sound waves of interest is passed through appropriately placed pinholes in the spatial filter and is reimaged on the screen. Thus there is a one-to-one correspondence between the local acoustic

[18] M. J. J. Zuliani, V. M. Ristic, P. J. Vella, and G. I. Stegeman, J. Appl. Phys. **44**, 2964 (1973).

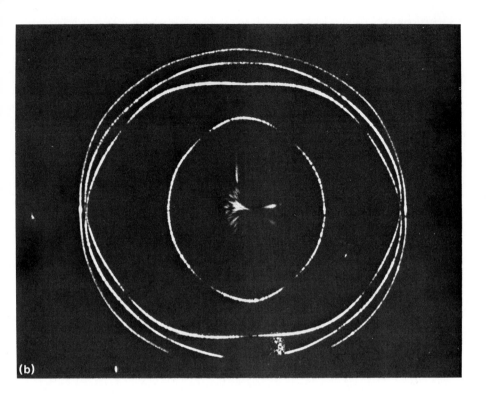

(b)

wave intensity in the sample and an illuminated area on the screen. The light transmitted without scattering through the material (or specularly reflected from the crystal surface) is focused as shown in Fig. 6 and blocked by the filter. Most of the stray light due to surface or material imperfections, etc., is also blocked since the pinholes subtend only a very small angle. (A small amount of this stray light is useful, however, for producing a weak image of the crystal as a reference for the acoustic power flow visualization.)

Typical results[17,18] are reproduced in Fig. 7. For Fig. 7a, the geometry of Fig. 6a was used to visualize surface waves traveling along the z axis on the y-cut face of lithium niobate. The waves are generated at one end of the surface by a bidirectional interdigital transducer (surface waves emanating from both sides) and are terminated by a piece of absorbing tape at the other end. The evolution of distinct maxima and minima along the propagation path due to acoustic beam diffraction is clearly visible. Shown in Fig. 7b are the acoustic modes present on reflection of yz-polarized shear waves at the $y = 0$ surface of lithium niobate: the incident shear wave is generated at the lower left-hand corner by a transducer and

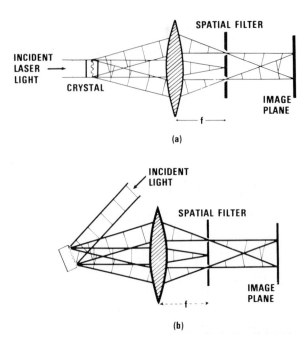

FIG. 6. Schlieren apparatus for visualizing sound waves on: (a) transmission; (b) reflection. (From Stegeman[6]).

both a specularly reflected shear wave and a mode-converted longitudinal wave (upper beam) are created on reflection from the upper sample surface. In this case the spatial filter is constructed to pass the appropriate three sets of diffraction spots.

This Schlieren technique is used primarily for diagnostic purposes, i.e., for evaluating acoustic beam diffraction, transducer generation, reflections, etc. The limitation is that large laser powers are needed since the light is spread out over a substantial cross section of the crystal. If a few square centimeters of sample area are illuminated, typically 10–100 mW of optical power are required to make readily visible acoustic beams with powers of a few tens of milliwatts.

9.4.3. Photoelectric Detection

Experiments in this classification are useful for evaluating the acoustic power. Typically, a photoelectric device is used to measure the acousto-optically scattered light $I_{\pm1}$ (and the unscattered light I_0). This can be achieved either in the far field (separated orders), in the Fourier plane of the first lens (Fig. 4), or after the spatial filter in Fig. 6.

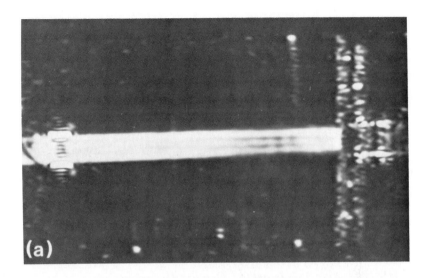

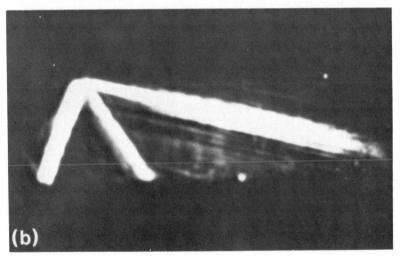

FIG. 7. Schlieren photographs of: (a) surface waves propagating on $y-z$ lithium niobate (taken from Zuliani et al.[18]); (b) shear wave reflection off a surface with partial mode conversion to longitudinal waves. (From Goruk et al.[17] Copyright, The Institute of Physics.)

A typical[19] apparatus useful for high-sensitivity measurements is shown in Fig. 8. This particular system was used for investigating surface waves using the corrugation scattering mechanism on the propagation surface.

[19] G. Cambon, M. Rouzeyre, and G. Simon, Appl. Phys. Lett. **18**, 295 (1971).

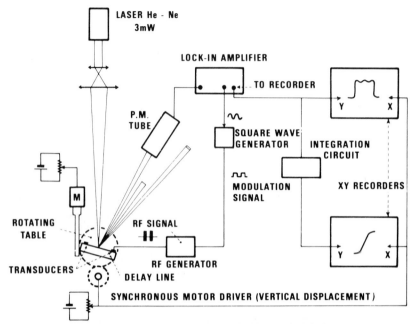

FIG. 8. Typical apparatus for photoelectric detection of the scattered light (from Cambon[19]).

(For studying bulk waves, scattering on transmission would be analyzed.) The acoustic signal at rf frequencies is modulated by a square-wave generator synchronized by a reference signal (100 Hz to 10 kHz) from the lock-in amplifier. Phase-sensitive detection is used here to subtract away the contributions to the measured detector output of both the stray light and the phototube dark noise. A small-diameter light probe is used to sample the acoustic wave front at different positions on the crystal, and the rest of the apparatus is used to move the sample relative to the laser and to synchronize this motion with an $X-Y$ recorder. (The sample is moved rather than the laser in order to maintain optical alignment.) The local acoustic amplitude is then evaluated from the local measurement of the ratio $I_{\pm1}/I_0$.

A typical result[19] is shown in Fig. 9 for the diffraction of an acoustic beam as it propagates away from a square-aperture surface wave transducer. Other parameters normally measured are attenuation, harmonic generation, parametric mixing, and beam steering. In the present form, this apparatus is only useful for measuring unidirectional sound waves. The sensitivity is typically 0.1 Å surface displacement.

Time resolution of the acoustic and/or optical waves is necessary for

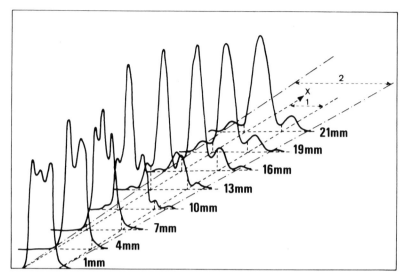

FIG. 9. Acousto-optic measurement of the acoustic beam generated by a transducer at various distances from the source: 1—transducer width (1 mm); 2—beam path (from Cambon[19]).

acoustic measurements in a standing wave geometry. Since acoustic velocities are typically less than 1 cm/μsec, sound waves take more than 1 μsec to traverse a sample of centimeter dimensions. Since forward traveling acoustic pulses of duration 1 μsec or less do not, in general, overlap

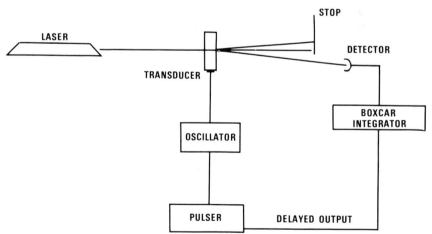

FIG. 10. Apparatus for resolving in time the light diffracted by acoustic pulses.

in space with reflected pulses, the light scattered into the ± 1 orders at any given time is due to only one of the sound waves. The laser source is sometimes also modulated so that it illuminates only the forward (or backward) traveling wave. Alternatively, the scattered light can be modulated or the detector gated on and off. This type of apparatus is shown in Fig. 10.

9.5. Fabry–Perot Interferometry

The Fabry–Perot interferometry approach[20] uses high-resolution optical interferometry to isolate the various spectral components (i.e., at ω_0, $\omega_0 + \Omega$, and $\omega_0 - \Omega$) produced by the acousto-optic interaction. This is achieved by frequency filtering the light by a Fabry–Perot interferometer which acts essentially as a very high-resolution, tunable, optical filter. Bandwidths as small as a megahertz and tuning ranges of many gigahertz are possible. The light transmitted by the interferometer is detected in a dc fashion by a photoelectric device and the experimental configuration is similar to that of Fig. 8 with the interferometer placed between the sample and the detector.

9.5.1. Basic Principles

The operating characteristics of a Fabry–Perot are well documented in a number of textbooks[21] on optics, and only the salient features will be reviewed here. A planar instrument consists of two flat parallel plates separated by a variable distance d with adjacent surfaces coated to a high reflectivity (typically $R > 95\%$, where R is the intensity reflection coefficient). The filter is tuned by varying the spacing between the plates over an optical wavelength.

A Fabry–Perot is basically an interference device. A transmission maximum occurs when $q\lambda = 2d$, where q is an integer and λ the optical wavelength, i.e., the maxima correspond to standing wave resonances in the optical cavity. Successive maxima occur whenever q changes by ± 1. Each wavelength present in the field incident on the Fabry–Perot produces its own series of maxima.

The bandwidth varies with the plate reflectivity, flatness, and parallelness. The reflectivity finesse is defined[21] by

$$\tilde{f}_R = \pi R^{1/2}/(1 - R) \tag{9.5.1}$$

[20] P. J. Vella, G. I. Stegeman, M. Zuliani, and V. M. Ristic, *J. Appl. Phys.* **44**, 1 (1973).
[21] M. Born and E. Wolfe, "Principles of Optics," 2nd ed., p. 323. Macmillan, New York, 1964.

and for plates with a rms deviation from perfect flatness of λ/m over the illuminated area, the flatness finesse is $\tilde{f}_F \simeq m/2$. The net finesse is approximated by

$$\frac{1}{\tilde{f}} = \frac{1}{\tilde{f}_R} + \frac{1}{\tilde{f}_F} \qquad (9.5.2)$$

and the instrumental bandwidth is $\delta\omega = \pi c/\tilde{f}d$. For $\lambda/200$ plates, the finesse is usually limited to less than 100. For example, if $d = 5$ cm and $R = 0.98$, then for $\lambda/200$ plates, $\delta\omega/2\pi = 50$ MHz.

The detectability of a spectral line depends on its frequency separation from other spectral components and its relative intensity. Signals of comparable intensity must be separated by at least $\delta\omega/2$. For weaker signals the separation must be larger and the contrast $c \; (=[1 + R]^2/[1 - R]^2)$ becomes important. (The contrast is defined as the ratio of the maximum to minimum transmitted intensities for a given wavelength.) Thus, at the midpoint between two successive maxima of an intense line, a signal of intensity of $\sim 1/c$ of the stronger line can just be measured. For $R = 0.98$, $c \sim 10^4$, which makes this instrument useful in the presence of intense stray light. (The contrast can be increased[4] to more than 10^{10} by passing the light more than once through the same inter-ferometer.)

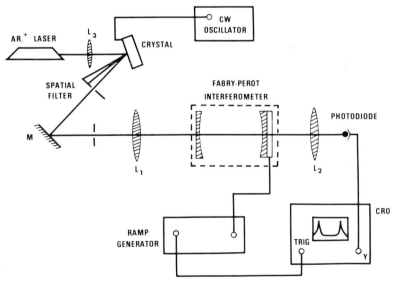

FIG. 11. Typical apparatus for frequency analyzing the scattered light with Fabry–Perot interferometers (from Vella *et al.*[20]).

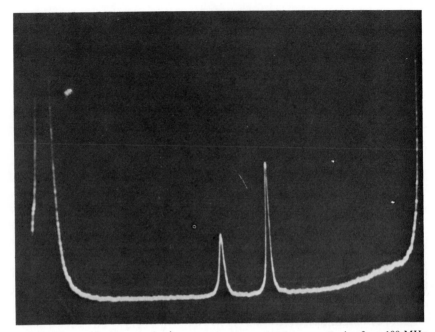

FIG. 12. Frequency spectrum of the +1 diffraction order due to scattering from 100-MHz surface waves traveling under an interdigital transducer on a SAW device. The two spectral lines occur at $\omega_0 \pm 100$ MHz and the measurement was done with a 10-cm confocal Fabry–Perot interferometer.

If large plate separations ($d > 5$ cm) are required for very high resolution, it is easier to use a confocal resonator geometry,[22] i.e., plates with a radius of curvature r are separated approximately by this same distance. Successive peaks occur when d changes by $\lambda/4$ and the reflectivity finesse is reduced by a factor of two from that of the parallel plate geometry. Spacings up to 1 m are used with resolutions of less than 1 MHz.

9.5.2. Experimental Geometry

A representative[20] experimental apparatus is shown in Fig. 11. The plates are mounted on small piezoelectric feet, which are used both to maintain optimum alignment of the plates as well as to vary the plate spacing. A voltage ramp that is linear in time is used to scan the interferometer. Lenses L_1 and L_2 are used to ensure a parallel beam at the interferometer input and to focus the light onto the detector, respectively.

[22] M. Hercher, *Appl. Opt.* **7**, 951 (1968).

The light source is a single-frequency laser, for example, a helium–neon or argon ion laser. This type of apparatus,[3-6] but with a cooled, selected, low-noise, photomultiplier tube, photon-counting electronics, repetitive scanning and data storage in multichannel memories, is used to study Brillouin scattering from thermally excited sound waves, i.e., the acoustic waves that constitute the "white noise" in a sample.

This technique is most useful for experimental conditions in which: (a) there is a great deal of stray light, and/or (b) there are sound waves traveling in both directions. A typical oscilloscope trace[20] of one sweep is shown in Fig. 12 for which the stray light is $\sim 10^3$ times more intense than the two signals of interest. The standing wave ratio is simply the ratio of the two acousto-optically scattered lines at $\omega_0 \pm \Omega$. This technique can easily detect surface displacement of 10^{-3} Å.

9.6. Heterodyne Techniques

The techniques[23] described here constitute the most powerful acousto-optic methods available for studying generated sound waves. In brief, they are based on electronically processing the difference frequency beat signals obtained when the acousto-optically scattered light is mixed by a square-law detector with a reference optical signal derived from the same laser source. Typically, both the acoustic amplitude and phase are measured as a function of position in the acoustic medium. In turn other parameters such as velocity and propagation direction can also be evaluated to a high degree of precision.

9.6.1. General Principles

The rationale behind the heterodyne technique is common to all versions. An oscillator of frequency Ω excites a transducer which launches the sound waves down the material. The light probe interacts with the sound wave at a distance x_0 measured from the transducer to the optical probe, which is assumed to be small in diameter relative to an acoustic wavelength. Thus the local time-dependent acoustic field has the form $u_0 \cos(\Omega t + \psi)$, where $\psi = kx_0$ is the local phase of the sound wave relative to the oscillator. The diffracted field is described by Eq. (9.2.5), and when it is mixed with a reference signal at the incident light frequency, a beat signal proportional to $u_0 \cos(\Omega t + \psi)$ is obtained as discussed in Chapter 9.3. When this beat signal is electronically mixed with a sample of

[23] R. M. De La Rue, R. F. Humphryes, I. M. Mason, and E. A. Ash, *Proc. IEEE* **119**, 117 (1972).

the oscillator output, the final signal is proportional to $u_0 \cos \psi$. Therefore, the local field amplitude and phase relative to the oscillator are recovered by scanning the optical probe across the surface.

Additional features common to all heterodyne systems are the restrictions on the wave fronts of the light beams incident on the detector.[23] The waves to be mixed must physically overlap on the detector photosensitive surface. In most cases the spot size of the light probe on the sample is less than one acoustic wavelength ensuring that the zeroth order does overlap with the ± 1 orders. (The usual diffraction spreading of the light is larger than the scattering angles due to the acousto-optic interaction.) Otherwise, lenses, etc., are required to recombine the various beams. It is also desirable that the mixing wave fronts be as parallel as possible in order to maximize the detector output signal. For example, if the detector surface corresponds to the $x'-y'$ plane (i.e., located at $z' = 0$), then two almost-parallel beams at near normal incidence would have the form $\exp\{i[(\omega_0 \pm \Omega)t - \kappa \sin \xi x' - \kappa \cos \xi z' + \psi]\}$ and $\exp\{i[\omega_0 t - \kappa z']\}$. The resulting beat signal is proportional to $\int \cos(\kappa \sin \xi x') \, dx' \, dy'$, which is the integral over the illuminated surface of the detector. The signal is a maximum only if ξ, the angle between the two incident optical wave vectors, is zero.

There are two basic variations on heterodyne techniques. In the first class, the reference field is essentially a fraction of the incident laser beam and the beat signals at the acoustic frequency Ω are analyzed. In the second class, superheterodyne systems produce signals at other frequencies by further modulating the incident and/or scattered light prior to detection.

9.6.2. Heterodyne Systems

The most versatile experimental configurations in heterodyne systems are essentially of two types.[6,23] In both types, the optical probe is focused down to a spot size smaller than an acoustic wavelength and the reference beam is at the laser frequency. The experimental geometries are shown in Figs. 13 and 14, respectively.

9.6.2.1. Simple Heterodyne Systems. Consider the configuration in Fig. 13. The reference beam can be obtained in a number of ways, and the prime objective is to produce as much overlap of the mixing beams on the detector as possible. For example, as indicated by path (a) in Fig. 13, the laser light can be sampled prior to incidence on the sample and redirected onto the detector via beamsplitters. One advantage of this approach is that the mixing beams can be independently adjusted to be parallel on the photodetector face. However, since the optical paths are different, fluctuations in the path length of the order of an optical wavelength

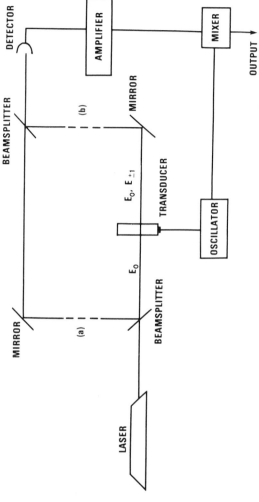

FIG. 13. Schematic of a simple heterodyne apparatus using (a) a fraction of the incident light or (b) the transmitted unscattered light as the reference beam.

(of either the reference or scattered beams) can seriously deteriorate the output signal. This factor severely limits the usefulness of this approach. In the second configuration, indicated by path (b) in Fig. 13, the reference beam is derived from a region of the sample near the acousto-optic interaction. If the unscattered (transmitted or specularly reflected) light is used, the 0 and ± 1 orders cannot, by definition, be exactly parallel at the detector face which results in a reduction in the signal as discussed before. Stray light at the sample surface and from sample imperfections in the volume can also be used, but the resulting amplitude and phase of such a reference beam will vary with position in the sample. The sensitivity to optical path length fluctuations is greatly reduced for both of these geometries since the optical paths are almost identical, but the disadvantages just mentioned have resulted in limited use of this technique.

The electronic processing subsequent to detection is the same for both configurations. The detector bandwidth must be at least as large as the acoustic frequency Ω being isolated. The beat signal from the phototube is mixed electronically with the oscillator thus yielding both the acoustic amplitude and phase. Typical sensitivities are of the order 0.1 Å.

9.6.2.2. Knife-Edge Method. This particular technique has been developed[24,25] to a very high degree of sophistication. In its most advanced form the acoustic phase fronts associated with a low-frequency standing wave are displayed on a television monitor. The features that distinguish this system from those discussed previously are the parallel superposition of the 0 and ± 1 orders at the detector and the use of a knife edge to block part or all of either the + 1 or − 1 orders from the detector.

The optical layout is shown in Fig. 14. A scanner systematically sweeps an incident laser beam across the surface (for scattering on reflection) or volume (for scattering on transmission) of the medium. The first lens (focal length f_1) focuses the light onto the acousto-optic interaction region to a spot size of the order of $2d = 2\lambda f_1/W$ (less than the acoustic wavelength $\Lambda = 2\pi/k$), where W is the cross-sectional dimension of the incident parallel beam (prior to L_1). The second lens (focal length f_2) recollimates the scattered light into parallel beams. Thus the various scattered orders have a size given by Wf_2/f_1 with parallel wave fronts at the knife edge. Note that this lens train ensures that the various diffraction orders always arrive at the same position in the plane of the knife edge, independent of the position of the probe on the surface. However, the

[24] R. Adler, A. Korpel, and P. Desmares, *IEEE Trans. Sonics Ultrason.* **SU-15**, 157(1968); R. L. Whitman and A. Korpel, *Appl. Opt.* **8**, 1567 (1969).
[25] E. Bridoux, J. M. Rouvaen, M. Moriamez, R. Torguet, and P. Hartemann, *J. Appl. Phys.* **45**, 5156 (1974).

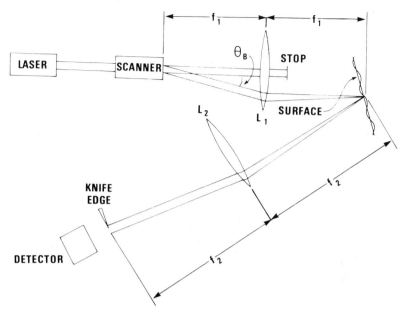

FIG. 14. Optical layout for the knife-edge method (from Adler *et al.*[24]).

centers of the ± 1 orders are displaced laterally from the zero order by $\pm k f_2/\kappa$, respectively.

The beat signal obtained depends on the degree of overlap of the zero and ± 1 orders on the detector surface. The knife edge blocks more of one order than the other. Assuming parallel wave fronts, the maximum detector signal is proportional to[24]

$$2 dk (u_0/\lambda) \cos(\Omega t + \psi) \qquad (9.6.1)$$

for $\lambda/2d \geqslant \lambda/\Lambda$ and to

$$4\pi[1 - d/\Lambda](u_0/\lambda) \cos(\Omega t + \psi) \qquad (9.6.2)$$

for $\lambda/\Lambda > \lambda/2d$ until $d = \Lambda$, after which the signal effectively remains zero. (The last condition corresponds to no overlap of the diffraction orders.) This configuration can also be analyzed in terms of specular reflection from a sinusoidally tilting surface with the results that the light is considered to be swept back and forth across the knife edge.

The electronics required here is essentially the same as shown (subsequent to the detector) in Fig. 13. The optical scanner is linked to the sweeping circuitry of a television monitor and the output is displayed on the screen. In the typical result[25] shown in Fig. 15 the wave fronts due to

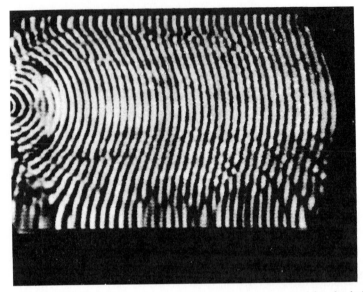

FIG. 15. Acoustic wave fronts due to a point source as measured by the knife-edge method (from Bridoux *et al.*[25]).

acoustic emission from a point source are displayed. This particular technique has proven to be very useful[23-25] in device diagnostics, nondestructive testing, and automated data acquisition.

There is, however, one major disadvantage to this approach. The sensitivity (which is typically 10^{-2} Å) decreases with the angle between the acoustic wave vector and the line of the knife edge, i.e., it is not sensitive to sound waves traveling parallel to the edge. On the other hand, this configuration is largely insensitive to random optical path length variations since the overlapped 0 and ± 1 orders follow the same trajectory.

This system can easily be modified to display standing acoustic waves if the scanning speed of the light beam across the sample is fast enough. When the scanning velocity is V along the acoustic propagation direction, the effective frequency shifts of the scattered light are given by $\Omega \pm \omega_0 V/c = \Omega \pm \Delta\Omega$ for waves traveling in the same and opposite directions, respectively, as the scanning beam, i.e., the light is additionally Doppler shifted due to the motion of the incident light relative to the sound waves. Typically, $\Delta\Omega \sim 1$ MHz and the two electrical signals from the detector at $\Omega + \Delta\Omega$ and $\Omega - \Delta\Omega$ are electronically separated as indicated in Fig. 16. Thus either component of the standing wave can be examined separately or both can be overlapped on the same display. This technique works well at low acoustic frequencies but it becomes difficult to separate $\Omega + \Delta\Omega$ from $\Omega - \Delta\Omega$ electronically if $\Omega \gg \Delta\Omega$.

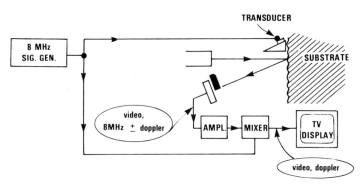

FIG. 16. Schematic of system used to separate the signals due to oppositely traveling sound waves (from Adler *et al.*[24]).

9.6.3. Superheterodyne Techniques

Three very powerful systems will be discussed in the category of super-heterodyne techniques. The beat signals of interest are not at the acoustic frequency Ω since the incident light and the scattered beams are further modulated, in one case by a Bragg cell and in the other two cases by an electro-optic modulator. These systems have greatly reduced sensitivity to random optical path variations.

9.6.3.1. Bragg Cell Modulation. The configuration of the optical components shown in Fig. 17 is reproduced from the work of Wickramasinghe and Ash.[26] The geometry corresponds basically to a Michelson interferometer with the Bragg cell playing the role of the beamsplitter. Each "reflection" is accompanied by a frequency shift of $\pm \Omega_m$, depending on whether the deflection is in the same or in the opposite direction to the acoustic wave vector in the Bragg cell. Light from the laser is partially transmitted to the sample region and partially deflected (at the frequency $\omega_0 + \Omega_m$) into the interferometer reference arm. (The light is incident at the appropriate Bragg angle and the "reflected" light corresponds to the $+1$ diffraction order of the Bragg cell.) The light in the reference arm is reflected back upon itself by a mirror and partially transmitted to the detector; the deflected part at the frequency $\omega_0 + 2\Omega_m$ is directed back toward the laser. A fraction of the light scattered by the sound waves is "reflected" into the -1 order of the Bragg cell and also illuminates the detector surface. In order to isolate the laser from the multiple, frequency-shifted, returning beams, an isolator is inserted before the laser. (For example, this could be achieved by a polarizer and a $\pi/4$ polarization rotator.)

---

[26] H. K. Wickramasinghe and E. A. Ash, *Proc. Symp. Opt. Acoust. Micro-Electron.* (J. Fox, ed.). Polytechnic Press, New York, 1974.

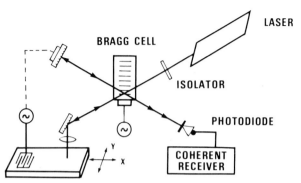

FIG. 17. Superheterodyne apparatus using a Bragg cell modulator (from Wickramasinghe and Ash[26]).

There are a large number of frequencies present in the beat signal at the detector output. Of particular interest in this technique are the signals obtained by mixing the light from the reference arm (at $\omega_0 + \Omega_m$) with light specularly (at $\omega_0 - \Omega_m$) and acousto-optically (at $\omega_0 - \Omega_m + \Omega$) reflected from the sample surface. As indicated in Fig. 18, the frequency components $2\Omega_m$ and $2\Omega_m - \Omega$ are electronically isolated and amplified. Further mixing of these two signals yields a signal at the frequency Ω, which is proportional to the acoustic amplitude and phase. (Since both the signals at $2\Omega_m$ and $2\Omega_m - \Omega$ contain the same phase variations due to optical path and laser instabilities, beating of the two together reduces the errors resulting from these problems substantially.) Finally, the amplitude and phase information is recovered by comparing the signal with one from the oscillator via a vector voltmeter.

This technique has been applied[26] to a variety of problems, such as one-dimensional holography, device diagnostics, and nondestructive testing, with great success. Furthermore, the acoustic phase velocity has been evaluated to a high precision by counting the number of acoustic cycles over a measured propagation distance, i.e., by measuring the acoustic wavelength very accurately. Accuracies of better than several parts per ten thousand have been achieved.[26] The sensitivity is better than 0.1 Å.

9.6.3.2. Electro-optic Modulation. The electro-optic modulation technique first reported by Parker[27] is very attractive since the only high-frequency component required is an electro-optic modulator. The beat signal of interest actually occurs at zero frequency.

The apparatus is shown in Fig. 19. An electro-optic modulator is

[27] T. E. Parker, in IEEE Ultrason. Symp. Proc. p. 365. IEEE, New York, 1974.

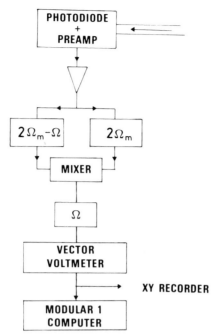

FIG. 18. Electronics used for processing the detector signals in the superheterodyne apparatus using a Bragg cell (from Wickramasinghe and Ash[26]).

driven by the same oscillator that excites the sound waves; thus the incident light is of the form $E_0[1 + \cos(\Omega t)] \exp(i\omega_0 t)$, i.e., it contains frequency components at ω_0 and $\omega_0 + \Omega$. Therefore, the acousto-optically scattered light will have components at ω_0. A lens system focuses the incident light into the acousto-optic interaction region, collects both the scattered and unscattered orders, and produces a magnified image of the crystal surface. (Included in this image is the phase grating created by the sound wave, but stationary in time.) This image is then sampled by an apertured photodetector, which limits the scattered light detected to an area with dimensions of less than one acoustic wavelength.

In contrast to the other heterodyne schemes described previously, the bandwidth of the detector can be as small as a few kilohertz. The goal is essentially to retain only the dc terms and reject the beat frequency signals. These terms are proportional to E_0^2, $2E_0^2 H' u_0 \kappa \cos(kx_0)$, and $E_0^2 u_0^2 H'^2$. Since usually $1 \gg u_0 H'$, the last term is negligible compared to the first two. The first dc term is removed by slowly modulating the oscillator (which excites the sound waves) and using phase-sensitive detection of the detector signal. Thus the term proportional to

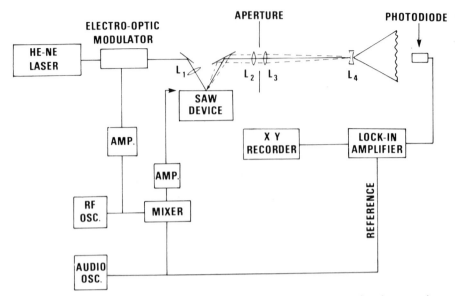

FIG. 19. Experimental apparatus for Parker's superheterodyne system using electro-optic modulation (from Parker[27]).

$u_0\kappa \cos(kx_0)$ is isolated and yields a measurement of the acoustic amplitude and phase as a function of position x_0.

This technique has a number of attractive features (other than the dc detection aspect). The system is simple to align since alignment is obtained by producing a sharp, focused image of the surface in the plane of the detector. Since the various diffraction orders follow essentially the same optical path, the effects of vibrations, etc., are minimal. It appears that this apparatus is limited only by the electro-optic modulator bandwidth, and therefore gigahertz acoustic frequencies can be investigated.

This apparatus is capable of all the measurements discussed previously for the other acousto-optic techniques. The sensitivity for measuring surface corrugations is better than 0.1 Å.

9.6.3.3. Engan's Method. The most advanced superheterodyne technique, reported recently by Engan,[28] combines the best features of the knife-edge method and electro-optic modulation. The experimental apparatus is identical to that shown in Fig. 19 except for the components between the sample and the lock-in amplifier: the appropriate sample geometry and detection scheme are illustrated in Fig. 20. As discussed in Section 9.6.3.2, the scattered signals of interest occur at the laser fre-

[28] H. Engan, *IEEE Trans. Sonics Ultrason.* **SU-25,** 372 (1978).

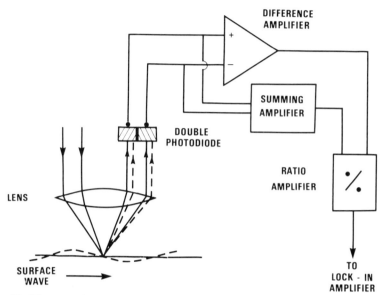

FIG. 20. The scattering geometry and detection electronics for the combined knife-edge and electro-optic modulation technique (from Engan[28]).

quency and the detector output is dc in nature. Therefore, neither high-speed detectors nor electronics are required. In this system the knife edge is replaced by a double photodiode which is centered very precisely on the zeroth diffraction order. The difference signal between the two detectors depends on the local acoustic amplitude and phase at the position of the laser probe on the surface. Since the double photodiode is centered on the unscattered light beam, the dc signal in the detectors' output due to the unscattered light is eliminated by the subtraction process. Note also that the double photodiode arrangement produces twice the signal obtained with a knife edge. The outputs of the two photodetectors are also summed and the ratio of the difference-to-sum signals is evaluated electronically: this allows the ratio $E_{\pm 1}/E_0$ and, therefore, the acoustic amplitude and phase to be measured directly, independent of the local surface reflectivity. By moving the laser probe over the acoustic propagation medium, the amplitude and phase of the sound wave can be evaluated, even in a standing wave geometry.[28,29]

This new technique is probably the most powerful and versatile one currently available in acousto-optics. It has been used to study acoustic

[29] R. M. De La Rue, *IEEE Trans. Sonics Ultrason.* **SU-24**, 407 (1977).

propagation on free surfaces, at boundaries, and inside surface acoustic wave devices. This apparatus has been operated at the highest frequencies reported to date for superheterodyne systems, i.e., 450 MHz and sensitivities of $\sim 10^{-3}$ Å have been demonstrated under the previously stated conditions.

9.7. Summary

The acousto-optic interaction is a very powerful tool for studying and evaluating acoustic phenomena. Essentially every acoustic parameter of interest, namely, the local amplitude, phase, frequency, wave vector, direction, phase velocity, and group velocity, can be measured.

The experimentalist has essentially a choice of two different techniques. The simplest version consists of separating out in space the various diffraction orders created by the acousto-optic interaction and then observing (or measuring) them individually. On the other hand, heterodyne systems rely on superimposing various scattered and unscattered beams on a square-law detector and isolating beat frequencies that are characteristic of the sound waves.

Which technique is appropriate for a given problem? If phase or very accurate velocity information is required, then heterodyne systems must be used. However, these methods are more complex than either of the dc detection techniques. Therefore, if only the acoustic amplitude must be measured, then the systems outlined in Chapters 9.4 and 9.5 are appropriate.

There are two other important factors that must also be considered. Since the signal in a heterodyne approach is proportional to $u_0 \kappa$ rather than to $u_0^2 \kappa^2$, which is characteristic of separate order dc detection, then for $1 \gg u_0 \kappa$ (the usual case), the heterodyne systems have a superior sensitivity. On the other hand, heterodyne methods require either a high-frequency modulator or a detector with a fast response time (and thus large bandwidth) and are, therefore, limited to acoustic frequencies less than a gigahertz. The dc techniques are useful at all acoustic frequencies.

The treatment presented here has not exhausted the available acousto-optic measurement techniques. Additional dc systems are discussed in Part 2, specifically with reference to precision velocity measurements. Other less widely used heterodyne techniques are summarized by Stegeman[6] and De La Rue et al.[23]

The theoretical analysis of the scattering presented here has, of necessity, been brief and incomplete. The scattering efficiency was described by a single parameter H, the calculation of which can be complicated, especially for anisotropic materials. Further information can be found in the review articles by Stegeman,[6] Chang,[7] Schmidt,[8] and Damon et al.[9] and the references contained therein.

10. SURFACE ELASTIC WAVES

By Richard M. White

List of Symbols

$\bar{a}$	mean surface roughness
A	displacement amplitude coefficient
A_1, A_2	amplitudes of fundamental, second harmonic
C	elastic stiffness
C_1, C_2, C_3	partial wave amplitude coefficients
D	electric displacement
e	piezoelectric constant
E	electric field
f	frequency; trapping factor
f_m	frequency of mth harmonic of interdigital transducers
h	distance from free surface
I_{AE}	acoustoelectric current density
k	propagation constant
K_S	electromechanical coupling constant for surface waves
L	path length
n	number of periods of interdigital transducer
N	integer
N_{opt}	optimum number of array electrode pairs
p	periodic distance of interdigital transducer
P_0	power density
R	distance from scattering object
S	strain
t	time
T	stress
u_1, u_2, u_3	components of particle displacement
$\hat{u}_1, \hat{u}_2$	peak values of particle displacement components
v	phase velocity
v_S	surface wave velocity
v_W	phase velocity of bulk wave in wedge
x, x_1, x_2, x_3	Cartesian coordinates
α_1, α_2	surface wave delay constants
ϵ	dielectric permittivity
θ	angle of bulk wave incidence on wedge surface
λ	wavelength
μ	mobility
ω	angular frequency

495

10.1. Introduction

Surface elastic waves—first analyzed by Lord Rayleigh and later the subject of extensive study by geophysicists—have been used since the 1950s in nondestructive testing of manufactured parts and during the 1970s in electronic signal-processing devices.[1] Surface elastic waves will propagate along the surface of any semi-infinite homogeneous or layered solid, whether the solid be elastically isotropic or anisotropic, homogeneous or inhomogeneous, electrically conducting or insulating, piezoelectric or nonpiezoelectric. In this part we shall consider a surface wave to be an organized propagating disturbance having a wavelike character (that is, one whose associated particle displacement, stress, and strain are solutions of a wave equation) whose energy is concentrated near the surface of the medium. Because of this energy concentration, the behavior of a surface elastic wave is particularly dependent on surface properties, so these waves offer unique measurement opportunities. Further, the recent development of electronic signal-processing devices employing surface elastic waves has led to a greatly increased understanding of surface wave properties and means of efficient transduction; in some instances these surface wave devices can themselves be used for measurement purposes.

10.2. Surface Waves in Semi-Infinite and Layered Media

Because our interest is in measurements that can be made with surface waves, we begin by asking what characteristics of the waves are observable quantities. If we could see the infinitesimal volumes of a solid along whose plane boundary a surface wave propagated, we would find particles being displaced from their equilibrium positions in directions normal and parallel to the boundary of the solid. Figure 1 suggests this but with an exaggerated vertical scale: for a typical surface wave experiment at 100 MHz, the wavelength would be 30 μm in a medium having the typical phase velocity 3000 m/sec; if the wave had 10 mW average power in a beam 1 cm wide, the peak vertical displacement might be about 1 Å or only 1/300,000 wavelengths. The particle motions (Fig. 2) are largest

---

[1] W. M. Ewing, W. S. Jardetsky, and F. Press, "Elastic Waves in Layered Media." McGraw-Hill, New York, 1957; I. A. Viktorov, "Rayleigh and Lamb Waves." Plenum, New York, 1967; B. A. Auld, "Acoustic Fields and Waves in Solids." Wiley (Interscience), New York, 1973; E. Dieulesaint and D. Royer, "Ondes Elastiques dans les Solides." Masson, Paris, 1974; A. A. Oliner (ed.), "Acoustic Surface Waves." Springer-Verlag, Berlin and New York, 1978. See also *Proc. IEEE Ultrason. Symp.* (1972–1978).

FIG. 1. Surface elastic or Rayleigh wave. A section through a semi-infinite solid is suggested, with dots showing the instantaneous positions of the elementary volumes of the solid as the wave propagates in the direction shown by the arrow. Note that the amplitude of motion decreases with the distance from the surface and the particles move both in the direction of propagation and perpendicular to it.

near the surface, and they virtually disappear a few wavelengths below the surface. The particle displacement can be sensed by optical means, and it can be inferred from the transport of small particles along the surface in a direction opposite to that of wave propagation owing to the retrograde elliptical motion of the surface atoms.

10.2.1. Analysis of Waves Guided along a Surface

We proceed much as in the case of bulk waves when deriving the components of a plane surface elastic wave propagating along the free surface of a semi-infinite homogeneous nonpiezoelectric solid: we write Hooke's law as

$$[\mathbf{T}] = [\mathbf{c}][\mathbf{S}], \qquad (10.2.1)$$

relating the components of the stress tensor $[\mathbf{T}]$ linearly through the elastic stiffness tensor $[\mathbf{c}]$ to the strain tensor $[\mathbf{S}]$. We also equate the inertial force on an elemental volume to the forces exerted by the medium via the stresses on the boundaries of the volume. In the surface wave case we also impose the additional condition that the stresses normal and tangential to the free surface plane must be zero. Numerical methods are then used to solve the resulting equations.

In the simplest case—that of an elastically isotropic solid— components of the particle displacement parallel (u_1) and perpendicular (u_2) to the surface have the form

$$u_1 = A[\exp(-\alpha_1 x_3) + C_1 \exp(-\alpha_2 x_3)] \cos(\omega t - kx_1)$$

$$= u_1 \cos(\omega t - kx_1), \qquad (10.2.2)$$

$$u_2 = A[C_2 \exp(-\alpha_1 x_3) + C_3 \exp(-\alpha_2 x_3)] \sin(\omega t - kx_1)$$

$$= u_2 \sin(\omega t - kx_1), \qquad (10.2.3)$$

for a wave of angular frequency ω and propagation constant k traveling in the x_1 direction. The constants α_1 and α_2 govern the rate of decay of am-

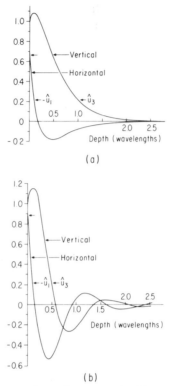

FIG. 2. Particle displacements for surface waves in: (a) an elastically isotropic medium having Poisson's ratio 0.29; (b) the cubic crystal nickel with propagation along the [100] direction on the (001) plane. The subscript 1 refers to the component in the direction of propagation and 3 to the component normal to the surface (Farnell,[2] with permission).

plitude with depth x_3 below the surface; they and the factors C_1, C_2, and C_3 are real functions of the elastic constants and the density. The factor A depends on the wave amplitude or, equivalently, its power. The peak amplitude terms are sketched in Fig. 2 for a representative isotropic material. Thus this simplest, Rayleigh, surface wave is a plane, nonuniform wave having components of displacement parallel and normal to the surface and lying in a plane parallel to the direction of propagation.

For a piezoelectric half-space we employ the piezoelectric equations of state, as in the case of bulk waves in a piezoelectric medium, instead of using Hooke's law. Thus the stress tensor is related to the strain tensor and to the electric field vector [E] by

$$[\mathbf{T}] = [\mathbf{c}][\mathbf{S}] - [\mathbf{e}][\mathbf{E}], \tag{10.2.4}$$

where [e] is the piezoelectric constant tensor. Correspondingly the electric displacement vector [D] in the solid is given by

$$[D] = [e][S] + [\epsilon][E], \qquad (10.2.5)$$

where [ϵ] is the permittivity tensor. Electrical boundary conditions at the surface provide a connection between the electric field inside the solid resulting from the deformation and the associated field outside. At a nonconducting boundary plane the components of electric displacement tangential to the boundary must be continuous. If instead an infinitesimally thin, perfect electrical conductor covers the surface, the tangential component of electrical displacement just inside the solid must be zero, and there is no electric field outside the solid.

Still more complicated media and structures are often of interest. If the medium is a piezoelectric semiconducting half-space, one must include electrical conductivity within the medium and require that there be no current flow through the free bounding surface. If there is more than one boundary, other surface or plate waves can propagate. The case of a thin, piezoelectric layer on a nonpiezoelectric substrate is of particular importance experimentally as transducers can be formed to launch and receive surface waves with such a layer, as we shall see later. Table I identifies many of the types of elastic waves that can propagate along the surfaces of bounded solids. Some of these waves are characterized schematically in Fig. 3.

10.2.2. Propagation Characteristics

The algebraic complexity of the surface wave problem forces one to rely heavily on computer solutions for wave characteristics. The principles of such analysis have been described by Farnell[2] and others. Fortunately, because of the interest in electronic and optical applications of surface waves, computer studies have already been made of many insulators, piezoelectrics, and semiconductors.[3]

Properties usually listed as functions of direction for propagation in a given homogeneous material include phase velocity, some factor relating wave amplitude (and electric field, in piezoelectrics) to total power flow, and strength of piezoelectric coupling. This last quantity is sometimes in-

---

[2] G. W. Farnell, Properties of elastic surface waves, *in* "Physical Acoustics," (W. P. Mason and R. N. Thurston, eds.), Vol. 6, pp. 109–166. Academic Press, New York, 1970.

[3] A. J. Slobodnik and E. D. Conway, "Microwave Acoustics Handbook," Vol. 1, Surface Wave Velocities. Physical Sciences Research Papers No. 414, Air Force Cambridge Research Laboratories, Bedford, Massachusetts, 1970; B. A. Auld, "Acoustic Fields and Waves in Solids," Vol. 2, Appendix 4. Wiley, New York, 1973.

TABLE I. Waves Associated with Boundaries for Selected Media[a]

Structure	Medium	Designation of wave	Components of particle displacement	Wave characteristics
Semi-infinite solid having plane stress-free boundary	Isotropic insulator	Rayleigh	u_1, u_3	Nondispersive propagation. Elastic fields (displacements, stress, strain, particle velocity) are sums of two decaying exponentials. Particle motion has two components, one parallel to propagation direction (u_1) and one normal to the surface plane (u_3)
	Anisotropic, nonpiezo-electric insulator	All are nondispersive. Fields are sums of up to three decaying exponentials. Type of solution depends on medium and direction of propagation relative to crystallographic axes		
		Rayleigh-type	u_1, u_3	All real decay constants. These waves occur only for particular directions in certain anisotropic materials
		Generalized Rayleigh waves	u_1, u_2, u_3	Oscillatory decay with distance below surface
		Leaky (or psuedo-) surface waves	u_1, u_2, u_3	Slow decay of wave amplitude with distance traveled as energy leaks into bulk. Solution contains a bulk partial wave
	Piezoelectric insulator	All are nondispersive.		
		Rayleigh-type, generalized Rayleigh, or leaky waves, as above	u_1, u_2, u_3	Mechanical wave accompanied by traveling electric fields inside and outside of solid for propagation in most directions in these anisotropic materials
		Bleustein–Gulyaev wave	u_2	Particle motion parallel to surface and normal to direction of propagation. Decay with depth much more gradual than for other surface waves

Medium		Wave	Displacement components	Characteristics
Infinitely broad plate of finite thickness in vacuum	Isotropic (can generalize to anisotropic as well)	Shear horizontal (SH waves)	u_2	Particle motion normal to propagation direction and parallel to plane of plate. Only lowest mode having no variation of displacement amplitude through transverse cross section is nondispersive. Modes can be characterized by symmetry with respect to median plane of plate; symmetric (or dilatational) modes and antisymmetric (or flexural) modes
		Lamb waves	u_1, u_2, u_3	Dispersive if wavelength is comparable with plate thickness. Particle motion parallel to direction of propagation and normal to plane of plate. One of these modes is also called the Lamé wave
Layered solid (layer of finite thickness bonded to semi-infinite solid)	Isotropic (can generalize to anisotropic as well)			Dispersive if wavelength is comparable with thickness of layer.
		Love wave	u_2	Particle motion parallel to plane of layer and normal to direction of propagation (like SH wave in free plate)
		Generalized Lamb waves	u_1, u_2, u_3	Particle motion components are parallel to direction of propagation and normal to plane of plate and possibly perpendicular to these as well. One of these modes is called the Sezawa wave
Two different semi-infinite media bonded at plane interface	Isotropic (can generalize to anisotropic as well with either different media or two identical media having different orientations)	Stoneley wave	u_1, u_2, u_3	Nondispersive. Wave is guided along interface only for particular combinations of elastic constants and densities

[a] For further details on these waves consult sources listed in text.[1] Displacement components u_1 and u_3 are parallel to propagation vector and surface normal respectively, while u_2 is normal to both u_1 and u_3.

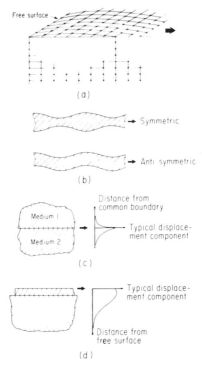

FIG. 3. Other modes in bounded solids: (a) an electroacoustic or Bleustein–Gulyaev wave which can propagate in piezoelectric solids. The isometric sketch showes the free surface and the medium below it. The particles move parallel to the free surface and normal to the propagation vector; the amplitude decreases with the distance from the surface and is quite sensitive to the electrical properties of the medium; (b) plate or Lamb waves. Particle motions are similar to those for the Rayleigh wave for plate thicknesses large compared with wavelength; (c) Stoneley wave at welded boundary of two dissimilar solids. The wave can exist only for pairs of isotropic or anisotropic solids having relative elastic properties which fall within a narrow range of values; (d) layered solid. The modes are similar to plate modes, the one like the symmetric plate mode being like the Rayleigh surface wave and the antisymmetric modes being known as Sezawa modes. Particle displacements for these modes in isotropic solids have components both parallel and perpendicular to the propagation vector. Love waves, having particle motion entirely transverse to the propagation vector, may also exist in layered solids.

dicated by the factor $\Delta v/v$, which represents the fractional decrease in phase velocity resulting when an infinitely thin, perfectly conducting layer is placed on the free surface of the piezoelectric medium, thereby reducing the piezoelectric stiffening of the wave by eliminating some stored electrical energy. Alternatively, the effective surface wave electromechanical coupling constant K_S may be given, where $K_S^2 \approx 2(\Delta v/v)$ for weakly or moderately piezoelectric substances. It is customary in describing surface waves to specify both the normal to the surface plane and the direction of propagation, relative to the crystallographic axes. For example, a frequently used, strongly piezoelectric crystal and orientation is denoted $YZ-LiNbO_3$, meaning propagation along the z direction in a lithium niobate crystal whose surface plane is normal to the y axis. The angle between the propagation vector and the power flow vector in anisotropic materials is another computed characteristic of experimental importance. Of course, for isotropic materials or for propagation normal to the symmetry direction in transversely isotropic solids, these two vectors are collinear and the power flow angle between them is zero. Studies of propagation in layered media have also been published.[4] Table II lists some surface wave properties of selected media.

Although numerical methods must be used to derive the properties of almost all surface waves, some general statements can be made about these waves. We shall consider first the primary characteristics that might describe any wave.

10.2.2.1. Primary Characteristics. In Table III we list and comment on surface wave characteristics, their accessibility to experimental determination, and typical ranges of their values. Of particular importance in using surface waves to probe a physical system are attenuation and phase velocity.

Attenuation may be determined by measuring the decrease in wave amplitude with distance traveled along the surface. The attenuation so defined may arise dissipatively from the local conversion of organized wave energy to heat[5] or it may result from the coupling of surface waves to bulk waves propagating into the solid (leaky surface waves) or even coupling to the atmosphere outside the solid. These two effects are usually small, amounting to small fractions of a decibel loss per wavelength of travel.[6] Nondissipative attenuation may result from surface roughness or defects or from scattering at grain boundaries in polycrystalline solids. The latter effect limits surface wave applications of ferroelectric ceramics such as lead–zirconate–titanate to frequencies below 50 MHz. Apparent attenuation in single-frequency measurements may also originate from nonlinear coupling to waves at other frequencies, which will be discussed later. As with bulk waves, surface wave attenuation or amplification can also occur

---

[4] G. W. Farnell and E. L. Adler, Elastic wave propagation in thin layers, *in* "Physical Acoustics" (W. P. Mason and R. N. Thurston, eds.), Vol. **9**, p. 35ff. Academic Press, New York, 1972.

[5] A. J. Slobodnik, P. H. Carr, and A. J. Budreau, *J. Appl. Phys.* **41**, 4380 (1970); P. J. King and F. W. Sheard, *ibid.* **40**, 5189 (1969).

[6] A. J. Slobodnik, *Proc. IEEE* **64**, 585 (1976).

TABLE II. Surface Wave Properties of Selected Media[a]

Material and orientation	Phase velocity (m/sec)	Magnitude of normalized components of particle velocity at surface $[10^{-6} \, (\text{m/sec})(\text{W/m})^{-1/2}]$			Magnitude of normalized potential at surface $[10^3 \, \text{V} \, (\text{W/m})^{-1/2}]$	Magnitude of normalized electric displacement component normal to surface $[10^{-12} \, (\text{C/m}^2)(\text{W/m})^{1/2}]$
		$\dot{u}_n$	$\dot{u}_p$	$\dot{u}_t$		
Bismuth germanium oxide $Bi_{12}GeO_{20}$ ([001]-cut,[110]-prop)	1680.7	$4.163\omega^{1/2}$	$2.535\omega^{1/2}$	—	$8.698\omega^{1/2}$	$45.80\omega^{1/2}$
Cadmium sulfide CdS(Z-cut, X-prop)	1728.9	$5.324\omega^{1/2}$	$2.861\omega^{1/2}$	—	$10.04\omega^{1/2}$	$51.41\omega^{1/2}$
Diamond C ([110]-cut,[001]-prop)	11063.0	$1.363\omega^{1/2}$	$1.123\omega^{1/2}$	—	—	—
Gallium arsenide GaAs([111]-cut, 30°[110]-prop)	2605.2	$3.627\omega^{1/2}$	$2.201\omega^{1/2}$	—	$1.803\omega^{1/2}$	$6.132\omega^{1/2}$
Lithium niobate $LiNbO_3$(Y-cut, Z-prop)	3487.7	$2.625\omega^{1/2}$	$1.777\omega^{1/2}$	—	$14.50\omega^{1/2}$	$36.79\omega^{1/2}$
Lithium tantalate $LiTaO_3$(Y-cut, Z-prop)	3229.9	$2.907\omega^{1/2}$	$2.045\omega^{1/2}$	—	$5.548\omega^{1/2}$	$15.21\omega^{1/2}$
Quartz SiO_2 (Y-cut, X-prop)	3159.3	$4.337\omega^{1/2}$	$2.897\omega^{1/2}$	$1.436\omega^{1/2}$	$8.726\omega^{1/2}$	$24.44\omega^{1/2}$
Sapphire Al_2O_3 (Z-cut,30°X-prop)	5706.4	$2.045\omega^{1/2}$	$1.265\omega^{1/2}$	—	—	—
Silicon Si (Z-cut, X-prop)	4921.2	$3.512\omega^{1/2}$	$2.862\omega^{1/2}$	—	—	—

[a] The power flow angle is zero for all the entries here, but it will, in general, be nonzero for other orientations in the anisotropic media listed. Values listed for piezoelectrics are for a free, unelectroded surface. Particle velocity, electrical potential, and electric displacement components are normalized with respect to the magnitude of the total mechanical power flow per unit width of the acoustic wave beam. Angular frequency ω is in radians per second. Subscripts n (normal) and p (parallel) refer to the directions perpendicular to the surface and parallel to the propagation vector, respectively. The subscript t (transverse) denotes the direction normal to the plane containing the direction of propagation and the normal to the surface. (Values taken from B. A. Auld, "Acoustic Fields and Waves in Solids," Vol. II, Appendix 4 Wiley (Interscience), 1973.)

TABLE III. Characteristics of Surface Waves and Their Measurement Applications

Characteristics	Comments and uses
Primary characteristics	
Frequency	Usually set by experimenter, ranging from a few megahertz (lower limit set by crystal size) to a few gigahertz (loss and transducer fabrication limits)
Particle displacement	Depends on strength of excitation. Absolute value can be measured optically. Typically a few angstroms or less
Stress	Depends on strength of excitation. Seldom measured directly
Strain	Depends on strength of excitation. Seldom measured directly. Typically 10^{-4} or less
Phase velocity	Important measurable characteristic. Depends on elastic constants, density, piezoelectric and dielectric constants, and temperature as well as on direction in anisotropic media. Frequency dependent in layered media at frequencies for which wavelength is comparable with layer thickness. Typical phase velocity is 3000 m/s
Particle velocity	Depends on strength of excitation. Seldom measured directly
Attenuation	Important measurable characteristic. Change in wave amplitude with distance traveled may be due to conversion of organized wave energy to heat (dissipation), geometrical effects such as scattering, diffraction, or leakage of waves from surface into bulk, or to nonlinear effects causing diversion of energy to other frequencies
Secondary characteristics	
Diffraction	Anisotropic media. Beam may broaden or narrow
Nonlinear effects	Depends on strength of excitation. Production of harmonics or sum and difference frequencies permit filtering for sensitive detection
Power flow angle	May be nonzero in anisotropic media

in a piezoelectric semiconductor in which there are mobile charge carriers.[7] Figure 4 shows the dissipative attenuation as a function of frequency for several important substances.[6]

Phase velocity varies as the square root of the ratio of elastic stiffness to mass density, as with bulk waves. For piezoelectrically active surface waves, the phase velocity depends as well on the piezoelectric and dielectric properties. Hence any factors that affect those properties, such as temperature and pressure, can perhaps be inferred from measurements of surface wave phase velocity. Figure 5 shows the temperature sensitivity

---

[7] K. M. Lakin and H. J. Shaw, *IEEE Trans. Microwave Theory Tech.* **MTT-17**, 912 (1969).

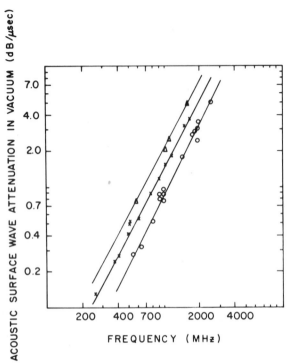

ACOUSTIC SURFACE WAVE ATTENUATION IN VACUUM (dB/μsec)

FREQUENCY (MHz)

FIG. 4. SAW attenuation in a vacuum as a function of frequency for YZ–LiNbO$_3$ (O), [001]–[110] and [111]–[110] Bi$_{12}$GeO$_{20}$ (×), and YX–quartz (Δ). Experimental slopes are all approximately proportional to the frequency squared (Slobodnik,[8] with permission).

of phase velocity for a number of piezoelectrics,[8] with velocities ranging from around 1600 m/sec in bismuth germanium oxide to more than 11,000 m/sec in diamond. Because these velocities are so much lower than the velocity of electromagnetic waves, an essentially quasi-static electric field having a decay distance of only $1/k = \lambda/2\pi$ exists just outside the surface of a piezoelectric supporting a surface wave. With this field, media placed quite close but not in actual contact with the piezoelectric can be probed.

The phase velocity of surface waves in a homogeneous semi-infinite solid is independent of the wave frequency—the medium is nondispersive—but variations of phase velocity with frequency arise in layered solids. Hence variations of the phase velocity can be used to sense changes in the thickness of layers such as might be produced by evaporation.

[8] R. M. O'Connell and P. H. Carr, *Proc. Ann. Symp. Frequency Control 32nd,* 189 (1978).

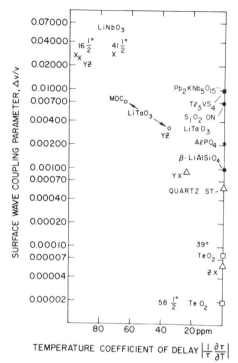

FIG. 5. Strength of the coupling factor versus the temperature coefficient of delay for various SAW crystals and orientations (O'Connell and Carr,[8] with permission).

10.2.2.2. Secondary Characteristics.

Diffraction occurs in surface wave experiments with finite-aperture transducers on anisotropic media because of the variation of phase velocity with the direction of propagation.[9] Diffraction causes wave intensities to differ from those expected from simple geometrical considerations and so must be taken into account when making attenuation measurements or when designing transducers.

Nonlinear effects can arise in insulating media because of the high concentration of elastic energy near the surface. In addition to this elastic nonlinearity, strong surface wave nonlinearities can arise in piezoelectric semiconductors because of the coupling of the traveling electric field with the bunched charge.[10] Harmonic generation and frequency mixing are observed and can be used to advantage experimentally, as we shall see (Section 10.3.3). In addition, it may be possible to determine the static

[9] M. S. Kharusi and G. W. Farnell, *IEEE Trans. Sonics Ultrason.* **SU-18**, 35 (1977); T. L. Szabo and A. J. Slobodnik, *ibid.* **SU-20**, 240 (1973).
[10] G. S. Kino, *Proc. IEEE* **64**, 724 (1976).

stress in a body from measurement of the surface wave velocity, since the elastic constants will be affected by the stress.[11]

The direction of surface wave power flow differs significantly from the propagation direction for all but the most symmetric directions in anisotropic media. Experimentally this can mean that energy from an input transducer may entirely miss an output transducer situated some distance away, so directions for which the power flow angle is zero are usually employed to avoid this so-called "beam steering" phenomenon.

10.2.3. Summary Comparison of Surface with Bulk Waves

The principal differences between surface and bulk elastic waves are:

(1) Plane surface waves are nonuniform with energy concentrated at the free surface. Thus the wave is accessible at the surface along its path of propagation, it is sensitive to surface conditions, and nonlinear effects may occur at moderate total power levels.

(2) The phase velocity of a surface wave propagating in a given direction is usually slightly less than that of the slowest bulk shear wave in the medium.

(3) A traveling electric field exists outside a piezoelectric solid in a vacuum supporting surface waves, decaying with distance h from the surface as $\exp(-kh)$, where k is the propagation constant for the surface wave.

(4) The fractional decrease of phase velocity $\Delta v/v$ when a perfectly conducting plane is put on the surface is a measure of the strength of piezoelectric coupling. The effective electromechanical coupling constant for surface waves is generally less than that for piezoelectrically active bulk modes in the medium because shorting the surface eliminates only that part of the electric energy stored quite near or outside the surface.

(5) Analytical expressions are seldom available for surface waves; instead compilations list properties such as phase velocity, decay constants, power flow angle, and strength of piezoelectric coupling as functions of direction with respect to crystallographic axes.

10.3. Transduction

Surface waves can be generated and received in many different ways, as summarized in Table IV (and shown in the accompanying figures in White[12]). When choosing a transducer one should consider the following:

[11] A. L. Nalamwar and M. Epstein, *J. Appl. Phys.* **47**, 43 (1976).
[12] R. M. White, *Proc. IEEE* **58**, 1236 (1970).

TABLE IV. Design Parameters for Interdigital Transducers on Various Piezoelectrics[a]

Piezo-electric	Cut and propagation direction	$\Delta v/v$ (%)	N_{opt}	M_{opt}	$\Delta\omega/\omega_0$	T ($\mu sec/cm$)	$T \cdot \Delta\omega/\omega_0$
$Bi_{12}GeO_{20}$	[110],[$\bar{1}$10]	1.15	6	183	0.17	6.33	1.08
CdS	XZ	0.31	12	54	0.09	5.82	0.52
$LiNbO_3$	YZ	2.46	4	108	0.24	2.88	0.69
$LiTaO_3$	ZY	0.82	7	31	0.14	2.86	0.40
PZT	Poled normal to surface	2.15	4	—	0.23	4.55	1.04
Quartz	YX	0.11	19	53	0.053	3.06	0.16
ZnO	XZ	0.56	8	99	0.12	3.74	0.45

[a] N_{opt} is the number of finger pairs for optimum bandwidth and M_{opt} the width in wavelengths of a transducer which would have 50-Ω radiation resistance at its center frequency. The last three columns are fractional bandwidth, time delay per centimeter of path traveled, and the time–bandwidth product. PZT is the designation of one commercially available form of the ferroelectric ceramic lead-zirconate-titanate. [From W. R. Smith *et al.*, *IEEE Trans. Microwave Theory Tech.* **MTT-17**, 865 (1969) with permission © 1969 IEEE.]

(1) Can piezoelectric media be used, either as semi-infinite solids or in thin-film form? If so, one can use the interdigital transducer (IDT), a reversible transducer employed in many surface wave, signal-processing devices (see Figs. 6 and 7).

(2) Is very wideband transduction required? If so, use of the wedge with a wideband, bulk wave transducer may be indicated (Fig. 8).

(3) Are wave amplitudes to be measured at various distances from the source? If so, optical detection may be desirable (Fig. 9).

(4) Are rapid measurements of many different samples to be made? If so, mode conversion in a fluid bath may be suitable (Fig. 10[13]).

(5) Are mere indications desired of the presence or absence of a wave on a piezoelectric system? Direct pickup of piezoelectrically coupled fields is possible (Fig. 11[14]).

(6) Are noncontacting transducers needed? Several means exist, including optical detection (Section 10.3.3), reversible piezoelectric or Lorentz force coupling to electric or magnetic fields produced by spatially periodic conductors located near the solid,[15] and thermoelastic generation of waves by surface absorption of transient electromagnetic radiation.[16]

[13] F. R. Rollins, Jr., *Appl. Phys. Lett.* **7**, 212 (1965).

[14] B. A. Richardson and G. S. Kino, *Appl. Phys. Lett.* **16**, 82 (1970).

[15] R. B. Thompson, *Proc. IEEE Ultrason. Symp., Phoenix, Arizona*, October 26–28, p. 74 (1977).

[16] R. E. Lee and R. M. White, *Appl. Phys. Lett.* **12**, 12 (1968); R. J. von Gutfeld and R. L. Melcher, *ibid.* **30**, 257 (1977).

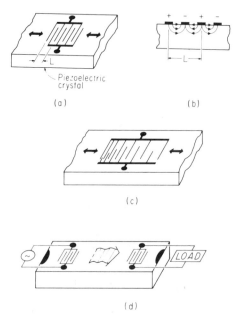

Fig. 6. Interdigital electrode array transducers: (a) array of constant periodic length L on piezoelectric crystal. (b) approximate distribution of electric fields in crystal at one instant; (c) array having variable periodicity for broadband operation and heavier relative weighting of midband frequency components because of the greater degree of finger overlap in the center of the transducer; (d) schematic of a typical SAW filter or delay line.

X-ray detection of surface waves is also possible because passage of the waves modulates both the Bragg angle and the distance between Bragg planes in a crystal.[17] We shall discuss the most useful of these means next.

10.3.1. Interdigital Transducers

The most frequently used transducer for electronic or optical work with surface waves is the interdigital transducer (IDT). The typical IDT consists of aluminum electrodes several thousand angstroms thick evaporated onto the surface of a piezoelectric substrate. For work with non-piezoelectric substrates, an IDT is placed beneath or on top of a piezoelectric thin film, such as ZnO sputtered onto the substrate. The electrodes are usually produced by conventional photolithography, although

[17] O. Berolo and D. Butler, *IEEE Ultrason. Symp. Proc., Phoenix, Arizona*, October 26–28, p. 98 (1977).

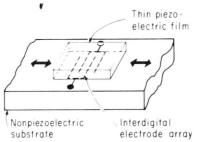

FIG. 7. Thin piezoelectric film with interdigital electrode array for launching or receiving surface waves on nonpiezoelectric substrate. Optimum location of IDT electrodes and an opposing uniform conducting field plate, if used, depends upon thickness of piezoelectric film, as indicated in Fig. 13.

electron beam techniques have been used to make IDTs for gigahertz frequencies.

At the fundamental frequency of operation, the surface wavelength at an input IDT will equal the periodic length L of the electrode array, so that waves generated by the elemental sources at the edges of each electrode interfere constructively, as in an end-fire electromagnetic antenna. At frequencies near the fundamental resonance or near those multiples of it for which constructive interference also occurs (usually the fifth, ninth, etc.), the amplitude of the wave excited by a given driving voltage decreases from its maximum at resonance in a $\sin(nx)/(x)$ fashion, where $x = (kp/2)(1 - f/f_m)$, with n the number of periods in the IDT, p the periodic distance of the transducers, f the operating frequency, and f_m the frequency of the mth harmonic.[18]

The simple IDT launches waves of equal amplitude in both directions on a surface, leading to an unavoidable minimum bidirectional loss of 6

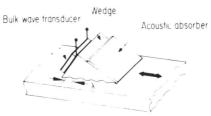

FIG. 8. Wedge transducer. The angle of the wedge is set so that the wavelength of the bulk wave, as measured along the contact surface, is approximately equal to the SAW wavelength.

[18] W. R. Smith, H. M. Gerard, J. H. Collins, T. M. Reeder, and H. J. Shaw, *IEEE Trans. Microwave Theory Tech.* **MTT-17**, 856, 865 (1969).

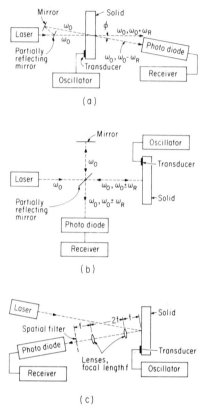

FIG. 9. Optical detection of surface waves: (a) detection in the far zone of the transmitted diffracted light. The reference light beam ω_0 transmitted through the solid mixes with the diffracted frequency-shifted beam ω_R in the photodiode. The diffraction angle depends on wave velocity, so either the surface or bulk waves can be detected; (b) Michelson interferometer arrangement for detection of diffracted light reflected from the solid. The partially reflecting mirror can be replaced with a Bragg cell and the fixed mirror can be vibrated to give phase sensitivity; (c) near-zone scheme with reflected light. The spatial filter is located in the imaged plane of amplitude-modulated light. Not shown are beam chopping and phase-sensitive detection often employed in these experiments.

dB for transduction in and out of a pair of transducers. (Unidirectional transducers can be made but they are fairly narrowband.) To this minimum loss must be added several other losses:

(a) conversion inefficiency resulting from electrical mismatching of transducer to electrical source or load. This loss term is usually minimized by connecting an inductor in series with the IDT to resonate with its electrical capacitance at the frequency of operation. Further, the real

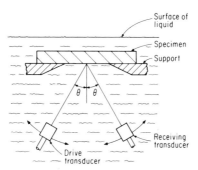

FIG. 10. Mode conversion at the liquid–solid boundary. The pressure wave from the drive transducer is partially converted to the surface elastic wave when set at the angle of incidence $\theta_C = \sin^{-1}(v_{\text{liquid}}/v_{\text{surface}})$. The arrangement shown here was used by Rollins[13] to determine surface wave velocities; the level of the reflected pressure wave at the receiving transducer dropped when the angle for maximum mode conversion was reached.

input resistance at resonance for an IDT is usually set equal to the internal resistance of the source or the load resistance by properly dimensioning the width (aperture) of the IDT to values obtained by computer analysis and listed as M_{opt} in Table IV.

(b) propagation loss in the medium;

(c) resistive losses in the electrodes of the IDT (usually a minor source of inefficiency). For each piezoelectric material an optimum number of array electrode pairs N_{opt} yield the maximum product of the bandwidth of operation times the fraction of power converted at center frequency. These values of N_{opt} are listed in Table IV along with the strength of coupling factor $\Delta v/v$.

Figure 12 shows the insertion loss of a delay line made on YZ–$LiNbO_3$, employing a 19-finger IDT having a 108-wavelength-wide aperture and

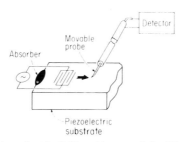

FIG. 11. Coupling to piezoelectric fields with a small flexible tip attached to a coaxial cable (see Richardson and Kino[14]).

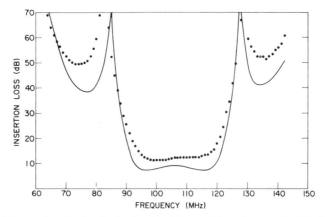

FIG. 12. Measured (points) and calculated (solid curve) insertion loss for YZ–lithium niobate delay line with two interdigital transducers (Smith *et al.*,[18] with permission). A single series inductor was used for tuning.

using tuning inductances on both input and output transducers.[18] Here, by symmetry, equal conversion losses occur at the identical input and output IDTs and account for most of the insertion loss because the propagation loss is fairly low. Variations on the simple IDT to adjust the frequency response include changing interelectrode spacings through the transducer and using different amounts of finger overlap particularly near the ends of the IDT ("apodization"). For more detailed design information, the reader should consult papers on IDT design.[19]

The coupling to surface waves in a piezoelectric layer on a semi-infinite substrate is plotted in Fig. 13 for the particular case of ZnO on semi-insulating silicon.[20] While highest coupling occurs for layers about half a wavelength thick and for the configuration having the interdigital electrodes on the interface and no opposing electrode or field plate, it may be difficult to make such thick films, so a compromise might be made with a thickness around 0.05 wavelengths and use of the configuration with the IDT on top and the continuous metal plate at the interface, corresponding to the smaller relative maximum in coupling.

[19] R. Tancrell, *IEEE Trans. Sonics Ultrason.* **SU-21,** 12 (1974); R. F. Mitchell and D. W. Parker, *Electron. Lett.* **10,** 512 (1974); A. J. Slobodnik, Surface Acoustic Wave Filters at UHF: Design and Analysis, AFCRL-TR-75-0311. Physical Sciences Research Papers No. 634, Air Force Cambridge Research Laboratories, Hanscom AFC, Massachusetts 01731 (3 June 1975); W. R. Smith and W. T. Pedler, *IEEE Trans. Microwave Theory Tech.* **MTT-23,** 853 (1975); W. R. Smith, *Wave Electron,* **2,** 25 (1976); R. F. Mitchell, *ibid.* **2,** 111 (1976); H. A. Matthews (ed.), "Surface Wave Filters." Wiley (Interscience), New York, 1977.
[20] G. S. Kino and R. S. Wagers, *J. Appl. Phys.* **44,** 1480 (1973).

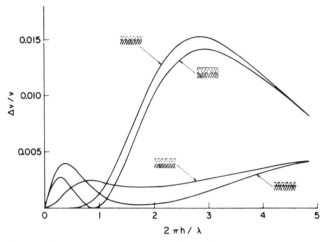

FIG. 13. Coupling factor versus piezoelectric film thickness for ZX-oriented piezoelectric ZnO on semi-insulating, 111-cut silicon (after Kino and Wagers[20] with permission). The propagation direction is $\overline{1}\overline{1}2$. Different locations of IDT electrodes are shown, together with the use of a uniform conducting field plate opposite the IDT electrodes in two of the configurations. The film thickness is h and the wavelength λ.

10.3.2. Wedge Transducers

In the wedge transducer[21] a bulk elastic wave from a piezoelectric source transducer impinges on the bottom surface of the wedge at the angle θ to the surface normal, where $\theta = \sin^{-1}(v_W/v_S)$, with v_W the phase velocity of the bulk wave in the wedge and v_S the surface wave velocity in the underlying solid. Plastic wedges are commonly chosen because of their low bulk wave velocities. The wedge transducer can have a bandwidth as large as that of the bulk wave transducer mounted on it, and the transducers can be used on any type of medium. Calculations[22] predict an overall efficiency as high as 80% of the efficiency of the bulk wave transducer used on the wedge.

A recent variation on the wedge technique involves coupling from a surface wave, rather than a bulk wave, using as couplant a low-velocity substance such as liquid ethylene glycol between a short surface wave delay line and the substrate onto which the waves are to be coupled. When set at the proper velocity-matching angle, efficient coupling through the interposed liquid occurs.[23] For example, a coupling loss

[21] I. A. Viktorov, "Rayleigh and Lamb Waves," p. 8. Plenum, New York, 1967.
[22] H. L. Bertoni and T. Tamir, *IEEE Trans. Sonics Ultrason.* **SU-22**, 415 (1975); H. L. Bertoni, *ibid.* **SU-22**, 421 (1975).
[23] B. T. Khuri-Yakub and G. S. Kino, *Appl. Phys. Lett.* **32**, No. 9, 513 (1978).

from surface wave to surface wave of only 3.5 dB was inferred from measurements at 100 MHz in an experiment in which two LiNbO$_3$ transducers
coupled energy into and out of a sample of Si$_3$N$_4$ ceramic.

10.3.3. Optical Detection

Light interacts with surface waves propagating along a free surface because of local changes in the index of refraction caused by the photo-
elastic effect (strain affects the dielectric permittivity) or, in piezoelectrics,
by the electro-optic effect (electric field affects the dielectric permittivity). Also, as suggested by Fig. 14, surface corrugations cause temporal
variations in the reflection of an optical beam incident upon a surface; the

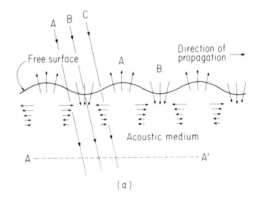

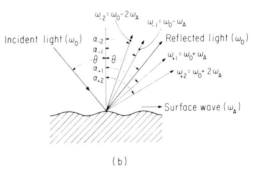

FIG. 14. Optical interaction with surface waves: (a) section through the surface shows the
instantaneous position of the rippled surface and particle motion (represented by short
arrows). Light ray A which passes through more material than ray B in reaching the interior
plane $A–A'$ also passes through the rarefied region underneath the surface and hence these
two diffracting effects tend to oppose each other. (b) light beam incident on the surface at
the angle produces the diffracted orders shown at angles α_i. The subscript O denotes optical frequency and subscript A acoustic wave frequency.

phase of the transmitted light at plane $A-A'$ will also differ locally from that of the incident light owing to the local variations in the amount of the transparent solid through which the beam has passed. The moving phase gratings or mirrors resulting from these three different phenomena cause diffraction and modulation of the light beam and can be used as a basis for the optical detection of surface waves. This subject has been treated thoroughly by Stegeman, both in this volume (Part 9) and in Stegeman.[24]

Stegeman classifies the techniques under the headings simple probing, Fabry–Perot techniques, and heterodyne techniques. In simple probing the amplitude of the ultrasonic wave as a function of position is to be determined. For example, if one wishes to measure the attenuation in a sample, one might probe using a focused laser beam which reflects from the surface over which the surface waves travel.[25] Since the reflected light is simultaneously diffracted and frequency modulated, the intensity of the surface waves may be measured either by using an optical stop to intercept the undiffracted beam and a simple photodetector to determine the intensity of the diffracted beam or by detecting the component modulated at the surface wave frequency in the output current from a fast photodetector. As the angle of diffraction depends on frequency, detectors can be employed at different locations to probe harmonics in nonlinear interactions; Lean and Powell[26] used this technique to measure attenuation caused by nonlinear elastic coupling.

A more sensitive approach involves using the solid on which the waves propagate as one mirror in a Fabry–Perot cavity. Because of the multiple reflections in the cavity, this technique permits measuring much smaller surface displacements than does simple probing, with values as low as 10^{-7} Å being theoretically possible with a 99% reflective surface.[27]

In the heterodyne techniques one employs spatial or temporal interference between optical fields detected by a square law device to produce an electrical output signal from which surface wave amplitude and phase information can be retrieved. Figure 15 shows schematically one heterodyne scheme in which a laser beam scans the surface in synchronism with the scan of a television display.[28] Electrical mixing of the video signal from the photodetector with an electrical signal from the generator driving the surface wave transducer permits phase information to be exhibited as gradations of intensity on the display. With such a display one can visu-

[24] G. I. Stegeman, *IEEE Trans. Sonics Ultrason.* **SU-23**, 33 (1976).
[25] A. J. Slobodnik, P. H. Carr, and A. J. Budreau, *J. Appl. Phys.* **41**, 4380 (1970).
[26] E. G. Lean and C. G. Powell, *Proc. IEEE* **58**, 1939 (1970).
[27] B. J. Hunsinger, *Appl. Opt.* **10**, 390 (1971).
[28] R. Adler, A. Korpel, and P. Desmares, *IEEE Trans. Sonics Ultrason.* **SU-15**, 157 (1968); R. L. Whitman and A. Korpel, *Appl. Opt.* **8**, 1567 (1969).

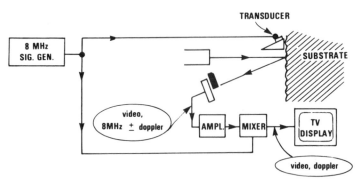

FIG. 15. Schematic diagram of an arrangement for the knife-edge technique for visualizing surface waves [R. Adler, A. Korpel, and P. Desmares, *IEEE Trans. Sonics Ultrason.* **SU-15**, 157 (1968)].

alize a surface wave beam quite clearly and easily observe the influences of surface structures on wave propagation. Commercial equipment for such display is available.[29] Velocity determinations to better than 1 part in 1000 are possible with such means for displaying phase as a function of position.

The advantages of these optical techniques are that the body is not disturbed by the probe as it moves about over the surface studied, the detection sensitivity can be high, absolute measurements of amplitude can be made, and virtually all types of media can be probed by choosing optical transmission or reflection as appropriate.

10.3.4. Electromagnetic Noncontacting Transducers

Useful sending and receiving transducers that need not physically contact the specimen studied can be based on the body forces produced in electrical conductors by the simultaneous presence of a static and a time-varying magnetic field. Figure 16 shows one such means for generating surface waves.[30] Consider first a single wire leg of the flat meanderline coil shown. If the wire carries an alternating current, it will produce an alternating magnetic field in the conductor beneath it. If in addition there is in the conductor a static magnetic field parallel to the surface and to the direction of wave propagation, Lorentz forces on the electrons

---

[29] A. Madeyski and L. W. Kessler, *IEEE Trans. Sonics Ultrason.* **SU-23**, 363 (1976).
[30] R. B. Thompson, *Proc. IEEE Ultrason. Symp., Phoenix, Arizona* p. 74 (1977); C. F. Vasile and R. B. Thompson, *ibid.* p. 84 (1977).

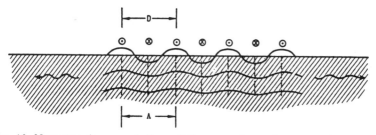

FIG. 16. Noncontacting meanderline coil for generating surface waves in a conducting medium. At the operating frequency, the periodic distance of the electromagnetic transducer ("EMAT") equals the surface wavelength in the medium (Thompson[30] with permission).

will couple to the lattice and act as sources of elastic waves. As with the interdigital transducer, the elementary sources will add coherently if the periodic distance of the transducer equals the elastic wavelength at the frequency of operation.

Analysis of these transducers[30] shows that the efficiency of coupling is proportional to the square of the static magnetic field, which might be produced by an electromagnet or by small permanent magnets made from a lightweight material such as samarium cobalt. The radiation resistance of these electromagnetic transducers (EMAT) is typically only about 50 $\mu\Omega$ and, since this is so small compared with the usual 50-Ω source or load impedances, the conversion loss on generation or reception is generally as large as 60 or 80 dB. The upper operating frequency of 10 MHz or so is set by limits on the attainable small periodicity of the meanderline and by eddy current losses. Nevertheless, the noncontacting feature makes this a useful transducer when waves are to be studied in bodies which have rough surfaces, are moving, or are at cryogenic or elevated temperatures. For example, these transducers have been used in rotation sensing[31] and in the inspection of welds in pipe (see Section 10.4.5).

[31] H. M. Frost, J. C. Sethares, and T. L. Szabo, *J. Appl. Phys.* **48**, 52 (1977).

10.3.5. Comments on Some Experimental Problems

Owing to the finite length of most surface wave transducers, the path length "between" transducers is seldom well defined, making the use of comparison methods with two adjacent identical paths attractive. Direct electromagnetic coupling of the electrical drive signal between input and output transducers is often more severe than with bulk transducers owing to the open nature of surface wave transducers. Keeping the connecting wires short and shaping them so they have a low profile (small mutual capacitance) can minimize this feedthrough. It is also helpful for shielding IDTs to place a broad electrode at ground potential between transducers on which the stray electric fields can terminate.

In piezoelectrics, the incidental excitation of bulk modes along with the surface mode can cause confusion as the bulk waves may reach the output transducer, for example, after reflecting from the bottom of the crystal. An IDT will usually excite bulk waves when operated above its fundamental surface wave frequency; bulk waves may also be launched by drive electric fields extending from top to bottom through the input end of the crystal. These well-understood problems[32] can be alleviated by one or more of the following: avoid establishing electric fields from the IDT to the bottom of the crystal by not putting a ground plane on the bottom of the crystal; roughen the bottom to scatter bulk wave energy; tilt the bottom so bulk waves reflect to the side rather than to the output end of the crystal; and put absorptive material such as room-temperature vulcanizing silicone rubber (RTV) or soft wax on the bottom of the crystal. Surface waves can also produce experimental surprises by propagating over rounded ends of a crystal and returning via the bottom surface. And, as mode coupling theory predicts, a wave launched on top of a crystal can convert quite efficiently to a wave traveling along the bottom surface if the crystal is too thin.[33]

10.4 Surface Wave Applications

Surface waves are used instead of bulk waves where one can exploit the unique features of the waves—the concentration of energy near the surface, the ability to tap or sample them as they propagate, and the existence of a traveling electric field accompanying the wave in piezoelectric media—or because of the ease of transduction with IDTs. In this section we shall discuss five types of applications: nonmeasurement applications,

[32] R. W. Wagers, *Proc. IEEE* **64**, 699 (1976).
[33] I. A. Viktorov, "Rayleigh and Lamb Waves," p. 93ff. Plenum, New York, 1967.

the study of thin films, measuring fields, determining semiconductor properties, and measuring elastic properties at surfaces.

10.4.1. Nonmeasurement Applications

Surface waves have been used to sense the position of a stylus or a human finger on a graphical input tablet for a computer system, and they can move small particles along a surface. The interaction of surface waves with light has interesting applications that will mentioned later, but the richest domain of nonmeasurement applications has been in making electronic surface acoustic wave (SAW) signal-processing devices. In them, an electrical input signal applied to an IDT produces an ultrasonic signal which ultimately propagates to an output IDT. The electrical output differs from the input because of the inherent propagation delay and the waveform or spectrum alterations produced by the input and output transducers and any intervening structures. Since these devices lie outside the subject matter of this volume we merely list the types of devices that have been made successfully to date: delay lines and recirculating memory stores; filters (bandpass, notch, matched, inverse, programmable, pulse compression, comb); oscillators; resonators; coders; convolvers; and correlators.[34]

Returning to optical interactions, the phenomena discussed in Section 10.3.3 can be used to make surface wave optical deflectors, modulators, and tunable filters which have properties comparable with or superior to the corresponding bulk wave devices. With the advent of surface-guided optical beams it has been possible to control dispersion and maintain high power densities over long interaction lengths and so to economize on the surface wave power required to achieve a given level of performance.[35] In the thin-film interactions optical energy can also be converted between TE and TM modes of propagation with surface elastic waves.

Figure 17 shows an arrangement used for optical modulation and deflection.[36] Here an optical beam is coupled into and out of a thin

---

[34] A wide range of SAW devices is discussed in the seven-article series: J. H. Collins and P. M. Grant, *Ultrasonics* **10**, 59 (1972); D. P. Morgan, *ibid.* **11**, 121 (1973); J. D. Maines and J. N. Johnston, *ibid.* **11**, 211 (1973); **12**, 29 (1974); D. P. Morgan, *ibid.* **12**, 74 (1974); M. F. Lewis, *ibid.* **12**, 115 (1974). See also *Proc. IEEE* **64**, Special Issue on Surface Elastic Waves (May 1976) and H. A. Matthews (ed.), "Surface Wave Filters." Wiley (Interscience), New York, 1977. The *Proc. IEEE Ultrason. Symp.* from 1972 through 1978 contain many articles on SAW devices.

[35] E. G. Lean, J. M. White, and C. D. W. Wilkinson, *Proc. IEEE* **64**, 779 (1976).

[36] C. S. Tsai, L. T. Nguyen, S. K. Yao, and M. A. Alhaider, *Appl. Phys. Lett.* **26**, 140 (1975).

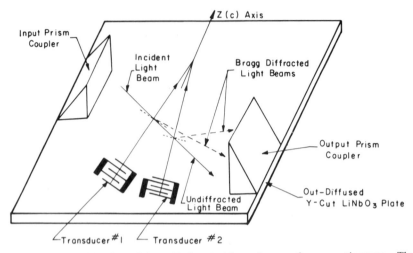

FIG. 17. Bragg diffraction of the guided optical beam by a surface acoustic wave. The two IDTs, having different center frequencies of operation, are tilted so that the Bragg condition is met over a wide range of SAW frequencies (Tsai *et al.*,[39] with permission).

optical-guiding film with prism couplers whose operation is analogous to that of the wedge ultrasonic surface wave transducer. Surface acoustic waves from the IDTs cause periodic changes in the index of refraction and hence partial reflection of the light beam. In order to obtain efficient deflection over a wide range of SAW frequencies, several transducers set at different angles have been used in this example so that the Bragg reflection conditions are satisfied in each portion of the spectrum. With such an arrangement, Tsai *et al.*[36] have reported deflection of a laser beam propagating in an optical waveguide made by titanium outdiffusion from lithium niobate where the 3-dB bandwidth was 175 MHz and only 200 mW electrical drive power was required for 50% diffraction efficiency. Lithium niobate was chosen because of its superior surface wave coupling properties, and yet the performance achieved was comparable with that of bulk wave devices employing materials having higher acousto-optic figures of merit. Schemes for using surface waves to scan optical images have been demonstrated,[37] but these techniques do not appear superior to other evolving solid-state approaches, such as scanning based on the charge-transfer principle.[38]

---

[37] See G. S. Kino, *Proc. IEEE* **64**, 734ff (1976).

[38] C. H. Sequin and M. F. Thompsett, 'Charge Transfer Devices,'' Chapter V. Academic Press, New York, 1975.

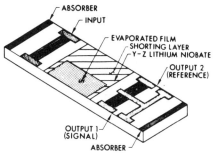

FIG. 18. Surface wave delay line for measurement of the phase shift induced by a thin evaporated film. The reference channel with its shorting layer serves as a comparison channel for reducing the effect of temperature changes (Hemphill[39] with permission).

10.4.2. Properties of Thin Films

Several different properties, or changes of properties, of thin films can be inferred from the measured phase velocity or attenuation of surface waves propagating over a thin film on a massive substrate. A typical arrangement is shown in Fig. 18, where the properties of a thin film being deposited are determined from electrical measurements made at the terminals of the surface wave transducers.[39] On a pulse-to-pulse basis we may determine the delay time or amplitude of an rf signal that is amplitude modulated by a rectangular pulse whose duration is less than the transit time through the delay line shown. Another approach to determining phase velocity is to connect an amplifier from output to input IDT and measure the frequency of the surface wave oscillator thus formed. The device will oscillate at a frequency within the operating frequency bands of the IDTs for which the round-trip phase shift is $2N$, where N is an integer. Since frequencies can be measured with great precision, such an arrangement is convenient for obtaining a value of phase velocity or change of phase velocity. One can also form a high-Q surface wave resonator incorporating the film to be examined within the resonator structure; shifts in the resonant frequency or changes in the electrical impedance measured at the input to a single transducer in such a resonator provide information about the phase velocity of the region. Intrinsic acoustic Qs as high as 25,000 have been obtained at 157 MHz in resonators formed by positioning an IDT between two arrays of 600 parallel reflecting grooves in an ST (surface wave, temperature-stabilized) quartz

[39] R. B. Hemphill, *Proc. IEEE Ultrason. Symp., Boston, Massachusetts* p. 340 (1972); R. B. Hemphill, *Proc. IEEE Ultrason. Symp., Monterey, California* p. 525 (1973).

crystal.[40] With such high Q values, small shifts in resonant frequency are measurable. The oscillator and resonator techniques are similar to bulk wave thickness monitors used in vacuum deposition systems, where the resonant frequency of a bulk crystal oscillator is measured as the medium is deposited on both it and the desired substrate. The surface wave measurement sensitivity can be high because of the very large number of wavelengths in the path.

We may also relate the phase velocity to material properties through computer solutions of the phase velocity of surface waves in layered media, provided the properties of the deposited medium are known, or through a one-time calibration. Values of from 100 to 2000° phase shift at 20 MHz are predicted for 1-μm-thick layers of such metallic evaporants as Al, Cu, and Au.[39]

An interesting variation useful for electrically conducting materials involves monitoring the surface wave attenuation as the film is deposited on a piezoelectric substrate. One observes a maximum attenuation when the electrical conductivity of the layer is such that the product of dielectric relaxation time in it times the angular frequency of the surface wave is roughly unity. This maximum results because for those conditions the piezoelectric fields are most effective at bunching and dragging electrons through the resistive medium and so transferring maximum energy from the wave.[41] For gold and aluminum the attenuation maxima for 20-MHz waves occur for film thicknesses between about 30 and 100 Å.[39] Harnik and Sader[42] have recently employed the SAW oscillator to measure changes of both phase velocity and attenuation in gold films being deposited on lithium niobate, obtaining the product of carrier mobility and Fermi energy with good accuracy.

Examples of using surface waves to study thin-film properties include that of superconducting films of indium[43] and lead,[44] where attenuation measured at 316 and up to 500 MHz, respectively, correlated well with predictions of the Bardeen–Cooper–Schrieffer theory for bulk superconductivity as the temperature was varied to cycle the films between superconducting and normal states. Another example of studies employing attenuation is the work of Feng *et al.*[45] on attenuation at 618 MHz caused by changes in applied magnetic field on 200-Å-thick nickel films. Attenu-

[40] D. T. Bell and R. C. M. Li, *Proc. IEEE* **64**, 711 (1976).

[41] R. A. Adler, *IEEE Trans. Sonics Ultrason.* **SU-18**, 115 (1971).

[42] E. Harnik and E. Sader, *Appl. Phys. Lett.* **33**, No. 12, 979 (1978).

[43] F. Akao, *Phys. Lett.* **409** (1969).

[44] E. Kraetzig, K. Walther, and W. Schulz, *Phys. Lett.* **30A**, 411 (1969).

[45] J. Feng, H. Fredricksen, C. Krischer, M. Tachiki, and M. Levy, *Proc. IEEE Ultrason. Symp. Phoenix, Arizona,* October 26–28, p. 328 (1977).

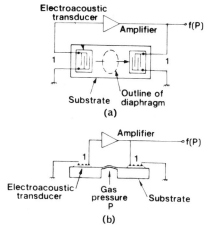

FIG. 19. Top (a) and cutaway side (b) views of SAW diaphragm pressure sensor (Reeder *et al.*[46] with permission). A second channel can be added to provide temperature compensation.

ation changes of 25 dB/cm were produced by magnetic field changes of 100 Oe as a result of the changes in the interaction of the internal magnetic moment of the film with the surface wave.

10.4.3. Sensing Fields

Pressure monitors have been made[46] which employ the change in wave velocity and path length to affect the frequency of a surface wave oscillator. Figure 19 shows a pressure sensor having a 7- or 8-mil-thick quartz region over which the waves propagate. Pressure applied to the underside of the membrane causes it to stretch, changing the path length and hence the oscillation frequency. Changes of from 3 to 15 ppm of the oscillation frequency (80 or 170 MHz) per psi applied have been observed, with a linearity from 0.1 to 2% for pressures in the range 0–20 psi and a temperature sensitivity below 0.3 ppm/°C. The same principle can be applied to sensing temperature, where both the change of elastic properties with temperature (usually weakening of interatomic bonds with increasing temperature) and the change in path (thermal expansion) are significant.

We can imagine employing surface waves to sense magnetic fields (ferromagnetic films) or electric fields (depleting a semiconducting film), but it is likely that existing techniques would be more effective and economical.

[46] T. M. Reeder, D. E. Cullen, and M. Gilden, *Proc. IEEE Ultrason. Symp., Los Angeles, California* p. 26 (1975).

10.4.4. Semiconductor Surface Properties

If a semiconductor is placed close enough to a piezoelectric plate on which a surface wave is propagating, charge carriers in the semiconductor can interact with the wave through the traveling electric field, changing both attenuation and phase velocity of the wave. One can also detect this interaction by observing the so-called acoustoelectric current flowing in the semiconductor, that is, the electric current flow represented by the charge carriers that are dragged along by the electric field of the surface wave in the piezoelectric. Techniques have been developed for obtaining the very close spacings required[47] (the electric field associated with a 200-MHz wave in a medium with phase velocity 3000 m/sec in vacuum decays to 37% of its value at the surface in a distance of only 2.4 μm).

The arrangement used in the surface measurement is shown schematically in Fig. 20.[48] The surface wave is generated and propagates in a lithium niobate plate. A semiconductor, which in this case is a gallium arsenide wafer having an epitaxial GaAs layer on its lower surface, is placed quite close to the lithium niobate. Spacer rails are typically used so that a uniform gap between the semiconductor and the lithium niobate of from 500 Å to a few micrometers is maintained; if the two solids were actually in contact, the wave energy would tend to leave the surface and propagate into the bulk of one crystal or the other. As Fig. 20 suggests, one may shine light into the semiconductor through the transparent lithium niobate to excite electrons and holes optically, and one may apply a dc bias to the contact on the semiconductor to control the density of carriers near its surface. If measurements of acoustoelectric current are also to be made, ohmic contacts are required at the left and right sides of the semiconductor.

The following quantities have been studied by Das, Gilboa, Motamedi, and co-workers[45] using such arrangements:

(1)   location of surface states within the energy band of epitaxial GaAs and in CdS. Monochromatic light of different wavelengths is used to illuminate the semiconductor while attenuation is measured. Attenuation from carriers within a Debye length of the surface exhibits a maximum when the wavelength of the light corresponds to energies of the surface

[47] J. H. Cafarella, W. M. Brown, E. Stern, and J. A. Auslow, *Proc. IEEE* **64,** 756 (1976).
[48] M. E. Motamedi, R. T. Webster, and P. Das, *Proc. IEEE Ultrason. Symp., Los Angeles, California* p. 668 (1975); H. Gilboa, M. E. Motamedi, and P. Das, *Appl. Phys. Lett.* **27,** 641 (1975); H. Gilboa, M. E. Motamedi, and P. Das, *Proc. IEEE Ultrason. Symp., Los Angeles, California* p. 633 (1975); P. Das, M. E. Motamedi, H. Gilboa, and E. T. Webster, *J. Vac. Sci. Technol.* **13,** 948 (1976); M. E. Motamedi, P. Das, and R. Bharat, *Proc. IEEE Ultrason. Symp., Annapolis, Maryland* 205 (1976).

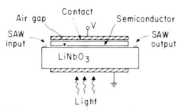

FIG. 20. Sketch of a typical arrangement for measuring the interaction of surface waves in piezoelectrics with the charge carriers in a nearby semiconductor. The bottom electrode on the lithium niobate is transparent to incident light (Motamedi *et al.*,[48] with permission).

states or to band-to-band transitions. These wavelengths for the GaAs studied were 0.86, 1.3, and 1.74 μm, in agreement with independent determinations of the surface state energies;

(2) variation of surface conductivity as a function of surface potential in silicon;

(3) type of carriers and variations due to illumination;

(4) lifetime associated with emission and absorption of carriers by surface traps. The temporal history (decay) of the surface wave attenuation is observed after charge has accumulated at the semiconductor surface in response to the application of a voltage pulse;

(5) Surface state density and charge carrier density at the surface of ion-implanted silicon. These measurements were made at frequencies from 45 to 230 MHz.

The acoustoelectric current measurements[49] are made as indicated in Fig. 20, but ohmic contacts on the left and right ends of the semiconductor permit the current flow through it to be measured. The acoustoelectric current density caused by attenuation of a power density P_0 over length L in a medium having phase velocity v_S and mobility μ is $I_{AE} = \mu P_0 Re(f)/v_S L$, where f represents the fraction of the charge that is free to move (i.e., is not in traps). In this surface wave measurement the surface mobility is obtained directly without need for an independent determination of the carrier density, as would be required if surface conductivity were measured instead.

It is not clear whether these measurement techniques will be adopted widely by workers outside the ultrasonic surface wave community, to whom the techniques appear manageable. It should be noted that although these are in principle noncontacting techniques, it is not easy to

---

[49] J. H. Cafarella, A. Bers, and E. Stern, *Proc. IEEE Ultrason. Symp., Milwaukee, Wisconsin* p. 216 (1974).

achieve the very close and uniform spacings of piezoelectric and semiconductor required by the method.

10.4.5. Elastic Surface Properties

Finally, we turn to the most natural measurement use of surface waves, determining elastic properties near the surfaces of material bodies. Because of the energy concentration near the free boundary, the phase velocity and the attenuation of surface waves are sensitive to elastic conditions at the surface, and because the depth to which particle motion occurs is comparable with a wavelength and so is inversely proportional to the frequency of the wave, measurements made at different frequencies can be used to probe to different depths below the surface and yield information about the gradient of properties near the surface.

Viktorov[50] summarizes the uses of Rayleigh and Lamb waves for detecting surface defects, such as cracks, surface pitting, voids, segregates, and foreign inclusions. Strong reflections and decreases in transmission are observed with surface wave beams incident upon cracks having depths of a wavelength or so, permitting surface inspection of manufactured parts during their lifetimes. Since the Rayleigh waves are confined in one direction by the guiding property of the surface, wave amplitude decreases more slowly than with bulk waves, varying as $R^{1/2}$, where R is the distance from a scattering object. Rayleigh waves propagate over convex surfaces having radii larger than a few wavelengths essentially as they do over plane surfaces, while on concave surfaces of small radius excess attenuation of the waves occurs because of the leakage of energy to bulk waves. Surface roughness is evidenced in measurements over a range of frequencies by the onset of attenuation in excess of ordinary lattice loss starting at a frequency for which the mean surface roughness $k\bar{a} \approx 1$, where $\bar{a}$ is the mean surface roughness.[51]

While surface waves have been used for three decades for surface inspection, only recently has there been a concerted theoretical and experimental effort made to quantify ultrasonic defect evaluation.[52] The methods and analyses of transduction, propagation, and reflections at discontinuities[53] carried out for development of SAW signal-processing

[50] I. A. Viktorov, "Rayleigh and Lamb Waves," pp. 57–65. Plenum, New York, 1967.

[51] R. M. White, *Proc. IEEE* **58**, 1266 (and Fig. 38) (1970).

[52] See annual reports on Interdisciplinary Program for Quantitative Flaw Definition. Rockwell International Science Center, Thousand Oaks, California, Contract No. F33615-74-C5180. Air Force Materials Laboratory, Wright–Patterson AFB, Ohio.

[53] For example, see Z. Alterman and D. Loewenthal, *Geophys. J. R. Astron. Soc.* **20**, 101 (1970); M. Mungsinghe and G. W. Farnell, *J. Appl. Phys.* **44**, 2025 (1973); F. C. Cuozzo, E. L. Cambiaggio, J.-P. Damiano, and E. Rivier, *IEEE Trans. Sonics Ultrason.* **SU-24**, 280 (1977).

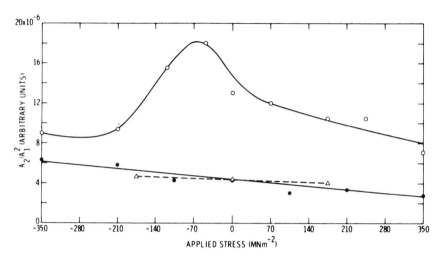

FIG. 21. Test results on the fatiguing of Al 2024 aluminum alloy in the presence of applied static tensile or compressional stress. The ordinate is the ratio of the second harmonic amplitude to the square of the 5-MHz fundamental wave amplitude. The top curve (○) for 5000 fatigue cycles (cracks initiated) shows generally more second harmonic production, with a peak at small values of static applied stress. Also shown are zero fatigue cycle (△) and 1920 fatigue cycles (●) with no cracks initiated (Buck et al.,[54] with permission).

devices contribute to understanding and innovation in this measurement application of surface waves.

One interesting new technique exploits harmonic generation to detect surface cracks. It has been observed that propagation over a surface containing microcracks is linear and little harmonic production is observed if a static applied tensile or compressive stress keeps the cracks either open or closed, but significant harmonic production occurs when a slight compressive stress is applied so that the cracks are open during part of the surface wave period and closed during the rest of it.[54] Figure 21 shows results of tests at 5-MHz fundamental frequency on an Al 2024 aluminum alloy that underwent up to 5000 fatigue cycles, resulting in the production of surface microcracks whose average lengths were approximately 15–20 μm and whose density was around 100 cm^{-2}. The relative amplitude of the 10-MHz second harmonic A_2/A_1^2 rises with the number of fatigue cycles, suggesting that the technique might measure the onset of fatigue and so be useful in predicting end-of-life for manufactured parts.

Gradients of some physical properties that might be reflected in gradients of elastic properties are listed in Table V. Measurements of surface wave velocity at different frequencies yield dispersion data from

[54] O. Buck, W. L. Morris, and J. M. Richardson, *Appl. Phys. Lett.* **33**, No. 5, 371 (1978).

TABLE V. Physical Property Gradients of Technological Importance[a]

Surface treatment	Property gradient	Example
Gas exposure	Hydrogen concentration	Hydrogen embrittlement of storage vessels
	Oxygen concentration	Surface alteration in titanium alloys
Quenching of steel	Microstructure gradient	Bearing surfaces of steel
Deposited layers	Stress gradient	Electroformed metals; prestressed safety glass; electrochemical charging of hydrogen in steel
Mechanical deformation	Cracks and cold work	Fatigue damage; machining damage
Ion implantation	Doping concentration in semiconductors	Boron in silicon
	Amorphous surface layer	Helium in crystalline quartz[b]
Liquid phase epitaxy (LPE)	Concentration gradients	LPE growth of III–IV compounds
Solar winds	Implanted hydrogen	Satellites and space vehicles; lunar rocks

[a] After Tittmann et al.[55]
[b] P. Hartemann, P. Doussineau, and A. Levelut, *Appl. Phys. Lett.* **33**, No. 3, 219 (1978); P. Hartemann, *J. Appl. Phys.* **49**, 5334 (1978).

which one may obtain profiles of elastic property versus depth. Inversion of the dispersion data has been done by a parametric approach, in which possible subsurface profiles are represented by functions containing finite sets of parameters which are adjusted to give the best fit to experimental data,[55] as well as by nonparametric approaches.[56] Figure 22 shows results for a test structure consisting of a Ni–Cu–Ni sandwich formed by electrodeposition on a solid nickel block. Phase–velocity measurements made at different frequencies and hence wavelengths (Fig. 22b) first depart from the value for nickel itself at wavelengths comparable with the depth of the copper layer, as the waves begin to "sense" the presence of the layer having the different elastic properties.

10.5. Summary and Conclusions

Surface waves have played a part in many different sorts of experiments. They are particularly well suited to optical deflection and modu-

[55] B. R. Tittmann, G. A. Alers, R. B. Thompson, and R. A. Young, *Proc. IEEE Ultrason. Symp., Milwaukee, Wisconsin* p. 561 (1974).
[56] J. M. Richardson, *J. Appl. Phys.* **48**, 498 (1977); J. M. Richardson and B. R. Tittmann, *ibid.* **48**, 5111 (1977); B. R. Tittmann and J. M. Richardson, *ibid.* **49**, 5242 (1978).

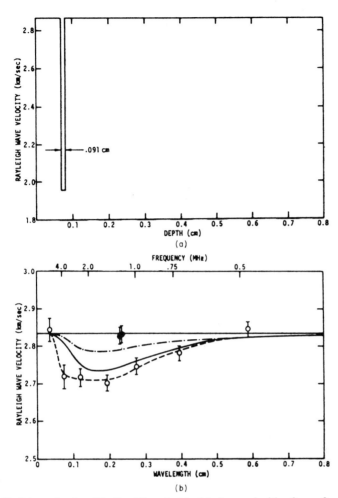

FIG. 22. Dispersion in a Ni–Cu–Ni sandwich: (a) phase velocities for surface waves on bulk nickel ($v_R = 2.835 \times 10^5$ cm/sec) and copper ($v_R = 1.96 \times 10^5$ cm/sec) versus the depths at which those constitutents appear for p = 8.9 gm/cc; (b) experimental data (○) obtained with the sandwich structure, together with the results of an exact theory (—) and a perturbation theory (– · – ·). Solid circles are the surface wave velocity measured on the solid nickel block (Tittman et al.,[55] with permission).

lation of light beams guided by thin dielectric layers, as they provide wideband interaction at frequencies of many hundred megahertz and require very low power to achieve deflection and modulation. There have been many applications in signal-processing devices, an area outside the scope of this volume. Some physical properties of surfaces have been measured with surface waves in a fashion parallel to that done earlier with bulk elastic waves; an example is the study of the superconducting–normal transition of thin, metallic films. In determining the locations and characteristics of cracks, defects, and voids at the surfaces of materials, surface waves are an excellent probe, and operation in a reflection mode or in a nonlinear mode is possible.

In some of these measurements it is convenient to use as a test bed a surface wave electronic device such as a high-Q resonator or a delay-line oscillator as this yields an output frequency which can be measured with great precision.

Transducers for use on piezoelectric insulators can readily be made provided even a modest photoresist facility is available; the upper attainable frequency increases with the quality of the facility. With optical probing, surface motions as small as a thousandth of an angstrom can be measured, albeit with quite a lot of equipment and precise alignment.

As understanding of surface wave technique becomes more widespread, one may expect to see increased use made of these waves because of their unique properties.

11. ACOUSTIC HOLOGRAPHY

By B. P. Hildebrand

11.1. Introduction

11.1.1. Phenomenological Foundation

Holography, a technique for recording and displaying three-dimensional information, is based on two venerable branches of optics: interference and diffraction. The word "hologram" and the technique were invented in 1948 by Dennis Gabor in connection with a problem in electron microscopy. It is noteworthy, however, that holography was first demonstrated in the optical regime, for this shaped both the course of events and the technique.

Ordinarily, an optical event is recorded by focusing an image into a photographic recording device by means of a lens. This is a reasonably good way to capture a scene except that the distance dimension is lost. This is because the film can record only intensity, and distance information is encoded in the phase of the light wave.† If no lens were used, the result would be a nondescript blob of white; the lens performs the task of equalizing the path lengths from a plane in the object space to the plane of the film, resulting in an image of that plane. Other planes, however, do not have their paths to the film equalized and so appear out of focus.

Gabor's stroke of genius came from the realization that there was a way to record phase on a phase-insensitive medium. Furthermore, there was a way to retrieve an image from such a record. The phase could be preserved by the simple expedient of interfering the light scattered or reflected by the object with a coherent background wave. Those parts of the two waves that had traveled paths differing by multiples of one-half wavelength would destructively interfere, leaving unexposed places on the film, while those differing by multiples of a wavelength would leave exposed places. A spherical wave radiated by a point object, when added to a colinear plane wave, would yield a circular interference pattern known as a *Fresnel-zone pattern*. Thus each point on the object would

---

† Of course, distance information can be encoded stereoscopically, as it is in normal vision in which only intensity is recorded also.

METHODS OF EXPERIMENTAL PHYSICS, VOL. 19

FIG. 1. Fresnel-zone pattern generated by interfering a plane ultrasonic wave with a spherical ultrasonic wave. The frequency was 2.5 MHz.

yield such a pattern and the complete record would be a superposition of them. Lateral displacement of points results in lateral displacement of Fresnel zones, and axial displacement of points results in a size change of the Fresnel patterns. Thus each point in object space is coded with its own Fresnel zone pattern in the hologram plane. An example of such a pattern is shown in Fig. 1. Note the ever-decreasing ring separation toward the edge of the pattern.

Retrieval of an image from the hologram can be accomplished by invoking diffraction theory. Huygens's principle states that a wave front can be considered to be the sum of wavelets from point sources placed on a preceding wave front. Thus if a Fresnel-zone pattern is placed in a beam of collimated light, the transparent regions in the pattern can be replaced with point sources. The wave front downstream then can be found by drawing tangents to the wavelets as shown in Fig. 2. Thus a Fresnel-zone pattern is seen to produce a converging beam. Note, however, that a diverging wave can also be drawn; so a hologram produces an image of the object (as a virtual image) situated exactly where the object was and a conjugate image (real). Additional light passes through the hologram undiffracted. A viewer then sees a superposition of light from these three beams.

If the coherent background reference beam is not colinear with the ob-

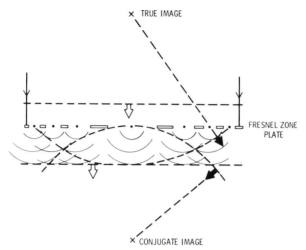

FIG. 2. Huygens's construction of a wave front from a Fresnel-zone pattern.

ject, the Fresnel-zone patterns are decentered. That is, the center of the circular pattern is off the film and thus is not recorded. An example of such a decentered zone pattern is shown in Fig. 3. When such a pattern is subjected to Huygens's principle, focal spots again appear, but now

FIG. 3. Fresnel-zone pattern formed by the interference of a spherical wave and noncolinear plane wave.

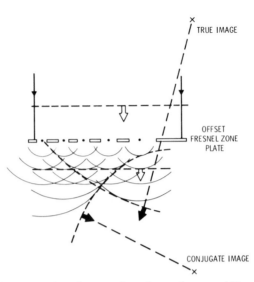

FIG. 4. Huygens's construction of a wave front from a decentered Fresnel-zone pattern.

they appear in positions such that the beams do not superpose, as shown in Fig. 4.

In this brief description we have attempted to convey the physical concepts on which holography is based, how strongly it is based on optics, and the possibility of measuring phase by interference. In the long-wavelength regime of acoustics, electronic methods for measuring phase have always been available. However, since humans see by light, holography was developed by people working in optics even though it would have been easier in the ultrasonic regime.

11.1.2. Historical Survey

As mentioned in Section 11.1.1, holography is a synthesis of interference and diffraction theory. Of these two subjects of classical optics, diffraction is the oldest, having first been discussed by Grimaldi in 1665.[1] Huygens put Grimaldi's theory on a firm mathematical foundation in 1678.[2] Young first demonstrated interference and proposed the wave theory to explain his findings.[3] Fresnel, in 1816, explained Young's inter-

[1] F. M. Grimaldi, "Physico-Mathesis de lumine, coloribus, et iride." Bologna, 1665.

[2] Chr. Huygens, "Traité de la lumière." (completed 1678, published in Leyden, 1690).

[3] Th. Young, *Philos. Trans. R. Soc. London* 12, 387 (1802); "Young's Works," Vol. 1, p. 202.

ference and Huygens's diffraction by means of wave theory, thus forcing the corpuscular theory to be abandoned.[4]

Interferometry rapidly became a major tool for precision measurement. Michelson, for example, used it to disprove the ether drift theory.[5] It was also used to test for imperfections in lenses and mirrors. It is the latter application that comes closest to holography.

In the late 1940s Gabor proposed the two-step process of imagery described in Section 11.1.1.[6] He proposed dispensing with the electromagnetic lens in the electron microscope by recording the electron diffraction pattern of the object and then using it to reconstruct an optical image. In this way he could avoid the large aberrations present in electromagnetic lenses of the time. Although this application was never pursued to its conclusion, Gabor's demonstration of holography triggered worldwide interest in the scientific community.

As indicated in Section 11.1.1, Gabor's holography resulted in two images, a virtual "true" and a real "conjugate" image. In addition, much of the incident light passes through the hologram undiffracted. Thus, as seen in Fig. 2, any attempt to view or record the true image will have to contend with undesired light. Many early researchers attempted to correct this defect with varying degrees of success.[7] In 1962 Leith, using his experience in communication theory, succeeded in separating the three beams by the simple expedient of imposing an angular separation between the light diffracted or reflected by the object and the background wave.[8] As shown in Fig. 3, this results in complete separation of the true and conjugate images and the reconstruction beam. The fortuitous invention of the laser at the same time provided the perfect coherent source that allowed truly spectacular holograms to be made. It was at this stage (1964) that holography went public. A great deal of excitement was generated by these spectacular displays, producing an immediate surge in research and a search for applications.

Since Gabor's original intent was to use holography as a means of visualizing nonvisible radiation, it is not surprising that this application received immediate attention. Holograms have been made with x rays, microwaves, electron beams, and ultrasound. It is the latter radiation with

[4] A. J. Fresnel, *Ann. Chim. Phys.* 1(2), 239 (1816); "Oeuvres," Vol. 1, pp. 89, 129.

[5] A. A. Michelson, "Light Waves and Their Uses." Univ. of Chicago Press, Chicago, Illinois, 1902.

[6] D. Gabor, A new microscope principle, *Nature (London)* **161**, (1948).

[7] A. Lohmann, Optical single-sideband transmission applied to the Gabor microscope, *Opt. Acta* **3**, 97 (1956).

[8] E. N. Leith and J. Upatnieks, Reconstructed wavefronts and communication theory, *J. Opt. Soc. Am.* **52**, 1123 (1962).

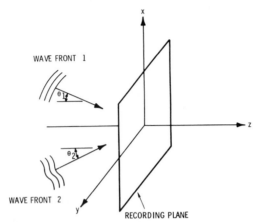

FIG. 5. Recording an optical hologram by interference.

which we are concerned in this part. The first published reference to ultrasonic holography is due to Greguss[9] in 1965, followed closely by Thurstone[10] in 1966. The main problem in developing this technique is the lack of a sensitive area detector, such as film. However, the traditional types of ultrasonic detectors have properties that make for very sensitive, convenient forms of holography not possible with intensity sensors.

11.2. Fundamental Concepts

11.2.1. Hologram Generation

Consider two beams of radiation intersecting on a plane as shown in Fig. 5. They are derived from the same source by some kind of interferometric arrangement as discussed in Chapter 11.1 or, in the case of acoustics, from an oscillator driving two transducers. The radiation at a point (x, y) on the detection plane can be described by the equation

$$S(x, y) = S_1(x, y) + S_2(x, y), \qquad (11.2.1)$$

where

$$S_1(x, y) = A_1(x, y) \cos[\omega t + \theta_1(x, y)],$$

$$S_2(x, y) = A_2(x, y) \cos[\omega t + \theta_2(x, y)],$$

[9] P. Greguss, Ultrasonics hologram, *Res, Film,* **5**(4), (1965).
[10] F. L. Thurstone, Ultrasound holography and visual reconstruction, *Proc. Symp. Biomed. Eng.* **1**, 12 (1966).

$A(x, y)$ is the amplitude, $\theta(x, y)$ the phase, and ω the radian frequency of the wave.

For optical radiation the detector is phase insensitive because of the extremely high frequency. Thus the recorded quantity is

$$I(x, y) = \langle [S(x, y)]^2 \rangle_t \qquad (11.2.2)$$

where $\langle \; \rangle_t$ denotes a time average.

Substitution of Eq. (11.2.1) into Eq. (11.2.2) yields

$$I(x, y) = S_1^2 + S_2^2 + (S_1 S_2)_+ + (S_1 S_2)_-, \qquad (11.2.3)$$

where

$$S_1^2 = \tfrac{1}{2} A_1^2(x, y)\langle \{1 + \cos 2[\omega t + \theta_1(x, y)]\} \rangle_t,$$

$$S_2^2 = \tfrac{1}{2} A_2^2(x, y)\langle \{1 + \cos 2[\omega t + \theta_2(x, y)]\} \rangle_t,$$

$$(S_1 S_2)_+ = \tfrac{1}{2} A_1(x, y) A_2(x, y) \langle \cos[2\omega t + \theta_1(x, y) + \theta_2(x, y)] \rangle_t,$$

$$(S_1 S_2)_- = \tfrac{1}{2} A_1(x, y) A_2(x, y) \langle \cos[\theta_1(x, y) - \theta_2(x, y)] \rangle_t.$$

Clearly, those terms containing ωt will average to zero, leaving

$$I(x, y) = \tfrac{1}{2}\{A_1^2(x, y) + A_2^2(x, y)\}$$
$$+ A_1(x, y) A_2(x, y) \cos[\theta_1(x, y) - \theta_2(x, y)]. \qquad (11.2.4)$$

The significance of this equation lies in the third term; note that the phase of both beams has been preserved. If either of the beams had not been present, all reference to phase and, hence, shape of the wave front would have been lost. This is a remarkable accomplishment in optics but not in acoustics. There are any number of ways this can be done electronically at longer wavelengths. For example, consider the system shown in Fig. 6. The signal reflected by the object and received at the detection plane is again

$$S_1(x, y) = A_1(x, y) \cos[\omega t + \theta_1(x, y)]. \qquad (11.2.5)$$

According to Fig. 6, this signal is multiplied by a reference signal derived from the oscillator to yield

$$S'(x, y) = A_2 A_1(x, y)\{\cos[2\omega t + \theta_1(x, y)] + \cos[\theta_1(x, y) - \theta_2]\}, \qquad (11.2.6)$$

where A_2 is the constant amplitude and θ_2 the constant phase of the reference signal. The low-pass filter rejects the oscillatory part of the signal, leaving

$$S''(x, y) = A_2 A_1(x, y) \cos[\theta_1(x, y) - \theta_2]. \qquad (11.2.7)$$

Thus phase is preserved without the requirement of an acoustic reference signal (but with the requirement of an electronic reference signal).

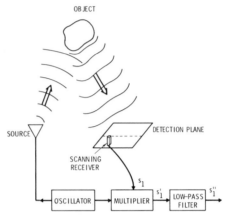

OBJECT

SOURCE

DETECTION PLANE

SCANNING
RECEIVER

S_1

OSCILLATOR — MULTIPLIER $\frac{S_1'}{}$ LOW-PASS FILTER $\frac{S_1''}{}$

FIG. 6. Recording an acoustical hologram by electronic phase detection.

11.2.2. Wave Reconstruction

In Chapter 11.1 we described how the wave front could be reconstructed from the hologram by diffraction. Suppose the hologram described by Eq. (11.2.4) had been written on a photographic plate and was illuminated by the wave front described by Beam 2 in Fig. 5:

$$S_2(x, y) = A_2(x, y) \cos[\omega t + \theta_2(x, y)]. \qquad (11.2.8)$$

After propagating through the hologram, the wave front becomes

$$I(x, y)S_2(x, y) = \tfrac{1}{2}\{A_1^2(x, y) + A_2^2(x, y)\}A_2(x, y) \cos[\omega t + \theta_2(x, y)]$$
$$+ \tfrac{1}{4}A_1(x, y)A_2^2(x, y) \cos[\omega t + \theta_1(x, y)]$$
$$+ \tfrac{1}{4}A_1(x, y)A_2^2(x, y) \cos[\omega t - \theta_1(x, y) + \theta_2(x, y)]. $$
$$(11.2.9)$$

The first term is basically the illumination beam modified in amplitude. The second term replicates Beam 1 and thus will produce an image of the object giving rise to it. The last term contains a phase conjugate to Beam 1 with twice the phase of the reconstruction beam.

If we use a reconstruction beam with a phase conjugate to that used previously, that is, $A_2(x, y) \cos[\omega t - \theta_2(x, y)]$, the second term becomes

$$\tfrac{1}{4}A_1(x, y)A_2^2(x, y) \cos[\omega t + \theta_1(x, y) - 2\theta_2(x, y)] \qquad (11.2.10)$$

and the third

$$\tfrac{1}{4}A_1(x, y)A_2^2(x, y) \cos[\omega t - \theta_1(x, y)].$$

Consequently, a conjugate wave front is reconstructed. This means that if $A_1(x, y) \cos[\omega t + \theta_1(x, y)]$ is a diverging wave, the conjugate wave will be converging.

The influence of an off-axis reference beam can be demonstrated by expressing the general phase terms θ_1, θ_2 as deviations from a plane wave; that is,

$$\theta_1(x, y) = \alpha_1 x + \Delta\theta_1(x, y), \qquad \theta_2(x, y) = \alpha_2 x + \Delta\theta_2(x, y) \quad (11.2.11)$$

where $\alpha_1 = (\sin \phi_1)/\lambda$, $\alpha_2 = (\sin \phi_2)/\lambda$, and ϕ_1, ϕ_2 are the directions of propagation.

Substituting Eq. (11.2.11) into Eq. (11.2.10) yields the result

$$
\begin{aligned}
I(x, y)S_2(x, y) = \ & \tfrac{1}{2}\{A_1{}^2(x, y) + A_2{}^2(x, y)\}A_2(x, y) \cos[\omega t + \alpha_2 x \\
& + \Delta\theta_2(x, y)] \\
& + \tfrac{1}{4}\{A_1(x, y)A_2{}^2(x, y)\} \cos[\omega t + \alpha_1 x + \Delta\theta_1(x, y)] \\
& + \tfrac{1}{4}A_1(x, y)A_2{}^2(x, y)\{\cos[\omega t + (2\alpha_2 - \alpha_1)x - \Delta\theta_1(x, y) \\
& + 2\,\Delta\theta_2(x, y)]\}.
\end{aligned}
\quad (11.2.12)
$$

This form shows clearly that the three beams are separated in space. The first term represents the undisturbed reconstruction beam propagating at the angle ϕ_2; the second term is a replica of the object beam propagating at the angle ϕ_1; and the third term represents a distorted conjugate object

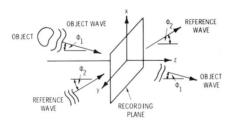

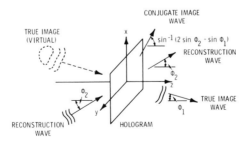

FIG. 7. Recording and reconstructing an off-axis hologram. Note the complete separation of diffracted waves.

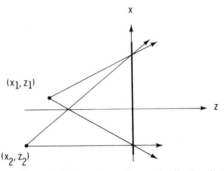

FIG. 8. Geometry for calculation of the phase distribution in the hologram.

beam propagating at $\sin^{-1}[2 \sin \phi_2 - \sin \phi_1]$, which, for small angles, becomes $2\phi_1 - \phi_2$. Hence the two beams bearing information about the object are diffracted symmetrically about the reconstruction beam. Figure 7 provides a schematic representation of this discussion.

11.2.3. Image Parameters

The preceding analysis provides a qualitative explanation of holographic imaging. It does not, however, provide quantitative answers to image parameters. For this, we launch into the type of analysis used for many years for determining image parameters of lens systems. Consider the system shown in Fig. 8. The phase on the hologram plane may be written

$$\theta_0(x) = \theta_1(x) - \theta_2(x) = (2\pi/\lambda_1)\{[(x - x_1)^2 + z_1^2]^{1/2} - [(x - x_2)^2 + z_2^2]^{1/2}\}. \tag{11.2.13}$$

This phase term is contained in the hologram as well as its conjugate. What must be done now is to consider the reconstruction system shown in Fig. 9. The phase of the wave after passing through the hologram will be needed:

$$\theta_i(x, y) = \pm \theta_0(x, y) - \theta_3(x, y)$$

$$= \pm (2\pi/\lambda_1)\{[(x - x_1)^2 + z_1^2]^{1/2} - [(x - x_2)^2 + z_2^2]^{1/2}\}$$

$$- (2\pi/\lambda_2)[(x - x_3)^2 + z_3^2]^{1/2}. \tag{11.2.14}$$

11.2.3.1. Image Position. In order for $\theta_i(x, y)$ to be representative of an image of the point (x_1, z_1), it must be expressed as

$$\theta_i(x, y) = (2\pi/\lambda_2)[(x - x_i)^2 + z_i^2]^{1/2}. \tag{11.2.15}$$

In order to equate Eqs. (11.2.14) and (11.2.15) it is necessary to expand

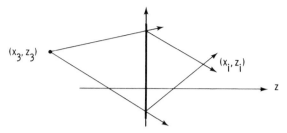

FIG. 9. Geometry for calculation of the parameters of the hologram image.

the square roots in the binomial series and equate terms:

$$\frac{2\pi}{\lambda_2}\left\{z_i + \frac{(x - x_i)^2}{2z_i} + \frac{(x - x_i)^4}{8z_i^3} + \cdots\right\}$$

$$= \pm\frac{2\pi}{\lambda_1}\left\{z_1 + \frac{(x - x_1)^2}{2z_1^2} + \frac{(x - x_1)^4}{8z_1^3} + \cdots\right.$$

$$\left. - z_2 - \frac{(x - x_2)^2}{2z_2} - \frac{(x - x_2)^4}{8z_2^3} + \cdots\right\}$$

$$- \frac{2\pi}{\lambda_2}\left\{z_3 + \frac{(x - x_3)^2}{2z_3} + \frac{(x - x_3)^4}{8z_3^3} + \cdots\right\}. \qquad (11.2.16)$$

The procedure used in lens imaging analysis is to compute imaging parameters by using the second-order terms and aberrations by using the third-order terms. Equating the coefficients of x^2 from the second-order terms, we obtain

$$\frac{1}{\lambda_2 z_i} = \pm\left(\frac{1}{\lambda_1 z_1} - \frac{1}{\lambda_1 z_2}\right) - \frac{1}{\lambda_2 z_3}, \qquad \frac{1}{z_i} = \pm\frac{\lambda_2}{\lambda_1}\left(\frac{1}{z_1} - \frac{1}{z_2}\right) - \frac{1}{z_3}.$$
$$(11.2.17)$$

This equation then provides the distance to both the true and conjugate images, the true being defined by the lower sign and the conjugate by the upper. Equating the coefficients of x yields

$$-\frac{x_i}{\lambda_2 z_i} = \pm\left(\frac{x_1}{\lambda_1 z_1} - \frac{x_2}{\lambda_1 z_2}\right) - \frac{x_3}{\lambda_2 z_3}, \qquad \frac{x_i}{z_i} = \pm\frac{\lambda_2}{\lambda_1}\left(\frac{x_1}{z_1} - \frac{x_2}{z_2}\right) - \frac{x_3}{z_3}.$$
$$(11.2.18)$$

This equation yields the lateral position of the image point. Thus Eqs. (11.2.17) and (11.2.18) provide image coordinates. In the terminology of optics, this is the position of the Gaussian image.

11.2.3.2. Aberrations. Third-order aberrations are obtained by combining all third-order terms in Eq. (11.2.16). Coefficients of x^4 represent spherical, of x^3 coma, of x^2 astigmatism, and of x field curvature.

Spherical:
$$\pm \frac{\lambda_2}{\lambda_1}\left(\frac{1}{z_1{}^3} - \frac{1}{z_2{}^3}\right) + \frac{1}{z_3{}^3} + \frac{1}{z_i{}^3},$$

Coma:
$$\pm \frac{\lambda_2}{\lambda_1}\left(\frac{x_1}{z_1{}^3} - \frac{x_2}{z_2{}^3}\right) + \frac{x_3}{z_3{}^3} + \frac{x_i}{z_i{}^3},$$

$$\pm \frac{\lambda_2}{\lambda_1}\left(\frac{y_1}{z_1{}^3} - \frac{y_2}{z_2{}^3}\right) + \frac{y_3}{z_3{}^2} + \frac{y_i}{z_i{}^3},$$

(11.2.19)

Astigmatism:
$$\pm \frac{\lambda_2}{\lambda_1}\left(\frac{x_1{}^2}{z_1{}^3} - \frac{x_2{}^2}{z_2{}^3}\right) + \frac{x_3{}^2}{z_3{}^3} + \frac{x_i{}^2}{z_i{}^3},$$

$$\pm \frac{\lambda_2}{\lambda_1}\left(\frac{y_1{}^2}{z_1{}^3} - \frac{y_2{}^2}{z_2{}^3}\right) + \frac{y_3{}^2}{z_3{}^2} + \frac{y_i{}^2}{z_i{}^2},$$

Field curvature:
$$\pm \frac{\lambda_2}{\lambda_1}\left(\frac{x_1 y_1}{z_1{}^3} - \frac{x_2 y_2}{z_2{}^3}\right) + \frac{x_3 y_3}{z_3{}^2} + \frac{x_i y_i}{z_i{}^3}.$$

11.2.3.3. Magnification. Magnification is defined as the ratio of the incremental change in image position to object position. That is, for lateral magnification

$$M_L = \partial x_i / \partial x_1 = \pm(\lambda_2/\lambda_1)z_i/z_1; \qquad (11.2.20)$$

for axial magnification

$$M_z = \partial z_i / \partial z_1 = \pm(\lambda_2/\lambda_1)(z_i/z_1)^2. \qquad (11.2.21)$$

Inspecting Eqs. (11.2.20) and (11.2.21) results in the expression

$$M_z = \pm(\lambda_1/\lambda_2)M_L{}^2. \qquad (11.2.22)$$

Equation (11.2.22) is the holographic equivalent to Maxwell's elongation formula for a lens.

11.2.3.4. Resolution. Resolution is the measure of the ability of an imaging system to distinguish between two adjacent points in an object. According to Hildebrand and Brenden, a simple way of computing the resolving capability of a hologram is to calculate how far the object point must be moved in order to change the number of fringes on the hologram by one.[11] This has been shown to be equivalent to the Rayleigh criterion.

From Eq. (11.2.13) we have the phase on the hologram plane

$$\theta_0(x) = (2\pi/\lambda_1)\{[(x - x_1)^2 + z_1{}^2]^{1/2} - [(x - x_2)^2 + z_2{}^2]^{1/2}\}.$$

---

[11] B. P. Hildebrand and B. B. Brenden, "An Introduction to Acoustical Holography," p. 47. Plenum, New York, 1972.

When the binomial expansion of both terms is made, and only the first two terms retained, this becomes

$$\theta_0(x) \cong \frac{2\pi}{\lambda_1} \left\{ z_1 - z_2 + \frac{x^2}{2}\left(\frac{1}{z_1} - \frac{1}{z_2}\right) - x\left(\frac{x_1}{z_1} - \frac{x_2}{z_2}\right) \right\}. \quad (11.2.23)$$

The phase at the extreme edges of the hologram is

$$\theta_0\left(+\frac{A}{2}\right) \cong \frac{2\pi}{\lambda_1} \left\{ (z_1 - z_2) + \frac{(A/2)^2}{2}\left(\frac{1}{z_1} - \frac{1}{z_2}\right) - \frac{A}{2}\left(\frac{x_1}{z_1} - \frac{x_2}{z_2}\right) \right\},$$

$$(11.2.24)$$

$$\theta_0\left(-\frac{A}{2}\right) \cong \frac{2\pi}{\lambda_1} \left\{ (z_1 - z_2) + \frac{(A/2)^2}{2}\left(\frac{1}{z_1} - \frac{1}{z_2}\right) + \frac{A}{2}\left(\frac{x_1}{z_1} - \frac{x_2}{z_2}\right) \right\},$$

where $\acute{A}$ is the width of the hologram. The difference between the two is

$$\delta\theta_0 = \frac{2\pi}{\lambda_1} A\left(\frac{x_1}{z_1} - \frac{x_2}{z_2}\right). \quad (11.2.25)$$

The incremental change of $\delta\theta_0$ due to an incremental change in x_1 is

$$\Delta\delta\theta_0 = (\partial\delta\theta_0/\partial x_1)\,\Delta x_1 = (2\pi A/\lambda_1 z_1)\,\Delta x_1. \quad (11.2.26)$$

Setting this quantity to 2π yields

$$\Delta x_1 = \lambda_1 z_1 / A. \quad (11.2.27)$$

Similarly,

$$\Delta z_1 = \lambda_1 z_1{}^2 / A x_1. \quad (11.2.28)$$

In order for the image to be outside the aperture we have $x_1 = A/2$. Hence

$$\Delta z_1 = 2\lambda(z_1/A)^2. \quad (11.2.29)$$

Often, in ultrasonic holography, the hologram itself is a different size than the area over which the data are taken. That is, a magnification has occurred. In this case the equations are altered as follows:

$$\frac{1}{z_i} = \pm \frac{\lambda_2}{m^2\lambda_1}\left(\frac{1}{z_1} - \frac{1}{z_2}\right) - \frac{1}{z_3},$$

$$\frac{x_i}{z_i} = \pm \frac{\lambda_2}{m\lambda_1}\left(\frac{x_1}{z_1} - \frac{x_2}{z_2}\right) - \frac{x_3}{z_3},$$

$$(11.2.30)$$

$$M_L = \pm \frac{\lambda_2}{m\lambda_1}\frac{z_i}{z_1},$$

$$M_z = \pm \frac{\lambda_2}{m^2\lambda_1}\left(\frac{z_i}{z_1}\right)^2,$$

where m is the magnification. Aberration expressions are similarly changed.[12] Resolution, of course, is unchanged.

11.3. Implementation

There are two basic methods for making an ultrasonic hologram, differentiated by the method of detection. The two methods also dictate the way in which the image is displayed.

11.3.1. Area Detector

The first method to be considered here is also the first practical method to be demonstrated.[13] It is the exact analog of optical holography in that two acoustical beams and an area detector are used. The area detector is the surface of a liquid. The two beams are directed to intersect at the surface where radiation pressure levitates the surface at those places where constructive interference takes place. Thus the surface becomes a hologram embossed on the surface of the liquid. A laser beam can then be reflected from the surface. The first-order diffraction from the surface constitutes the reconstructed object. Figure 10 shows a schematic diagram

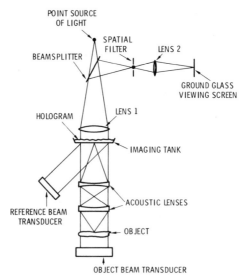

FIG. 10. Liquid surface levitation holography system.

[12] B. P. Hildebrand and B. B. Brenden, "An Introduction to Acoustical Holography," p. 38. Plenum, New York, 1972.
[13] R. K. Mueller and N. K. Sheridan, *Appl. Phys. Lett.* **9,** 328–329 (1966).

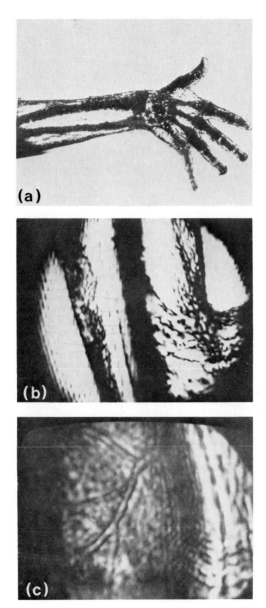

FIG. 11. Three examples of imaging with the liquid surface holography system: (a) a human hand and forearm showing details of bone structure; (b) the attachment of tendons to the humerus; and (c) branching vascular structure in the biceps.

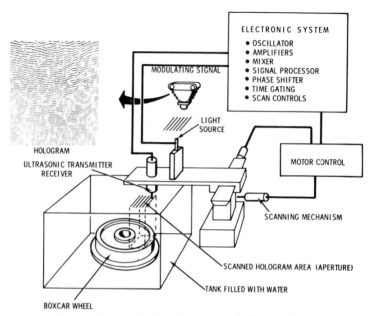

ELECTRONIC SYSTEM

• OSCILLATOR
• AMPLIFIERS
• MIXER
• SIGNAL PROCESSOR
• PHASE SHIFTER
• TIME GATING
• SCAN CONTROLS

MODULATING SIGNAL

LIGHT SOURCE

HOLOGRAM

ULTRASONIC TRANSMITTER RECEIVER

MOTOR CONTROL

SCANNING MECHANISM

SCANNED HOLOGRAM AREA (APERTURE)

TANK FILLED WITH WATER

BOXCAR WHEEL

FIG. 12. Scanned holography using a single transducer.

of such a system.   Note the ultrasonic lens in the object beam.   For certain practical reasons related to the power required to levitate the surface, the object is imaged into the surface, thus forming a so-called "focused" image hologram.[14]

This system operates in real time since the surface responds immediately.   The sound is pulsed at the rate of 60–100 Hz, thus providing motion capability.   A closed-circuit television camera and monitor are used as the display device.   This system showed great promise for medical imaging purposes.   However, for reasons related to the delicacy of adjustments required to keep the liquid surface leveled and vibration free, and the rapid developments in pulse-echo and array techniques, it seems to have been passed by.

This system was one of the first ultrasonic imaging devices to exhibit operation in real time.   It caused great initial excitement, and subsequent improvements have resulted in remarkably good images, as shown in Fig. 11.   It remains to be seen whether this technique will be a commercial success.

---

[14] B. P. Hildebrand and B. B. Brenden, "An Introduction to Acoustical Holography," p. 28.   Plenum, New York, 1972.

11.3.2. Sampling Detectors

The implementation that appears to have gained wide acceptance uses the well-developed technology of piezoelectric devices. Current holographic devices use a single transducer and mechanical scanning as shown in Fig. 12. The mechanical scanner transports the transducer in a series of translated linear scans. The transducer is pulsed at appropriate intervals; the echo is received and processed by a phase-sensitive detector. As discussed in Section 11.2.1 this method eliminates the need for a physical reference beam. The processed signal modulates a synchronously scanned light source, which is photographed to form the hologram. The hologram is then inserted into an optical system as shown in Fig. 13. The optical beam is usually made to be converging. The first-order diffracted beams will then focus to an image that may be photographed or cast onto a video camera for display on a monitor.

This description indicates that a single transducer is used both as the source and the receiver. This is a natural extension of ultrasonic holography stemming from long experience in pulse-echo work in nondestructive testing. It is also common practice in synthetic-aperture side-looking radar systems. This technique brings with it certain advantages, not the least of which is improved resolution. An analysis, similar to that outlined in Section 11.2.3 yields the following image parameter equations:

$$\frac{1}{z_i} = \pm \frac{\lambda_2}{m^2 \lambda_1} \left(\frac{2}{z_1} - \frac{1}{z_2} \right) - \frac{1}{z_3},$$

$$\frac{x_i}{z_i} = \pm \frac{\lambda_2}{m \lambda_1} \left(\frac{2x_1}{z_1} - \frac{x_2}{z_2} \right) - \frac{x_3}{z_3},$$

$$M_{\mathrm{L}} = \pm \frac{2\lambda_2}{m \lambda_1} \frac{z_i}{z_1},$$

$$M_z = \pm \frac{2\lambda_2}{m^2 \lambda_1} \left(\frac{z_i}{z_1} \right)^2, \qquad (11.3.1)$$

$$\Delta x_1 = \lambda_1 \frac{z_1}{2A},$$

$$\Delta z_1 = \frac{\lambda_1}{2} \left(\frac{z_1}{A} \right)^2.$$

Note that the image appears closer to the hologram by a factor of 2 and has doubled resolution. These factors of 2 result from changing the distances of both the source and the receiver. The rate of change of phase across the aperture is twice that seen when only the receiver moves.

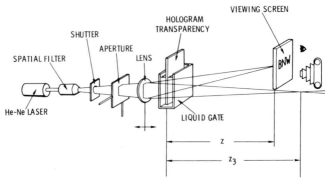

FIG. 13. Optical system for obtaining the image from a scanned hologram.

Thus the convenience of using a single transducer has brought with it significant advantages.

The advantage of an imaging system is that quantitative measurements on the image can be related to the dimensions of the object being imaged. Therefore, we now rearrange Eqs. (11.3.1) to reflect the actual geometry used in the optical system shown in Fig. 13. For practical reasons, the video camera will remain at a fixed distance from the hologram. In order to focus the image on the camera, then, the lens is moved, thereby moving the image. The distance z_2 is infinity since the electronic references simulates a plane wave. Hence, the equation for z_i can be written

$$\frac{1}{z_i} = \pm \frac{2\lambda_2}{m^2\lambda_1 z_1} - \frac{1}{z_3} = \frac{1}{z},$$

where z is the distance from hologram to camera.

Solving for z_1 we have

$$z_1 = \pm \frac{2\lambda_2}{m^2\lambda_1}\left(\frac{1}{z_3} + \frac{1}{z}\right)^{-1}. \tag{11.3.2}$$

Thus we move the lens until the image is in focus, measure the distance z_3, and solve for the object distance z_1. To find the size of the object we must work with the third of Eqs. (11.3.1):

$$M_L = m(1 + z/z_3). \tag{11.3.3}$$

In Eqs. (11.3.2) and (11.3.3) it must be remembered that z_3 is negative because the light beam is converging to a point to the right of the hologram. This was set up in the original equations according to the convention shown in Figs. 8 and 9, where the distance coordinate of the object

(z_1), reference (z_2), and reconstruction (z_3) points are considered positive to the left of the hologram plane and the image (z_i) point is positive to the right of the hologram. Using Eq. (11.3.3) we can estimate the size of the object from a measurement of the image.

The diffractive element in this case is the hologram written on film. This, of course, causes a delay between the time of scan and display of the image. Polaroid film is generally used, so the delay is no more than 2 or 3 min. However, mechanical scanning of a single transducer element is a rather slow process, usually in the neighborhood of 10–15 min. Piezoelectric arrays allow rapid electronic scanning of the hologram.[15] For example, a linear array electronically scanned along its length and mechanically translated can scan a 15×15 cm aperture in 50 sec (120×120 points in the aperture). In this case, film becomes the limiting component in the imaging system. A number of electro-optic light valves, such as the Xerox Ruticon, Itek PROM, and Hughes liquid crystal can replace film. Those light valves are electrically or optically addressable so the hologram signal can be written on them in real time. In this case, then, the image can be seen in real time but with a 50-sec time delay; that is, as the array translates to a new position, the hologram can be rolled up on the light valve and the first line deleted. Thus a continuously updated hologram and image can be displayed 50 sec behind the scanner.

The type of holography just described is essentially a sampling system. Single or multiple transducers are moved across an aperture and individual measurements of complex amplitude made at specific points in that aperture. This constitutes a sampled aperture and, as such, must obey the rules of sampling theory as described in many texts on radar theory. The major characteristic of a sampled hologram is a multiplicity of images (for uniform sampling).[16] The images will overlap if Nyquist's sampling criterion is violated. For distinct images, this means that we must sample the wave front at least twice in the shortest spatial wavelength in the wave front. Since the shortest possible spatial wavelength in the aperture is λ, the wavelength of the sound, an array of elements spaced by $\lambda/2$ is adequate. Much greater spacing is usually sufficient since the object is generally small compared to the aperture and is situated directly below it, resulting in a spatial wavelength much longer than λ. Nevertheless, if nothing is known of the object size or position, $\lambda/2$ spacing is indicated.

[15] F. L. Becker, V. L. Crow, J. C. Crowe, T. J. Davis, B. P. Hildebrand, and G. J. Posakony, Development of an ultrasonic imaging system for inspection of nuclear pressure vessels, *Proc. AIRAPT Int. High Pressure Conf., 6th Boulder, Colorado* (1977).

[16] B. P. Hildebrand and B. B. Brenden, "An Introduction to Acoustical Holography," p. 112. Plenum, New York, 1972.

11.4. Applications

Optical holography has found its greatest application in generalized interferometry, even though its imaging capability is certainly spectacular. Ultrasonic holography, however, is used mostly for imaging, with only a few experiments in interferometry having been reported.[17] This is because any method capable of yielding a visible image of an invisible object assumes great importance in medicine, nondestructive testing, and seismic prospecting, for example.

11.4.1. Imaging

As mentioned above, the capability to obtain images of objects immersed in opaque media has been the overriding motivation for developing ultrasonic holography. The liquid surface area detection system has the capability for real-time imaging and, hence, finds favor in medical applications. Because the liquid surface responds to the radiation pressure instantaneously, the sound can be pulsed at television frame rates. Thus moving objects can be imaged. The earliest example of this was a motion picture film of a moving goldfish. Several consecutive frames from this motion picture are shown in Fig. 14. Initial systems were developed for medical applications. Some examples of images of parts of the human anatomy appear in Fig. 11. It must be noted that the physician is able to manipulate the patient to obtain the optimum image. Experimental instruments have been evaluated for breast screening, obstetrics, and orthopedics. The medical jury is still out, although many complimentary reports have appeared.[18]

The scanning system has become a popular tool in the nondestructive field. It is being used quite extensively in the nuclear industry to provide accurate measurements of flaws in the pressure vessel. Some experimental images useful for evaluation of the capabilities of these systems are shown in Figs. 15 and 16. The first of these illustrates the resolving capability and its dependence on the wavelength of the sound. The object is a series of flat-bottomed holes in an arrangement convenient for resolution measurements. The holes are drilled 100 mm into a 200-mm block of aluminum. The holes in the three rows are separated by 2, 1, and 0.5 mm, respectively. The two images were made with 2.5- ($\lambda = 2$

[17] B. P. Hildebrand and K. Suzuki, Acoustical holographic interferometry, *Jpn. J. Appl. Phys.* **14**, 805–813 (1975).

[18] L. Weiss, Some pathobiological considerations of ultrasonic imaging and holography, *in* "Ultrasonic Imaging and Holography" (G. W. Stroke *et al.*, eds.), pp. 561–587. Plenum, New York, 1974.

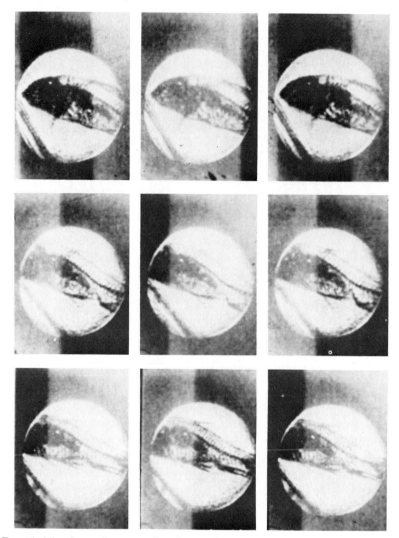

FIG. 14. Nine frames from a motion picture of a live goldfish imaged with the liquid surface system.

mm) and 5-MHz ($\lambda = 1$ mm) sound. Note that with 5 MHz all three rows are resolved, whereas with 2.5 MHz the last row is not. Thus the resolution appears to be in the neighborhood of $\lambda/2$. Figure 16 is an example of this technique on a simulated flaw (12-mm-square piece of graphite) imbedded in a 200-mm stainless-steel weld. This represents the type of flaw of interest in nuclear pressure vessel welds.

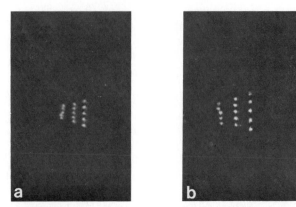

FIG. 15. Image of flat-bottomed holes through 100 mm of aluminum: (a) 3.1 MHz; (b) 5 MHz.

11.4.2. Interferometry

Holographic interferometry is carried out by superimposing two coherent images of the object taken in two states. That is, if the object is physically distorted between the two exposures of an optical hologram, the resulting image will show interference fringes indicative of the distortions. Another type of fringe pattern can be formed by providing two exposures of the object with a slight difference in wavelength. It has been shown that this results in an image containing contour fringes indicating elevation changes of $\lambda^2/2 \, \Delta\lambda$, where $\Delta\lambda$ is the change in wavelength.[19] Due to the very short wavelength of light, registration of the two holograms is possible only by double exposing the same photographic plate. In ultrasonic holography it is relatively easy to register two individual holograms. Thus an object may be compared with itself after a long interval of time in service to determine if any change has occurred.

Figure 17 is an example of the fringe pattern obtained by tilting a plane object. Figure 18 shows the pattern resulting from the distortion of a thin brass sheet. Figure 19 shows the contour fringe exhibiting the shape of a cylindrical surface. The frequency was changed by the amounts shown.

11.5. Computer Reconstruction

The development of faster and cheaper computers has revived interest in computer reconstruction of holograms. In fact, it seems certain that computers will replace the optical reconstruction techniques.

[19] B. P. Hildebrand, The role of coherence theory in holography with application to measurement, *in* "The Engineering Uses of Holography" (E. R. Robertson and J. M. Harvey, eds.). Cambridge Univ. Press, London and New York, 1969.

FIG. 16. Photograph and acoustical images of the simulated flaw: (a) photograph of test specimen clearly showing the clad surface; (b) image of the flaw taken through the smooth surface (bottom side in photograph); (c) image of flaw taken through clad surface (top side in photograph). The hologram was made at a frequency of 5.1 MHz yielding a lateral resolution of 1.2 mm and a lateral magnification of 0.2 in (b) and 0.366 in (c).

11.5.1. Computer Reconstruction of Holograms

In the previous description of holography we have not categorized the holograms but have performed a completely general analysis. It is now necessary to study a particular type of hologram called the "lensless Fourier-transform hologram."[20] The distinguishing characteristic of this type of hologram is the placement of the reference source coplanar with the object point. Physically, this means that the interference fringes are

[20] J. B. DeVelis and G. O. Reynolds, "Theory and Applications of Holography," p. 14. Addison-Wesley, Reading, Massachusetts, 1975.

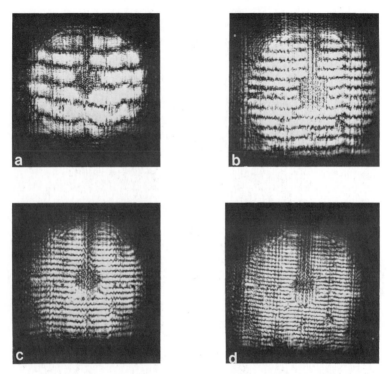

FIG. 17. Holographic interferometry by making two holograms of the object in two states and interfering the images optically. The plane object is tilted by (a) 1°; (b) 2°; (c) 3°; (d) 4°.

approximately straight lines oriented perpendicular to the line joining the two points and spaced inversely to their distance apart. The degree of linearity of these fringes depends on the distance of the hologram from the object plane, that is, the hologram must be in the far field of the object.

If we recall a fundamental observation about the Fourier transform, we can connect this type of hologram to it. This observation is that the one-dimensional Fourier transform of two delta functions is a sine wave. The two-dimensional Fourier transform of two delta functions is a sine-wave grating structure. Thus the hologram formed by placing the reference source next to and in the plane of the object yields a two-dimensional Fourier transform of the object. The mathematical demonstration of this follows. Consider Eq. (11.2.13), with $z_1 = z_2$:

$$\theta_0(x) = \frac{2\pi}{\lambda_1}\left\{[(x - x_1)^2 + z_1^2]^{1/2} - [(x - x_2)^2 + z_1^2]^{1/2}\right\} \quad (11.5.1)$$

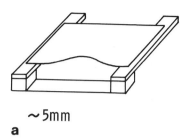

~5mm

a

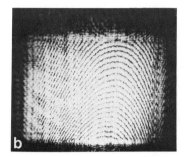

b

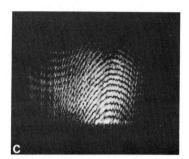

c

FIG. 18. Holographic interferometry of a thin brass sheet deformed between exposures: (a) brass plate with inclination; (b) reconstructed image from the double exposed hologram; (c) reconstructed image from the two holograms.

If we expand the square roots in a binomial series, we have

$$\theta_0(x) = \frac{2\pi}{\lambda_1} z_1 \left\{ \left(\frac{x_1{}^2 - x_2{}^2}{2z_1{}^2} \right) - \frac{x(x_1 - x_2)}{z_1{}^2} \right. $$
$$ \left. + \frac{(x - x_1)^4 - (x - x_2)^4}{8z_1{}^4} + \cdots \right\}. \qquad (11.5.2)$$

The complex amplitude on the hologram plane is, therefore,

$$\exp j\theta_0(x) \cong \exp j \left[\frac{2\pi}{\lambda_1} \left(\frac{x_1{}^2 - x_2{}^2}{2z_1} \right) \right] \exp j \left[-\frac{2\pi}{\lambda_1} \left(\frac{x_1 - x_2}{z_1} \right) x \right], \qquad (11.5.3)$$

where we have neglected higher-order terms. The first exponential is a constant and the second a sinusoidal phase grating with spatial frequency equal to

$$\frac{2\pi}{\lambda_1} \left(\frac{x_1 - x_2}{z_1} \right) = \omega_x. \qquad (11.5.4)$$

Thus, as discussed earlier, the closer the reference is to the object point,

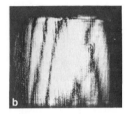
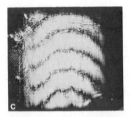

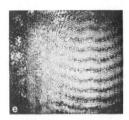

FIG. 19. Holographic interferometry by making two holograms of the object at different frequencies and interfering the images optically. In this case the frequency of the sound was changed to obtain the contour intervals shown: (a) curved aluminum slab; (b) without inclination $\Delta z = 1.28$ mm; (c) convex top surface with inclination $\Delta z = 2.56$ mm; (d) convex bottom surface with inclination $\Delta z = 1.28$ mm; (e) convex top surface with inclination $\Delta z = 1.28$ mm.

the lower the grating frequency and vice versa. Equation (11.5.3) will also be recognized as the Fourier-transform kernel.

Once it is realized that this type of hologram is a Fourier transform of the object, it becomes obvious that an image can be retrieved by inverse Fourier transformation of the hologram. Since ultrasonic holograms typically contain 100×100 to 300×300 sample points, modern computers are quite capable of performing Fourier transforms using fast algorithms.

It must be emphasized that the image is perfect only if the hologram is in the far field of the object and the object is planar. Object points not in the plane of the reference source produce curved rather than straight fringes. Thus a Fourier transform will yield an aberrated image of those points—basically a defect of focus. Points in the near field, however, will not be imaged well even if they are in the plane of the reference.

Three-dimensional objects must, therefore, be imaged by making Fourier-transform holograms for each plane of the object. Generally the planes will be separated by the axial resolution cell size as described by Eq. (11.2.28). Ordinarily this would require that a multiplicity of holograms be made with the reference source placed adjacent to the object in planes separated by the axial resolution cell size. Then images of planes through the depth of the object can be reconstructed by computer.

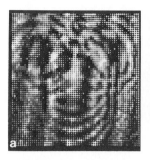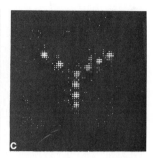

FIG. 20. Steps in the computer reconstruction of complex amplitude data: (a) real part of the complex amplitude plot; (b) synthetic Fourier transform hologram; (c) image by inverse Fourier transformation (Courtesy of G. L. Fitzpatrick, Holosonics, Inc., Richland, Wa).

A better alternative to this is to take only one set of data without any reference beam, simply by measuring phase and amplitude electronically. Then the computer adds the reference beam by adding the appropriate phase value to each measured sample and makes the multiple holograms from one set of measured data. Images are then produced by inverse Fourier transformation. Figure 20 shows the steps in this procedure. Figure 20a shows the real part of the complex amplitude distribution obtained by scanning a transducer over a Y pattern of flat-bottomed holes drilled in a block of aluminum. Figure 20b shows the computer-generated hologram made by adding the phase representing a point reference beam adjacent to the Y pattern. Figure 20c shows the image resulting by performing an inverse Fourier transform of the hologram. It should be noted that once the data are in digital form it is possible to apply various digital operations to improve the image. For example, the data may be smoothed and processed for noise removal before reconstruction. Alternatively, or in addition, the image may be processed to yield edge enhancement, intensity contouring, or even false-color encoding for improved appearance.

11.5.2. Imaging by Backward Wave Reconstruction

Having abandoned the idea of a physical reference beam and the idea of reconstruction by diffraction of light, it is possible to go back even further. That is, we can drop the concept of holography entirely and consider the problem from the basis of wave propagation.[21] The problem consists of predicting an image from a measurement of a wave field. The basic idea for imaging by backward wave propagation resides in the wave

---

[21] D. L. Van Rooy, Digital Ultrasonic Wavefront Reconstruction in the Near Field. IBM Report #320.2402 (1971).

equation

$$(\nabla^2 + k^2)f(x, y, z) = 0,$$ (11.5.5)

where $f(x, y, z) = g(x, y, z) \exp[j\theta(x, y, z)]$, the scalar wave field, and k is the wave number. Taking the two-dimensional Fourier transform of $f(x, y, z)$ yields

$$F(u, v, z) = \iiint f(x, y, z) \exp[-2\pi j(ux + vy)] \, dx \, dy.$$ (11.5.6)

Similarly

$$f(x, y, z) = \iint F(u, v, z) \exp[2\pi j(ux + vy)] \, du \, dv.$$ (11.5.7)

Substituting Eq. (11.5.7) into Eq. (11.5.5) yields the equation

$$\iint \left\{ \frac{\partial^2 F}{\partial z^2} (u, v, z) + [k^2 - (2\pi u)^2 - (2\pi v)^2] \right\}$$

$$\times F(u, v, z) \exp[2\pi j(ux + vy)] \, du \, dv = 0.$$ (11.5.8)

This equation is satisfied only if

$$\frac{\partial^2 F(u, v, z)}{\partial z^2} + [k^2 - (2\pi u)^2 - (2\pi v)^2]F(u, v, z) = 0.$$ (11.5.9)

The solution to this well-known differential equation is

$$F(u, v, z) = F^+(u, v) \exp jkz \left[1 - \left(\frac{2\pi u}{k}\right)^2 - \left(\frac{2\pi v}{k}\right)^2 \right]^{1/2}$$

$$+ F^-(u, v) \exp(-jkz) \left[1 - \left(\frac{2\pi u}{k}\right)^2 - \left(\frac{2\pi v}{k}\right)^2 \right]^{1/2},$$ (11.5.10)

where F^+ indicates forward and F^- backward propagation.

Now, consider a unit amplitude plane wave. It can be expressed by

$$B(x, y, z) = \exp[jk(\alpha x + \beta y + \gamma z)],$$ (11.5.11)

where $\gamma = (1 - \alpha^2 - \beta^2)^{1/2}$ and $\alpha, \beta,$ and γ are the direction cosines of the propagation vector.

The integrand of Eq. (11.5.7) can therefore be interpreted as a plane wave propagating with direction cosines

$$\alpha = 2\pi u/k = \lambda u,$$ (11.5.12)

$$\beta = 2\pi v/k = \lambda v,$$ (11.5.13)

$$\gamma = [1 - (\lambda u)^2 - (\lambda v)^2]^{1/2}$$ (11.5.14)

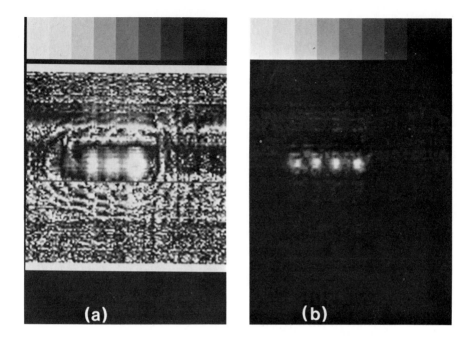

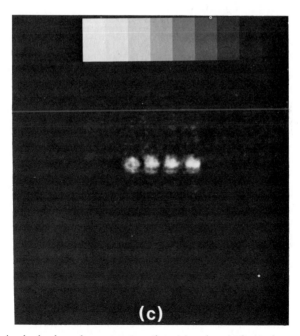

FIG. 21. Imaging by backward wave propagation: (a) real part of complex amplitude on data plane; (b) intensity in a plane not the plane of the object; (c) intensity in the plane of the object. The object is a row of flat-bottomed holes seen through 75 mm of steel.

and amplitude

$$F(u, v, z)\, du\, dv, \qquad \text{where} \quad u = \alpha/\lambda, \quad v = \beta/\lambda. \qquad (11.5.15)$$

Thus Eq. (11.5.7) says that the wave front in the data plane can be thought to consist of the sum of an angular distribution of plane waves.[22] Equation (11.5.10) states that the angular spectrum of plane waves on the plane $z = z$ can be found by multiplying the angular spectrum of plane waves on the plane $z = 0$ by an exponential factor that accounts for the phase change due to the distance propagated.

Rewriting Eq. (11.5.10) to reveal the angular dependence more clearly, we obtain

$$F\left(\frac{\alpha}{\lambda}, \frac{\beta}{\lambda}, z\right) = F\left(\frac{\alpha}{\lambda}, \frac{\beta}{\lambda}, 0\right) \exp[-2\pi jz(1 - \alpha^2 - \beta^2)^{1/2}], \qquad (11.5.16)$$

where we have used only the backward propagation term.

Thus the following procedure will result in an image generated from a measurement of a wave field:

(1)   Measure the complex value of the field $f(x, y, z)$ on the detection plane $z = 0$.

(2)   Take the two-dimensional Fourier transform of $f(x, y, 0)$ to obtain the spectrum of plane waves $F(\alpha/\lambda, \beta/\lambda, 0)$.

(3)   Multiply the spectrum of plane waves by the backward propagation factor $\exp[-jkz(1 - \alpha^2 - \beta^2)^{1/2}]$ to obtain the angular spectrum of plane waves $F(\alpha/\lambda, \beta/\lambda, z)$ on the desired image plane $z = z$.

(4)   Take the inverse Fourier transform of $F(\alpha/\lambda, \beta/\lambda, z)$ to obtain the image complex amplitude $f(x, y, z)$.

(5)   Plot magnitude $|f(x, y, z)|^2$ and/or phase $\theta(x, y, z)$.

When the image plane $z = z$ coincides with the plane of the object, the result is an image. This procedure is good at any distance; there is no far-field limitation. Furthermore, no conjugate or zero-order wave exists, since we use only the backward part of the wave expressed by Eq. (11.5.12). Since no off-axis reference beam is used, the spatial frequency in the data plane is less than half that of the Fourier-transform hologram. Thus the computer needs to work with a data set less than half as large. An illustration of this type of image reconstruction is shown in Fig. 21.

[22] J. W. Goodman, "Introduction to Fourier Optics," p. 48. McGraw-Hill, New York, 1968.

12. COMPUTERIZED TRANSMISSION TOMOGRAPHY

By James F. Greenleaf

12.1. Introduction: Implementation of Tomography Methods in Medicine

The recent use of computerized tomography methods in medical x rays has had a tremendous impact on the field of radiology.[1] For the first time in the history of roentgenography, the radiologist can obtain quantitative measures of the spatial distribution of roentgen linear absorption coefficients within a plane through the body without interference from overlying tissues. Medical applications of computerized tomographic reconstruction methods are now being developed in many areas of medical research, using forms of energy such as x radiation,[2,3] electrons,[4] gamma rays,[5,6] alpha particles,[7] nuclear magnetic resonance,[8] positrons[9] and, more recently, ultrasound.[10-19]

[1] H. L. Baker, J. K. Campbell, O. W. Houser, D. F. Reese, P. F. Sheedy, and C. B. Holman, *Mayo Clinic Proc.* **49,** 17 (1974).

[2] R. A. Brooks and G. DiChiro, *Phys. Med. Biol.* **21,** 684 (1976).

[3] A. M. Cormak, *J. Appl. Phys.* **34,** 2722 (1963).

[4] D. J. DeRosier and A. Klug, *Nature (London)* **217,** 130 (1968).

[5] D. E. Kuhl and R. Q. Edwards, *Radiology* **80,** 653 (1963).

[6] D. E. Kuhl, L. Hale, and W. L. Eaton, *Radiology* **87,** 278 (1966).

[7] K. M. Crowe, T. F. Budinger, J. L. Cahoon, V. P. Elischer, R. H. Huesman, and L. L. Kanstein, *IEEE Trans. Nucl. Sci.* **NS-22** (3), 1752 (1975).

[8] P. C. Lauterbur, W. V. House, D. M. Kramer, C. N. Chen, F. W. Porretto, and C. S. Dulcey, *Proc. Image Processing 2-D and 3-D Reconstruct. from Projections, August 4–7* p. MA10-1 Technical Digest, Optical Society of America, Stanford (1975).

[9] M. M. Ter-Pogossian, M. E. Phelps, E. J. Hoffman, and N. A. Mullani, *Radiology* **114,** 89 (1975).

[10] J. F. Greenleaf, S. A. Johnson, S. L. Lee, G. T. Herman, and E. H. Wood, *in* "Acoustical Holography" (P. S. Green, ed.), Vol. 5, p. 591. Plenum, New York, 1975.

[11] J. F. Greenleaf, S. A. Johnson, W. F. Samayoa, and F. A. Duck, *in* "Acoustical Holography" (N. Booth, ed.), Vol. 6, p. 71. Plenum, New York, 1975.

[12] J. F. Greenleaf and S. A. Johnson, *Proc. Seminar Ultrason. Tissue Characterizations* p. 109. Natl. Bur. Stand. Special Publ. 453 (1975).

[13] S. A. Johnson *et al. in* "Acoustical Holography" (L. Kessler, ed.), Vol. 7, p. 307. Plenum, New York, 1977.

METHODS OF EXPERIMENTAL PHYSICS, VOL. 19

Historically, ultrasound has been used to obtain tomographs using echo modes such as B-scans[20] and, more recently, echo mode C-scans.[21] These methods result in qualitative mappings of tissue interfaces and geometries. Images representing quantitative acoustic parameters[13,22-24] may be intrinsically more valuable than qualitative images representing only tissue interfaces and geometries, since fundamental mechanical tissue parameters, such as adiabatic compressibility and density, may be calculated from basic acoustic parameters.[25] Unlike those generated by current B-scan techniques, such images should be independent of the abilities of the technician or physician doing the scan and should depend only on the precision of the instrument involved. The disadvantage of the current quantitative methods is that very large apertures are necessary to obtain a large range of viewing angles required for current methods of reconstruction. This limits the methods to a few available organs such as the breast. However, new methods of tomography are being developed that may allow quantitative imaging methods to be developed from the echo mode of data acquisition in addition to the transmission modes.[26,27] In this part, methods will be derived for quantitatively reconstructing two-dimensional or three-dimensional spatial distributions of values of acoustic properties such as attenuation, speed of sound, absorption, and others related through line integrals to measurable and quantifiable attributes of the received signals.

[14] J. F. Greenleaf, S. A. Johnson, W. F. Samayoa, and C. R. Hansen, in "Acoustical Holography" (L. Kessler, ed.), Vol. 7, p. 263. Plenum, New York, 1977.

[15] G. H. Glover, Natl. Bur. Stand. Int. Symp. Ultrason. Tissue Characterization, 2nd, Gaithersburg, Maryland, June 13–15 (1977).

[16] P. L. Carson, L. Shabason, D. E. Dick, and W. Clayman, Natl. Bur. Stand. Int. Symp. Ultrason. Tissue Characterization, 2nd, Gaithersburg, Maryland, June 13–15 (1977).

[17] P. J. Carson, T. V. Oughton, W. R. Hendee, and A. S. Ahuja, Med. Phys. 4, 302 (1977).

[18] G. H. Glover, and L. C. Sharp, IEEE Trans. Sonics Ultrason. SU-24(4), 229 (1977).

[19] R. Dunlap, Ultrasonic CT imaging, Ph.D. Thesis, Christchurch, New Zealand (1978).

[20] G. Kossoff, Proc. R. Soc. Med. 67, 135 (1974).

[21] R. Mezrich, in "Acoustical Holography" (L. Kessler, ed.), Vol. 7, p. 51. Plenum, New York, 1977.

[22] J. P. Jones, Ultrason. Int., 1973 Conf. Proc., 1973 p. 214. PIC Science and Technology Press, London, 1974.

[23] A. C. Kak, L. R. Beaumont, and J. Wolfley, School of Electrical Engineering, Purdue Univ., TR-EE 75-7 (1974).

[24] A. C. Kak and F. J. Fry, Ultrasonic Tissue Characterization. Natl. Bur. Stand. Special Publ. 453 (1975).

[25] L. E. Kinsler and A. R. Frey, "Fundamentals of Acoustics," p. 115. Wiley, New York, 1950.

[26] F. A. Duck, in "Computer Aided Tomagraphy and Ultrasonics in Medicine" (Josef Raviv, ed.), p. 137. North-Holland Publ., Amsterdam, 1979.

[27] R. M. Lewitt, State Univ. of New York at Buffalo. Personal communication.

The problem of reconstructing a three-dimensional object from its two-dimensional projections (shadowgraphs) has arisen in a large number of scientific and medical areas. The basic mathematical problem was posed as early as 1917 by Radon,[28] who actually provided a closed-form solution assuming ideal mathematical conditions. In 1956 Bracewell[29] considered and provided a practical solution to this problem in relation to strip integration in radioastronomy. In 1963 Cormack[30] pointed out the importance of the problem in radiology, and in 1969 DeRosier and Klug[4] applied it to the field of electron microscopy. Many different methods of acquiring data and mathematically solving the reconstruction problem have been proposed in these and other fields; for a survey, see the work of Gordon and Herman[31] or Brooks and DeChiro.[2]

12.1.1. Notation

The word *tomography* refers to "slice imaging." Computerized tomography (CT), when used in ultrasonics, refers to computer-aided B-scanning, to echo C-scanning, or to computer-aided transmission tomography, which will be described in this part. The word *reconstruction* refers to the process of forming, from measured or simulated projection data, an image consisting of pixels (picture elements) or voxels (volume elements) whose brightnesses are proportional to the value of an internal material property.

A generalized coordinate system will be used to specify position in space with a vector $\mathbf{r}$, consisting of an n-tuple of elements, for example (x, y), where n is the dimensionality of the space (usually $n = 2$) in which the function to be obtained, usually denoted by F, is defined. The function F could represent the acoustic attenuation coefficient, acoustic speed, or even, for example, specific acoustic impedance. A ray is that path it can be assumed the acoustic energy traverses and often (in geometric acoustics) is taken as being everywhere orthogonal to the wave front (for a pulse of acoustic energy) or parallel to the wave vector $\mathbf{k}$ (for continuous waves).

For two dimensions, the integral of $F(\mathbf{r})$ along a ray is called the *ray sum* or the *ray projection;* that is, the projection or profile

$$p(t, \boldsymbol{\theta}) = \int_{\mathbf{r}\cdot\boldsymbol{\theta}\,=t} F(\mathbf{r})\,d\mathbf{r}), \qquad (12.1.1)$$

[28] J. Radon, *Berg. Verh. Sachs. Acad. Wiss.* **69**, 262 (1917).

[29] R. N. Bracewell, *Aust. J. Phys.* **9**, 198 (1956).

[30] A. M. Cormack, *J. Appl. Phys.* **34**, 2722 (1963).

[31] R. Gordon and G. T. Herman, *in* "International Review of Cytology" (G. H. Bourne and J. F. Danielli, ed.), Vol. 38, p. 111. Academic Press, New York, New York, 1974.

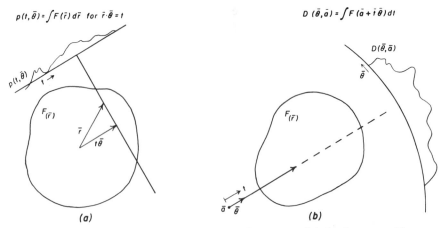

$p(t,\bar{\theta})=\int F(\bar{r})\,d\bar{r}$ for $\bar{r}\cdot\bar{\theta}=t$

$D(\bar{\theta},\bar{a})=\int F(\bar{a}+\bar{t}\bar{\theta})\,dt$

Fig. 1. Definition of geometry for projection profiles: (a) parallel; (b) divergent. The parallel rays are denoted by direction $\hat{\theta}$ orthogonal to the detector line or plane and by distance t from the center of the object region. Divergent rays are denoted by direction $\hat{\theta}$ and position of source **a**.

where $\boldsymbol{\theta}$ is the unit vector in the direction orthogonal to the ray. The operator in Eq. (12.1.1) can be called the *parallel projection operator* since all lines satisfying $\mathbf{r} \cdot \boldsymbol{\theta} = t$ are parallel (Fig. 1a). We take $F(\mathbf{r})$ to be zero outside a disk of some suitable radius $|\mathbf{r}| = R$.

We can define the divergent beam operator as

$$D(\boldsymbol{\theta}, \mathbf{a}) = \int F(\mathbf{a} + t\boldsymbol{\theta})\, dt, \qquad (12.1.2)$$

where $\boldsymbol{\theta}$ is the unit direction vector for the ray, **a** is the position of the source (Fig. 1b), and the integration is performed in the direction $\boldsymbol{\theta}$. Again the limits of integration are determined by a disk $|\mathbf{r}| = R$ outside of which $F = 0$. Usually this disk is taken to be just inside the contour of sources of energy. The reconstruction problem consists of solving Eq. (12.1.1) or (12.1.2) for $F(\mathbf{r})$ given measurements of projection profiles $p(t, \boldsymbol{\theta})$ or $D(\boldsymbol{\theta}, \mathbf{a})$.

A particularly eloquent formulation of an inversion technique, apparently due to Bracewell,[28] is based on the projection theorem stated as follows:

Projection Theorem.   Given the parallel projections $p(t, \boldsymbol{\theta})$ of a material property $F(\mathbf{r})$, defined on a two-dimensional disk, the Fourier transform of F is related to the Fourier transforms of the projection profiles $p(t, \boldsymbol{\theta})$ in the following way:

$$\mathcal{F}_{x,y}[F](u, v) = \mathcal{F}_t[p](w, \boldsymbol{\theta}), \qquad (12.1.3)$$

where $\boldsymbol{\theta} = (\cos\theta, \sin\theta)$, $u = w\cos\theta$, $v = w\sin\theta$, $\mathcal{F}_{x,y}$ is a two-dimensional Fourier-transform operation with respect to x and y, and $\mathscr{F}_t$ the one-dimensional Fourier-transform operation with respect to t. This equation states that the one-dimensional Fourier transform of a projection profile obtained at an angle θ is equal to the values of the two-dimensional Fourier transform of F along a line at an angle θ through the origin of the $u-v$ plane.

An inversion procedure directly based on the projection theorem is applicable only to parallel projections and for straight rays. However, it will be shown in Section 12.3.2 to be a special case of a more general problem in acoustics that involves curved rays. Inversion methods for divergent geometry have been developed by several investigators (see Brooks and DiChiro[2] for a review) but will not be described here.

12.1.2. List of Symbols

A	c_0/c; refractive index if c is real
$\mathbf{a}$	position of transmitter for assumed divergent beam geometry
$c(\mathbf{r})$	local acoustic speed, isotropic
D_f	root-mean-square duration of tissue transfer function
D_t	root-mean-square duration of tissue impulse response
$D(\boldsymbol{\theta},\mathbf{a})$	divergent ray sum of F
F	"density" function to be reconstructed
f	perturbation of c_0, i.e., $c = c_0(1 + f)$
$g(\mathbf{r},w)$	complex tissue transfer function
$\mathbf{K}$	wave vector for continuous waves
k_0	wave number in unperturbed medium $k_0 = 2\pi/\lambda_0$
$L(w)$	measured log amplitude ratios of tissue layer
N	real index of refraction
N_0	uniform isotropic refractive index (real)
N_1	perturbed refractive index
N_P	number of parallel channels
N_R	number of rays required to obtain reconstruction
n_Z	perturbation on refractive index
n_r	real part of n
n_i	imaginary part of n
$P_{\text{sum}}(\mathbf{r},t)$	total excess pressure
$p(t,\boldsymbol{\theta})$	parallel ray sum of F
$\mathbf{r}$	position vector
T	temperature in degrees celsius
t	scalar measure of distance
$\mathbf{t}$	ray tangent vector in laboratory coordinate system
$\mathbf{t}_m$	ray tangent vector in fluid coordinate system
t_c	arbitrary reference point in time
$U(\mathbf{r})$	time-independent solution of wave equation
U	coordinate in Fourier domain (abscissa)
$\mathbf{V}(\mathbf{r})$	fluid velocity (local)
V_x	fluid velocity component in x direction
V_y	fluid velocity component in y direction

V	coordinate in Fourier domain (ordinate)
X_0	position of receiver
α	acoustic amplitude attenuation constant
α_0	acoustic amplitude attenuation coefficient for linear attenuation
$\beta(\mathbf{r},t)$	phase term of complex tissue transfer function
Δf	differential frequency
Δt	sample period for calculation of frequency
$\boldsymbol{\theta}$	unit vector for direction of ray (straight rays)
λ_0	wavelength in unperturbed medium
τ_r	reverberation time of acoustic enclosure
$\phi(\mathbf{r})$	eikonal or acoustic path length
ϕ_0	solution to unperturbed eikonal equation
ϕ_1	part of eikonal solution due to perturbation
ϕ_{1r}	real part of ϕ_1
ϕ_{1I}	imaginary part of ϕ_1
Ψ	complex function, real part of which is excess local pressure
ω	radian frequency of signal or pressure
$\mathscr{F}_{x,y}[F](U,V)$	two-dimensional Fourier transform of $F(x,y)$
$\mathscr{F}_t[p](w,\boldsymbol{\theta})$	one-dimensional Fourier transform of $p(t,\boldsymbol{\theta})$

12.2. Straight Ray Transmission Tomography

In this section the transmission acoustic tomography problem will be formulated as a reconstruction problem for the case of energy traveling in a straight line. The aim is to derive relationships between measurable properties of acoustic energy received after transmission through biological tissue and basic acoustic properties within the tissue. These relationships should be recognizable as a form of Eq. (12.1.1) or (12.1.2) in order to solve for the material properties using established methods of reconstruction for straight ray projections. More complex methods of modeling acoustic energy interactions, which include curved ray effects due to refraction and diffraction, will be described in Chapter 12.3.

In the following sections, the physical basis for applying the straight-line reconstruction methods to acoustic energy propagation is derived under a set of assumptions and conditions that, when applied to the acoustic wave equation, result in mathematical relationships for the real and imaginary parts of acoustic refractive index in terms of parameters measurable from the acoustic signal received from transmission scans. These derivations are for continuous functions. Discretization would be required for computer implementation of the methods of solution.

12.2.1. Single-Frequency (Narrowband) Method

The simplest case for illustrating the relationship between acoustic energy propagation and the reconstruction equations is that in which variations in the refractive index of the material under investigation are small enough that we may assume that all second-order terms in the wave equation can be deleted. This case will first be considered for a single frequency and subsequently for wideband signals.

Under conditions in which tissue density varies only slightly within the organ under study,[32] the propagation of ultrasound at a single frequency ω can be described by the wave equation

$$\nabla^2 \Psi(\mathbf{r}, t) - \frac{1}{c(\mathbf{r})^2} \frac{\partial^2 \Psi(\mathbf{r}, t)}{\partial t^2} = 0, \tag{12.2.1}$$

where c is the local complex velocity, which may depend on position $\mathbf{r}$. The excess local pressure P is given by the real part of $\Psi(\mathbf{r}, t)$.

To remove time dependence, we assume sinusoidal time dependence; that is,

$$\Psi(\mathbf{r}, t) = \exp(-i\omega t)U(\mathbf{r}), \tag{12.2.2}$$

where $U(\mathbf{r})$ is a complex-valued time-independent function. This substitution transforms Eq. (12.2.1) into a Helmholtz equation

$$\nabla^2 U + [\omega^2/c^2]U = 0. \tag{12.2.3}$$

Let the factor in the square brackets be replaced by $[k_0 A(\mathbf{r})]^2$, where $k_0 = 2\pi/\lambda_0 = \omega/c_0$ and c_0 is a fixed propagation speed such as that in water. In this case the factor A^2 becomes c_0^2/c^2. If c is real, then $A = N$, where N is the real index of refraction.

Equation (12.2.3) is transformed by the substitution

$$U(\mathbf{r}) = U_0 \exp[ik_0\phi(\mathbf{r})], \tag{12.2.3a}$$

into

$$ik_0^{-1}\nabla^2\phi - \nabla\phi \cdot \nabla\phi + A^2 = 0, \tag{12.2.4}$$

where $\phi(\mathbf{r})$ is the eikonal or acoustic path length function. If we let λ_0 approach zero (i.e., no diffraction), then Eq. (12.2.4) becomes

$$\nabla\phi \cdot \nabla\phi = A^2. \tag{12.2.5}$$

Assuming the unperturbed medium has no dissipation, we let $A(\mathbf{r}) = N_0 + n(\mathbf{r})$, where N_0 is the real (uniform and isotropic) index of refraction of the unperturbed medium. Then

$$\phi = \phi_0 + \phi_1, \tag{12.2.6}$$

where ϕ_0 is the solution in the unperturbed medium and ϕ_1 is due to the perturbation; that is,

$$\nabla\phi_0 \cdot \nabla\phi_0 = N_0^2. \tag{12.2.7}$$

[32] L. M. Brekhovskikh, "Waves in Layered Media," p. 170. Academic Press, New York, 1960.

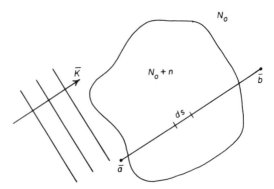

FIG. 2. Geometry for diffraction tomography. A continuous wave with vector $\mathbf{K}$ travels in a medium with constant refractive index N_0 and enters an object with refractive index $N_0 + n$. Rays have initial direction $\mathbf{K}$, incremental length ds, and may or may not be straight depending on the propagation model being considered.

If the terms in n^2 and $\nabla \phi_1 \cdot \nabla \phi_1$ are assumed to be negligible, Eq. (12.2.5) becomes

$$\nabla \theta_0 \cdot \nabla \theta_1 = N_0 n. \qquad (12.2.8)$$

An unperturbed plane wave solution of Eq. (12.2.8) is $\phi_0 = N_0 \mathbf{K} \cdot \mathbf{r}$, giving

$$\hat{\mathbf{K}} \cdot \nabla \phi_1 = n, \qquad (12.2.9)$$

where $\hat{\mathbf{K}}$ is a unit vector in the direction of propagation of the plane wave (see Fig. 2). Assume n has real and imaginary parts n_r and n_i, respectively, and ϕ_1 has real and imaginary parts ϕ_{1r} and ϕ_{1i}, respectively. Also note from Eq. (12.2.3a) that $\phi_{1i}(\mathbf{r}) = k_0^{-1} \ln|U_0/U(\mathbf{r})|$.

Using these new variables Eq. (12.2.9) can be integrated along a path between $\mathbf{a}$ and $\mathbf{b}$. Notice that the path $\mathbf{b} - \mathbf{a}$ must be parallel to $\mathbf{K}$ (Fig. 2). Letting ds be the element of length along that path, we obtain for the real part

$$\phi_{1r}(\mathbf{b}) - \phi_{1r}(\mathbf{a}) = \int_0^{|\mathbf{b}-\mathbf{a}|} n_r(s)\, ds. \qquad (12.2.10)$$

Both sides of Eq. (12.2.10), when divided by c_0, become the perturbation in the time of flight of the acoustic wave from $\mathbf{a}$ to $\mathbf{b}$.

Integrating Eq. (12.2.9) for the imaginary part gives

$$\phi_{1i}(\mathbf{b}) - \phi_{1i}(\mathbf{a}) = \int_0^{|\mathbf{b}-\mathbf{a}|} n_i(s)\, ds \qquad (12.2.11)$$

or

$$\ln|U(\mathbf{b})/U(\mathbf{a})| = -k_0 \int_0^{|\mathbf{b}-\mathbf{a}|} n_{\mathrm{i}}(s)\,ds. \qquad (12.2.12)$$

This is in the same form as the Lambert–Beer law for linear attenuation in optical or x-ray radiation. In this case, it is valid for a single frequency only. Both Eqs. (12.2.10) and (12.2.12) can now be written in the form of either Eq. (12.1.1) or Eq. (12.1.2) and may be used for reconstruction of images of n_{i} or n_{r} from appropriate data using narrowband acoustic signals.

Since, typically, $n_{\mathrm{r}} \leq 0.10 N_0$ in tissues and since the attenuation in water is small relative to other path integrals, the assumptions that higher-order terms can be ignored in Eq. (12.2.5) after perturbation of ϕ leads to the straight-line approximation, which is reasonable and produces fairly good images.[31]

12.2.2. Multiple-Frequency (Broadband) Method

The derivation just described was for monochromatic signals only. The following is a derivation of the relationship of the amplitude and phase of a broadband pulse received after transmission through tissue to the complex refractive index of the tissue as a function of frequency. This derivation again assumes the refractive index deviates only slightly from a constant value.

A plane wave traveling in the $\mathbf{K}$ direction can be approximated by

$$\Psi(\mathbf{r},\,t) = U_0 \exp\{i[-\omega t + k_0(N_0\mathbf{K}\cdot\mathbf{r} + \phi_1(\mathbf{r},\,\omega))]\}, \qquad (12.2.13)$$

where we have set $N_0\mathbf{K}\cdot\mathbf{r} = \phi_0$ and $\phi_1(\mathbf{r},\,\omega) = \phi_1$, to be consistent with Eq. (12.2.6). Note that ϕ_1 is complex, i.e., $\phi_1 = \phi_{1\mathrm{r}} + i\phi_{1\mathrm{i}}$.

Summing the pressures at a point $\mathbf{r}$ for all positive frequencies, we obtain

$$P_{\mathrm{sum}}(\mathbf{r},\,t) = \mathrm{Re}\int_0^\infty U(\omega)\exp\left(\frac{i\omega}{c_0}\phi_1\right)\exp\left\{i\omega\left[-t + \frac{1}{c_0}(N_0\mathbf{K}\cdot\mathbf{r})\right]\right\}d\omega,$$

$$(12.2.14)$$

where the complex function $U(\omega)$ is the phase and amplitude weight for the pressure at each frequency. The second exponential term,

$$\exp\{i\omega[-t + (1/c_0)(N_0\mathbf{K}\cdot\mathbf{r})]\},$$

is the unperturbed contribution for the immersing fluid, which is assumed to have refractive index N_0. The term $\exp[(i\omega/c_0)\phi_1]$ is the modification of the unperturbed term caused by perturbations in the index of refraction.

We can now determine $U(\omega)$ from experimentally measured data in the following manner. Examining Eqs. (12.2.10) and (12.2.12), we note that at $\mathbf{r} = 0$ ($\mathbf{a} = \mathbf{b}$), $\phi_1 = 0$. This reduces Eq. (12.2.14) to

$$P_{\text{sum}}(0, t) = \text{Re} \int_0^\infty U(\omega) \exp(-i\omega t) \, d\omega. \qquad (12.2.15)$$

Note that $\mathbf{r} = 0$ can be some convenient reference point at which P_{sum} can be measured on the transmitter side of the material under study. To determine $U(\omega)$ we merely take the Fourier transform of Eq. (12.2.15), using the transform pair

$$\mathcal{F}_t[h](\omega) = \int_{-\infty}^\infty h(t) \exp(i\omega t) \, dt = H(\omega),$$

$$\mathcal{F}_w[H](t) = \frac{1}{2\pi} \int_{-\infty}^\infty H(\omega) \exp(-i\omega t) \, d\omega = h(t). \qquad (12.2.16)$$

Hence

$$\mathcal{F}_t[P_{\text{sum}}](0, w) = \pi U(\omega), \qquad \omega \geqslant 0. \qquad (12.2.17)$$

The reconstruction problem consists of determining the refractive indices from Eqs. (12.2.10) and (12.2.12). Therefore, we must determine ϕ_1 in Eq. (12.2.14) given measurements of P_{sum} at many points in space and time. In principle, ϕ_1 can be determined in the following manner. Using the transform pair in Eqs. (12.2.16) to transform (12.2.14) with respect to t, we obtain

$$\mathcal{F}_t[P_{\text{sum}}](\mathbf{r}, w) = \pi \exp\left(\frac{i\omega}{c_0} \phi_1\right) \exp\left(i \frac{\omega}{c_0} N_0 \mathbf{K} \cdot \mathbf{r}\right) U(\omega), \qquad \omega \geqslant 0.$$

$$(12.2.18)$$

The term $(\omega/c_0)N_0\mathbf{K} \cdot \mathbf{r}$ represents the acoustic path length for frequency ω through the unperturbed medium and can be determined experimentally by recording the arrival time (phase velocity) of all frequencies through, say, water. Rewriting Eq. (12.2.18), we have

$$\exp\left(i \frac{\omega}{c_0} \phi_{1r}\right) \exp\left(-\frac{\omega}{c_0} \phi_{1i}\right)$$
$$= \frac{\mathcal{F}_t[P_{\text{sum}}][(\mathbf{r}, w)] \exp[-(i\omega/c_0)N_0\mathbf{K} \cdot \mathbf{r}]}{\pi U(\omega)}. \qquad (12.2.19)$$

The term on the right-hand side can be experimentally determined for those values of ω that give a nonvanishing $U(\omega)$. Let us denote it by $g(\mathbf{r}, \omega)$. Then

$$g(\mathbf{r}, \omega) = |g(\mathbf{r}, \omega)| \exp[i\beta(\mathbf{r}, \omega)], \qquad (12.2.20)$$

where β is the phase term. Substituting Eq. (12.2.20) into (12.2.19) and equating phase and amplitude components, we have

$$\frac{\omega}{c_0} \phi_{1r}(\mathbf{r}, \omega) = \beta(\mathbf{r}, \omega),$$

$$-\left[\frac{\omega}{c_0} \phi_{1i}(\mathbf{r}, \omega)\right] = \ln|g(\mathbf{r}, \omega)|.$$

(12.2.21)

We note that $g(\omega)$ must be obtained through a division by $U(\omega)$ representing a process of deconvolution in the time domain, which can be an unstable process. Several authors have investigated such problems: Dines and Kak deal with the acoustic cases[33], while general spectral analyses signals are covered by Childers.[34]

12.2.3. Reconstruction of Nonscalar Parameters Assuming Straight Rays

The previous derivations have been confined to reconstructing *scalar quantities*. Reconstruction of *vectors*, for straight rays, is a different problem related to fluid-flow measurement.

Fluid-flow velocity within a measurement region may be determined by transmitting and receiving acoustic energy through the measurement region along a plurality of rays such that each volume element is traversed by a set of rays having nonzero direction cosines in each direction for which flow components are to be reconstructed.[35] The propagation times of the acoustic energy along the plurality of rays constitute the only measurements required by the method.

Each measurement of ray propagation time is an integral of a function of acoustic speed and fluid velocity along the ray, and the set of such measurements around a plurality of directions constitutes a simultaneous set of integral equations that may be solved to obtain the unknown fluid velocity vector field. The equation that relates the propagation time along each ray from source $\mathbf{a}$ to receiver $\mathbf{b}$ is given by

$$t_{ab} = \int_a^b ds/|\mathbf{t}_m c(\mathbf{r}) + \mathbf{V}(\mathbf{r})|,$$

(12.2.22)

where $\mathbf{t}_m$ is the unit tangent vector along the acoustic ray as seen from the moving medium, $c(\mathbf{r})$ the local acoustic speed as seen in the moving

[33] K. A. Dines and A. C. Kak, *IEEE Trans. Acous. Speech Signal Processing* **ASSP-25** (4), 346 (1977).

[34] D. G. Childers (ed.), "Modern Spectral Analysis." IEEE Press, New York, 1978 (distributed by Wiley, New York).

[35] S. A. Johnson, J. F. Greenleaf, M. Tanaka, and G. Flandro, *Instrum. Soc. Am. Trans.* **6**, 3 (1975).

medium, and $\mathbf{V}(\mathbf{r})$ the fluid velocity vector as measured in the coordinate system.

In general, the actual ray path taken by acoustic energy from the source $\mathbf{a}$ to the receiver $\mathbf{b}$ is not known a priori, even though the time of propagation may be measured quite accurately. The actual ray paths often differ only slightly from straight lines as shown in Section 12.2.1. In this section we shall consider the case of low fluid velocities, i.e., fluid velocities less than 10% of the speed of sound.

When the velocity of the fluid is everywhere much less than the speed of sound, the denominator in Eq. (12.2.22) can be approximated by the expression $c + \mathbf{t}_m \cdot \mathbf{V}$, where it is assumed that the ray tangent vector $\mathbf{t}_m$ as seen from a coordinate system embedded in the fluid is almost identical to the ray tangent vector $\mathbf{t}$ as seen in the laboratory coordinate system. In most fluids, when $|\mathbf{V}| \ll c$, the variations in $c(\mathbf{r})$ caused by flow are also correspondingly small. With these assumptions, Eq. (12.2.22) can be approximated by

$$t_{ab} = \int_{\mathbf{a}}^{\mathbf{b}} \frac{1}{c} \left(1 - \frac{\mathbf{t} \cdot \mathbf{V}}{c} \right) ds. \qquad (12.2.23)$$

The assumption that $\mathbf{t}_m$ is nearly equal to $\mathbf{t}$ allows the further assumption that the acoustic rays are nearly straight lines.

Thus the ray paths may be found in terms of the known source point $\mathbf{a}$ and receiver point $\mathbf{b}$. The use of straight-line rays provides a further benefit, namely, a simple method for separating the dependence of the time t_{ab} on both $c(\mathbf{r})$ and $\mathbf{V}(\mathbf{r})$. This separation is obtained by forming the linear combinations $(t_{ab} + t_{ba})$ and $(t_{ab} - t_{ba})$, where t_{ba} refers to the propagation time between $\mathbf{b}$ and $\mathbf{a}$ when $\mathbf{b}$ serves as the source and $\mathbf{a}$ as the receiver and where $\mathbf{t}$ is in the direction $\mathbf{b} - \mathbf{a}$. That is,

$$t_{ab} + t_{ba} = 2 \int_{\mathbf{a}}^{\mathbf{b}} \frac{1}{c(\mathbf{r})} ds, \qquad (12.2.24)$$

$$t_{ab} - t_{ba} = -2 \int_{\mathbf{a}}^{\mathbf{b}} \frac{\mathbf{t} \cdot \mathbf{V}}{c^2(\mathbf{r})} ds. \qquad (12.2.25)$$

Thus even if $c(\mathbf{r})$ is a function of position $\mathbf{r}$, it may be found by the inversion of the multiple set of Eq. (12.2.24), assuming that an independent set of ray paths through the flow channel can be obtained. This solution may be used to define c in Eq. (12.2.24) so that $\mathbf{V}$ may be found from the same or a similar set of ray measurements by solving Eq. (12.2.25).

An expression for $\mathbf{t} \cdot \mathbf{r}$ may be written in terms of components V_x, V_y of $\mathbf{V}$ and the direction cosines of the vector $\mathbf{t}$. With this substitution for $\mathbf{t} \cdot \mathbf{V}$, Eq. (12.2.25) becomes (note the symbol $\equiv$ indicates a definition)

$$I_0(a_ib_j) \equiv t_{a_ib_j} - t_{b_ja_i} = -2\int_a^b \frac{1}{c^2}(V_x\cos\theta + V_y\sin\theta)\,ds. \quad (12.2.26)$$

With I_0 determined for each ray in a sufficiently large set of rays, it is possible for Eq. (12.2.26) to be inverted.

The integral in Eq. (12.2.26) may be approximated as a discrete sum by subdividing the plane within which fluid velocity is being measured into finite elements or pixels. Then ds corresponds to the length of the ray in each pixel. Using this notation, Eq. (12.2.26) becomes

$$I_1(a,i,j)_s = -2\sum_k^M \frac{1}{c^2}(V_x(k)\cos\theta_s + V_y(k)\sin\theta_s)L_{sk}. \quad (12.2.27)$$

where L_{sk} is the length of ray s in pixel k and M the total number of pixels. The case of nonconstant speed of sound c is also described by Eq. (12.2.24). Thus both V and c can be reconstructed even in those cases where c is not a constant.

Algebraic reconstruction technique (ART) is a method commonly used for reconstructing *scalar* quantities from their projections. In the implementation of the fluid-flow technique just described, reconstruction of vector quantities is required. This is accomplished by writing $(t \cdot V)$ as $(V_x\cos\theta_s + V_y\sin\theta_s)$. At each point along the ray, $\cos\theta_s$ and $\sin\theta_s$ are known and are included as a part of the weight L_{sk}. The quantities V_x and V_y are each to be calculated for the N^2 pixels (picture elements) in a square array with N pixels on a side. One way to accomplish this task is to modify the ART algorithm to reconstruct scalars V_x and V_y as separate arrays of N^2 pixels.[35] Thus two images V_x and V_y would be reconstructed from the double set of data (t_{ab} and t_{ba} for each ray) using the same iterative methods of reconstruction used for one set of data.[27]

12.2.4. Reconstruction of Temperature Fields

The theory developed in the preceding sections may be applied under certain conditions to the determination of the three-dimensional distribution of temperature. For example, if it is the case that the material to be probed by the measuring acoustic fields is homogeneous, then the reconstruction of the acoustic speed may be related to the temperature of the material by a method similar to that proposed by Sweeney,[36] who used optical rather than acoustic properties of matter. For example, for pure degassed water, the velocity of sound c (in meters per second) in the

---

[36] D. W. Sweeney, D. T. Attwood, and L. W. Coleman, *Proc. Image Processing 2-D and 3-D Reconstruct. from Project.* August 4–7, p. MA7-1 Technical Digest, Optical Society of America, Stanford, (1975).

neighborhood of 19°C can be given by[37]:

$$c = 1402 + 5.037T + 5.809 \times 10^{-2}T^2, \qquad (12.2.28)$$

where T is measured in degrees Celsius. The inverse function giving T as a function of c is the required mapping function. A difference of 2.0°C for water produces about the same percent change in velocity of sound as the difference between striated muscle and water. Reconstruction accuracy may be increased if the *difference* in temperature is required. Thus circulation and metabolic heat generation may possibly be reconstructed. A map of mixing of two identical fluids of different temperatures may also be reconstructed.[35] In this way, a reconstruction of c can be transformed to a reconstruction of T. The reconstruction of c is obtained by solving the system of Eq. (12.2.12).

12.3. Reconstruction Using Curved Rays

Chapter 12.2 formulated imaging problems in ways that can be solved using the reconstruction methods developed in x-ray tomography. . The derivations were based on perturbation methods for solving the wave equation [Eq. (12.2.1)] that deleted higher-order terms containing significant ray-bending effects. Reconstruction methods based on straight-ray path assumptions result in images that are distorted and have decreased resolution when applied to media in which significant ray bending occurs.

When ray bending occurs, several strategies are available for improving images obtained in ultrasonic transmission tomography. Methods of geometric acoustics (i.e., excluding diffraction) lead to the development of ray tracing techniques to be described in Section 12.3.1. Methods of physical acoustics, which include diffraction effects, will be developed in Section 12.3.2. Methods of solution of the physical acoustic problem in tomography, which include effects of complex refractive index and acoustic impedance, have not yet been developed. A currently unknown but important factor in such a development is detailed knowledge of the scattering characteristics of the biomaterials under investigation. As more knowledge is gained about the scattering characteristics of tissue, more realistic models of interaction between acoustic energy and tissue can be developed, resulting in more meaningful solutions or images.

[37] F. J. Miller and T. Julinski, *J. Acoust. Soc. Am.* **57**, 312 (1975).

12.3.1. Ray Tracing Methods

Glover,[15] Greenleaf and Johnson,[12] and Schomberg[38] have suggested that perhaps an iterative procedure could be derived, which used the distribution of refractive index reconstructed from straight line methods to provide an estimate for the curved paths of the rays. These curved paths could then be used to reconstruct a more accurate estimate of the refractive index. One would continue the process in an iterative fashion until a sufficiently accurate solution was obtained. Such a technique would require the calculation of ray paths through each current estimate of the distribution of acoustic speed. Jakowatz and Kak[39] and Johnson *et al.*[13] have derived digital ray-tracing methods for acoustics using the eikonal equation and assumptions of geometric acoustics.

Ignoring diffraction, the eikonal [Eq. (12.2.5)] is given by

$$\nabla\phi \cdot \nabla\phi = N^2, \tag{12.3.1}$$

where N is the real, isotropic relative refractive index c_0/c. We define a vector $\mathbf{R}(s)$, which is the position vector along a ray as a function of length s along the ray. Then, since geometric optics requires rays to be perpendicular to the wave fronts ($\theta = $ const), we have

$$d\mathbf{R}/ds = \text{grad } \phi. \tag{12.3.2}$$

Now let $\phi = \phi_0 + \phi_1$, where ϕ_0 is the eikonal for the constant reference refractive index and ϕ_1 is due to the perturbation in refractive index. Dropping quadratic terms in $\nabla\phi$, a linearized eikonal equation can be written

$$2(\nabla\phi_1 \cdot \nabla\phi_0) = 2N_0N_1 + N_1^2, \tag{12.3.3}$$

where N_0 and N_1 are the unperturbed and perturbed refractive indices, respectively. This equation can be integrated over ray paths through the tissue. These can then be used to reconstruct a more accurate version of the refractive index.

Schomberg[38] has embedded such a ray-tracing method in the ART[31] reconstruction algorithm. In computer simulations, improved images of refraction indices were obtained. However, Mueller *et al.*,[40] using a model with typical biomedical geometries, refractive indices and wavelengths,

[38] H. Schomberg, *J. Phys.* D **11**, L181, L185 (1978).

[39] C. V. Jakowatz and A. C. Kak, Computed Tomographic Imaging Using X-Rays and Ultrasound. Purdue Univ. Tech. Rep. TR-EE76-26 (1976).

[40] R. K. Mueller, M. Kaveh, and R. D. Iverson, A new approach to acoustic tomography using diffraction tomography, *in* "Acoustical Imaging" (A. Metherell, ed.), Vol. 8, pp. 615–628. Plenum, New York, 1978.

and methods of geometric acoustics for ray tracing, have shown that the best expected resolution is in the order of ten wavelengths. Simulations by Johnson *et al.*[35] have shown some improvement using ray tracing. However, the simulated data were themselves obtained by approximate digital ray tracing methods.

Because of the small size of scatterers ($\ll 1$ mm), diffraction effects may have to be included in the tissue models to obtain images of resolution on the order of a few wavelengths. Such methods have been called "diffraction tomography" by Mueller and Kaveh.[40,41]

12.3.2. Diffraction Tomography: An Inversion (Reconstruction) Method for Low Attenuation That Includes Curved Rays

Since the wave fronts exiting from the tissue undergo diffraction and have also diffracted within the tissue, the assumption that the arrival times and amplitudes are affected only by refraction within the tissue is not completely valid. Under assumptions of low attenuation, one can use a method, first proposed by Iwata and Nagata,[42] that solves for the distribution of the refractive index given the measured distribution of the phase and amplitude of the scattered acoustic wave for narrowband insonification. The following development gives a solution for the distribution of acoustic refractive index under assumptions of low attenuation and includes the effects of diffraction.

12.3.2.1. Rytov's Approximation. Iwata and Nagata[42] have pointed out that the Rytov approximation[43] may have use in obtaining refraction index measurements in cases where the variation in refraction index is small locally but may extend over an area of many wavelengths. We begin with

$$\Psi = \exp(ik_0\,\phi). \qquad (12.3.4)$$

Substituting Eq. (12.3.4) into Eq. (12.2.3) gives

$$ik_0^{-1}\nabla^2\phi - (\nabla\phi)^2 + N^2 = 0. \qquad (12.3.5)$$

We divide N and ϕ into unperturbed and perturbed parts, giving

$$N = 1 + N_1 \quad \text{and} \quad \phi = \phi_0 + \phi_1. \qquad (12.3.6)$$

[41] M. Kaveh, R. K. Mueller, and R. D. Iverson, *Comput. Graphics Image Process.* **9**, 105–116 (1979).

[42] K. Iwata and R. Nagata, *Jpn. J. Appl. Phys.* **14**, 379 (1975).

[43] J. W. Strohbehn *in* "Progress in Optics" (L. Wolf, ed.), p. 85. North-Holland Publ., Amsterdam, 1971.

Note that ϕ_0 is the eikonal in the absence of the object. Substituting Eq. (12.3.6) into Eq. (12.3.5) gives

$$ik^{-1}\nabla^2\phi_0 + ik_0^{-1}\nabla^2\phi_1 - (\nabla\phi_0)^2 - 2(\nabla\phi_0\nabla\phi_1)$$
$$- (\nabla\phi_1)^2 + 1 + 2N_1 + N_1^2 = 0. \qquad (12.3.7)$$

Neglecting N_1^2 and $(\nabla N_1)^2$ and noting that $ik_0^{-1}\nabla^2\phi_0 - (\nabla\phi_0)^2 + 1 = 0$ gives

$$\nabla^2[\phi_1 \exp(ik_0\phi_0)] + k_0^2[\phi_1 \exp(ik_0\phi_0)] = i2k_0N_1 \exp(ik_0\phi_0). \qquad (12.3.8)$$

Using the geometry to be shown in Fig. 4, the solution of this equation for N_1 is[38]

$$\mathscr{F}_{x,y}[N_1](U,\,V) = \frac{1}{2\pi}\frac{u}{k_0} \exp[i(k_0 - u)X_0]\mathscr{F}_y\phi_1^{(n)}(v,\,\theta), \qquad (12.3.9)$$

where

$$U = (u - k_0) \cos\theta_0 - v \sin\theta_0, \qquad (12.3.10)$$

$$V = (u - k_0) \sin\theta_0 + v \cos\theta_0, \qquad (12.3.11)$$

$$u^2 + v^2 = k_0^2, \qquad (12.3.12)$$

$\mathscr{F}_{x,y}[f](U,\,V)$ is the two-dimensional Fourier transform of $f(\mathbf{r})$ and $\mathscr{F}_y[\phi_1](v,\,\theta)$ the Fourier transform with respect to y of the measured values of $\log(\Psi/\Psi_0)$ along a line $x = X_0$ on the far side of the disturbance $f(\mathbf{r})$ (see Fig. 3).

Figures 3 and 4 illustrate the procedure for obtaining $F(U,\,V) = \mathscr{F}_{x,y}(f)(U,\,V)$. At angle $\theta^{(n)}$ one obtains measurements of $\Psi(x,\,y)^{(n)}$, the pressure amplitude, and phase along the receiver line $x = X_0$, where

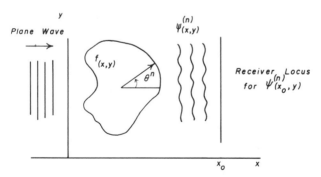

FIG. 3. Geometry for data acquisition in Rytov approximation. The plane wave traveling in $+x$ direction enters an object having refractive index $c_0(1 + f(x,\,y))$. The scattered wave $\Psi_{(X_0,y)}^{(n)}$ is measured along a line at X_0. The object is rotated at angles $\theta^{(n)}$ and additional measurements are taken for each viewing angle.

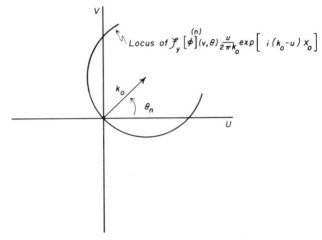

FIG. 4. Locus of Fourier transform of acoustic pressure profile in UV plane. The Fourier transform of measured pressure $\phi_1 = \ln|\Psi/\Psi_0|$ is multiplied by the phase factor and distributed over a circle having its center at the wave vector K_0 within the UV plane. Acquisition of circles at each of many $\theta^{(n)}$ allows the UV plane to be filled out, thus reconstructing the Fourier transform of N_1 [Eq. (12.3.9)].

$\Psi_0(X_0, y)$ is the pressure measured with no object in the field. The discrete Fourier transform of the measurements of $\log(\Psi/\Psi_0)$ then represents samples of $F(U, V)$ (after multiplying by the phase factor $(u/2\pi k_0)$ $\exp[-i(u - k_0)X_0]$) along a circle (see Fig. 4) in the UV plane centered at

$$U = k_0 \cos \theta^{(n)} \quad \text{and} \quad V = k_0 \sin \theta^{(n)}. \quad (12.3.13)$$

For each angle of view $\theta^{(n)}$, a separate circle is obtained and for enough angles of view, the entire UV plane is filled out. After interpolation of the values from the circles onto the points of a grid in the Cartesian plane, an inverse discrete Fourier transform gives $f(\mathbf{r})$.[40,41] Note that as $k_0 \rightarrow \infty$ we obtain the projection theorem of Section 12.1.1 since the radii of the circles go to infinity and the arcs become straight lines through the UV plane.

12.4. Data Acquisition and Signal Analysis

The procedure for obtaining solutions of the real and imaginary parts of n from Eqs. (12.2.10) and (12.2.12) requires the obtaining of sufficient measurements (i.e., independent equations) to solve for all the unknowns. This is done by measuring propagation time or amplitude for as many rays through the tissue as possible, given the available geometry. This is accomplished by scanning the tissue (obtaining "profile data") at each of several angles of rotation about an axis perpendicular to the scan plane.

12.4.1. Methods for Straight Line Reconstruction

Ultrasound computer-assisted tomographs of acoustic speed and of attenuation at several frequencies can be obtained with scanners of various geometries. Ultrasound computerized tomography profile data consist of measurements of time-of-flight or of amplitude at several frequencies. These measurements are made from the received acoustic signals after the acoustic energy traverses the object along many different directions. The data can be obtained by divergent beam scanning with a single pair of transducers, one on each side of the region of interest. A schematic diagram of such a data collection geometry is shown in Fig. 5. The scan angle θ is incremented at equiangular intervals using a stepping motor. After the scan, the transmitter–receiver arm is rotated to a new θ and another scan is obtained. Thus, for each θ, there is one profile of data. A group of profiles separated in view by equiangular increments gathered over a 360° range result in a set of data used for reconstruction.

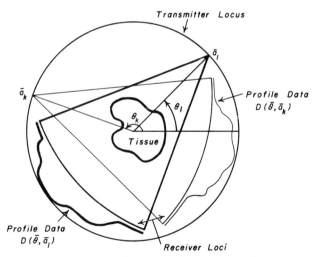

FIG. 5. Data collection geometry for ultrasonic computerized tomography. The divergent beam collection geometry in which transmitter loci are at a_1, a_2, . . . , a_n and the receivers are in directions θ from the transmitter. The profiles $D(\theta, a_i)$ are used for reconstruction and represent pressure amplitude and phase measurements for ultrasonic reconstruction [Reproduced with permission from J. F. Greenleaf *et al.*, *Proc. Int. Conf. Rev. Informat. Processing in Medical Imaging, 5th.* ORNL/BCTIC-2-103-115 (1978)].

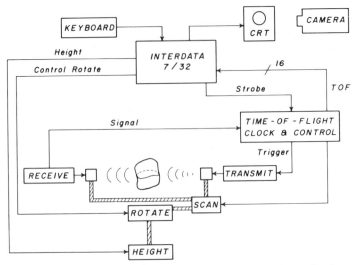

FIG. 6. Example of ultrasound computed tomography system: data collection and control. Computer controls transmit delay and transducer position. The time-of-flight signal is measured with a 16-bit, 100-MHz clock that is read by the counter at up to 400 points along a scan profile with each point separated by 0.15°. There are 120 views, each containing 400 samples, and each sample containing up to 8 channels of 16-bit time-of-flight data can be obtained in a period of 2.0 min [reproduced with permission from J. F. Greenleaf *et al., Proc. Int. Conf. Rev. Informat. Processing in Medical Imaging, 5th* ORNL/BCTIC-2-103-115 (1978)].

A schematic of an experimental arrangement for collecting computerized tomography data is shown in Fig. 6. The system is driven by a minicomputer that controls the height and rotation angle θ of the scanner arm and collects the time-of-flight data from the digital output of the time-of-flight (TOF) clock and control box. This TOF clock and control box is interfaced to the digital input–output bus of the minicomputer and, upon a command from the computer, collects time-of-flight data $\beta(\mathbf{r}, \omega)$ and/or attenuation data $|g(\mathbf{r}, \omega)|$ [Eq. (12.2.21)]. The data are then input to the computer as the scanner arm steps to the next orientation in the scan, repeating this process until one profile of data is collected. At the end of the scan the scanner arm is rotated to the new θ position by a command from the minicomputer, and the process of collecting profile data is initiated again by a trigger to the TOF clock and control circuit. Data for attenuation can be obtained simultaneously by measuring the pulse energy at several frequencies. Methods of analysis of such data were described in Chapter 12.2. These time-of-flight and/or attenuation profiles, representing propagation delay or amplitude attenuation of acoustic

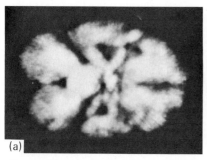
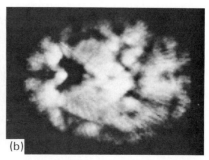

(a) (b)

FIG. 7. Ultrasound transmission tomographs of acoustic speed through transverse planes within fresh excised human brain obtained with straight line reconstructions using time-of-flight measurements at 400 points along each of 120 profiles separated by 3°. The image in (b) is 1–2 cm superior to that in (a). Attenuation is low in the brain, allowing very accurate measurements of pulse arrival time. The range of speed represented in the image is approximately from 1545 to 1555 m/sec for brain and about 1520 m/sec for water background.

pulses along many rays through the tissue under examination, are input to reconstruction programs to obtain quantitative distributions of acoustic speed or attenuation [via Eqs. (12.2.21)] in a transverse plane through the organ.

Examples of ultrasonic transmission tomographs of acoustic speed through transverse planes of a fresh, excised human brain are shown in Fig. 7. The striking detail is present because of the small range of acoustic speeds of the brain (1540–1560 m/sec). Figure 8 illustrates reconstruction of acoustic speed and attenuation in the breast of a 68-year-old woman. The region of high acoustic speed and high attenuation at 9:00 in the right breast is cancer. Techniques for obtaining data for these images have been described elsewhere.[44]

12.4.2. Signal Analysis Methods for Pulse Arrival Time

Measurement of the phase term $\beta(\mathbf{r}, \omega)$ in Eq. (12.2.20) is required for calculation of the distribution of acoustic speed using Eqs. (12.2.19) and (12.2.20). In addition, Eq. (12.2.10) requires that one measure the transit time of the acoustic path from the transmitter at $\mathbf{a}$ to the receiver at $\mathbf{b}$ (Fig. 2) for each ray through the object. This requires the measurement of phase velocity of the signal. Since there are many signals traversing multiple paths and arriving at the receiver at nearly the same time, it is very difficult to define the phase velocity for an individual path.

[44] J. F. Greenleaf, S. A. Johnson, and A. H. Lent, *Ultrasound Med. Biol.* **3**, 216 (1978).

RIGHT   LEFT

SPEED

ATTENUATION

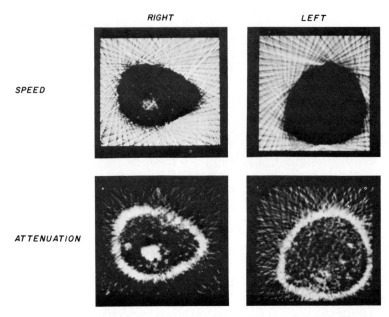

FIG. 8. Ultrasound transmission tomographs of acoustic speed and attenuation in coronal planes of right and left breasts of a 68-year-old woman. Cancer is seen in the right breast at 9:00, consisting of a circular region of high acoustic speed and high attenuation. The image was reconstructed from 60 views around 360°, each profile having 400 samples.

One approach is to measure the time of arrival of the signal energy using threshold detectors after high-gain receivers. In this way, one defines as the arrival time that point in time at which the signal goes above (or below) some preset threshold. The technique of detecting earliest arrival time of the pulses using threshold detectors results in measurements of "signal velocity" rather than group or phase velocity as a measure of time of flight. However, the probability is high that the energy has been received over the least curved path through the tissue, since the earliest arriving energy probably traveled the least curved-ray paths if acoustic speed within the specimens is assumed to be nearly homogeneous. This, of course, is not true in situations where, adjacent to a straight line path between transmitter and receiver, there is a region of very high acoustic speed. In this case, the fastest path may indeed be the curved path through the region of high acoustic speed and not the straight path between the transducers. Therefore the measured geometric extent of regions of high acoustic speed is exaggerated. Resolution is also affected by measuring the earliest arrival time using threshold detection methods. For example, energy received from small regions of high acoustic speed can be easily detected since these regions will cause some energy to arrive and be detected earlier than the bulk of the energy through other pathways; on the other hand, energy received from small regions of low acoustic speed will be masked by the bulk of energy arriving earlier through nearby higher acoustic speed paths.

12.4.3. Analysis of Amplitude of Received Signals for Reconstruction of Acoustic Attenuation

The availability of high-speed analog-to-digital (A/D) converters allows acquisition of digitized transmitted and received pulse waveforms, making possible several methods for determining the absorption and attenuation of ultrasound in tissues. Let

$$\Psi(\omega, X) = \Psi_0(\omega) \, R \exp \left\{ - \int_0^X [\alpha(\omega, x) + i(x\beta(\omega) - t)] \, dx \right\} \quad (12.4.1)$$

be the plane wave amplitude expression for an idealized pressure wave of ultrasound propagating in the x direction and received at $x = X$, where R is the loss due to reflection, α the attenuation coefficient, β a function describing velocity dispersion, which is ω/c for no dispersion, $\Psi_0(\omega)$ input signal amplitude, and $\Psi(\omega, x)$ the received signal. The relationship between α and ϕ can be obtained from Eqs. (12.2.18) and (12.4.1) where we obtain

$$\int_0^X \alpha(\omega, x) \, dx - \ln R = \frac{\omega}{c_0} \phi_{1i} \bigg|_0^X . \quad (12.4.2)$$

This indicates, from Eq. (12.2.21), that

$$\ln|g(X, \omega)| - \ln|g(0, \omega)| = \int_0^X \alpha(\omega, x) \, dx - \ln R. \quad (12.4.3)$$

We know that α is a linear function of frequency for most tissues,[45] thus we can model

$$\alpha(\omega, x) = \alpha_0(x)f, \quad (12.4.4)$$

where f is frequency.
This gives, since $\ln|g(0, \omega)| = 0$,

$$\ln|g(X, \omega)| = \int_0^X \alpha_0(x)f \, dx - \ln R. \quad (12.4.5)$$

Hence

$$\frac{d}{df} (\ln|g(X, \omega)|) = \int_0^X \alpha_0(x) \, dx \quad (12.4.6)$$

---

[45] D. E. Goldman and T. F. Hueter, *J. Acoust. Soc. Am.* **28**, 35 (1956).

and

$$\frac{d}{df}[\ln|g(x, \omega)|] - \ln|g(X, \omega)| = \ln R. \qquad (12.4.7)$$

Therefore, we can reconstruct images of the frequency-dependent attenuation α and the frequency-independent attenuation R if we know $g(X, \omega)$ at enough frequencies with enough signal-to-noise ratio to obtain its derivative.

Kak and Dines[46] have pointed out that if one assumes a linear dependence of attenuation on frequency such as

$$\alpha(f) = \alpha_0|f|, \qquad (12.4.8)$$

one can show the following relationship between α_0 and D_t, D_t being the root-mean-square (rms) duration of the tissue layer impulse response $G(\mathbf{r}, t)$, where G is the inverse Fourier transform of $g(\mathbf{r}, \omega)$, defined in Eq. (12.2.20).

Kak and Dines showed that

$$\int_a^b \alpha \, ds = 2\pi D_t, \qquad (12.4.9)$$

where $|\mathbf{b} - \mathbf{a}|$ is the length of the path through the tissue and

$$D_t = \left[\int_{-\infty}^{\infty} (t - t_c)^2 |G(\mathbf{r}, t)|^2 \, dt / \int_{-\infty}^{\infty} |G(\mathbf{r}, t)|^2 \, dt \right]^{1/2}, \qquad (12.4.10)$$

where t_c is some reference point in time, such as the peak of $G(\mathbf{r}, t)$. Under such assumptions a simple calculation on the impulse response of the tissue $G(\mathbf{r}, t)$ will give a measure of the attenuation coefficient path length product needed for the reconstruction equations. Kak has also pointed out that the rms bandwidth of the tissue transfer function $g(\mathbf{r}, \omega)$ [Eq. (12.2.20)], defined as

$$D_f = \left[\int_{-\infty}^{\infty} (\omega - \omega_c)^2 |g(\mathbf{r}, \omega)|^2 \, d\omega / \int_{-\infty}^{\infty} |g(\mathbf{r}, \omega)|^2 \, d\omega \right]^{1/2}, \qquad (12.4.11)$$

can be related to the attenuation coefficient of the tissue by

$$\int_a^b \alpha \, ds = \frac{D_f}{2\pi}. \qquad (12.4.12)$$

These measures are difficult to use since one is first required to obtain the transfer function or impulse response of the tissue layer between the transmitter and receiver. Nevertheless, this "deconvolution" process is

---

[46] A. C. Kak and K. A. Dines, *IEEE Trans. Biomed. Eng.* **BME-25** (No. 4), 321 (1978).

also required to solve Eq. (12.2.19), previously derived for wideband straight-ray reconstruction methods. For review of methods for obtaining impulse responses of tissues, see Kak and Dines.[46]

12.4.4. Acquisition of Data for Diffraction Tomography

The equations for diffraction tomography derived in Section 12.3.1 necessitate the use of plane wave insonification. One could accomplish this with a very large planar transmitter on one side of the tissue and a scanning point receiver or an array of receivers on the opposite side of the tissue. The geometry of Section 12.3.1 also requires the receiver locus to be a straight line in the plane of the cross section to be imaged. The measurements necessary at each receiver position would apparently be the phase and the amplitude of the received fixed-frequency energy relative to the values obtained with the specimen removed.

A possible problem would be the effect of standing waves on the measurement, since continuous waves are required. One might be able to transmit a burst of sine waves long enough to insonify the entire object at once but short enough to generate few standing waves within the enclosure containing the specimen.

It remains to be seen whether solutions to the wave equation can be obtained for point source insonification with or without including attenuation and impedance.[40] It seems clear, however, that a divergent beam geometry using point sources and receivers rather than planar sources would be simpler to implement experimentally.

12.4.5. Phase Interference

The arrival at the transducer of wave fronts from multiple directions causes phase interference over the surface of the transducer. Since piezoelectric transducers integrate the amplitude of the signal over their surface, an inaccurate estimate for intensity (the integral of the squared amplitude) would be obtained using such transducers, limiting the practicality of accurate amplitude reconstructions. This effect of phase cancellation is very detrimental when using transducers with extended apertures.[47,48] The use of small apertures, on the order of the size of a half-wavelength, while decreasing the phase sensitivity of the transducer,

[47] E. J. Farrell, *Proc. IEEE Conf. Pattern Recognition and Image Processing*, RC 7064 (#30271), p. 1, Chicago, Illinois (May 31–June 2, 1978).

[48] J. G. Miller, M. O'Donnell, J. W. Mimbs, and B. E. Sobel, *Natl. Bur. Stand. Int. Symp. Ultrason. Tissue Characterization, 2nd, Gaithersburg, Maryland, June 13–15* (1977).

also greatly decreases the sensitivity of the receiving system. Miller et al.[48] have studied the phase cancellation effect for extended piezoelectric transducers and have constructed a phase-insensitive receiver using cadmium sulfide, which detects the *intensity* of the signal, as described in Part 1. This transducer is less sensitive than piezoelectric transducers but it is apparently extremely useful for measuring the intensity of signals. Measurement of intensity at separate frequencies using this transducer apparently requires sequential transmission of individual bursts of narrowband energy but may solve the phase interference problem.

12.4.6. Measurement of Amplitude versus Frequency

If Δt and Δf are measures of signal duration and frequency bandwidth, respectively, and k is a constant depending on these measures, the uncertainty relation $\Delta t \Delta f \geq k$ comes into effect when measuring bursts of energy and determining amplitude versus frequency to obtain data for frequency-dependent attenuation.

High-resolution measurements of amplitude versus frequency require long samples of signal (Δt). The result is that very accurate measurements of amplitude versus frequency in the received pulse cannot be obtained from the early portion of the signal. Therefore, reconstructions of attenuation, a strong and nonlinear function of frequency, require long sample periods of the signal to define amplitude accurately at a given frequency. However, as the length of the signal required for analysis is increased, the geometric distribution of accepted path lengths increases. Thus the effective beam width of the transmission scanning system is broadened. This causes a loss of spatial resolution of such measurements through beam widening.

Lens–transducer combinations for maintaining narrow spatial beam width combined with narrowband frequency measurements are difficult to use in systems using parallel data acquisition from arrays of transducers, as required in high-speed scanners, since many receiver elements must be insonified simultaneously.

12.4.7. Speed of Data Acquisition

The period of time required for data acquisition is limited by the reverberation time τ_r of the tank in which the specimens are being scanned. The highest available pulse repetition frequency is $1/\tau_r$, where τ_r is the time for multiple echoes in the tank to decrease to an amplitude below a specified level, which depends on the signal-processing system being used. The number of parallel receiver channels N_P inversely affects the time of data acquisition required for each reconstruction. If the number of rays required for a reconstruction is N_R, then the time T for data acquisition of data required for one cross-sectional reconstruction is $T = N_R \tau_r / N_P$. The use of fan-shaped beams allows the exposure of more than one receiver transducer at a time and thus provides possibility of parallel data acquisition. However, parallel receiver exposure also increases the probability of multiple-path detection for narrowband measurement of amplitude, as discussed in Section 12.4.6. Thus the tradeoff is between narrowband focused beam measurement of amplitude with slow scanning and fan beam insonification with parallel acquisition of data for fast scans.

One technique for increasing the pulse repetition frequency is to encode the transmitted pulse allowing more than one pulse at a time to be in transit across the object. The multiple encoded signals can be separated from one another by suitable signal processing, especially if the separate signals are temporally orthogonal to one another. The processing for such a method, however, is extremely complex and under the conditions of scanning acoustically complex objects, the correlation methods become extremely difficult since the expected waveforms are greatly distorted by the variable-frequency filter represented by varying lengths and characteristics of tissue through which the acoustic energy has passed.

12.5. Summary

Computer-assisted ultrasound tomography is a useful method for obtaining quantitative measurements of two- and three-dimensional distributions of material properties such as acoustic attenuation and speed. Resulting images of these parameters should provide a noninvasive method of studying organ structure and function in health and disease. The development of transmission tomography methods for use in accessible organs such as the breast should not only provide a diagnostic tool but lead to a better understanding of reflection tomography techniques having applicability to a broader range of organs in biology and medicine.

AUTHOR INDEX

Numbers in parentheses are reference numbers and indicate that an author's work is referred to although the name is not cited in the text.

A

Abdulla, U., 341
Abeles, B., 432
Abraham, B., 448, 449, 450
Adams, C. E., 326
Adams, S., 431
Adithan, M., 352
Adler, E. L., 503
Adler, L., 77, 109, 110
Adler, R., 484, 485, 486(24), 487, 517, 518, 524
Ahlers, G., 437
Ahuja, A. S., 563(17), 564
Airy, G. B., 300
Akao, F., 524
Akulichev, V. A., 369, 387, 395(84), 398
Alers, G. A., 256, 270, 530, 531(55)
Alfrey, T., 144
Alhaider, M. A., 521, 522(36)
Al-Temini, C. A., 303
Alterman, Z., 528, 530(53)
Amdur, I., 208
Anderson, G. 108, 257, 258(39)
Andreae, J. H., 108, 131
Andreatch, P., Jr., 94, 151, 157(31, 32), 258
Apfel, R. E., 315, 316, 362, 366, 367, 368, 371, 381, 387
Applegate, K. R., 202(48), 204
Asay, J. R., 107
Ash, E. A., 481, 482(23), 486(23), 487, 488, 489, 492(23)
Asono, T., 202(52), 204
Atkins, K. R., 421
Attwood, D. T., 575

B

Auld, B. A., 34, 35(3), 48, 54, 59(3), 62, 65 496, 499, 501(1), 504
Auslow, J. A., 526

Bailey, E. D., 168
Bains, E. M., 120
Bajons, P., 347
Baker, H. L., 563
Baker, N. V., 341
Baker, W. O., 106, 147
Bamber, J. C., 13, 14
Baranskii, K. N., 39
Barger, J. E., 13, 14, 367, 371(25), 393
Barksdale, A., 233
Barlow, A. J., 24, 106, 147, 150, 151, 153, 154, 156, 168(33), 170, 171, 172, 173, 174(66), 176(34), 177
Barmatz, M., 429, 437, 453
Barnard, G. A., 303
Barnes, R. P., Jr., 80, 301
Barone, A., 133
Barrett, C. S., 237, 251, 276, 284(1), 292, 293
Barshauskas, K., 76
Bass, R., 72, 108
Bassett, J. D., 337
Basurmonova, O. K., 327, 330(135)
Bateman, T. B., 159
Baxter, K., 315
Bauer, H. J., 192
Beams, J. W., 368
Beasley, J. D., 339
Beaumont, L. R., 564
Becker, F. L., 551
Bell, D. T., 524

591

SUBJECT INDEX

A

Absorption
  in biological tissues, 13
  by chemically reacting systems, 183, 188
  definition of, 20, 107
  introductory theory, 20–27
Absorption coefficient, 18, 20–21, 24–26
  due to single relaxation process, 183,
    188
  of various materials, 10–13
AC-cut quartz, 52
Acetic acid, 193, 202
Acetone, 189, 190, 195
Acoustic calorimetry, 192
Acoustic cavitation, see Cavitation
Acoustic devices, acoustooptic, characteri-
    zation of, 477, 480, 486–492
Acoustic dispersion, 41, 183, 187
Acoustic drying, 342–345
Acoustic gain, 192, 217
Acoustic holography, see Holography
Acoustic impedance, see Impedance
Acoustic oscillation, 192
Acoustic streaming, see Streaming
Acoustic streaming method, attenuation
    measurement, 132–133
Acoustooptic figures of merit, 463–464
Acoustooptic measurement of sound wave
  amplitude and intensity, 476, 478–481,
    484–492
  attenuation, 474–481, 487
  beam diffraction, 475–477, 484–492
  beam propagation, 475–477, 484–492
  phase, 484–492
  power, 476–492
  presence of, 457, 471–473, 482
  propagation, 475–477, 484–492
  reflection, 475–477, 484–487, 490
  standing wave ratio, 480, 484–487, 490
  velocity, 471–473, 487
Acoustooptic phenomena, 302, 326, 455–
    493

Acoustooptic techniques, 467–492
  classification of, 467–469
  dc detection of light, 467–469
  diffraction orders, observation of, 470–
    471
  Fabry–Perot interferometry, 478–481
  heterodyne, 469, 481–492
  knife edge, 484–487, 490–492
  photoelectric detection, 474–478
  Schlieren studies, 472–474
Activation energy, 221
Activity, 201
Activity coefficient, 201–207
Adenosine, 203
Adenosine-5'-phosphate, 202
Advancement of chemical reaction, 184
Affinity of chemical reaction, 184
Agglomeration, 340–342
Algebraic reconstruction technique, 575
Alloys, polycrystalline, attenuation in,
    267–290
Aminobenzoic acids, 225–229
Amplitude, see acoustooptic measurement
    of sound wave, amplitude and inten-
    sity
Anisotropic specimens, errors due to, 291–
    294
Anisotropy
  and beam spreading, 292
  calculation for martensite, 254–255
  and elastic moduli, 293–294
  macroscopic, in polycrystalline materi-
    als, 275–277, 292–294
  of pearlite, 267, 284–288
  preferred orientation as cause, 251–252,
    275–277, 292–294
  in pyrolytic graphite, 292–293
  reduction by phase transformations,
    253–256, 267, 280–290
  in scattering by grains, 241–244, 253–
    256, 267–290
  in tool steel, 293–294
  of velocity, 277